Ecotoxicology and Genotoxicology

Non-traditional Aquatic Models

Issues in Toxicology

Series Editors:
Diana Anderson, *University of Bradford, UK*
Michael D. Waters, *Michael Waters Consulting, USA*
Timothy C. Marrs, *Edentox Associates, UK*

Editorial Advisor:
Alok Dhawan, *CSIR-Indian Institute of Toxicology Research, Lucknow, India*

Titles in the Series:

1: Hair in Toxicology: An Important Bio-Monitor
2: Male-mediated Developmental Toxicity
3: Cytochrome P450: Role in the Metabolism and Toxicity of Drugs and other Xenobiotics
4: Bile Acids: Toxicology and Bioactivity
5: The Comet Assay in Toxicology
6: Silver in Healthcare
7: *In Silico* Toxicology: Principles and Applications
8: Environmental Cardiology
9: Biomarkers and Human Biomonitoring, Volume 1: Ongoing Programs and Exposures
10: Biomarkers and Human Biomonitoring, Volume 2: Selected Biomarkers of Current Interest
11: Hormone-Disruptive Chemical Contaminants in Food
12: Mammalian Toxicology of Insecticides
13: The Cellular Response to the Genotoxic Insult: The Question of Threshold for Genotoxic Carcinogens
14: Toxicological Effects of Veterinary Medicinal Products in Humans: Volume 1
15: Toxicological Effects of Veterinary Medicinal Products in Humans: Volume 2
16: Aging and Vulnerability to Environmental Chemicals: Age-related Disorders and their Origins in Environmental Exposures
17: Chemical Toxicity Prediction: Category Formation and Read-Across
18: The Carcinogenicity of Metals: Human Risk Through Occupational and Environmental Exposure
19: Reducing, Refining and Replacing the Use of Animals in Toxicity Testing
20: Advances in Dermatological Sciences
21: Metabolic Profiling: Disease and Xenobiotics
22: Manganese in Health and Disease
23: Toxicology, Survival and Health Hazards of Combustion Products
24: Masked Mycotoxins in Food: Formation, Occurrence and Toxicological Relevance

25: Aerobiology: The Toxicology of Airborne Pathogens and Toxins
26: Chemical Warfare Toxicology, Volume 1: Fundamental Aspects
27: Chemical Warfare Toxicology, Volume 2: Management of Poisoning
28: Toxicogenomics in Predictive Carcinogenicity
29: Human Stem Cell Toxicology
30: The Comet Assay in Toxicology, 2nd edition
31: Computational Systems Pharmacology and Toxicology
32: Ecotoxicology and Genotoxicology: Non-traditional Terrestrial Models
33: Ecotoxicology and Genotoxicology: Non-traditional Aquatic Models

How to obtain future titles on publication:
A standing order plan is available for this series. A standing order will bring delivery of each new volume immediately on publication.

For further information please contact:
Book Sales Department, Royal Society of Chemistry, Thomas Graham House, Science Park, Milton Road, Cambridge, CB4 0WF, UK
Telephone: +44 (0)1223 420066, Fax: +44 (0)1223 420247
Email: booksales@rsc.org
Visit our website at www.rsc.org/books

Ecotoxicology and Genotoxicology
Non-traditional Aquatic Models

Edited by

Marcelo L. Larramendy
National University of La Plata, Argentina
Email: marcelo.larramendy@gmail.com

ROYAL SOCIETY OF CHEMISTRY

THE QUEEN'S AWARDS FOR ENTERPRISE: INTERNATIONAL TRADE 2013

Issues in Toxicology No. 33

Print ISBN: 978-1-78262-781-4
Two-volume set print ISBN: 978-1-78801-169-3
PDF eISBN: 978-1-78262-988-7
EPUB eISBN: 978-1-78801-179-2
ISSN: 1757-7179

A catalogue record for this book is available from the British Library

The Royal Society of Chemistry is a charity, registered in England and Wales, Number 207890, and a company incorporated in England by Royal Charter (Registered No. RC000524), registered office: Burlington House, Piccadilly, London W1J 0BA, UK, Telephone: +44 (0) 207 4378 6556.

For further information see our web site at www.rsc.org

Printed in the United Kingdom by CPI Group (UK) Ltd, Croydon, CR0 4YY, UK

Preface

Many important ecosystems around the world are being constantly challenged owing to the growing human and industrial pressure exerted upon them. The use of various biomarkers in local, easily available species can be applied to evaluate the response of the biota to such pollutants. Several biological parameters mirror the interactions between toxic agents and biotic matrices. These are powerful tools that can be applied to environmental monitoring tests and studies. Their responses may reveal general deleterious effects to the organism, pinpointing alterations at a cellular, biochemical and molecular level, as well as higher levels of organisation.

Our global society needs to table down actions and set rules to evaluate and considerably reduce the real and potentially hazardous factors in the environment that can, as previously stated, result in health risks for all forms of life (including *Homo sapiens sapiens*). Despite major positive contributions in the field of health, owing to the immense progress achieved in science, technology and industrialization, the interaction between environmental risk and health is an often intricate equation, not self-evident, that involves a variety of not only social, political and economic, but also lifestyle factors. This cannot be emphasized enough. Health depends on the good quality of environmental "basic ingredients", such as air, water, soil and food, among others. We believe that the ultimate challenge in this matter is to weigh-in short-term positive gains, while, at the same time, taking into account long-term effects of substances used. Available information about the toxic effects of heterogeneous xenobiotics, continuously released into human habitats, inadvertently, deliberately, or by non-regulated industrial discharges on biological components of the environment, is inconclusive.

There is not a clear-cut definition of the concept of Environmental Health. Various openings help us in the understanding of this concept. According to

Issues in Toxicology No. 33
Ecotoxicology and Genotoxicology: Non-traditional Aquatic Models
Edited by Marcelo L. Larramendy

Published by the Royal Society of Chemistry, www.rsc.org

the World Health Organization, it is defined by "all the physical, chemical and biological factors external to a person and all the related factors impacting upon behaviours. It encompasses the assessment and control of those environmental factors that can potentially affect health. It is targeted towards preventing disease and creating health-supportive environments. . ." For the National Environmental Health Association, this concept refers to "the protection against environmental factors that may adversely impact human health or the ecological balances essential to long-term human health and environmental quality, whether in the natural or man-made environment." A third definition by the National Institute of Environmental Health Science also involves the criteria that "the social environment encompasses lifestyle factors like diet and exercise, socioeconomic status, and other societal influences that may affect health."

In general terms, our health and the health of many other species are negatively affected by five broad categories of environmental hazards, namely, electromagnetic fields (produced by high power lines, electrical wiring, appliances, mobile phones, computers, and TV sets, *etc.*), radiation (including nuclear fallout from weapons testing, fission materials from nuclear power plants and their respective accidents, leaking radioactive disposal sites, air travel and x-rays), toxic chemicals (some organochlorines, phthalates, polybrominated flame retardants, perfluorinated substances, bisphenol-A) and several toxic metals, among others, which have been shown to have endocrine-disrupting properties, and finally, soil mineral depletion as a complex environmental hazard.

By definition, health risk assessment in its quantitative and/or qualitative determinations includes variants such as the type of risk involved and the severity of response, within or without a probabilistic context. In this regard, risk-based methods of analysis play a strategic role in identifying and ranking adverse responses or the structure of the effects of exposure *vis-à-vis* environmental factors.

Many compounds can be hazardous if not used appropriately and may present a real risk to the environment, contaminating soil, water and air. Most of the pollutants in the different environmental compartments exert their effects through cytotoxic, genotoxic and metabolically toxic mechanisms. In pollution studies, there is an increasing interest in biomonitoring markers of biological exposure to pollutants. To achieve this goal, several end-points for the three above-mentioned factors have been used in aquatic and terrestrial invertebrate and vertebrate species on contaminated areas (*in situ* assays) and to screen for xenobiotics after direct or indirect exposure (*in vivo* assays).

The use of invertebrate and vertebrate autochthonous species as indicators for monitoring pollutant-induced deleterious environmental effects will raise the current awareness of real and potential hazards. It is also known that most of the environmental pollutants not only affect target organisms, but concomitantly exert negative effects on non-target species as well.

Invertebrate and vertebrate animal models have been used for decades in acute and chronic toxicity tests for hazard identification. They can be very efficient screening systems that have a major role to play in toxicity research, because certain aspects of their biology, physiology and genetic characteristics make them suitable models in ecotoxicological and genotoxicological studies.

These two books intend to provide an overview of the use of non-conventional, locally available, invertebrate and vertebrate species as experimental models for the study of different toxicological aspects induced by environmental pollutants in both aquatic and terrestrial ecosystems. Volume One, *Ecotoxicology and Genotoxicology: Non-traditional Aquatic Models* includes examples of the use of aquatic species or aquatic stages of terrestrial species and Volume Two, *Ecotoxicology and Genotoxicology: Non-traditional Terrestrial Models,* is committed to terrestrial non-conventional animal models.

Both volumes aim to shed some light on the matter, whilst offering relevant tools for evaluating risk and to provide a framework for practical discussions. These will foster decisions and actions required to reduce environmental health risk against environmental factors. This piece of work has been systematized for the sake of clarity, presenting some real-life examples and extending concepts (of hazardous factors) to living species that may stimulate new research ideas and trends in the relevant fields.

Available information has been compiled from a diversity of sources, trying to achieve a representative global and geographical balance, as far as possible, whilst at the same time aiming at high-quality studies. We believe that this piece of work is unique in this sense.

Many researchers from different parts of the world have contributed to the publication of this book. Given the fast pace of new scientific publications shedding more light on the matter, these books will probably be outdated very soon. We regard this as a positive and healthy fact. We hope that these books will meet the expectations and needs of all those interested in the environmental risk assessment field of study by the use of widely available species worldwide. Finally, we also hope that the examples included in the different chapters of these books will awaken the ability to search for new organisms in local and regional ecosystems to pursue further studies in ecotoxicology and genotoxicology. If our wishes are granted, we shall be happy to oblige and edit the next edition of this series.

Prof. Dr Marcelo L. Larramendy
and Dr Guillermo Eli Liwszyc

Contents

Issues in Toxicology No. 33
Ecotoxicology and Genotoxicology: Non-traditional Aquatic Models
Edited by Marcelo L. Larramendy

Published by the Royal Society of Chemistry, www.rsc.org

Section I: Aquatic Invertebrates as Experimental Models

CHAPTER 1

The Comet Assay in Aquatic (Eco)genotoxicology Using Non-conventional Model Organisms: Relevance, Constraints and Prospects

MARTA MARTINS[a,b] AND PEDRO M. COSTA*[a]

[a] MARE – Marine and Environmental Sciences Centre, Departamento de Ciências e Engenharia do Ambiente, Faculdade de Ciências e Tecnologia, Universidade Nova de Lisboa, 2829-516 Caparica, Portugal; [b] UCIBIO, REQUIMTE, Departamento de Química, Faculdade de Ciências e Tecnologia, Universidade Nova de Lisboa, 2829-516 Caparica, Portugal
*Email: pmcosta@fct.unl.pt

1.1 Introduction

The integrity of the genome is the very foundation of the organism and all the complex downstream events that mediate the various levels of biological organization, from gene to protein, then cell and tissue, and from there to individual, population and ecosystem. Not surprisingly, the "success of the fittest" involves the ability to cope with agents that may interfere with the genome and its transcription. When this ability is overwhelmed (or led into malfunction) by any given agent, such as chemical or radiation, the genetic

Issues in Toxicology No. 33
Ecotoxicology and Genotoxicology: Non-traditional Aquatic Models
Edited by Marcelo L. Larramendy

Published by the Royal Society of Chemistry, www.rsc.org

material accumulates lesions that lead to metabolic dysfunction and then to cell death, or to the fixation of mutations if the cell does survive, thus propagating altered genetic material in somatic or germline cells. The latter case implies severe implications not only for the individual but also for the entire population since it may cause reproductive impairment, teratogenesis and, very importantly, tumorigenesis. Genotoxicity is therefore a phenomenon that affects all aspects of ecosystem functioning and may determine populational and species fitness in their changing habitat, rendering paramount the determination of its effects in ecologically relevant organisms outside the scope of the acknowledged laboratory model. The range of such "unconventional" models is increasingly wide, with particular respect to aquatic organisms, comprising many species of fish to molluscs (especially bivalves), to crustaceans, annelids and even echinoderms, cnidarians or macrophytes, whose exemplificative applications will be detailed in subsequent sections. On the contrary, the range of acknowledged model organisms holding some degree of ecological relevance is rather narrow. These include wild-type or genetically modified strains of the freshwater teleost zebrafish (*Danio rerio*) or the cladoceran crustacean *Daphnia magna*. In spite of their value in many fields of research (including biomedical, in the case of the former), these models are entirely laboratorial and cannot provide an entirely realistic insight into ecosystem function impairment by pollutants nor ensure the much needed long-term monitoring programs.

Not surprisingly, on account of the basic Paracelsian principle "it is only the dose that separates benefit from poison", toxicologists have long tried to understand, quantify and predict the effects of substances that may damage the genome of both humans and wildlife. As such, the first methods to detect and quantify DNA damage were adopted, adapted and improved from mammalian models (mostly *in vitro*) by environmental scientists from the start and successfully applied to a wide range of vertebrate and invertebrate organisms holding ecological and even economical relevance, thus giving birth to the domain of ecogenotoxicology. This implied, nonetheless, not only changes in protocols to harvest tissue and cells, for instance, but also in the interpretation of the findings *per se*, since genotoxicity is a complex biological phenomenon that depends on multiple pathways that likely differ between distinct taxa.

The methods to detect and quantify genotoxicity first focused on whole-chromosome changes, such as micronuclei and other nuclear abnormalities, which can be expeditiously scored, for instance, in whole-blood samples of non-mammalian vertebrates (since erythrocytes are nucleated), or the sister chromatid assay. These methods detect large-scale, irreparable, lesions that derive from clastogenic and aneugenic events. To this is added the widespread ^{32}P-postlabelling method for detecting DNA–xenobiotic adducts. Other methods, such as the Ames test, address mutagenesis by itself by detecting the reversion of his-mutant *Salmonella* strains back to bacteria able to synthetize this amino acid, by the action of mutagens. Even though the adequacy and value of these methods is still beyond dispute, there was still a lack of a protocol that could efficiently detect alterations to the genome at the DNA strand level. A revolution thus took place when the single cell gel electrophoresis (SCGE) assay, or simply the "Comet" assay, was developed and rapidly incorporated within toxicological sciences, with emphasis on ecotoxicology and environmental toxicology. The common alkaline variant of the Comet assay, which stands as the workhorse of the protocol, originally settled by Singh *et al.*[1] and based on the "neutral" version developed by Östling and Johanson,[2] is nowadays little used. In fact, alkaline Comet assay or simply Comet assay are terms that are used almost interchangeably.

The principle of the assay is simple. Since DNA, like many organic molecules, is charged, when subjected to an electric field the smaller fragments will migrate faster towards one of the poles in a strong alkali environment, preceded by DNA denaturation in the same alkali buffer (~pH 13). Thus, the DNA of individual cells is exposed after embedding in an agarose matrix and the amount of fragment DNA migrating towards the positive electrode, *i.e.* the anode (since oligonucleotides are, essentially, anions), can be determined after staining and scoring using microscopy and imaging tools. The term "Comet" results from the typical shape of DNA after cell lysis and electrophoresis (the "nucleoid"), since large oligonucleotides, *i.e.*, little or not at all fragmented, will be retained in the head whereas the smaller move toward the anode, forming the "tail". The migration of fragments, however, depends on several aspects that often tend to be overlooked. First, DNA is itself a supercoiled molecule formed by two oligonucleotide strands. Second, genotoxicants may or may not lead to direct strand breakage. In fact, one of the most critical factors of the assay is DNA denaturation under alkaline conditions since this will permit separation of the two strains and therefore allows the expression of single-strand breaks (if any), and the expression of double-strand breaks (if any) that were transformed into the single-strand after denaturation. To this is added the relaxation of altered DNA segments (loops) and expression of the so-called alkali-labile sites that consist essentially of altered nucleobases that, when DNA is loosened, may break upon electrophoresis (see for instance Tice *et al.*[3]). The intensity of the staining between the head and tail can then be extrapolated as the relative proportion between fragmented and unfragmented DNA as a simple metric among the several direct or derived measures that can be retrieved from analysis and that will be debated further on.

In other words, in spite of the many types of DNA damage that might occur, the alkaline Comet assay indiscriminately detects damage that may either result in strand breakage or contribute to relaxing the DNA molecule to the point of favouring migration towards the anode. As such, the Comet assay has been used to provide a measure of "total strand breakage", which, in spite of some bias, is evidently more accurate than "total DNA damage". There are, however, variants of the alkaline Comet assay that permit some discrimination of damage by type, which will be addressed later on.

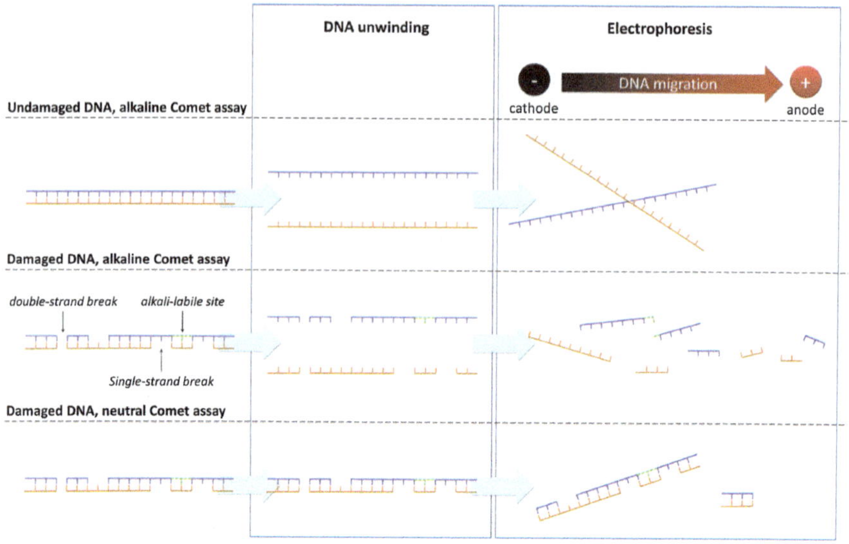

Figure 1.1 Illustration of the basic principle of the Comet assay. When subjected to a strongly alkaline buffer, DNA tends not only to unwind but also to separate the two chains, then exposing single- and double-strand breaks, as well as eventually some alkali-labile sites (*e.g.* altered nucleotides) that may break under the alkali electrophoresis. Compared to the "neutral" version, the alkaline Comet assay yields not only more fragments but also lower molecular weight fragments, whose speed of migration is therefore higher, thus producing longer "tails".

By comparison, the scantly used "neutral" Comet assay follows the exact same principle. However, the denaturation/electrophoresis buffer has a lower pH (10), therefore failing to efficiently separate the DNA strands, favouring only the migration of double-strand fragments (Figure 1.1).

There is thus an important difference in the meaning of DNA strand-breakage when compared to other genotoxicity assessment methods, especially the micronucleus test and its variants, often referred to as nuclear abnormalities (NAs), since the latter refers to whole-chromosome damage such as aneugenesis (chromosomes that are not integrated within the nucleus of a daughter cell) or clastogenesis (chromosomal fragmentation), which is commonly associated with faulty cell division. Unlike DNA strand-breakage, lesions at the chromosome level are most unlikely to be repaired (Fenech *et al.*[4]). Even though the relevance of scoring NAs in aquatic ecotoxicology is undisputable,[5] it has been shown, even in studies with non-model marine fish like sole and bass, that the two measures may not necessarily be correlated.[6,7] However, unlike assessing NAs, the Comet assay is not yet used regular biomonitoring approaches in aquatic ecotoxicology. It could be argued that the lack of standardization of protocols and its multiplicity could be hindering the value of the SCGE assay; however, in most cases, the logistics of field sampling greatly favour the high cost-effectiveness of preparing blood smears when compared to a molecular method whose accurateness greatly relies on avoiding accessory strand breakage. Still, the Comet assay has been widely employed in both *in situ* and *ex situ* (laboratory) bioassays and, although to a lesser extent, in passive biomonitoring of marine and freshwater ecosystems, thus involving surveys with a broad range of unconventional model organisms, as debated below. Still, there are many technical aspects that render aquatic ecogenotoxicity with these species particularly challenging. As previously highlighted in the few reviews specifically dedicated to the topic, the application of the Comet assay cannot be based on the same assumptions of biomedical research and human-oriented toxicology that, to date, still dictate most protocols and guidelines.[8–10]

1.2 The Comet Assay in Aquatic Ecotoxicology: Role of Unconventional Models

1.2.1 Aquatic Ecosystems as the Ultimate Fate for Pollutants

When translated to ecotoxicology, the principle with which Paracelsus gave birth to toxicology simply stands as "contamination does not necessarily mean pollution". In other words, hazard and risk are two distinct concepts. Whereas some substances may be more hazardous than others (*e.g.*, we can compare the metals cadmium and copper), risk is the probability of adverse effects occurring. This means that the dose or concentration can turn a scarcely hazardous agent into a high-risk pollutant.[11] The ecotoxicologist must keep in mind that contamination occurs when the levels of one or more given agents surpass baseline environmental concentrations. If these concentrations cause deleterious effects to biota, then pollution is indeed occurring. The main challenges are, first, to detect deleterious effects and, second, to determine causality. In fact, ecosystem complexity is one of the major factors hindering the establishment of cause–effect relationships in this field of research. On the other hand, dealing with non-model organisms, quite often from "unconventional" taxa, poses additional challenges, albeit being crucial to understand how the ecosystem, and not just a species or a population, is affected by pollutants. Altogether, aquatic ecosystems hold many characteristics that render aquatic ecotoxicology as complex as it is important: (i) the aquatic environment is invariably the ultimate fate of environmental toxicants; (ii) areas adjacent to marine and freshwater ecosystems have always received the highest anthropogenic pressure; (iii) the sources of toxicants are multiple, natural or anthropogenic, and include aquatic transport, direct discharge, atmosphere, urban drainage and maritime/fluvial transport; and (iv) aquatic ecosystems, especially those of transitional waters, have peculiar characteristics that render them ideal for accumulation, transformation and long-term storage of hazardous substances, particularly in sediments.[11–13] Altogether, surveying the effects of pollutants on aquatic organisms is paramount as a tool for the diagnosis of ecological status and as a means to understand how a toxicant can affect the functioning of an entire ecosystem. In other words, surveying aquatic organisms plays an important role in Environmental Risk Assessment (ERA), whether as a measure of exposure (effects-oriented research) or as a means to understand why and how a substance becomes toxic to aquatic biota (mechanism-oriented research). In either case, model organisms, such as laboratory strains of the zebrafish or *Daphnia,* are mere surrogates and are not realistic representatives of wildlife. Even though clear advantages of these model strains, such as reduced intraspecific variability and high genomic annotation, permit important basic toxicological research, extrapolation towards wild organisms must be cautionary.

While mechanistic research in ecologically relevant organisms is far from being as developed as in human toxicology, effect-oriented studies are of upmost importance to quantify exposure in these "models" since, unless the concentrations of toxicants are either too high or too low, chemical analyses of sediments, waters and biota may be insufficient. Furthermore, it has long been acknowledged that ERA should not rely on a single Line-of-Evidence (LOE), such as a on a single biomarker or chemical determination of toxicants, but rather it should be an integrative approach comprising several LOEs, often referred to as the Weight-of-Evidence (WOE) approach. Determining genotoxicity has been proposed as an active component for these approaches as a biomarker of effect. The reader may refer to the excellent reviews by van der Oost *et al.*,[14] Martín-Díaz *et al.*[15] and Chapman *et al.*[16] for a definition of biomarker practices in Aquatic Ecotoxicology.

The number of genotoxicant substances present in the aquatic milieu keeps increasing. Many of these pollutants are acknowledged to be effective or potential carcinogens to humans, as classified by the International Agency for Research on Cancer,[17] but their effects (including those from mixtures) on wildlife are mostly unknown. Among these substances are "classical" genotoxicants, such as many polycyclic aromatic hydrocarbons (PAHs), dioxins [such as tetrachlorodibenzodioxin (TCDD)], pesticides, and As and Cr compounds, just to state a few other examples. However, novel, "emerging", compounds are springing up, such as nanomaterials like carbon nanotubes and metal nanoparticles, whose genotoxic effects have been tested in freshwater snails.[18,19] In addition, complex mixtures of organic and inorganic sediment pollutants have been found to cause DNA strand breakage in marine fish even when the individual concentrations of the compounds would indicate reduced or null risk.[6,20] In a similar example from freshwater environments, crucian carp (*Carassius carassius*) exposed to flood water from agricultural grounds yielded significant DNA strand breakage in whole-blood cells, albeit yielding null effects when fish were exposed to ecologically relevant concentrations of pesticides found in the area, confirming not only the complexity of environmental samples but also the need to safeguard some measure of realism in ERA.[21]

It is also of upmost importance to understand that DNA damage may have multiple causes, depending on the substance (or mixtures), organism, route of exposure, concentration and even tissue and organ, if applicable. On the other hand, DNA damage results from the balance between aggression and repair, since evolution has favoured metabolic pathways that protect the integrity of the species' genetic heritage, which is far from being fully understood, especially in invertebrates, rendering data interpretation cautionary (Figure 1.2). Many organic compounds, like PAHs, which are hydrophobic and metabolically inert, are bioactivated by CYP (cytochrome P450) mixed-function oxidases (MFOs) and other enzymes into highly genotoxic metabolites that form bulky adducts with the DNA molecule. It is

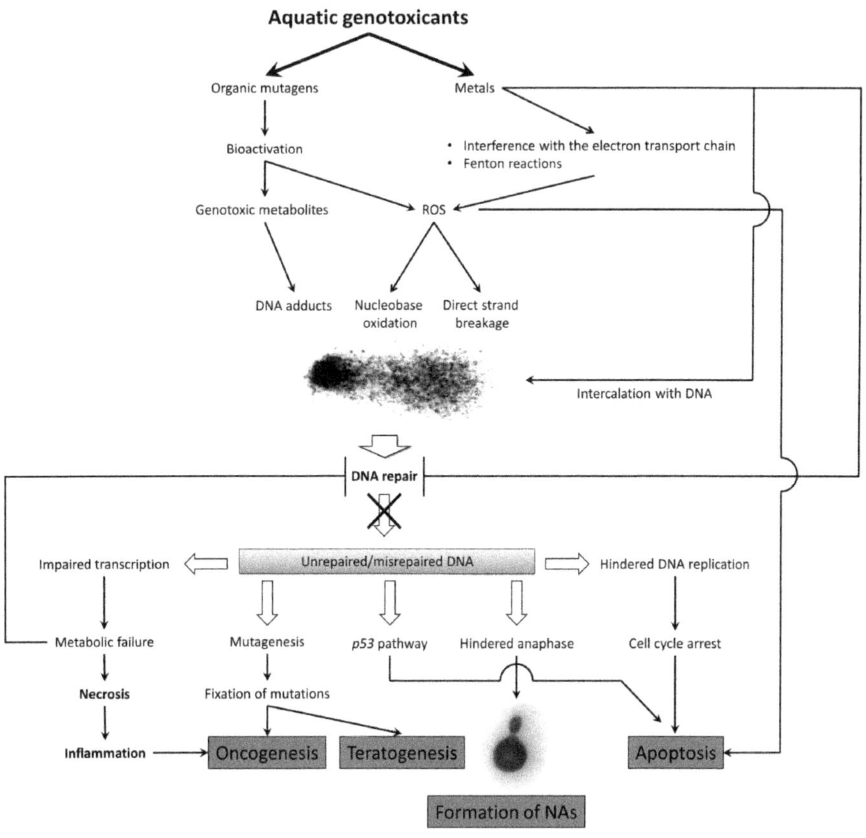

Figure 1.2 A simplified overview of the molecular pathways that link genotoxicity detectable by the Comet and nuclear abnormalities (NA) assays with oncogenesis, teratogenesis and other pathological alterations as expected to occur in vertebrates, exemplified here as in co-exposure to organic mutagens (such as some PAHs and dioxins) and metals to highlight some of the potential interactions between toxicants. These pathways, albeit originally described in higher-order vertebrates, have also been described in fish.

the case for diol-epoxides generated after activation of the potent mutagen and carcinogen benzo[*a*]pyrene (B[*a*]P). Nowadays, it known that the binding of B[*a*]P metabolites to DNA is not random as there is affinity towards sequences of proto-oncogenes (such as those of the *ras* family in vertebrates, with known homologues in fish), leading to their activation, overexpression and triggering of anaplastic transformation of cells. The reader may refer to the review by Xue and Warshawsky[22] for further details. The bioactivation process may generate reactive oxygen species (ROS) that, among other effects, can oxidise nucleobases, especially the highly reactive hydroxyl (•OH) radical (see Cadet *et al.*[23]). These single-strand lesions may be repaired by the nucleotide (NER) and base (BER) excision repair pathways, respectively. Furthermore, NER can be global genome (GG-NER) or transcription-coupled (TC-NER), involved in the removal of DNA and RNA polymerase blocking adducts, respectively, such as the bulky adducts formed by PAH metabolites.[24] Double strand breakage can also be repaired, albeit by the more complex processes of homologous recombination (HR) and non-homologous end-joining (NHEJ). The aforementioned pathways are the most familiar to toxicologists. Nonetheless, to these are added mismatch repair and interstrand crosslink repair. These repair mechanisms are well described in mammals and even in fish.[25] Nevertheless, little is known about DNA repair in invertebrates. In turn, metals (and some metalloids, like As) hold a very distinct mode-of-action as genotoxicants, as they enter the nucleus and intercalate with DNA only at high concentrations. The mechanism is then mostly indirect, through, for instance, the interference with DNA repair mechanisms, generation of reactive oxygen species (ROS) or general metabolic impairment. In the first case, metals like Cd (toxic) and Cu (essential) are believed to displace Zn from the active sites of Zn-finger enzymes involved in DNA damage detection and repair.[26,27] On the other hand, Se and its derived compounds may ameliorate damage, presumably though an anti-oxidative effect.[28] Interestingly, even apoptosis, which can be triggered intrinsically through the *p53* pathway when DNA repair is overwhelmed, can be blocked by toxic metals like Cd even in the presence of potent mutagens like B[*a*]P, which has been found to occur even in marine fish.[29] It may be inferred, therefore, that dealing with environmental samples and field-collected (or tested) animals will most likely imply interpreting results from interaction of toxicants. Moreover, mutagenesis and tumorigenesis are chronic effects that not only take time to occur as the fixation of the mutation may not necessarily correlate to the extent of DNA lesions determined by the Comet assay.[30] Still, in spite of its urgent demand, establishing cause–effect relationships between genotoxicants and neoplasia-related disease in wild aquatic organisms seldom occurs, with few exceptions, such as the notorious work by Myers *et al.*[31] that related PAH exposure to neoplasia-related disease in English sole (*Pleuronectes vetulus*) from the Puget Sound. This work resulted, nonetheless, from extensive sampling campaigns and statistical modelling. Genotoxicity was hitherto not assessed.

1.2.2 Unconventional Models in the Biomonitoring of Aquatic Ecosystems

In the European Union, similarly to other industrialised nations, the need for monitoring the aquatic environment has been translated into policy and regulations, which implies more than mere substance testing, since it is critical that developed programmes consider passive sampling of local species for ERA. In the United States, for instance, monitoring programmes such as NOAA's (National Atmospheric and Oceanic Administration) National Status and Trends programmes (such as Mussel Watch) and the US Geological Survey's Biomonitoring of Status and Trends have been running for many decades for the purpose of ERA in marine and river basins, respectively, surveying a wide range of ecologically and economically relevant species, especially fish (*e.g.* salmonids) and shellfish (like mussels, cockles and clams). However, these approaches do not yet include genotoxicity assessment, favouring long-established methods such as histopathology and toxicant burden analyses. In the European Union, both the Marine Strategy Framework Directive (Directive 2008/56/EC) and the updated Water Framework Directive (Directive 2008/56/EC) mandate the development of effective monitoring programmes for marine and continental waterbodies, which include, as in the famous Descriptor 8 of the first, the need to ascertain the occurrence of deleterious effects to biota as a consequence of anthropogenic action (besides determining the contaminant burden in living marine resources of relevance for human consumption, as stated in Descriptor 9). This means, for an environmental toxicologist, to determine if contamination is rising towards pollution.[32,33] Still, some EU countries have their own long-established monitoring programmes for the aquatic milieu, such as the United Kingdom's Clean Seas Environmental Programme (CSEMP) and the transnational Biological Effects Quality Assurance in Monitoring Programmes (BEQUALM) to standardize biomarker methods. However, as with the aforementioned programmes, neither includes genotoxicity assessment, which is an important gap. In fact, some studies have disclosed the occurrence of genotoxicity in wildlife long after the removal of the stressor or accident clearance. It is the example of mussels (*Mytilus trossulus*) and clams (*Protothaca staminea*) collected from the Exxon Valdez accident area, in Alaska, that revealed an association between DNA damage (determined though the Comet assay) and PAHs.[34] In fact, the Comet assay has already been proposed as one of the main techniques to detect genotoxic effects in marine wildlife after oil spills, in large part owing to the known genotoxicity of many aromatic hydrocarbons.[35] It must be noticed, though, that the Comet assay has reduced or null specificity towards a specific toxicant of class or toxicant, mostly owing to the aforesaid complexity of genotoxicant action, which mandates caution when interpreting data obtained *in situ* or *ex situ* with intricate matrices such as natural waters and sediments. As such, careful planning of experiments, choice of biological model, complementary analyses (*e.g.* chemical determinations) and objective data interpretation are mandatory.[16,36]

1.3 Application of the Comet Assay in Non-conventional Aquatic Models

Non-conventional models are mostly employed as surrogates for wildlife or as indicators of ecosystem status in effects-oriented studies. Since different taxa have distinct molecular and physiological pathways to deal with a pollutant, the "one measure fits all" concept cannot possibly be applied to the biomarker approach in ERA. The choice of model or sentinel/indicator organism, and approach, is thus paramount and results from a careful balance between ecological relevance and the need to circumvent noise variables. It must be noted that the diversity of biological models in studies involving the Comet assay in marine or estuarine organisms is wide. However, the diversity of freshwater "models" keeps increasing, leading to applications in perhaps unsuspected organisms, from fish and molluscs to flatworms, leeches, and reptiles—without neglecting macrophytes, which have been receiving important attention in pesticide-related risk assessment.[37–40]

1.3.1 Fish and Other Vertebrates

Fish are unquestionably one of the most important sentinel organisms in biomonitoring, owing to their ecological and economical relevance and their similarity towards high-order vertebrates, *i.e.* mammals, for which toxicological mechanisms are far better described.[41,42] It must be noted that the zebrafish is, in fact, the only true acknowledged piscine biological model, holding high importance in many fields of research, from toxicology to cancer research, benefitting from high genomic annotation and availability of wild-type and genetically modified strains for high-profile biomedical research.[43] This model is thus out-of-scope of the present work.

Besides their abundance in the wild or availability from aquaculture facilities, the ease of collecting blood samples and performing the Comet assay on whole-blood (since all cells are nucleated, unlike in mammals) renders these models particularly appealing. Moreover, blood cells (more than cell lines) have been found to hold high responsiveness to the Comet assay. Kilemade *et al.*,[44] for instance, disclosed that blood, liver and gill cells were more sensitive than epidermis and spleen in turbots (*Scophthalmus maximus*) exposed *ex situ* to contaminated marine sediments. Still, gills may be an important target owing to the permanent contact with contaminants in waters or fine sediment particles. Della Torre *et al.*[45] revealed DNA strand breakage in conger eel (*Conger conger*) gills (but not liver, kidney, muscle or intestine) from a chemical weapon dumping site in the Mediterranean Sea, likely to be impacted by the old chemical warfare agent yperite ("mustard gas").

In marine and estuarine environments, flatfish (Pleuronectiformes) have particular relevance in the ERA of ecogenotoxicants owing to their economical and ecological relevance and to their close contact with sediments. Furthermore, these fish tend to have relatively slow growth and attain prolonged age, therefore being more prone to acquire chronic toxicopathological disease when subjected to moderate-low levels of toxicants throughout their lifetime. Although far from meeting the requirements as true biological models, their availability in the environment and from mariculture facilities is also an advantage for bioassay-based assessments. Another important aspect regarding these animals, especially when passive biomonitoring (field sampling) is involved, is that flatfish tend to be gregarious and relatively loyal to their habitats.[46] In addition, these animals often inhabit transition ecosystems, such as estuaries, which tend to be some of the most impacted aquatic habitats. The relevance of flatfish yielded attempts to develop specific Comet assay protocols for these animals.[47] Applications of the Comet assay on flatfish are wide and range from passive biomonitoring to field and laboratory assays, with mixed results, albeit with a tendency to yield a good link with the global pattern of environmental (sediment) contamination, especially metals and PAHs, which tend to be the main pollutants of concern in the marine environment. Species are often chosen accordingly to their regional relevance. As such, while soleids like *Solea senegalensis* (Figure 1.3B) tend to be sampled or tested in SW Europe and the Mediterranean,[10,20] while species like the dab (*Limanda limanda*), the English sole or the turbot (*Scophthalmus* spp.) are important targets in northern European countries, Canada and the USA.[43,48–50] Still, among coastal species of interest are also included bass (*Dicentrarchus labrax*), butterfish (*Pholis gunnellus*), wrass (*Symphodus melops*) and even eel (*Anguilla anguilla*).[7,51–53] Interestingly, common species like the gilthead seabream, *Sparus aurata* (Figure 1.3B), for instance, although well-known to ecotoxicologists more dedicated to substance testing[54] in SW Europe and the Mediterranean (mostly owing to its availability from mariculture facilities), have yet to be put to test in real biomonitoring scenarios using the Comet assay. Overall, virtually all of these species are able to produce convincing results in monitoring and substance-testing results and in the integration of the Comet assay with other biomarker techniques, even though the overall appraisal of these works yields the notion that there is much interspecific

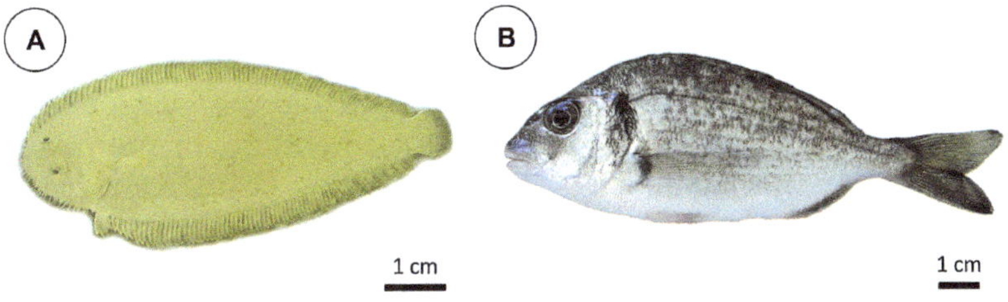

Figure 1.3 Examples of marine teleosts employed in ecotoxicology-related studies that used the Comet assay to determine genotoxicity. (A) Juvenile *Solea senegalensis* Kaup, 1858 (Pleuronectiformes: Soleidae). (B) Juvenile *Sparus aurata* L. 1758 (Perciformes: Sparidae). See [ref. 8, 20, 54], for examples.

variability, which calls for caution when selecting the target organism. Refer to Martins and Costa[10] for a detailed listing of published literature on marine fish.

In freshwater environments, the variety of target species for genotoxicity assessment is too wide to list here in full detail. Not surprisingly, freshwater fish are become increasingly important in the monitoring of tropical ecosystems, whose preservation is a priority worldwide. Tropical fish tend to be robust and easy to breed. More conventional models like the zebrafish, but also tilapia, carp and goldfish (all tropical cyprinids), for instance, greatly benefit from these characteristics. In addition, even in non-tropical countries worldwide these animals are deployed as laboratory models for a variety of subjects, toxicology (environmental or pharmacological) included. It is also the case of the tilapia (*Oreochromis* spp.), whose robustness improves the logistics of long term-assays like the one reported by Lima *et al.*[55] for the testing of chronic genotoxicity caused by animal farm effluents, which showed an increase in both oxidative stress biomarkers and DNA strand breakage. Southern American Characiforms (like *Prochilodus lineatus*) are also gaining ground, such as *Channa punctata* (=*Channa punctatus*) in SE Asia, for both ERA and substance testing.[56–58] The type of stressors that are studied are wide but there is a clear concern with the quality of effluents and the effects of pesticides upon the integrity of DNA (and other endpoints), thus reflecting the priorities of developing rural areas.

Amphibians, on the other hand, are far from being unconventional models in many fields of life science research, including substance or pollutant testing. Comet assay protocols are indeed available for model species such as the frog *Xenopus laevis*. The reader is thus diverted, for instance, to the review by de Lapuente *et al.*[9] However, biomonitoring approaches for ecogenotoxicants in freshwater ecosystems with wild amphibians are less common. These animals are very sensitive to pollutants and many are endangered species, which, in spite of increasing their relevance, may render extensive campaigns prohibitive. Still, there are a few works that should be mentioned as important examples. It is the case of the work by Gonçalves *et al.*[59] with tadpoles of the tropical frog *Dendropsophus minutus*. This work integrated both Comet assay and the micronuclei test (in blood, similarly to fish) and revealed the good sensitivity of both methods by comparing animals from impacted and reference sites. Maselli *et al.*[60] also stated the sensitivity of undisclosed anuran species collected in the wild for ecogenotoxicity assessment with the Comet assay. Furthermore, in the past there have been successful attempts to develop bioassays *in situ* with caged tadpoles, which may be an interesting alternative to passive biomonitoring.[61]

The application of the Comet assay in higher-order aquatic vertebrates is uncommon. As one of the few examples, Lee *et al.*[62] applied the Comet assay in the bottlenose dolphin (*Tursiops truncatus*) lymphocytes of dolphins from an impacted and a reference location, but yielded a stronger relation between susceptibility to acquire infection than environmental status. Caliani *et al.*[63] successfully surveyed DNA strand breakage in whole-blood of the loggerhead turtle (*Caretta caretta*) collected in the Mediterranean Sea but might have neglected factors such as migration or signs of trauma in animals. Similarly, Zapata *et al.*[64] studied the application of the Comet assay and the micronucleus test in blood cells of the freshwater turtle *Trachemys callirostris* collected from several sites in Colombia. The authors claimed technical success and established a baseline of DNA damage for future endeavours. It must be noted, though, that these works with high-order vertebrates almost invariably suffer from constraints such as reduced number of specimens and high interspecific variability, which cannot thus be overcome by stratifying sampling into male/female or adult/juvenile, for instance. Furthermore, in spite of the many attempts to standardize the protocol (including the recent OECD guidelines for *in vivo* testing with mammals), the Comet assay is essentially comparative in biomonitoring studies, which requires adequate reference areas, some assurance that migration is limited and considering other noise factors such as natural disease.

1.3.2 Molluscs

Bivalves offer special advantages in ERA, from abundance and ecological relevance to the fact that they are easy to collect, handle and often even breed. In addition, these are fixed organisms and therefore they reflect the conditions of a given area throughout their lifecycle. The metabolization of organic toxicants like PAHs into genotoxic metabolites is, however, thought to be reduced, in comparison with vertebrates.[65] However, CYP-like proteins are known to exist in marine and freshwater bivalves and metabolization occurs that is able to generate PAH metabolite–DNA adducts.[66–69] Still, information regarding these pathways and their relation to DNA in molluscs and invertebrates in general is overall scarce, which further complicates the interpretation of results from the Comet assay since it is possible that lower or null metabolization may yield false negatives. On the other hand, differences in DNA repair, compared to their vertebrate counterparts, may result in increased sensitivity to genotoxicants and/or raising of the baseline levels of DNA damage by natural accumulation of lesions.

Marine mussels (*Mytilus* spp.) are one of the most important sentinel organisms for marine ERA for being abundant and sensitive, being applied in bioassays and passive sampling.[70,71] Nonetheless, sediment-burrowing bivalves like cockles and clams have also been surveyed through the Comet assay. As an example, Dallas *et al.*[71] disclosed good sensitivity towards DNA strand breakage in haemocytes of both mussels and cockles collected from an impacted estuary. However, differential sensitivity of marine bivalves has been reported, with sediment-burrowers potentially being more able to cope with ecogenotoxicants.[72] In another example, Martins *et al.*[73] disclosed DNA strand-breakage to occur in the marine clam *Ruditapes decussata* (=*R. decussatus*) subjected to sediment bioassays, inclusively in animals exposed to realistic concentrations of phenanthrene, a PAH regarded as non-carcinogenic, in a study involving the Comet assay adapted to gill cells.

Oysters are also of particular importance in biomonitoring programmes and have been deployed as target organisms in a range of studies, from basic substance testing to passive biomonitoring, as in the work by Bisset *et al.*,[74] who performed geospatial analysis of DNA strand breakage in the oyster *Crassostrea virginica* collected from an impacted bay in southern USA and disclosed good correlation with the proximity to industrial areas.

There are many species of freshwater bivalves that have been successfully applied in ecogenotoxicological studies with the Comet assay. There is, however, a trend favouring the zebra mussel (*Perna viridis*) for laboratory testing, whereas passive or active biomonitoring involves a multiplicity of species of regional significance, including invasive species such as the Asian clam, *Corbicula fluminea*.[75–78] It must be noted, though, that the heterogeneity of freshwater environments and the higher effects of seasonality are often noted to have a significant impact on Comet assay data.[76,77] Yet among aquatic molluscs, gastropods, especially freshwater, such as *Lymnaea* spp., have also been gaining some attention, albeit almost invariably in works related to the testing of various substances and even nanomaterials and not so much in true biomonitoring procedures.[18,19,79] Interestingly, Vincent-Hubert *et al.*[79] hypothesized that Cd may inhibit DNA repair enzymes in the freshwater snail *Potamopyrgus antipodarum*, similarly to what has been discussed for vertebrates. Overall, the full potential of these organisms for the ERA of ecogenotoxicants remains to be ascertained. Still, their ease to collect in the environment, breed and handle in the laboratory renders these animals interesting non-conventional models in the field of research. Among the rarest examples within molluscs are the application of the Comet assay in wild *Octopus vulgaris*, revealing differential baseline levels of DNA strand breakage between distinct organs, with the gonads yielding the lowest levels, followed by the "kidney", gills and digestive gland.[80]

1.3.3 Other Organisms: From Crustaceans to Algae

There are few reports on the use of the Comet assay to measure DNA strand breaks in crustaceans and they almost exclusively deal with toxicity testing rather than biomonitoring. The freshwater crustacean *Daphnia magna*, an acknowledged model species, may be considered the main target and has its own protocols for performing the Comet assay on the haemolymph-derived cells[81] that may be a good basis for other small-sized species. Among the few examples with unconventional species are works such as that from Hook and Lee,[82] who exposed embryos of the marine shrimp *Palaemonetes pugio* to Cr and B[*a*]P to study DNA repair among different developmental stages, disclosing no differences during exposure but showing that later stage embryos could recover faster from insult. In one of the scarce examples of biomonitoring approaches with crustaceans, Roberts *et al.*[83] investigated DNA strand breakage through the Comet assay in wild *Corophium volutator* (Amphipoda) to address the interaction between ocean acidification and metal toxicity.

Although widely employed in ecotoxicology, there are few applications of the Comet assay on Polychaeta, for which coelomocytes are the preferred target for the assay (as for their terrestrial counterparts, the Oligochaeta), most of which are related to substance testing.[84–86] Interestingly, there is some evidence in deposit-feeding Polychaeta of the ability to metabolize PAHs and generate DNA strand breakage in parallel.[87] Even though the mechanisms remain obscure, this makes these benthic animals interesting for ERA of aquatic sediments. Finally, a word must be provided for taxa that have been almost neglected from ecogenotoxicity, such as echinoderms and cnidarians. These organisms, so far subjected only to model toxicants in attempts to optimize protocols, have already yielded promising results with the Comet assay, revealing at least comparable sensitivity to other aquatic organisms such as mussels.[88,89]

Even though aquatic plants (marine included), have been proposed as indicator and sentinel organisms for ERA,[90] only a few studies have, so far, been conducted on ecogenotoxicity assessment. Overall, aquatic macrophytes and algae pose important problems for the Comet assay owing to the existence of a cell wall and subsequent impairment of lysis. However, there are a few interesting studies with freshwater macrophytes to address the effect of pesticides, which is an urgent problem at least in wetlands and other lentic ecosystems surrounding agricultural areas.[40] These promising works, which employ mechanical exposure of nuclei through tissue splicing, may indicate a novel path in ecogenotoxicity testing. Conversely, phytoplanktonic microalgae, in spite of the convenience of analysing cell suspensions, have produced null or scant results, in most part due to problems with cell lysis.[91] Table 1.1 summarizes some of the most significant examples on the application of the Comet assay in non-conventional aquatic model organisms.

1.4 Methods

1.4.1 The Comet Assay and its Modifications: Discriminating Type of Damage and Addressing DNA Repair in Unconventional Aquatic Models

The Comet assay has been widely applied *in vitro* and *in vivo* in model and non-model eukaryotic lifeforms. From the same basic protocol, several adaptations have been derived to fit cell cultures, whole blood or haemolymph, and cells harvested from solid tissues of diverse model and non-model organisms. However, the standard Comet assay is composed of basic eight steps: (i) collecting samples from live specimens; (ii) suspending cells in adequate buffer; (iii) embedding cells in a gel matrix; (iv) spreading the embedded cells onto pre-coated microscope slides; (v) lysing cells with detergents in a hypersaline buffer; (vi) promoting DNA alkali unwinding; (vii) subjecting the nucleoids to electrophoresis in an alkaline moiety and (viii) neutralizing, staining and scoring the slides. Besides proper collection

Table 1.1 Summary of example applications of the Comet assay with non-conventional models discriminated by main taxa and habitat of the species.

Organism	Environment	Approach	Toxicant(s)	Reference
Amphibians				
Bufo americanus	Freshwater	*In situ* bioassay	Undisclosed	61
Dendropsophus minutus	Freshwater	Passive sampling	Likely agricultural	59
Rana clamitans	Freshwater	*In situ* bioassay	Undisclosed	61
Annelids				
Arenicola marina	Marine	*Ex situ* bioassay	B[*a*]P	85
Capitella capitata	Marine	*Ex situ* bioassay	PAH (fluoranthene)	84
Hirudo verbana	Freshwater	*Ex situ* bioassay	Al compounds (water and sediments)	38
Nereis diversicolor	Marine	*Ex situ* bioassay	B[*a*]P, Ag nanoparticles	85,86
Nereis virens	Marine	*Ex situ* bioassay	PAH (fluoranthene)	84
Cnidarians				
Anthopleura elegantissima	Marine	*Ex situ* bioassay	B[*a*]P	88
Crustaceans				
Corophium volutator	Marine	*Ex situ* bioassay	Metals (sediment-bound)	83
Palaemonetes pugio	Marine	*Ex situ* bioassay	B[*a*]P and Cr	82
Echinoderms				
Asterias rubens	Marine	*Ex situ* bioassay	Model genotoxins	89
Fish				
Anguilla anguilla	Catadromous	*Ex situ* bioassay	Yperite	45
Carassius carassius	Freshwater	*Ex situ* bioassay	Pesticides (flood water)	21
Conger conger	Marine	Passive sampling	Yperite	45
Channa punctatus	Freshwater	*Ex situ* bioassay	Pendimethalin	56
Dicentrarchus labrax	Marine	*Ex situ* bioassay	PAHs (sediment-bound)	7
Limanda limanda	Marine	Passive sampling	Likely PAHs	48,50
Oreochromis niloticus	Freshwater	*Ex situ* bioassay	Livestock industry effluents	55
Pholis gunnellus	Marine	Passive sampling	Undisclosed	52
Platichthys flesus	Marine	Passive sampling	Mixed	92
Pleuronectes vetulus	Marine	Passive sampling	Mixed (potential endocrine disruptors)	49

Pleuronichthys verticalis	Marine	Passive sampling	Mixed (potential endocrine disruptors)	49
Prochilodus lineatus	Freshwater	Passive sampling, *ex situ* bioassay	Mixed (agricultural, urban and industrial), Cu	57,58
Scophthalmus maximus	Marine	*Ex situ* bioassay	Mixed (sediment-bound)	44
Solea senegalensis	Marine	*Ex situ/in situ* bioassay	Mixed (sediment-bound)	6,20
Sparus aurata	Marine	*Ex situ*	Cu	54
Symphodus melops	Marine	*Ex situ* bioassay	Styrene	51
Zoarces viviparus	Marine	Passive sampling	Mixed (oil spill)	93
Flatworms				
Dugesia schubarti	Freshwater	*Ex situ* bioassay	$CuSO_4$	37
Macrophytes				
Myriophyllum quitense	Freshwater	*Ex situ* bioassay	Azoxystrobin	40
Mammal				
Tursiops truncatus	Marine	Passive sampling	Unknown	62
Molluscs				
Cerastoderma edule	Marine	Passive sampling	Metals	71
Corbicula fluminea	Freshwater	*Ex situ* bioassay	Atrazine and Roundup	78
Crassostrea virginica	Marine	Passive sampling	Mixed	74
Dreissena polymorpha	Freshwater	*Ex situ/in situ* bioassay	Cd and B[*a*]P, PAHs	75,77
Lymnea luteola	Freshwater	*Ex situ* bioassay	Ag nanoparticles, carbon nanotubes	18,19
Mytilus edulis	Marine	*Ex situ* bioassay, passive sampling	Styrene, undisclosed environmental pollutants, PAHs (sediment-bound), metals	51,68,70,71
Mytilus trossulus	Marine	Passive sampling	PAHs (oil spill)	34
Octopus vulgaris	Marine	Passive sampling	Likely metals	80
Potamopyrgus antipodarum	Freshwater	*Ex situ* bioassay	Cd and bisphenol	79
Protothaca staminea	Marine	Passive sampling	PAHs (oil spill)	34
Ruditapes decussata	Marine	*Ex situ* bioassay	PAHs (sediment-bound)	73
Scapharca inaequivalvis	Marine	*Ex situ*	Cu	54
Scrobicularia plana	Marine	*Ex situ* bioassay	Ag nanoparticles	86
Sinanodonta woodiana	Freshwater	Passive sampling	Phosphates and metals	76
Reptiles				
Caretta caretta	Marine	Passive sampling	Unknown	63
Trachemys callirostris	Freshwater	Passive sampling	Unknown	64

of tissue and cells (a high percentage of viable cells should be achieved, typically 70% or more), one of the main points for the success of the Comet assay is to ensure that accessory DNA damage is avoided, *e.g.* by working under dim light and in the cold. Even though some cryopreservation protocols for cell suspensions have been developed, the Comet assay should be performed immediately after harvesting biological material to avoid DNA degradation. In addition, it is of great importance that the researcher keeps the protocol constant, once this is set, to safeguard the comparability of results. Table 1.2 provides a general protocol suitable for most circumstances.

Peripheral fluids such as blood and haemolymph from aquatic vertebrates and invertebrates, respectively, are suitable and logistics-friendly. In addition, collection may be performed in a non-destructive way. In fish, good quality blood samples may be collected from the caudal peduncle or immediately above the lateral line with a syringe treated (pre-washed) with an anticoagulant such as heparin or EDTA. Immediately after collection, the samples should be diluted in cold phosphate-buffered saline (PBS), typically between 1:100 and 1:1000. Too few or too many cells in the Comet field may pose problems upon scoring. On the other hand, haemolymph is usually more difficult to obtain, depending on species and location of puncture. In mussels and many other bivalves, haemolymph can be collected by an expert hand from the adductor muscles. Kenny's Balanced Salt Solution (KBSS) is the commonly recommended buffer to suspend cells harvested from most invertebrates, especially molluscs. In crustaceans, however, syringe puncturing may be more problematic, particularly in smaller animals. To circumvent this, Pellegri *et al.*[81] devised a protocol for *Daphnia magna* in which a quick step with an amalgamator device in presence of glass microspheres proved to be efficient for haemolymph extraction in this species and that PBS is the most appropriate buffer. Cells may be obtained from solid tissue by brief mechanical splicing (chopping) the material in adequate buffer, followed by soft pipetting.[73] Previous protocols for animal tissue included a collagenase step. However, this can usually be omitted, also avoiding accessory DNA damage. In order to precipitate the debris and damaged cells, centrifuging the cell suspension at low speed is an effective procedure in order to obtain the supernatant ready to be embedded in LMPA (low-melting point agarose). Plant material is more problematic owing to the presence of cell walls, but mechanical extraction of nuclei tends to be efficient in soft material.[90]

Recently, some modifications of the standard Comet protocol have been developed in order to detect different types of DNA damage and, eventually, to address mechanisms of DNA repair. These approaches are based on the excision of damage nucleotides by lesion-specific endonucleases after cell lysis, then generating breaks that are detectable after alkali unwinding and electrophoresis. As such, these enzyme-linked Comet assay approaches tend to improve the sensitivity of the method. Commercial forms of these restriction enzymes are now available, usually human forms obtained from recombinant bacteria. The most common are the BER enzymes oxoguanine

Table 1.2 Proposed consensus alkaline Comet assay protocol suitable for the majority of biological models. All steps should be conducted under controlled temperature and dim light or dark (immersion steps) whenever possible.

Step	Duration	Temperature	Solution/buffer	Notes
Cell suspensions	—	4 °C	PBS (vertebrates and plant material) KBSS (invertebrates)	Dilute whole blood or haemolymph 1 : 100 to 1 : 1000. If working with solid tissue, rapidly mince in cold buffer and add sample and buffer to a tube in the proportion of 1 : 10 or greater. Release the cells by soft pipetting. Centrifuge briefly at low speed and use only the supernatant for the assay and for cell counting. Typically, No enzymatic steps (*e.g.* collagenase) are needed. For difficult plant material, refer to Costa *et al.*[94] and references therein.
Dilution in LMPA	—	~37 °C	0.5–1% LMPA (in PBS or KBSS)	Dilute cell suspension in molten LMPA (1 : 100). Typically, the *in vitro* Comet assay employs lower concentrations of LMPA (0.5–0.75%), whereas 1% is suitable for *in vivo* studies.
Preparation of slides	—	Room (~20 °C)	—	Place 2×75–80 μL of cells in LMPA per pre-coated slide. Use frosted or single-frosted slides for labelling.
Place coverslip			—	
Solidification	15 min	4 °C	—	

Table 1.2 (*Continued*)

Step	Duration	Temperature	Solution/buffer	Notes
Remove coverslip			—	
Lysis	30–60 min	4 °C	0.45 M NaCl (m/v); 40 mM EDTA (m/v); 5 mM Tris (pH 10).	The buffer has a short shelf-life even in the cold (1–2 weeks). Add 10% v/v DMSO and ~1% v/v Triton X-100 before use. The buffer cannot be re-used.
DNA unwinding	40 min	4 °C	0.1 µM EDTA; 0.3 M NaOH (pH ≈ 13)	
Electrophoresis	30 min	4 °C	0.1 µM EDTA; 0.3 M NaOH (pH ≈ 13)	Run electrophoresis at 25 V. Buffer can be re-used once.[a]
Neutralizing	15 min	4 °C	0.2 M Tris-HCl buffer (pH 7.5)	
[Fixation]	*10 min*	*4 °C*	*Absolute methanol*	*[Optional]*
Stain	5–10 min	Room (~20 °C)	Ethidium bromide (0.02 mg L^{-1} in water) or suitable replacement, such as SYBR Green	The staining time (to be done in the dark) allows the dye to bind to DNA, otherwise bleaching of the dye will occur immediately with exposure to UV. If working with pre-fixed, dry, slides, rehydrate in cold distilled water for at least 15 min. The slides are not to be washed after staining. Mount with coverslip. Caution: ethidium bromide is a suspected mutagen.[b]

[a]If the electrophoresis speed is too low, check the volume of the buffer in the tank (should just cover the slides).
[b]The slides can be washed in absolute methanol or ethanol after scoring, stored dry and re-analysed.

glycosylase (OGG) and formamidopyrimidine glycosylase (FPG), which convert sites containing oxidized nucleobases (such as oxoguanine) to single-strand breaks.[95] By contrasting the findings from the enzyme-linked and the standard Comet assay, a proportion of DNA oxidative damage can be derived. Although almost routinely employed in clinical research using mammalian cell lines,[96] with applications in ecotoxicology studies with human cell cultures,[97] the enzyme-modified Comet assay is just starting to be applied in ecogenotoxicology studies with unconventional models. It is the case, for instance, in the work by Gielazyn *et al.*[98] to detect oxidative DNA lesions in the oyster *Crassostrea virginica* and the clam *Mercenaria mercenaria*. Additionally, there is a modification of the Comet assay to detect alterations to the organisms' ability to repair DNA *per se*, which is based on treating suspended cells with a DNA-damaging agent like ethylmethylsulfonate (EMS), followed by a recovery period and then running the Comet assay.[99] Nonetheless, this protocol is not yet practiced *in vivo* in ecotoxicology. It must be noted, though, that alterations to the mechanisms of DNA repair in fish cell lines have already been proposed as potential biomarkers of genotoxicity.[25]

1.4.2 Final Remarks on Analysis and Interpretation of Data

There are many factors that influence the Comet assay performance, such as variations in LMPA concentration, alkaline incubation time, electrophoresis voltage (and duration) plus the scoring method, which were pointed out by Azqueta and Collins[100] as issues that can increase the variability of the results. As an example, small variations in electrophoresis speed may increase the size of tails without meaning increased DNA fragmentation, therefore rendering Comet length highly biased as a metric when compared to relative metrics. In addition, performing the Comet assay *in vivo* yields considerable variations in the levels of DNA damage within the same slide, since distinct types of cells at different stages of their life cycle are being surveyed. However, this hindrance can be overcome if at least 50–100 intact nucleoids per slide/individual were scored. In general, employing relative metrics, such as the %DNA in the tail or Olive Tail Moment, may avoid constraints created by variations induced by protocol or something as simple as genome size.[10,101]

Statistics and sampling or experimental planning are also paramount to guarantee data quality and have been discussed elsewhere.[10] Very importantly, the researcher must be aware that non-conventional models, whether reared in the laboratory or collected in the wild, offer high intraspecific variability that will invariably dilute statistical significance if replication and stratification by factors like age and gender are not properly taken into account. In addition, proper controls and references are mandatory in ERA in order to achieve the most important yet the most challenging goal in ecotoxicological studies in such complex areas like aquatic ecosystems—causality.

Abbreviations

B[*a*]P	benzo[*a*]pyrene
BER	Base excision repair
CYP	Cytochrome P450
ERA	Environmental risk assessment
HR	Homologous recombination
KBSS	Kenny's balanced salt solution
LMPA	Low melting point agarose
MFO	Mixed-function oxidase
NA	Nuclear abnormality
NHEJ	Non-homologous end-joining
NER	Nucleotide excision repair
NMPA	Normal melting point agarose
PAH	Polycyclic aromatic hydrocarbon
PBS	Phosphate-buffered saline
ROS	Reactive oxygen species
SCGE	Single cell gel electrophoresis
TCDD	Tetrachlorodibenzodioxin

References

1. N. P. Singh, M. T. McCoy, R. R. Tice and E. L. Schneider, A simple technique for quantitation of low levels of DNA damage in individual cells, *Exp. Cell Res.*, 1988, **175**, 184–191.
2. O. Östling and K. J. Johanson, Microelectrophoretic study of radiation-induced DNA damages in individual mammalian cells, *Biochem. Biophys. Res. Commun.*, 1984, **123**, 291–298.
3. R. R. Tice, E. Agurell, D. Anderson, B. Burlinson, A. Hartmann, H. Kobayashi, Y. Miyamae, E. Rojas, J.-C. Ryu and Y. F. Sasaki, Single cell gel/comet assay: guidelines for *in vitro* and *in vivo* genetic toxicology testing, *Environ. Mol. Mutagen.*, 2000, **35**, 206–221.
4. M. Fenech, M. Kirsch-Volders, A. T. Natarajan, J. Surralles, J. W. Crott, J. Parry, H. Norppa, D. A. Eastmond, J. D. Tucker and P. Thomas, Molecular mechanisms of micronucleus, nucleoplasmic bridge and nuclear bud formation in mammalian and human cells, *Mutagenesis*, 2011, **26**, 125–132.
5. C. Bolognesi and M. Hayashi, Micronucleus assay in aquatic animals, *Mutagenesis*, 2011, **26**, 205–213.
6. P. M. Costa, J. Lobo, S. Caeiro, M. Martins, A. M. Ferreira, M. Caetano, C. Vale, T. A. Delvalls and M. H. Costa, Genotoxic damage in *Solea senegalensis* exposed to sediments from the Sado Estuary (Portugal): Effects of metallic and organic contaminants, *Mutat. Res.*, 2008, **654**, 29–37.
7. M. Martins, A. M. Ferreira, M. H. Costa and P. M. Costa, Comparing the genotoxicity of a potentially carcinogenic and a noncarcinogenic PAH,

singly, and in binary combination, on peripheral blood cells of the European sea bass, *Environ. Toxicol.*, 2016, **31**, 1307–1318.
8. C. L. Mitchelmore and J. K. Chipman, DNA strand breakage in aquatic organisms and the potential value of the comet assay in environmental monitoring, *Mutat. Res.*, 1998, **399**, 135–147.
9. J. de Lapuente, J. Lourenço, S. A. Mendo, M. Borràs, M. Martins, P. M. Costa and M. Pacheco, The Comet Assay and its application in the field of ecotoxicology: A mature tool that continues to expand its perspectives, *Front. Genet.*, 2015, **6**, 180.
10. M. Martins and P. M. Costa, The comet assay in Environmental Risk Assessment of marine pollutants: applications, assets and handicaps of surveying genotoxicity in non-model organisms, *Mutagenesis*, 2015, **30**, 89–106.
11. P. M. Chapman, Determining when contamination is pollution - weight of evidence determinations for sediments and effluents, *Environ. Int.*, 2007, **33**, 492–501.
12. G. Chen and P. A. White, The mutagenic hazards of aquatic sediments: a review, *Mutat. Res.*, 2004, **567**, 151–225.
13. T. Ohe, T. Watanabe and K. Wakabayashi, Mutagens in surface waters: a review, *Mutat. Res.*, 2004, **567**, 109–149.
14. R. van der Oost, J. Beyer and N. P. Vermeulen, Fish bioaccumulation and biomarkers in environmental risk assessment: A review, *Environ. Toxicol. Pharmacol.*, 2003, **13**, 57–149.
15. M. L. Martín-Díaz, J. Blasco, D. Sales and T. A. DelValls, Biomarkers as tools to assess sediment quality, *Trends Anal. Chem.*, 2004, **23**, 807–818.
16. P. M. Chapman, F. Wang and S. S. Caeiro, Assessing and managing sediment contamination in transitional waters, *Environ. Int.*, 2013, **55**, 71–91.
17. IARC, A review of human carcinogens - chemical agents and related occupations, *IARC Monogr.*, 2012, **100F**, 1–32.
18. D. Ali, Oxidative stress-mediated apoptosis and genotoxicity induced by silver nanoparticles in freshwater snail *Lymnea luteola* L., *Biol. Trace Elem. Res.*, 2014, **162**, 333–341.
19. D. Ali, M. Ahmed, S. Alarifi and H. Ali, Ecotoxicity of single-wall carbon nanotubes to freshwater snail *Lymnaea luteola* L.: Impacts on oxidative stress and genotoxicity, *Environ. Toxicol.*, 2015, **30**, 674–682.
20. P. M. Costa, T. Neuparth, S. Caeiro, J. Lobo, M. Martins, A. M. Ferreira, M. Caetano, C. Vale, T. A. DelValls and M. H. Costa, Assessment of the genotoxic potential of contaminated estuarine sediments in fish peripheral blood: laboratory *versus in situ* studies, *Environ. Res.*, 2011, **111**, 25–36.
21. T. Polard, S. Jean, L. Gauthier, C. Laplanche, G. Merlina, J. M. Sánchez-Pérez and E. Pinelli, Mutagenic impact on fish of runoff events in agricultural areas in south-west France, *Aquat. Toxicol.*, 2011, **101**, 126–134.

22. W. Xue and D. Warshawsky, Metabolic activation of polycyclic and heterocyclic aromatic hydrocarbons and DNA damage: A review, *Toxicol. Appl. Pharmacol.*, 2005, **206**, 73–93.
23. J. Cadet, T. Douki and J. L. Ravanat, Oxidatively generated base damage to cellular DNA, *Free Radical Biol. Med.*, 2010, **49**, 9–21.
24. M. Fousteri and L. H. F. Mullenders, Transcription-coupled nucleotide excision repair in mammalian cells: molecular mechanisms and biological effects, *Cell Res.*, 2008, **18**, 73–74.
25. A. Kienzler, S. Bony and A. Devaux, DNA repair activity in fish and interest in ecotoxicology: A review, *Aquat. Toxicol.*, 2013, **134–135**, 47–56.
26. M. Asmuss, L. H. Mullenders, A. Eker and A. Hartwig, Differential effects of toxic metal compounds on the activities of Fpg and XPA, two zinc finger proteins involved in DNA repair, *Carcinogenesis*, 2000, **21**, 2097–2104.
27. A. Hartwig, M. Asmuss, I. Ehleben, U. Herzer, D. Kostelac, A. Pelzer, T. Schwerdtle and A. Bürkle, Interference by toxic metal ions with DNA repair processes and cell cycle control: molecular mechanisms, *Environ. Health Perspect.*, 2002, **110**(S5), 797–799.
28. E. Soto-Reyes, L. M. Del Razo, M. Valverde and E. Rojas, Role of the alkali labile sites, reactive oxygen species and antioxidants in DNA damage induced by methylated trivalent metabolites of inorganic arsenic, *Biometals*, 2005, **18**, 493–506.
29. P. M. Costa, E. Chicano-Gálvez, J. López Barea, T. À. DelValls and M. H. Costa, Alterations to proteome and tissue recovery responses in fish liver caused by a short-term combination treatment with cadmium and benzo[a]pyrene, *Environ. Pollut.*, 2010, **158**, 3338–3346.
30. M. Šrut, A. Štambuk and G. I. Klobučar, What is Comet assay not telling us: AFLP reveals wider aspects of genotoxicity, *Toxicol. In Vitro*, 2013, **27**, 1226–1232.
31. M. S. Myers, L. L. Johnson and T. K. Collier, Establishing the causal relationship between polycyclic aromatic hydrocarbon (PAH) exposure and hepatic neoplasms and neoplasia-related liver lesions in English sole (*Pleuronectes vetulus*), *Hum. Ecol. Risk Assess.*, 2003, **9**, 67–94.
32. Á. Borja, M. Elliott, J. Carstensen, A.-S. Heiskanen and W. van de Bund, W. Marine management-towards an integrated implementation of the European Marine Strategy Framework and the Water Framework Directives, *Mar. Pollut. Bull.*, 2010, **60**, 2175–2186.
33. B. P. Lyons, J. E. Thain, G. D. Stentiford, K. Hylland, I. M. Davies and A. D. Vethaak, Using biological effects tools to define good environmental status under the European Union Marine Strategy Framework Directive, *Mar. Pollut. Bull.*, 2010, **60**, 1647–1651.
34. R. E. Thomas, M. Lindeberg, P. M. Harris and S. D. Rice, Induction of DNA strand breaks in the mussel (*Mytilus trossulus*) and clam (*Protothaca staminea*) following chronic field exposure to polycyclic aromatic

hydrocarbons from the Exxon Valdez spill, *Mar. Pollut. Bull.*, 2007, **54**, 726–732.
35. C. Martínez-Gómez, D. Vethaak, K. Hylland, T. Burgeot, A. Köhler, B. P. Lyons, J. Thain, M. J. Gubbins and I. M. Davies, A guide to toxicity assessment and monitoring effects at lower levels of biological organization following marine oil spills in European waters, *ICES J. Mar. Sci.*, 2010, **67**, 1105–1118.
36. J. A. Hagger, M. B. Jones, D. R. Leonard, R. Owen and T. S. Galloway, Biomarkers and integrated environmental risk assessment: are there more questions than answers? *Integr. Environ. Assess. Manage.*, 2006, **2**, 312–329.
37. T. Guecheva, J. A. P. Henriques and B. Erdtmann, Genotoxic effects of copper sulphate in freshwater planarian *in vivo*, studied with the single-cell gel test (comet assay), *Mutat. Res.*, 2001, **497**, 19–27.
38. Z. Mihaljević, I. Ternjej, I. Stanković, M. Kerovec and N. Kopjar, Application of the comet assay and detection of DNA damage in haemocytes of medicinal leech affected by aluminium pollution: A case study, *Environ. Pollut.*, 2009, **157**, 1565–1572.
39. G. L. Poletta, A. Larriera, E. Kleinsorge and M. D. Mudry, Genotoxicity of the herbicide formulation Roundup® (glyphosate) in broad-snouted caiman (*Caiman latirostris*) evidenced by the Comet assay and the Micronucleus test, *Mutat. Res.*, 2009, **672**, 95–102.
40. D. S. Garanzini and M. L. Menone, Azoxystrobin causes oxidative stress and DNA damage in the aquatic macrophyte *Myriophyllum quitense*, *Bull. Environ. Contam. Toxicol.*, 2015, **94**, 146–151.
41. C. L. Bolis, M. Piccolella, A. Z. Dalla Valle and J. C. Rankin, Fish as model in pharmacological and biological research, *Pharmacol. Res.*, 2001, **44**, 265–280.
42. A. Hallare, T.-B. Seiler and H. Hollert, The versatile, changing, and advancing roles of fish in sediment toxicity assessment - a review, *J. Soils Sediments*, 2001, **11**, 141–173.
43. H. Feistma and E. Cuppen, Zebrafish as a cancer model, *Mol. Cancer Res.*, 2008, **6**, 685–694.
44. M. F. Kilemade, M. G. J. Hartl, D. Sheehan, C. Mothersill, F. N. A. M. van Pelt, J. O'Halloran and N. M. O'Brien, Genotoxicity of field-collected inter-tidal sediments from Cork Harbor, Ireland, to juvenile turbot (*Scophthalmus maximus* L.) as measured by the Comet assay, *Environ. Mol. Mutagen.*, 2004, **44**, 56–64.
45. C. Della Torre, T. Petochi, C. Farchi, I. Corsi, M. M. Dinardo, V. Sannarini, L. Alcaro, L. Mechelli, S. Focardi, A. Tursi, G. Marino and E. Amato, Environmental hazard of yperite released at sea: Sublethal toxic effects on fish, *J. Hazard. Mater.*, 2013, **248–249**, 246–253.
46. C. Vinagre, J. Salgado, H. N. Cabral and M. J. Costa, Food web structure and habitat connectivity in fish estuarine nurseries - impact of river flow, *Estuaries Coasts*, 2011, **34**, 663–674.

47. K. Belpaeme, K. Cooreman and M. Kirsch-Volders, Development and validation of the *in vivo* alkaline comet assay for detecting genomic damage in marine flatfish, *Mutat. Res.*, 1998, **415**, 167–184.
48. F. Akcha, G. Leday and A. Pfohl-Leszkowicz, Measurement of DNA adducts and strand breaks in dab (*Limanda limanda*) collected in the field: effects of biotic (age, sex) and abiotic (sampling site and period) factors on the extent of DNA damage, *Mutat. Res.*, 2004, **552**, 197–207.
49. M. A. Rempel, J. Reyes, S. Steinert, W. Hwang, J. Amstrong, K. Sakamoto, K. Kelley and D. Schlenk, Evaluation of relationships between reproductive metrics, gender and vitellogenin expression in demersal flatfish collected near the municipal wastewater outfall of Orange County, California, USA, *Aquat. Toxicol.*, 2006, **77**, 241–249.
50. M.-H. Dévier, M. Le Dû-Lacoste, F. Akcha, B. Morin, L. Peluhet, K. Le Menach, T. Burgeot and H. Budzinsky, Biliary PAH metabolites, EROD activity and DNA damage in dab (*Limanda limanda*) from Seine Estuary (France), *Environ. Sci. Pollut. Res.*, 2013, **20**, 708–722.
51. E. Mamaca, R. K. Bechmann, S. Torgrimsen, E. Aas, A. Bjørnstad, T. Baussant and S. Le Floch, The neutral red lysosomal retention assay and Comet assay on haemolymph cells from mussels (*Mytilus edulis*) and fish (*Symphodus melops*) exposed to styrene, *Aquat. Toxicol.*, 2005, **75**, 191–201.
52. V. Bombail, D. Aw, E. Gordon and J. Batty, Application of the comet and micronucleus assays to butterfish (*Pholis gunnellus*) erythrocytes from the Firth of Forth, Scotland, *Chemosphere*, 2001, **44**, 383–392.
53. M. Benedetti, F. Ciaprini, F. Piva, F. Onorati, D. Fattorini, A. Notti, A. Ausili and F. Regoli, A multidisciplinary weight of evidence approach for classifying polluted sediments: Integrating sediment chemistry, bioavailability, biomarkers responses and bioassays, *Environ. Int.*, 2012, **38**, 17–28.
54. R. Gabrianelli, G. Lupidi, M. Villarini and G. Falconi, DNA damage induced by copper on erythrocytes of gilthead sea bream *Sparus aurata* and mollusk *Scapharca inaequivalvis*, *Arch. Environ. Contam. Toxicol.*, 2003, **45**, 350–360.
55. P. L. Lima, J. C. Benassi, R. C. Pedrosa, J. Dal Magro, T. B. Oliveira and D. Wilhelm Filho, Time-course variations of DNA damage and biomarkers of oxidative stress in tilapia (*Oreochromis niloticus*) exposed to effluents from a swine industry, *Arch. Environ. Contam. Toxicol.*, 2006, **50**, 23–30.
56. I. Ahmad and M. Ahmad, Fresh water fish, *Channa punctatus*, as a model for pendimethalin genotoxicity testing: A new approach toward aquatic environmental contaminants, *Environ. Toxicol.*, 2016, **31**, 1520–1529.
57. C. A. Freire, L. R. Souza-Bastos, J. Chiesse, F. H. Tincani, L. D. S. Piancini, M. A. F. Randi, V. Prodocimo, M. M. Cestari, H. C. Silva-de-Assis, V. Abilhoa, J. R. S. Vitule, L. P. Bastos and C. A. de Oliveira-Ribeiro, A multibiomarker evaluation of urban, industrial, and

agricultural exposure of small characins in a large freshwater basin in southern Brazil, *Environ. Sci. Pollut. Res.*, 2015, **22**, 13263–13277.
58. J. D. Simonato, M. Mela, H. B. Doria, I. C. Guiloski, M. A. Randi, P. S. Carvalho, P. C. Meletti, H. C. Silva de Assis, A. Bianchini and C. B. Martinez, Biomarkers of waterborne copper exposure in the Neotropical fish *Prochilodus lineatus*, *Aquat. Toxicol.*, 2016, **170**, 31–41.
59. M. W. Gonçalves, T. B. Vieira, N. M. Maciel, W. F. Carvalho, L. S. F. Lima, P. G. Gambale, A. D. da Cruz, F. Nomura, R. P. Bastos and D. M. Silva, Detecting genomic damages in the frog *Dendropsophus minutus*: Preserved *versus* perturbed areas, *Environ. Sci. Pollut. Res.*, 2015, **22**, 3947–3954.
60. V. Maselli, G. Polese, D. Rippa, R. Ligrone, R. K. Rastogi and D. Fulgione, Frogs, sentinels of DNA damage induced by pollution in Naples and the neighbouring Provinces, *Ecotoxicol. Environ. Saf.*, 2010, **73**, 1525–1529.
61. S. Ralph and M. Petras, Caged amphibian tadpoles and in situ genotoxicity monitoring of aquatic environments with the alkaline single cell gel electrophoresis (comet) assay, *Mutat. Res.*, 1998, **413**, 235–250.
62. R. F. Lee, K. Bulski, J. D. Adams, M. Peden-Adams, D. G. Bossart, L. King and P. A. Fair, DNA strand breaks (comet assay) in blood lymphocytes from wild bottlenose dolphins, *Mar. Pollut. Bull.*, 2013, **77**, 355–360.
63. I. Caliani, T. Campani, M. Giannetti, L. Marsili, S. Casini and M. C. Fossi, First application of comet assay in blood cells of Mediterranean loggerhead sea turtle (*Caretta caretta*), *Mar. Environ. Res.*, 2014, **96**, 68–72.
64. L. M. Zapata, B. C. Bock, L. Y. Orozco and J. A. Palacio, Application of the micronucleus test and comet assay in *Trachemys callirostris* erythrocytes as a model for *in situ* genotoxic monitoring, *Ecotoxicol. Environ. Saf.*, 2016, **122**, 108–116.
65. L. S. Peters, F. Telli-Karakoç, A. Hewer and D. H. Phillips, *In vitro* mechanistic differences in benzo[a]pyrene-DNA adduct formation using fish liver and mussel digestive gland microsomal activating systems, *Mar. Environ. Res.*, 2002, **54**, 499–503.
66. S. Canova, P. Degan, L. D. Peters, D. R. Livingstone, R. Voltan and P. Venier, Tissue dose, DNA adducts, oxidative DNA damage and CYP1A-immunopositiveproteins in mussels exposed to waterborne benzo(a)pyrene, *Mutat. Res.*, 1998, **399**, 17–30.
67. F. Akcha, C. Izuel, P. Venier, H. Budzinski and T. Burgeot, and J.-F. Narbonne Enzymatic biomarker measurement and study of DNA adduct formation in benzo[a]pyrene-contaminated mussels, Mytilus galloprovincialis, *Aquat. Toxicol.*, 2000, **49**, 269–287.
68. J. P. Shaw, A. T. Large, J. K. Chipman, D. R. Livingstone and L. D. Peters, Seasonal variation in mussel *Mytilus edulis* digestive gland cytochrome P4501A- and 2E-immunoidentifed protein levels and DNA strand breaks (Comet assay), *Mar. Environ. Res.*, 2000, **50**, 405–409.

69. A. Bourgeault and C. Gourlay-Francé, Monitoring PAH contamination in water: Comparison of biological and physico-chemical tools, *Sci. Total Environ.*, 2013, **454–455**, 328–336.
70. M. Martins, P. M. Costa, J. Raimundo, C. Vale, A. M. Ferreira and M. H. Costa, Impact of remobilized contaminants in *Mytilus edulis* during dredging operations in a harbour area: Bioaccumulation and biomarkers responses, *Ecotoxicol. Environ. Saf.*, 2012, **85**, 96–103.
71. L. J. Dallas, V. V. Cheung, A. S. Fisher and A. N. Jha, Relative sensitivity of two marine bivalves for detection of genotoxic and cytotoxic effects: a field assessment in the Tamar Estuary, South West England, *Environ. Monit. Assess.*, 2013, **185**, 3397–3412.
72. J. Fernández-Tajes, F. Flórez, S. Pereira, T. Rábade, B. Laffon and J. Méndez, Use of three bivalve species for biomonitoring a polluted estuarine environment, *Environ. Monit. Assess.*, 2011, **177**, 289–300.
73. M. Martins, P. M. Costa, A. M. Ferreira and M. H. Costa, Comparative DNA damage and oxidative effects of carcinogenic and non-carcinogenic sediment-bound PAHs in the gills of a bivalve, *Aquat. Toxicol.*, 2013, **142**, 85–95.
74. W. Bisset, (Jr.), L. Smith and J. A. Thompson, Geostatistical analysis of DNA damage in oysters, *Crassostrea virginica*, in Lavaca Bay, Texas, *Ecotoxicology*, 2009, **18**, 69–74.
75. F. Vincent-Hubert, A. Arini and C. Gourlay-Francé, Early genotoxic effects in gill cells and haemocytes of *Dreissena polymorpha* exposed to cadmium, B[a]P and a combination of B[a]P and Cd, *Mutat. Res.*, 2011, **723**, 26–35.
76. S. Kolarević, J. Knežević-Vukcevic, M. Paunivić, M. Kračun, B. Vasiljević, J. Tomović, B. Vuković-Gačić and Z. Gačić, Monitoring of DNA damage in haemocytes of freshwater mussel *Sinanodonta woodiana* sampled from the Velika Morava River in Serbia with the comet assay, *Chemosphere*, 2013, **93**, 243–251.
77. C. Michel, A. Bourgeault, C. Gourlay-Francé, F. Palais, A. Geffard and F. Vincent-Hubert, Seasonal and PAH impact on DNA strand-break levels in gills of transplanted zebra mussels, *Ecotoxicol. Environ. Saf.*, 2013, **92**, 18–26.
78. K. C. dos Santos and C. B. R. Martinez, Genotoxic and biochemical effects of Atrazine and Roundup, alone and in combination, on the Asian clam *Corbicula fluminea*, *Ecotoxicol. Environ. Saf.*, 2014, **100**, 7–14.
79. F. Vincent-Hubert, M. Revel and J. Garric, DNA strand breaks detected in embryos of the adult snails, *Potamopyrgus antipodarum*, and in neonates exposed to genotoxic chemicals, *Aquat. Toxicol.*, 2012, **122–123**, 1–8.
80. J. Raimundo, P. M. Costa, C. Vale, M. H. Costa and I. Moura, DNA damage and metal accumulation in four tissues of feral *Octopus vulgaris* from two coastal areas in Portugal, *Ecotoxicol. Environ. Saf.*, 2010, **73**, 1543–1547.
81. V. Pellegri, G. Gorbi and A. Buschini, Comet Assay on *Daphnia magna* in eco-genotoxicity testing, *Aquat. Toxicol.*, 2014, **155**, 261–268.

82. S. E. Hook and R. F. Lee, Genotoxicant induced DNA damage and repair in early and late developmental stages of the grass shrimp *Palaemonetes pugio* embryo as measured by the comet assay, *Aquat. Toxicol.*, 2004, **66**, 1–14.
83. D. A. Roberts, S. N. R. Birchenough, C. Lewis, M. B. Sanders, T. Bolam and D. Sheahan, Ocean acidification increases the toxicity of contaminated sediments, *Global Change Biol.*, 2013, **19**, 340–351.
84. L. Bach, A. Palmqvist, L. J. Rasmussen and V. E. Forbes, Differences in PAH tolerance between *Capitella* species: Underlying biochemical mechanisms, *Aquat. Toxicol.*, 2005, **74**, 307–319.
85. C. Lewis and T. Galloway, Genotoxic damage in polychaetes: A study of species and cell-type sensitivities, *Mutat. Res.*, 2008, **654**, 69–75.
86. P.-E. Buffet, A. Zalouk-Vergnoux, A. Châtel, B. Berthet, I. Métais, H. Perrein-Ettajani, L. Poirier, A. Luna-Acosta, H. Thomas-Guyon, C. Risso-de Faverney, M. Guibbolini, D. Gilliland, E. Valsami-Jones and C. Mouneyrac, A marine mesocosm study on the environmental fate of silver nanoparticles and toxicity effects on two endobenthic species: The ragworm *Hediste diversicolor* and the bivalve mollusc *Scrobicularia plana*, *Sci. Total Environ.*, 2014, **470–471**, 1151–1159.
87. A. Palmqvist, H. Selck, L. J. Rasmussen and V. E. Forbes, Biotransformation and genotoxicity of fluoranthene in the deposit feeding polychaete *Capitella* sp. I, *Environ. Toxicol. Chem.*, 2003, **22**, 158–166.
88. C. L. Mitchelmore and S. Hyatt, Assessing DNA damage in cnidarians using the Comet assay, *Mar. Environ. Res.*, 2004, **58**, 707–711.
89. M. N. Canty, T. H. Hutchinson, R. J. Brown, M. B. Jones and A. N. Jha, Linking genotoxic responses with cytotoxic and behavioural or physiological consequences: Differential sensitivity of echinoderms (*Asterias rubens*) and marine molluscs (*Mytilus edulis*), *Aquat. Toxicol.*, 2009, **94**, 68–76.
90. L. Ferrat, C. Pergent-Martini and M. Roméo, Assessment of the use of biomarkers in aquatic plants for the evaluation of environmental quality: application to seagrasses, *Aquat. Toxicol.*, 2003, **65**, 187–204.
91. F. Akcha, G. Arzul, S. Rousseau and M. Bardouil, Comet assay in phytoplankton as biomarker of genotoxic effects of environmental pollution, *Mar. Environ. Res.*, 2008, **66**, 59–61.
92. J. Laroche, O. Gauthier, L. Quiniou, A. Devaux, S. Bony, E. Evrard, J. Cachot, Y. Chérel, T. Larcher, R. Riso, V. Picherau, M. H. Devier and H. Budzinski, Variation patterns in individual fish responses to chemical stress among estuaries, seasons and genders: the case of the European flounder (*Platichthys flesus*) in the Bay of Biscay, *Environ. Sci. Pollut. Res.*, 2013, **20**, 738–748.
93. G. Frenzilli, V. Scarcelli, I. Del Barga, M. Nigro, L. Förlin, C. Bolognesi and J. Sturve, DNA damage in eelpout (*Zoarces viviparus*) from Göteborg harbour, *Mutat. Res.*, 2004, **552**, 187–195.
94. P. M. Costa, A. Milhinhos, M. Simões, L. Marum, A. M. Oliveira, M. H. Costa and C. Miguel, Determining DNA strand breakage from

embryogenic cell cultures of a conifer species using the single-cell gel electrophoresis assay, *Tree Genet. Genomes*, 2012, **8**, 425–430.
95. A. R. Collins, Investigating oxidative DNA damage and its repair using the comet assay, *Mutat. Res.*, 2009, **681**, 24–32.
96. A. Azqueta, Slyskova, J. Langie, S. A. S. Gaivão, I. O. and A. Collins, Comet assay to measure DNA repair: approach and applications, *Front. Genet.*, 2014, **5**, 288.
97. P. M. Costa, M. Pinto, A. M. Vicente, C. Gonçalves, A. P. Rodrigo, H. Louro, M. H. Costa, S. Caeiro and M. J. Silva, An integrative assessment to determine the genotoxic hazard of estuarine sediments: combining cell and whole-organism responses, *Front. Genet.*, 2014, **5**, 437.
98. M. L. Gielazyn, A. H. Ringwood, W. W. Piegorsch and S. E. Stancyk, Detection of oxidative DNA damage in isolated marine bivalve hemocytes using the comet assay and formamidopyrimidine glycosylase (Fpg), *Mutat. Res.*, 2003, **542**, 15–22.
99. M. F. Pinto, H. Louro, P. M. Costa, S. Caeiro and M. J. Silva, Exploring the potential interference of estuarine sediment contaminants with the DNA repair capacity of human hepatoma cells, *J. Toxicol. Environ. Health, Part A*, 2015, **78**, 559–570.
100. A. Azqueta and A. R. Collins, The essential comet assay: a comprehensive guide to measuring DNA damage and repair, *Arch. Toxicol.*, 2013, **87**, 949–968.
101. P. Duez, G. Dehon and J. Dubois, Validation of raw data measurements in the comet assay, *Talanta*, 2004, **63**, 879–886.

CHAPTER 2

Adverse Effects of Pharmaceutical Products in the Marine Environment: The Use of Non-target Species to Evaluate Water and Sediment Matrices

L. A. MARANHO,*[a] G. V. AGUIRRE-MARTÍNEZ[b,c,d] AND M. L. MARTÍN-DÍAZ[b,c]

[a] Department of Marine Science, São Paulo Federal University, Santos, São Paulo, Brazil; [b] Andalusian Centre of Marine Sciences and Technology (CACYTMAR), University of Cádiz, Polígono Río San Pedro, s/n, 11510, Puerto Real, Cádiz, Spain; [c] Physical Chemical Department, Faculty of Marine and Environmental Sciences, University of Cádiz, Polígono Río San Pedro, s/n, 11510, Puerto Real, Cádiz, Spain; [d] Facultad de Ciencias de la Salud. Universidad Arturo Prat, casilla 121, 1110939 Iquique, Chile
*Email: lmaranho@gmail.com

2.1 Pharmaceuticals in Environmental Matrices

The presence of pharmaceuticals in water and sediment matrices has become an environmental issue since the 1990s.[1] They have been detected

Issues in Toxicology No. 33
Ecotoxicology and Genotoxicology: Non-traditional Aquatic Models
Edited by Marcelo L. Larramendy

Published by the Royal Society of Chemistry, www.rsc.org

with increasing frequency in the aquatic environment in the $ng\,L^{-1}$ to $\mu g\,L^{-1}$ range. Pharmaceutical active compounds usually enter the aquatic environment through municipal effluents, with detoxification processes in wastewater treatment plants (WWTPs) not being sufficient to manage them. Removal efficiency of contaminants from WWTPs influents varies greatly with seasonality, quantities of individual active pharmaceutical ingredients being used by the population, properties of individual contaminants, antagonism/synergism behaviours of mixtures, and the type of wastewater treatment being applied.[2] Contaminants, including emergent compounds such as pharmaceuticals, can cause risk to human health and the environment associated with their presence and frequency of occurrence.[3]

Consequently, pharmaceuticals can reach surface water, groundwater and sediments.[3] Drug concentrations in marine sediments are in the $ng\,g^{-1}$ order, often greater than those detected in surface water,[2,4–8] indicating that this matrix can act as a long-term sink and/or source of such compounds to aquatic systems.[9] Sediments can be remobilized, and contaminants may be bioavailable to the water column affecting associated biota.[10,11]

One important factor that determines the adsorption and accumulation of pharmaceuticals in the environment is the polarity of such compounds. Many drugs are weak acids or basics, which may be ionisable, influenced by the pK_a (acidity constant) or the pH of the medium. This is related to the adsorption of drugs in organisms and sediments. Once ionized, a compound can lead to a number of sub-compounds and/or be adsorbed to particles suspended in the water column and deposited in the sediment. Solubility, organic matter, total organic carbon (TOC) and % of fine particles in the sediment are among the factors that differentiate environments in relation to pollution of pharmaceuticals. Pharmaceutical bioavailability also varies between organisms owing to their behaviour, feeding type, body composition and lifestyle in the environment.

Current regulatory requirements for environmental risk assessment (ERA) of "emerging compounds" and their inclusion within the regulatory framework for chemicals have been discussed in the United States, Canada, Japan, Australia, Switzerland and the European Union.[12] The European Medicines Agency[13] released a guideline describing how to evaluate the potential risks of pharmaceutical products entering the environment, describing a step-wise tiered procedure for ERA: pre-screening assessment (Tier 1) and environmental fate and effect analysis (Tier 2).

Scientists believe that research should be focused on developing an ERA methodology in which marine environment components are included. In contrast to other pollutants, pharmaceuticals are specifically designed to be bioactive at low concentrations, exerting pharmacological and physiological effects on humans or animals under veterinary treatment. Nevertheless, their effects on non-target species are difficult to predict and may often be detrimental.[14]

2.2 Adverse Effects of Pharmaceutical Products on Aquatic Biota

2.2.1 Laboratory Studies

Compared to chemical data, little information has been published regarding the aquatic ecotoxicology of pharmaceuticals, especially relevant data regarding exposure and the possible adverse effects of drugs and bioaccumulation in aquatic organisms. The majority of these studies have been focused on laboratory approaches. These approaches allow the use of negative toxicity water or sediment controls, they are less expensive to perform than field studies, and data obtained in the laboratory has less variability than data collected in bioassessment, field or mesocosm studies.[15]

2.2.2 Short-term Assays

During the last few decades, the impact of chemical pollution has been focused on the conventional "priority" pollutants (PAHs, PCBs, metals, pesticides, *etc.*), and especially those acutely toxic/carcinogenic pollutants that persist in the environment.[16] Standardized toxicity tests are recommended to screen the potential hazards of aquatic contaminants and to develop environmental quality criteria.[15] Guidelines for water and sediment quality assessment include the use of different endpoints, such as bioluminescence inhibition of the bacteria *Vibrio fischeri* (Microtox®), spermiotoxicity and embryotoxicity of sea urchins, growth inhibition of microalgae, and mortality of amphipods and *Daphnia magna*.[17–20]

However, the constant discharge of pharmaceuticals gives them a state of pseudo-persistence[6] and may produce many adverse effects in non-target organisms. The mixture of pharmaceutical products can work as synergist or antagonist substances in the presence of other contaminants, which can distort the real effect of a specific pharmaceutical product. In fact, there is a need to obtain information about the adverse effects of each pharmaceutical product separately, since they have different mechanisms of action.

For marine organisms, standardized toxicity tests such as Microtox®, microalgae growth inhibition and sea urchin spermiotoxicity applied to study conventional "priority" pollutants are not effective to evaluate pharmaceutical toxicity by itself in a water matrix.[21,22] For sediment spiked with pharmaceutical products, the use of *Ampelisca brevicornis* amphipod mortality (Figure 2.1), Microtox® Solid Phase Test (SPT), and sea urchin spermiotoxicity and embryotoxicity may be recommended to evaluate the toxicity of some pharmaceuticals, such as carbamazepine, ibuprofen, fluoxetine, 17α-ethynylestradiol, propanolol and caffeine, including environmental concentrations (ranging from 0.05 to 500 $ng\,g^{-1}$).[2]

The number of studies about the possible adverse effects owing to the exposure of invertebrates to pharmaceutical compounds has increased.[2,3,23–28] Previous studies demonstrated the suitability of short-term effects for

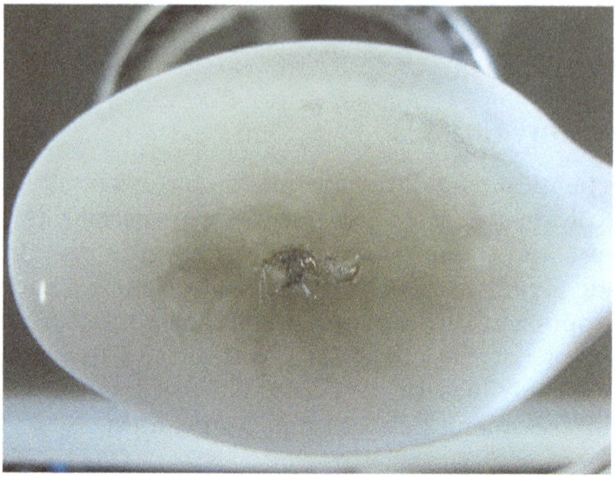

Figure 2.1 Amphipods *Ampelisca brevicornis*.

Figure 2.2 Sea urchins *Paracentrotus lividus*.

distinguishing changes in reproduction or early stage development. The most sensitive test to assess pharmaceutical impacts in aquatic ecosystems owing to short-term exposure showed that early stages of development (*i.e.*, sea urchin *Paracentrotus lividus* embryotoxicity test) (Figure 2.2) were more sensitive to pharmaceutical contamination.[2,23,24]

The importance of studying marine environments affected by pharmaceutical products contamination is related to WWTPs and the mixture of contaminants. In fact, the first sediment quality guidelines (SQG) taking into consideration emerging compounds as pharmaceuticals products and surfactants [secondary alkane sulfonates (SAS)] was published concerning the Bay of Cádiz (Southwest, Spain).[2] Nevertheless, acute toxicity effects on the biota associated with sediment affected by WWTPs may be applied to estimate toxicity in organisms and different endpoints, as with the use of

Ampelisca brevicornis amphipod mortality and Microtox® (SPT) for bulk sediment.[2,29] This fact demonstrated that policies for contamination control were ineffective for dealing with population expansion and economic activities, since contaminants continue to be released into the environment and cause toxicity. There was no environmental legislation about pharmaceutical products or SAS impacts; therefore, such study demonstrated that these compounds could be associated with adverse effects on the aquatic biota. The integrated analysis of lines of evidence (short-term assays and chemical analysis) in the sediment allowed differences to be observed between toxicity and contamination according to seasonality.

2.2.3 Biomarkers of Stress and Effect

Owing to the continued presence of low concentrations in the environment, persistence and continuous supply through WWTPs, drugs are more likely to cause chronic or subchronic effects. Toxic effects occur in individuals from the interaction of contaminants with biomolecules, providing structural and/or functional changes to essential cell activity. These changes affect different trophic levels represented by the "domino effect" from the cell biochemistry, organs and tissues up to the ecosystem and biosphere. In this context, "early warning tools" allow the assessment of adverse effects, providing information on the mechanism of action and anticipating the knowledge of the ecological effects that may occur on biota.[30]

Biomarkers of stress and effect are incorporated in environmental studies to assess toxic and metabolic effects of pollutants, individually or in combination, in increasingly smaller levels of biological organization. Biomarkers are recommended as tools by various international organizations such as UNEP, OSPAR, OECD, ICES and ICO.[31] The Canadian Agency has made previous studies on the acute and chronic effects in the laboratory and the field concerning pharmaceutical and wastewater discharge. Exposure to environmental concentrations of these compounds may induce oxidative stress and genetic damage in exposed laboratory organisms.[32] However, studies are mostly on water, and therefore further research is necessary related to sediment, more specifically on sublethal toxicity, using tools such as biomarkers.[33]

In this regard, the evaluation of pharmaceutical toxicity in marine organisms should include sensitive responses, such as biomarkers of general stress and biochemical biomarkers. The evaluation of lysosomal membrane stability (LMS) as a biomarker of general stress syndrome has been recommended and applied in fish, mussels, crabs and clams[23,24,33–38] and it has been employed in wide-scale Tier 1 bio-monitoring programs.[23,24,33] LMS measurements are currently included in the general guidelines for monitoring programs (JAMP by OSPAR) and proposed as a marine pollution index to evaluate stress responses in mollusks by the Mediterranean Pollution Programme (MEDPOL).[37]

Laboratory experiments with seawater spiked with caffeine, ibuprofen, carbamazepine, novobiocin and tamoxifen validated the sensitivity of this

Figure 2.3 Crab *Carcinus maenas.*

Figure 2.4 Clam *Ruditapes philippinarum.*

biomarker in invertebrates as crabs *Carcinus maenas* (Figure 2.3) and clams *Ruditapes philippinarum* (Figure 2.4), demonstrating significant stress when exposed to environmental concentrations.[23,24,39,40] In addition, polychaetes *Hediste diversicolor*[28,41] (Figure 2.5) and amphipods *Ampelisca brevicornis*[27] were considered good bioindicators for evaluating the chronic toxicity effects of the pharmaceuticals carbamazepine, ibuprofen, fluoxetine, 17α-ethynylestradiol, propranolol and caffeine spiked in marine sediments, including at environmental concentrations.

When the original bioavailable compound or its metabolites bind to cellular macromolecules, toxic effects are manifested. Ultimately, this can lead to rupture of the membrane, cell damage and/or genotoxic effects, which may subsequently result in the development and progression of disease (*e.g.* cancer). Metabolism is, therefore, an important determinant of the activity

Figure 2.5 Polychaetes *Hediste diversicolor*.

and half-life of the compound in the organism.[42] The first battery of biomarkers taken into consideration in this chapter is directly related to metabolism, antioxidant activities, oxidative stress and effects caused by xenobiotics on the health of non-target organisms.

Most xenobiotics are catalyzed by Phase I metabolism enzymatic reactions, also known as the mixed function oxidase (MFO) system. CYP 450, which comprises a large family of cytochromes, is a heme protein located in the endoplasmic reticulum[43] where the oxidation of lipophilic substrates is catalyzed using NADPH and O_2. In this chapter, two enzyme activities were considered for Phase I: ethoxyresorufin *O*-deethylase (EROD) and dibenzylfluorescein dealkylase (DBF). The enzymatic activity of EROD is related to CYP 450 1A-like, which is involved in reactions of mono-oxygenation of dioxins and PAHs[44] and drugs such as CBZ.[45] The enzyme activity is related to the DBF CYP 450 3A-like, which is involved in the metabolism of many pharmaceuticals.[44] Pharmaceuticals can activate or deactivate detoxification pathways (phase I and phase II).[27,28,46] Phase II metabolism [*e.g.* enzyme glutathione *S*-transferase (GST)] promotes the combination of endogenous compounds with the xenobiotic by tripeptide glutathione, making them soluble and more easily excreted. As for the antioxidant defense system, changes in enzyme activities may reflect an increase in the synthesis of reactive oxygen species (O_2^-, H_2O_2) induced by exposure to xenobiotics, associated with oxidative stress, defense mechanisms and overlapping pro-oxidant forces. The enzyme glutathione peroxidase (GPX) works in the degradation of H_2O_2 with a recognized role in the preservation of oxidative stress. Supplemental enzymes such as glutathione reductase (GR) produce GSH and NADPH to maintain the antioxidant defense mechanism in the cell.[47]

Since disturbances that occur from exposure to contaminants are demonstrated by different biochemical process changes, which can result in cell damage, such as lipid peroxidation (LPO) and mitochondrial DNA damage (strand breaks).

As for the action of pollutants on the neuronal system, special attention has been paid to the functions of the acetyl cholinesterase enzyme (AChE), which performs the function of decomposing the neurotransmitter acetylcholine at neuromuscular junctions and synapses, which prevents the continuous activation of the nerve, being vital to the proper functioning of sensory and neuromuscular systems of the body.[48]

The use of biomarkers of Phase I, Phase II, oxidative stress, genotoxicity and neurotoxicity as early warning tools for ERA has been proposed by several scientists.[33,49–51]

Bioassays or biomarkers should focus on the specific mechanisms of action of drugs on biota.[16] The second class of biomarkers cited in this chapter is related to the mechanism of action of the drug in question, with possible effects on survival, immune function and reproduction.[52] This group is related to the endocrine status of gametogenesis and energy levels: activity of monoamine oxidase (MAO), activity of cyclooxygenase (COX), total lipid content (TLP) and mitochondrial electron transport (MET). MAO activity plays a vital role in the inactivation of neurotransmitters (*e.g.* norepinephrine, serotonin, dopamine). COX activity is an enzymatic activity that is responsible for the formation of important biological mediators. It corresponds to neuroendocrine state and may also be a measure for immunotoxicity. Inhibition of this activity can lead to pain relief and inflammation. TLP is directly related to energy reserves and MET is related to energy expenditure, mainly related to breathing.

Previous studies indicated the chronic toxicity of pharmaceutical products in marine organisms when responses of detoxification pathways, oxidative stress, energy status and reproduction were significantly different compared to controls.[27,28,39–41,46]

Results from these studies have indicated the presence, availability, and toxicity of pharmaceuticals and validate the use of biomarkers when assessing the effects of pharmaceuticals within a marine environmental risk assessment framework, using as bioassay the model species *C. maenas, R. philippinarum*, and *P. lividus* for water matrix and the species *R. philippinarum, H. diversicolor* and *A. brevicornis* for sediment matrix. Furthermore, owing to their sensibility, easy procedure, and low cost, the use of biomarkers should be included not only in ERA but also in legislation.

2.2.4 *In situ* Studies

Ecotoxicological data obtained in laboratory studies are often difficult to translate into accurate predictions of potential adverse effects *in situ*. Since either overestimation or underestimation of the effects can occur, laboratory results are best validated by field research.[51] Transplantation of organisms to assess environmental quality avoids these drawbacks and allows the integration of field evaluation with the laboratory. Previous studies include analysis of biomarkers in different caged organisms, such as oysters,[53] clams,[36,45,54] mussels,[44,52,55] crabs[45] and polychaetes.[56]

Despite many studies on the toxic effect of WWTPs and agricultural chemicals, among other xenobiotics related to the harmful effects on aquatic organisms, little is known about the combined sublethal effects of these chemicals in the field.[57] *In situ* studies using transplantation of organisms from areas not contaminated to areas suspected of contamination have been widely applied.[36,53,58] This procedure can provide information on the action and bioaccumulation of contaminants under natural conditions, promoting typical levels of control experiments in the laboratory, but with the realism of field studies.[59]

In this context, bivalves have acquired global significance as bioindicators as they have been used in monitoring programs of coastal waters, lagoons,[60,61] estuaries[53] and freshwater environments.[44] Bivalves have broad geographical distribution, direct availability through aquaculture, and are effective for use in entry cage experiments along shorelines.[31]

The battery of biochemical responses and the LMS assay were suitable tools to evaluate the environmental quality of sediments directly affected by wastewater effluents *in situ*.[36] The chemical composition of sediment directly affected by wastewater effluents had seasonal changes, which were reflected in the health status of benthic organisms. Nevertheless, even with seasonal fluctuations, this methodology is recommended for the environmental evaluation of contaminated areas. In winter, urban effluents were detoxified by Phase I (DBF activity of the metabolism) and conjugated by Phase II (GST activity), antioxidant defenses increased (GR activity) and the exposure resulted in neurotoxicity (AChE activity) and lipid peroxidation (LPO). Urban effluents lead to detoxification metabolism (EROD activity), oxidative effects (LPO and DNA damage), neurotoxicity (AChE activity) and neuroendocrine disruption [COX activity and alkali-labile proteins (ALP) levels], which are involved in inflammation and spawning delay in clam populations in summer. Adverse effects varied according to the contamination level, and are also dependent on the reproductive cycle of the clams.

2.3 Final Considerations

Research of pharmaceutical toxicity should be focused on developing a risk assessment methodology in which marine environment components are included. In contrast to other pollutants, pharmaceuticals are specifically designed to have pharmacological and physiological effects on humans or animals under veterinary treatment.

Pharmaceutical toxicity cannot be evaluated using traditional tools that have been applied to study other compounds that are more toxic to marine organisms, such as PAHs, PCBs and pesticides. In the case of pharmaceuticals, acute tests should be avoided as they are not effective for evaluating water samples. For sediment, the use of *A. brevicornis* amphipod mortality, Microtox® SPT, and sea urchin spermiotoxicity and embryotoxicity may be recommended to evaluate the toxicity of pharmaceuticals. In this regard, the evaluation of pharmaceutical toxicity in marine organisms should include sensitive responses, such as embryotoxicity, biomarkers of general stress and effect biomarkers. Sublethal responses lead to more sensitivity and accuracy when determining the environmental risk of pharmaceuticals in marine environments.

Finally, it is remarkable to take into consideration the synergistic and antagonistic effects of pharmaceuticals when they are present in the environment in mixtures with other pharmaceuticals and/or other compounds. The study of effluent toxicity, acting as sources of pharmaceutical products to the environment, could bring insight to management strategies in order to prevent future effects at the ecosystem level.

Acknowledgements

Luciane A. Maranho thanks CNPq for the postdoctoral scholarship (process n. 166122/2015-7), CAPES/MEC-Brazil (BEX 0362/10-7) for the doctoral scholarship and for the partial financial support of this study. Gabriela Aguirre-Martínez would like to thank Consejería de Economía, Innovación y Ciencia (Regional Government of Andalusia, Spain) for the financial support for the postdoctoral and doctoral research. Additionally, these studies were funded by the project P09-RNM-5136 (Government of Andalusia, Spain) and by the European Regional Development Fund (FEDER).

References

1. N. A. Doerr-MacEwen and M. E. Haight, Expert stakeholders' views on the management of human pharmaceuticals in the environment, *Environ. Manage.*, 2006, **38**(5), 853–866.
2. L. A. Maranho, M. C. Garrido-Pérez, R. M. Baena-Nogueras, P. A. Lara-Martín, R. Antón-Martín, T. A. DelValls and M. L. Martín-Díaz, Are WWTPs effluents responsible for acute toxicity? Seasonal variations of sediment quality at the Bay of Cádiz (SW, Spain), *Ecotoxicology*, 2015, **24**(2), 368–380.
3. B. Ferrari, N. Paxeus, R. Lo Giudice, A. Pollio and J. Garric, Ecotoxicological impact of pharmaceuticals found in treated wastewaters: study of carbamazepine, clofibric acid, and diclofenac, *Ecotoxicol. Environ. Saf.*, 2003, **55**(3), 359–370.
4. T. A. Ternes, H. Andersen, D. Gilberg and M. Bonerz, Determination of estrogens in sludge and sediments by liquid extraction and GC/MS/MS, *Anal. Chem.*, 2002, **74**(14), 3498–3504.
5. D. Löffler and T. A. Ternes, Determination of acidic pharmaceuticals, antibiotics and ivermectin in river sediment using liquid chromatography-tandem mass spectrometry, *J. Chromatogr. A*, 2003, **1021**(1–2), 133–144.
6. M. D. Hernando, M. Mezcua, A. R. Fernández-Alba and D. Barceló, Environmental risk assessment of pharmaceutical residues in wastewater effluents, surface waters and sediments, *Talanta*, 2006, **69**(2), 334–342.

7. B. F. Da Silva, A. Jelic, R. López-Serna, A. A. Mozeto, M. Petrovic and D. Barceló, Occurrence and distribution of pharmaceuticals in surface water, suspended solids and sediments of the Ebro river basin, Spain, *Chemosphere*, 2011, **85**(8), 1331–1339.
8. M. G. Pintado-Herrera, E. González-Mazo and P. A. Lara-Martín, Environmentally friendly analysis of emerging contaminants by pressurized hot water extraction-stir bar sorptive extraction-derivatization and gas chromatography-mass spectrometry, *Anal. Bioanal. Chem.*, 2013, **405**(1), 401–411.
9. È. A. M. Gilroy, V. K. Balakrishnan, K. R. Solomon, E. Sverko and P. K. Sibley, Behaviour of pharmaceuticals in spiked lake sediments - Effects and interactions with benthic invertebrates, *Chemosphere*, 2011, **86**(6), 578–584.
10. A. Cesar, C. D. S. Pereira, A. R. Santos, D. M. S. Abessa, R. B. Choueri, I. Riba, M. C. Morales-Caselles and T. A. DelValls, Comparative sediment quality assessment in different littoral ecosystems from Spain (Gulf of Cadiz) and Brazil (Santos and São Vicente estuarine system), *Environ. Int.*, 2007, **33**(4), 429–435.
11. R. J. Torres, D. M. S. Abessa, F. C. Santos, L. A. Maranho, M. B. Davanso, M. R. L. Nascimento and A. A. Mozeto, Effects of dredging operations on sediment quality: contaminant mobilization in dredged sediments from the Port of Santos, SP, Brazil, *J. Soils Sediments*, 2009, **9**, 420–432.
12. J. O. Straub and T. H. Hutchinson, in *Human Pharmaceuticals in the Environment: Current and Future Perspectives*, ed. B. Brooks and D. Huggett, Springer Science + Business Media, 2012, pp. 17–49.
13. EMA/CHMP/SWP/4447/00. 2006. Available at: http://www.ema.europa.eu/docs/en_gb/document_library/scientific_guideline/2009/10/wc500003978.pdf.
14. M. Hampel, J. E. Bron, J. B. Taggart and M. J. Leaver, The antidepressant drug carbamazepine induces differential transcriptome expression in the brain of Atlantic salmon, *Salmo salar*, *Aquat. Toxicol.*, 2014, **151**, 114–123.
15. B. W. Brooks, E. M. Dzialowski, P. K. Turner, J. K. Stanley and E. A. Glidewell, in *Pharmaceutical effects on freshwater invertebrates*. Annual Meeting of the American Society of Limnology and Oceanography, Salt Lake City, UT, 2003.
16. C. G. Daughton and T. A. Ternes, Pharmaceuticals and personal care products in the environment: agents of subtle change?, *Environ. Health Perspect.*, 1999, **107**(6), 907–938.
17. CEDEX, Spanish Action Levels for dredged material management, CEDEX, 1994, Madrid, Spain.
18. USEPA - USEPA/USACE EPA-823-B-98- 004, Evaluation of dredged material proposed for discharge in waters of the U.S.—testing manual. 1998.
19. GIPME - Global Investigation of Pollution in the Marine Environment. Guidance on assessment of sediment quality. London, UK. International Maritime Organization. 2000.

20. SEDNET. The SedNet Strategy Paper. The opinion of SedNet on environmentally, socially and economically viable sediment management. 2003. pp. 22, www.sednet.org.
21. G. V. Aguirre-Martínez, M. A. Owuor, C. Garrido-Pérez, M. J. Salamanca, T. A. DelValls and M. L. Martín-Díaz, Are standard tests sensitive enough to evaluate effects of human pharmaceuticals in aquatic biota? Facing changes in research approaches when performing risk assessment of drugs, *Chemosphere*, 2015, **120**, 75–85.
22. G. V. Aguirre-Martínez, C. Okello, M. J. Salamanca, C. Garrido-Pérez, T. A. DelValls and M. L. Martín-Díaz, Is the step-wise tiered approach for ERA of pharmaceuticals useful for the assessment of cancer therapeutic drugs present in marine environment? *Environ. Res.*, 2016, **144**, 43–59.
23. G. V. Aguirre-Martínez, S. Buratti, E. Fabbri, T. A. DelValls and M. L. Martín-Díaz, Using lysosomal membrane stability of haemocytes in *Ruditapes philippinarum* as a biomarker of cellular stress to assess contamination by caffeine, ibuprofen, carbamazepine and novobiocin, *J. Environ. Sci.*, 2013, **25**(7), 1408–1418.
24. G. V. Aguirre-Martínez, S. Buratti, E. Fabbri, T. A. DelValls and M. L. Martín-Díaz, Stability of lysosomal membrane in *Carcinus maenas* acts as a biomarker of exposure to pharmaceuticals, *Environ. Monit. Assess.*, 2013, **185**(5), 3783–3793.
25. J. Damásio, M. Fernández-Sanjuan, J. Sánchez-Avila, S. Lacorte, N. Prat, M. Rieradevall, A. M. V. M. Soares and C. Barata, Multi-biochemical responses of benthic macroinvertebrate species as a complementary tool to diagnose the cause of community impairment in polluted rivers, *Water Res.*, 2011, **45**(12), 3599–3613.
26. S. Franzellitti, S. Buratti, P. Valbonesi and E. Fabbri, The mode of action (MOA) approach reveals interactive effects of environmental pharmaceuticals on *Mytilus galloprovincialis*, *Aquat. Toxicol.*, 2013, **140–141**, 249–256.
27. L. A. Maranho, L. B. Moreira, R. M. Baena-Nogueras, P. A. Lara-Martín, T. A. DelValls and M. L. Martín-Díaz, A candidate short-term toxicity test using *Ampelisca brevicornis* to assess sublethal responses to pharmaceuticals bound to marine sediments, *Arch. Environ. Contam. Toxicol.*, 2014, **68**(2), 237–258.
28. L. A. Maranho, R. M. Baena-Nogueras, P. A. Lara-Martín, T. A. DelValls and M. L. Martín-Díaz, Bioavailability, oxidative stress, neurotoxicity and genotoxicity of pharmaceuticals bound to marine sediments. The use of the polychaete *Hediste diversicolor* as bioindicator species, *Environ. Res.*, 2014, **134**, 353–365.
29. L. A. Maranho, M. C. Garrido-Pérez, T. A. DelValls and M. L. Martín-Díaz, Suitability of standardized acute toxicity tests for marine sediment assessment: pharmaceutical contamination, *Water, Air, Soil Pollut.*, 2015, **226**(3), 65.

30. R. D. Handy, T. S. Galloway and M. H. Depledge, A proposal for the use of biomarkers for the assessment of chronic pollution and in regulatory toxicology, *Ecotoxicology*, 2003, **12**(1–4), 331–343.
31. M. P. Cajaraville, M. J. Bebianno, J. Blasco, C. Porte, C. Sarasquete and A. Viarengo, The use of biomarkers to assess the impact of pollution in coastal environments of the Iberian Peninsula: a practical approach, *Sci. Total Environ.*, 2000, **247**(2–3), 295–311.
32. F. Gagné, C. Blaise, M. Fournier and P. D. Hansen, Effects of selected pharmaceutical products on phagocytic activity in *Elliptio complanata* mussels, *Comp. Biochem. Physiol., Part C: Toxicol. Pharmacol.*, 2006, **143**(2), 179–186.
33. A. Viarengo, D. Lowe, C. Bolognesi, E. Fabbri and A. Koehler, The use of biomarkers in biomonitoring: a 2-tier approach assessing the level of pollutant-induced stress syndrome in sentinel organisms, *Comp. Biochem. Physiol.*, 2007, **146**(3), 281–300.
34. S. Buratti, J. Ramos-Gómez, E. Fabbri, T. A. DelValls and M. L. Martín-Díaz, Application of neutral red retention assay to caged clams (*Ruditapes decussatus*) and crabs (*Carcinus maenas*) in the assessment of dredged material, *Ecotoxicology*, 2012, **21**(1), 75–86.
35. L. A. Maranho, T. A. DelValls and M. L. Martín-Díaz, Assessing potential risks of wastewater discharges to benthic biota: an integrated approach to biomarker responses in clams (*Ruditapes philippinarum*) exposed under controlled conditions, *Mar. Pollut. Bull.*, 2015, **92**(1–2), 11–24.
36. L. A. Maranho, C. André, T. A. DelValls, F. Gagné and M. L. Martín-Díaz, *In situ* evaluation of wastewater discharges and the bioavailability of contaminants to marine biota, *Sci. Total Environ.*, 2015, **538**, 876–887.
37. C. Martínez-Gómez, J. Benedicto, J. A. Campillo and M. Moore, Application and evaluation of the neutral red retention (NRR) assay for lysosomal stability in mussel populations along the Iberian Mediterranean coast, *J. Environ. Monit.*, 2008, **10**, 490–499.
38. M. N. Moore, M. H. Depledge, J. W. Readman and D. R. Leonard, An integrated biomarker-based strategy for ecotoxicological evaluation of risk in environmental management, *Mutat. Res.*, 2004, **552**(1–2), 247–268.
39. G. V. Aguirre-Martínez, T. A. DelValls and M. L. Martín-Díaz, Identification of biomarkers responsive to chronic exposure to pharmaceuticals in target tissues of *Carcinus maenas*, *Mar. Environ. Res.*, 2013, **87–88**, 1–11.
40. G. V. Aguirre-Martínez, T. A. DelValls and M. L. Martín-Díaz, Early responses measured in the brachyuran crab *Carcinus maenas* exposed to carbamazepine and novobiocin: application of a 2-tier approach, *Ecotoxicol. Environ. Saf.*, 2013, **97**, 47–58.
41. L. A. Maranho, C. André, T. A. DelValls, F. Gagné and M. L. Martín-Díaz, Toxicological evaluation of sediment samples spiked with human pharmaceutical products: Energy status and neuroendocrine effects in

marine polychaetes *Hediste diversicolor*, *Ecotoxicol. Environ. Saf.*, 2015, **118**, 27–36.
42. J. A. Timbrell, Biomarkers in toxicology, *Toxicology*, 1998, **129**(1), 1–12.
43. J. J. Stegeman, M. Brouwer, T. D. G. Richard, L. Forlin, B. A. Fowler, B. M. Sanders and P. A. van Veld, in *Biomarkers: Biochemical, Physiological and Histological Markers of Anthropogenic Stress*, ed. R. J. Huggett, R. A. Kimerly, P. M. Mehrle and H. L. Bergman Jr, Lewis Publishers, Chelsea, MI, USA, 1992, pp. 235–335.
44. F. Gagné, C. André, P. Cejka, C. Gagnon and C. Blaise, Toxicological effects of primary-treated urban wastewaters, before and after ozone treatment, on freshwater mussels (*Elliptio complanata*), *Comp. Biochem. Physiol., Part C: Toxicol. Pharmacol.*, 2007, **145**(4), 542–552.
45. M. L. Martín-Díaz, J. Blasco, D. Sales and T. A. DelValls, Biomarkers study for sediment quality assessment in Spanish ports using the crab *Carcinus maenas* and the clam *Ruditapes philippinarum*, *Arch. Environ. Contam. Toxicol.*, 2007, **53**(1), 66–76.
46. G. V. Aguirre-Martínez, T. A. DelValls and M. L. Martín-Díaz, General stress, detoxification pathways, neurotoxicity and genotoxicity evaluated in *Ruditapes philippinarum* exposed to human pharmaceuticals, *Ecotoxicol. Environ. Saf.*, 2016, **124**, 18–31.
47. D. J. Reed, Regulation of reductive processes by glutathione, *Biochem. Pharmacol.*, 1986, **35**, 7–13.
48. J. F. Payne, A. Mathieu, W. Melvin and L. L. Fancey, Acetylcholinesterase, and old biomarker with a new future? Field trials in association with two urban rivers and a paper mill in Newfoundland, *Mar. Pollut. Bull.*, 1996, **32**(2), 225–231.
49. K. Broeg, H. V. Westernhagen, S. Zander, W. Körting and A. Koehler, The "bioeffect assessment index" (BAI). A concept for the quantification of effects of marine pollution by an integrated biomarker approach, *Mar. Pollut. Bull.*, 2005, **50**(5), 495–503.
50. P. K. S. Lam, Use of biomarkers in environmental monitoring, *Ocean Coast. Manage.*, 2009, **52**, 348–354.
51. R. Van der Oost, J. Beyer and N. P. E. Vermeulen, Fish bioaccumulation and biomarkers in environmental risk assessment: a review, *Environ. Toxicol. Pharmacol.*, 2003, **13**, 57–149.
52. F. Gagné, C. Blaise, C. André, C. Gagnon and M. Salazar, Neuroendocrine disruption and health effects in *Elliptio complanata* mussels exposed to aeration lagoons for wastewater treatment, *Chemosphere*, 2007, **68**(4), 731–743.
53. L. A. Maranho, C. D. S. Pereira, R. B. Choueri, A. Cesar, P. K. Gusso-Choueri, R. J. Torres, D. M. S. Abessa, R. D. Morais, A. A. Mozeto, T. A. DelValls and M. L. Martín-Díaz, The application of biochemical responses to assess environmental quality of tropical estuaries: field surveys, *J. Environ. Monit.*, 2012, **14**(10), 2608–2615.
54. M. L. Martín-Díaz, J. Blasco, M. Gonçales de Canales, D. Sales and T. A. DelValls, Bioaccumulation and toxicity of dissolved heavy metals

from the Guadalquivir Estuary after the Aznalcóllar mining spill using *Ruditapes philippinarum*, *Arch. Environ. Contam. Toxicol.*, 2005, **48**(2), 233–241.

55. F. Gagné, B. Bouchard, C. André, E. Farcy and M. Fournier, Evidence of feminization in wild *Elliptio complanata* mussels in the receiving waters downstream of a municipal effluent outfall, *Comp. Biochem. Physiol., Part C: Toxicol. Pharmacol.*, 2011, **153**(1), 99–106.
56. J. Ramos-Gómez, M. Martins, J. Raimundo, C. Vale, M. L. Martín-Díaz and T. A. DelValls, Validation of *Arenicola marina* in field toxicity bioassays using benthic cages: biomarkers as tools for assessing sediment quality, *Mar. Pollut. Bull.*, 2011, **62**(7), 1538–1549.
57. A. J. O'Neill, T. S. Galloway, M. A. Browne, A. Dissanayake and M. H. Depledge, Evaluation of toxicity in tributaries of the Mersey estuary using the isopod *Asellus aquaticus* (L.), *Mar. Environ. Res.*, 2004, **58**(2–5), 327–331.
58. C. D. S. Pereira, D. M. S. Abessa, A. C. D. Bainy, L. P. Zaroni, M. R. Gasparro, M. C. Bicego, S. Taniguchi, T. H. Furley and E. C. P. M. Sousa, Integrated assessment of multilevel biomarker responses and chemical analysis in mussels from São Sebastião, São Paulo, Brazil, *Environ. Toxicol. Chem.*, 2007, **26**(3), 462–469.
59. D. L. Kalpaxis, C. Theos, M. A. Xaplanteri, G. P. Dínos, A. V. Catsiki and M. Leotsínidis, Biomonitoring of Gulf of Patras, N. Peloponnesus, Greece. Application of a biomarker suite including evaluation of translation efficiency in *Mytilus galloprovincialis* cells, *Environ. Res.*, 2004, **94**(2), 211–220.
60. Y. Lafontainey, F. Gagne, C. Blaise, G. Costan, P. Gagnon and H. M. Chan, Biomarkers in zebra mussels (*Dreissena polymorpha*) for the assessment and monitoring of water quality of the St Lawrence River (Canada), *Aquat. Toxicol.*, 2000, **50**(1–2), 51–71.
61. C. Nasci, N. Nesto, R. A. Monteduro and L. Da Ros, Field application of biochemical markers and a physiological index in the mussel, *Mytilus galloprovincialis:* transplantation and biomonitoring studies in the lagoon of Venice (NE Italy), *Mar. Environ. Res.*, 2002, **54**(3–5), 811–816.

CHAPTER 3

Rotifers as Models for Ecotoxicology and Genotoxicology

ROBERTO RICO-MARTÍNEZ,*[a] MARIO A. ARZATE-CÁRDENAS,[a] JESÚS ALVARADO-FLORES,[b] IGNACIO A. PÉREZ-LEGASPI[c] AND GUSTAVO E. SANTOS-MEDRANO[a]

[a] Universidad Autónoma de Aguascalientes, Centro de Ciencias Básicas, Departamento de Química, Avenida Universidad 940, C.P. 20131, Aguascalientes, Ags., Mexico; [b] Consejo Nacional de Ciencia y Tecnología. Unidad de Ciencias del Agua, Centro de Investigación Científica de Yucatán, Cancún, Quintana Roo, Mexico; [c] División de Estudios de Posgrado e Investigación, Instituto Tecnológico de Boca del Río, Km. 12 Carr, Veracruz-Córdoba, Boca del Río, Veracruz C.P. 94290, México
*Email: rrico@correo.uaa.mx

3.1 Rotifers as Models for Ecotoxicology

Rotifers have many characteristics that explain their use as test models for ecotoxicology: (a) ability to produce cysts, (b) ease of culture, (c) low maintenance costs, (d) rapid growth, (e) small size, and (f) sensitivity.[1] They have consistently been used to assess the environmental status of many rivers.[2] Rotifers have been used worldwide to assess industrial, rural, and urban discharges. Sarma *et al.*[3] showed that Mexico City urban wastewater affects the instantaneous growth rate of *Brachionus patulus*. In Southern Italy,

Issues in Toxicology No. 33
Ecotoxicology and Genotoxicology: Non-traditional Aquatic Models
Edited by Marcelo L. Larramendy

Published by the Royal Society of Chemistry, www.rsc.org

Isidori *et al.*[4] employed *B. plicatilis* in 24 h toxicity tests and found that all samples of municipal solid waste landfills showed acute toxicity. Park *et al.*[5] conducted toxicity testing with *B. plicatilis* to determine LC_{50} values for Korean wastewaters, finding acute toxicity from industrial, rural and urban wastewaters. In Aguascalientes, Mexico, *Lecane quadridentata* has been used to determine LC_{50} values for municipal, industrial, and agricultural sites, most of the sites studied showed acute toxicity.[2] José de Paggi and Devercelli[6] examined the influence of watershed land use on microzooplankton around the city of Santa Fe in Argentina. They found that river microzooplankton abundance and rotifer species assemblages were good indicators of land use, where *Brachionus* spp. were associated with saline waters in rural areas and *Keratella* spp. (except *K. tropica*) with urban water bodies.

Rotifers are welcome as model organisms for ecotoxicology and genotoxicology owing to their ecological relevance mainly as zooplanktonic species of great importance in many aquatic food webs.[7] Rotifers can be easily cultured, populations can be maintained in the laboratory easily, and clonal cultures can be established. Protocols to obtain DNA even from single individuals have been widely employed, whereby cysts can be obtained from several species; these cysts can be stored at room temperature and allow the development of toxicity kits which can then be widely used for assessing industrial wastewater when environmental agencies ask for particular analysis. At the same time, many ecotoxicological protocols have been developed that use parthenogenetic eggs.[1]

Modern toxicity studies using rotifers as model organisms started as early as 1964.[8] However, a significant increase in contributions using rotifers for ecotoxicological studies started in 1990.[9] Nowadays, 12 genera and 35 species of rotifers have been used to develop acute and chronic toxicity tests.[1] Assessment endpoints determined in rotifer toxicity studies are varied and have included the following: (a) mortality (with exposure times of 30 min, 24 h, 48 h or 96 h); (b) reproduction inhibition and behavior; (c) enzyme biomarkers; (d) mRNA biomarkers; (e) induction of stress proteins; and (f) predator–prey interactions.[1]

The database of substances that have been analyzed for toxicity exposure using rotifers includes 15 metals: Ag, Al, Cd, Cr, Cu, Ga, Fe, Hg, In, Mn, Ni, Pb, Ti, Tl, and Zn.[1]

Regarding pesticide exposure analyzed using rotifers, the database contains 13 organophosphate (OP), five organochlorines (OC), four herbicides (H), three pyrethroids (P), two carbamates (C), and one fungicide (F): azinphos-methyl (OP), carbaryl (C), 4-chloroaniline, chlorpyrifos (OP), cypermethrin (P), diazinon (OP), 3,4-dichloroaniline (OC), diesel 2, diesel 6, dimethoate (OP), endosulfan (OC), Faena® (H), fenitrothion (OP), fonofos (OP), glyphosate (H), isoprothiolane (OP), lindane (OC), malathion (OP), methoprene (OP), methyl parathion (OP), NaPCP (OC), 4-nitrophenol, omethoate (OP), parathion (OP), pendimethalin 60%, (H), pentachlorophenol (OC), phenol (F), permethrin (P), resmethrin (P), *S*-metolachlor

31.2% + terbuthylazine 18.8% (H), 2,3,4,6-tetrachlorophenol (OC), thiophanate-methyl (C), thiram 80% (F), and trichlorfon (OP).[1]

Finally, the database also includes 38 other organic compounds: acetone, amlodipine, amlodipine A1 (pyridine derivative), ammonium chloride, benzene, bezafibrate, bezafibrate B1, chloramines, chlorine, chlorodinitrobenzene, chloroform, crude oil, cyanide, dexamethasone and its photoproducts, 3,4-dichloroaniline, dichlorophenoxy acetic acid, diesel fuel, ethanol, ethyl acetate, ethylene glycol, fenitrothion, fenofibrate, fenofibrate F1, free ammonia, free chlorine, furosemide, furosemide F1 (dimer 2), gemfibrozil, gemfibrozil G1, hexane, methanol, phenol, prednisolone and its photoproducts, toluene, tributyltin, trichlorofon, vinyl acetate, and xylene; less than five inorganic compounds have been analyzed for exposure using rotifers.[1]

Rotifers have been used to study endocrine disruption caused by the fungicide vinclozolin.[10] They have been widely used in analysis of microcosms and mesocosms,[1] from small aquaria[11] to whole lake experiments.[12] Species sensitivity distributions have been also determined using rotifers.[13,14] Rotifers have been widely used to study eutrophication, acidification, metal and pesticide stress in natural populations.[1] Studies of bioconcentration of toxic substances in rotifers are limited to: (a) PCBs[15] and (b) selenium,[16] both for *B. calyciflorus*, (c) mono-, di-, and tributyltin with *B. plicatilis*,[17] and (d) lead biomagnification in the predator rotifer *Asplanchna brightwellii*,[18] which represents the first report of biomagnification for this metal, which in the past was considered in many textbooks to not be subject to biomagnification.

3.2 Rotifers as Models for Genotoxicology

3.2.1 Environmental Genomics

Walsh[19] described four major periods for the development of rotifer genetics and classified it as a long but sporadic history; at that time, PCR and protein electrophoresis were considered novel methods with the potential to elucidate, in conjunction with "classic" methods, the pathways and proteins (genes) involved in: (a) control of sexuality, where the real genetic studies come from the late 1980s and early 1990s, when the mate recognition pheromone was characterized;[20,21] (b) chromosome number, where cytological techniques unveiled the number and organization of rotifers karyotypes and its relation to life histories;[22–24] (c) physiological genetics, in which some genes are involved in maintenance and detoxification processes, like reactive oxygen species neutralization;[25] and (d) population and evolutionary genetics.[26]

Such approaches revealed that rotifers have been considered model organisms for a long time and that although genetics studies with these invertebrates are not as numerous as for other phyla, nowadays there is an increasing number of publications demonstrating that previous assumptions did indeed have a genetic component; for example, mate recognition in rotifers as in other animals needs pheromone participation; thus, Snell *et al.*[27] identified through cDNA libraries and bioinformatics approaches some genes that encode proteins with the expected properties of mate recognition pheromone (MRP); in addition, their functionality was assessed *via* RNAi knockdown, demonstrating its participation in MRP synthesis.

Important contributions to rotifer genetics have recently been made by studying rotifer genomes. Within the class Monogononta, *Brachionus* is the most genus since several species genomes are nowadays described and available, including those recommended in international guidelines like *B. calyciflorus* and *B. plicatilis*,[28,29] and two other species intended to become model organisms are *B. ibericus*[30] and *B. koreanus*.[31] Although bdelloid rotifers have been of great interest because of their loss of sexual reproduction and their ability to survive in anhydrobiosis,[32] the number of complete genomes is very limited in comparison to monogononts; *Adineta vaga* is the only bdelloid with its genome sequenced to date.[33] Moreover, complete mitochondrial genomes are available for *Rotaria rotatoria* (Bdelloidea)[34] and *B. koreanus*.[35]

Genomics and genetics have provided researchers with relevant information for developing tools to assess how environmental factors (including toxicant exposure) influence biochemical and physiological processes along with survival and fecundity. Calorie restriction in rotifers enhances life span, up-regulates genes related to carbohydrate metabolism and improves the expression of manganese superoxide dismutase (Mn-SOD) and catalase (CAT), which are protective enzymes against oxidative stress.[36–39]

The protective role of certain enzymes has been tested in rotifers through gene expression profiles. Jung and Lee[40] described the role played by heat shock proteins genes (*hsp*) in *B. koreanus* when exposed to cadmium or copper, and from their expression profiles (up-regulation) concluded that *hsp27, hsp70*, and *hsp90α-1* could serve as molecular markers of exposure to heavy metals. Furthermore, copper and cadmium altered the expression profile of glutathione S-transferase (GST) genes and the activity of glutathione cycle enzymes (glutathione reductase and glutathione peroxidase) in *B. koreanus*, which could be implemented as biomarkers of an early oxidative stress process like GST.[41,42]

In oxidative stress not only metals participate but also organic compounds, in which biocides are included. For instance, *B. koreanus* is the species with more publications regarding the effect of pesticides on expression levels of antioxidant enzymes like GST and SOD, cytochrome P450 (CYP), and *hsp*, whose patterns depend on the toxicant chemical properties and time of exposure.[43,44] Besides enzymes that neutralize reactive oxygen species or detoxify chemical compounds, DNA repair mechanisms in the same *B. koreanus* have been assessed in organisms exposed to UV-B or gamma radiation through real-time RT-PCR to evaluate the relative amounts of transcripts, finding higher mRNA levels in radiation-exposed animals in comparison to the control groups because of alterations of the redox steady state.[45,46]

In recent years, novel toxicants have emerged owing to health and care products, which reach water bodies since the conventional wastewater treatments cannot remove them. Therefore, there is a need to investigate the effects of pharmaceuticals (*i.e.*, antibiotics, anti-inflammatory drugs) on aquatic organisms. Thus, Rhee *et al.*[47] performed acetylcholinesterase (AChE) expression analysis in *B. koreanus* exposed to six different pharmaceuticals: acetaminophen, atenolol, carbamazepine, oxytetracycline, sulfamethoxazole, and trimethoprim. In addition, the AChE activity was assessed in the same treatments. Enzymatic activity and mRNA expression were significantly inhibited by exposure to acetaminophen, carbamazepine, and trimethoprim; moreover, oxytetracycline and sulfamethoxazole barely decreased AChE mRNA expression.

In summary, transcriptomics provides relevant information regarding the effects of diverse factors on specific genes, which may serve as molecular markers or even as fingerprints to identify the most likely toxicant to which individuals in natural environments could have been exposed. Nonetheless, there is still the need for further research in environmental transcriptomics with rotifers as model organisms.

3.2.2 Single Cell Electrophoresis (Comet Assay)

Secondary DNA modifications include strand breaks, changes in base composition or increased activity of DNA repair mechanisms. Several chemicals cause DNA strand breaks as part of their toxicity mechanisms, which can be evaluated either *in vivo* (animal models) or *in vitro* (cell line cultures) using organisms isolated from natural environments or model organisms raised in laboratory conditions.[48] For instance, methodologies that assess DNA integrity in cells from toxicant-exposed individuals include Single Cell Gel Electrophoresis (SCGE), which is commonly known as the Comet assay since intact and broken DNA are visualized as the nucleus and tail of a comet, respectively.[49] Image analysis methods are currently applied to quantify the amount of DNA in the comet tail and classify the shape into one out of five categories after staining DNA with fluorescent dyes as ethidium bromide or SYBR Green I.[50]

Despite the broad application of SCGE in aquatic organisms, both marine and freshwater animals, it is important to point out that some considerations should be made when running this test; for instance, there could be high variability in comet shapes and quantities because of intrinsic tissue characteristics (*i.e.*, cell cycle stage) or even caused by manipulation.[51] Moreover, higher variability has been identified when using complete organisms (*i.e.*, invertebrates like cladocerans) in comparison to single cell culture types, likely because entire organism homogenates contain several sorts of cell lineages instead of a single one. Therefore, genotoxicity research requires working with a specific tissue like blood, hepatocytes, leukocytes and coelomocytes.[51]

Despite the fact that SCGE has been successfully performed in annelids, fish, mollusks, mussels, oysters, polychaetes, *etc.*, zooplanktonic species represent a challenge to obtain cells from the same tissue. Thus, Pellegri *et al.*[52] developed and optimized a method for acquiring cells from *Daphnia magna* hemolymph, in which tissue disruption was carried out with a short shock that included the use of glass microspheres, giving a significant improvement to the current methodologies.

Rotifers, as model organisms, are included in international guidelines for acute, chronic, and sublethal toxicity evaluation, but despite their broad use in environmental toxicology there are not yet any report regarding SCGE in rotifers. This is probably owing to the limited biomass that can be harvested from these zooplankters or the difficulty to isolate cells from a single tissue. However, scientific literature describes the use of embryos from different sources for Comet assay protocols;[49] therefore, instead of struggling with the optimization of tissue separation, research might focus on using rotifer eggs (embryos) that can be easily collected in laboratory conditions, or cysts from sediment samples or commercially available samples,[53] which in the end could represent the advantage of evaluating toxicant-induced DNA alterations on rotifers in early life stages. Finally, integrative analysis might include Comet assay results and correlate them with other cellular processes, like apoptosis induction, carbohydrate metabolism modification, cell division (mitosis) alterations, delayed organogenesis, and hatchability reduction, therefore presenting a complete landscape of how toxicants affect rotifers.

3.2.3 Cell Cycle Parameters

Rotifer cell cycle and gene transcription are valuable tools in toxicity tests to evaluate the adverse effects of chemicals, where the gene transcription of biomarkers for xenobiotic detoxification and general stress response can be highly informative. The rotifers have sexual (meiosis) and asexual reproduction (mitosis). The induction of mictic reproduction in rotifers is a highly adaptive mechanism of great ecological relevance. Induction of sexual reproduction involves the production of amphoteric females that produce eggs by simultaneous oogenesis *via* meiosis (male ova) and mitosis (female ova): a female carries both male and female offspring.[54] The males copulate the amphoteric females to produce diploid eggs, or cysts (called resting eggs by some authors). However, in the Rotifera there is a taxonomic group called the Bdelloidea that lacks sex; their reproduction is strictly parthenogenetic and they are able to produce eggs that are highly tolerant to desiccation using a mechanism named anhydrobiosis. In general, the embryonic development of parthenogenetic eggs in rotifers involves different stages: (1) first cell progenitors (macromers and micromers) form the blastula, (2) which originates a gastrula by epiboly and (3) organogenesis.[55]

During the embryologic development there are different genetic patterns of expression in parthenogenetic eggs and resting eggs: the expression of gen *vasa* and *nanos* [*vasa* encodes an adenosine triphosphate (ATP)-dependent RNA helicase that is a member of the DEAD box protein family and *nanos* genes have been implicated in specification of both germline and somatic cell fate, although their roles in the formation and/or maintenance of PGCs are more broadly conserved] are evident in the vitellarium and eggs of *B. plicatilis*; this denotes an epigenetic origin of germinal cells.[56] During the hatching of cysts the following facts have been observed: (1) high expression of LEA proteins and heat shock protein family, (2) medium expression of genes ferritin, glutathione-6-tranferase and hsp70, and (3) low expression of trehalose-6-phosphate synthase, fatty-acid binding protein, lipoprotein lipase and the aquaporins gene family.[57] We emphasize the anhydrobiosis mechanism of Bdelloidea and the absence of meiosis as tools to study evolution and genotoxicology because they are an interesting group with peculiarities in their genome: degenerate tetraploidy, extreme compartmentalization of mobile elements, and the presence of large amounts of foreign DNA in subtelomeric regions.[58]

Several characteristics make rotifers good candidates for genotoxicological analysis: (a) cellular development in rotifers, (b) patterns of expression during formation of embryos, (c) the ability to produce cysts, (d) conserved genomes of populations, (e) the alternation of asexual and sexual reproduction by induction, and (f) multiple generations in a short time. It is important to investigate in detail the alterations, adverse effects, and patterns of gene expression in the vitellarium, a syncytial gland responsible for producing eggs in rotifers. Rotifers have many potential genes to study and in fact several genes have already been detected in *B. plicatilis*: genes related to cell structure and motility (6.8%), gene and protein expression (25.1%), metabolism (18.9%), and cell signaling and communication (10.6%), using transcript abundance by functional class of predicted protein product analysis,[59] and the whole transcriptome of *B. koreanus* using RNA-seq technology reported a total 13 784 genes identified from Blast analysis.[60]

3.2.4 Micronucleus, Nuclear Abnormalities, Microsomal and Mutagenic Assays

Micronucleus and nuclear abnormalities have emerged as one of the preferred methods for assessing chromosome damage because they enable both chromosome loss and chromosome breakage to be measured reliably. These nuclear abnormalities can only be expressed in cells that complete nuclear division.[61] Their presence in cells is a reflection of structural and/or numerical chromosomal aberrations arising during mitosis.[62,63]

Information about micronucleus and nuclear damage is scarce in rotifers. However, a few examples exist on the numbers of chromosomes in embryo nuclei, for example: *Phylodina roseola* had 13 chromosomes, *Macrotrachela quadricornifera* had 10 chromosomes, and *Habrotrocha constricta* and *A. vaga* had 12 chromosomes.[64] Interestingly, the bdelloids are extraordinarily resistant to ionizing radiation and desiccation. It is almost certain that this fact is the result of evolutionary adaptation to repair DNA breaks and to prevent or repair other damage caused by the episodes of desiccation to which bdelloids are exposed in their ephemerally aquatic habitats. This ability to remain fertile after experiencing high levels of radiation is a characteristic of the bdelloid lifestyle, and offers an explanation for why bdelloid genomes are made up of collinear chromosomes.[65] For this reason, bdelloids have been used to study the effects of gamma radiation in *Adineta vaga*, which was irradiated with gamma rays (1120 and 5000 Gy), causing low fecundity and DNA breakage.[66] The monogonont rotifer *B. koreanus* has been irradiated with gamma rays from 0 to 7000 Gy, and LD_{50}s were 2900 and 2300 Gy at 24 h (LD_{50}-24 h) and 96 h (LD_{50}-96 h) after irradiation, respectively.[67] Bdelloid rotifers seem to be more resistant than other groups of rotifers. This characteristic makes rotifers excellent models to measure the DNA damage and abnormalities in nuclear cells, or embryos, when they are exposed to mutagenic substances.

Recently, and for the first time, microsomal studies used as a biomarker in rotifers have been documented. For example, identification of stress granules expression in the rotifer *B. manjavacas* in response to heat stress, osmotic stress and nutrient deprivation.[68] In the vitellarium of rotifers we can observe a lot of inclusion droplets (lipids and lysosomes). These microsomal inclusions release enzymes that cause cellular death. There is also a relationship of increasing inclusion bodies and vacuole autophagy structures as aging increases; in fact, these autophagy vacuoles stop the reproduction of parthenogenetic eggs[69] (Figure 3.1). In *A. brightwelli* the rate of nuclear division in the gastric glands and vitellarium of the rotifer, as determined by daily nuclear counts, was retarded by dietary restriction.[70]

Finally, to measure mutagenesis in rotifers we can examine the infinite asexual populations, which can accumulate an increasing number of deletions and mutations by a process known as Muller's ratchet, which consists of successive losses of the fittest or least-loaded classes of individuals in the populations.[71] Bdelloidea are a strictly parthenogenetic group with asexual reproduction, without meiosis: the oocytes are formed through mitotic divisions, with no reduction of chromosome number and no indication of chromosome pairing,[72] they have gene conversion, which may limit the accumulation of deleterious mutations in the absence of meiosis.[73] However, in monogonont rotifers, induction of sexuality increases the genetic diversity of populations producing diapausing embryos.[74] For example, in a study of the genetic population of a *B. plicatilis* species complex, they were shown to have a high level of genetic diversity consisting of 43 multilocus genotypes, which were clustered into six well differentiated groups, often found in sympatry, with few or no hybrids between them,[75] as a consequence of the sexual reproduction. However, in asexual rotifers like bdelloids, the homogenizing and diversifying roles of sex may have been replaced by gene conversion and horizontal gene transfer in an unexpected convergence of evolutionary strategy with prokaryotes.[76] The mutagenic alterations as an endpoint for toxicology assays using rotifers, including micronucleus, nuclear abnormalities, and microsomal test, are a new area for future investigations. For example, the rotifer *B. calyciflorus* exposed to vinclozolin

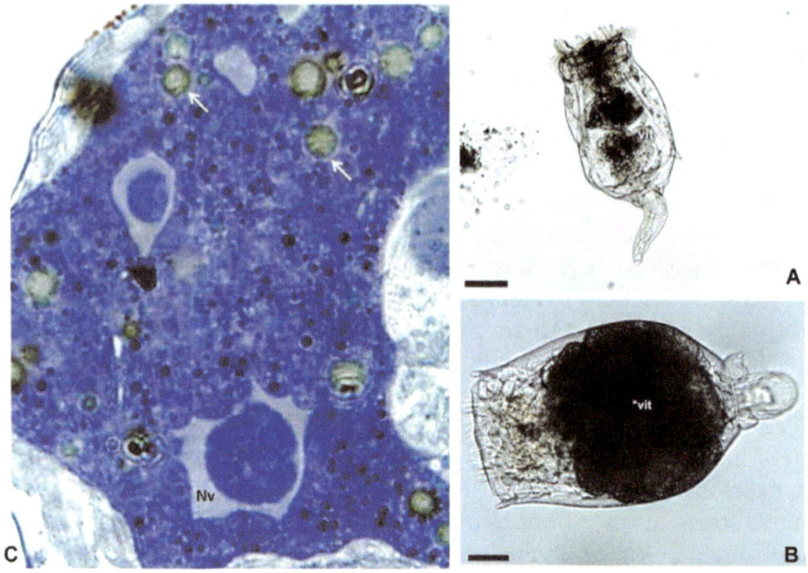

Figure 3.1 Inclusion bodies in the vitellarium of *Brachionus calyciflorus*. (A) Normal vitellarium size in the rotifer. (B) Abnormal vitellarium size in rotifers after chronic exposure to vinclozolin. (C) Transversal cut, 300 nm, inclusion in LR white resin, using an Ultramicrotome LEICA, observed at 100× using an Axio Scope 40 Zeiss microscope. Vit, vitellarium; Nv, nuclei of vitellarium.[22]

is able to pass on the adverse effects to subsequent generations.[74] Rotifers can be easily cultured, reproduce quickly, and occur at high levels of clonal, genetic diversity in nature: for these reasons, they are excellent subjects for the study of macro and micro evolutionary changes[77] in populations with toxic chronic exposure that induce adaptations that make the rotifer more tolerant to toxicity.[78]

3.3 Cytogenetics

In the last four decades a wide variety of biological methods have been developed to investigate different endpoints, sub-lethal effects, and biochemical responses, evaluating: acute toxicity, cytotoxicity and genotoxicity in marine and freshwater organisms.[77]

Cytotoxicity can be defined as the interference of a chemical compound with structures and/or functions essential for the survival and reproduction of almost any cell: integrity of membranes, metabolism, degradation of cellular constituents (DNA, RNA), ion regulation, and cell division.[78,79] The Multicentre Evaluation of *In Vitro* Cytotoxicity (MEIC) study was a programme that was designed and organised by the Scandinavian Society for Cell Toxicology to evaluate the relevance of in vitro toxicity tests for predicting human toxicity. MEIC was a voluntary effort by international laboratories to test the same 50 reference chemicals in their own *in vitro* toxicity systems. These *in vitro* results were compared with human lethal blood concentrations and other relevant acute systemic toxicity data.[80] The first ten chemicals (acetylsalicylic acid, amitriptyline, diazepam, digoxin, ethanol, ethylene glycol, ferrous sulfate, isopropanol, methanol, paracetamol) were tested for acute toxicity in four standardized cyst-based aquatic invertebrate tests, consisting of two rotifer species (the estuarine *B. plicatilis* and the freshwater *B. calyciflorus*) and two crustacean species (the halophilic anostracan *Artemia salina* and the freshwater anostracan *Streptocephalus proboscideus*). Mortality was the test criterion and toxic effects, expressed as 24 hour LC_{50} values, were correlated with rodent and human acute oral toxicity data. Generally, a good correlation was obtained between any of the invertebrate tests and the rodent data. Likewise, the predictive screening potential of the aquatic invertebrate tests for acute oral toxicity in man was slightly better than the rat test for eight (excluding diazepam and digoxin) and nine (including diazepam, excluding digoxin) of the ten substances. The aquatic test systems, however, appear to be more suitable for compounds that are soluble in water.[81]

3.3.1 Chromosome Damage

One type of aquatic organism with more studies on chromosomal damage is fish, with about 40 species that have been evaluated with different toxicants, like metals and organic compounds.[82] In the species of rotifers *Adineta vaga* (carbonylation assayed by ELISA),[65] *Philodina roseola*[65] and *Euchlanis dilatata* (both species by examination of DNA breakage),[83] the effects of ionizing radiation were investigated at the chromosome level; the researchers found significantly higher resistance in response to gamma radiation even at several hundreds to thousands of gray with a high dose rate (dozens or hundreds of Gy min^{-1}).

3.3.2 Nuclear Morphology

Rao[84] evaluated the toxic effects of monocrotophos on the ciliates *Paramecium caudatum* and *Oxytricha fallax;* he found macronuclear changes like unevenly divided, vacuolated, fragmented and rod shaped at concentrations of 100 mg L^{-1} of monocrotophos. Lal and Saxena[85] found that continuous treatment with 100 mg L^{-1} DDT produced several nuclear abnormalities in the ciliate *Stylonychia notophora;* however, comparing these values with the rotifer *B. calyciflorus* for chronic toxicity, concentrations higher than 0.16 mg L^{-1} significantly decreased the mictic rate of the rotifers.[86] The rotifer *Brachionus* sp. was exposed to a lipophilic extract of *Vicicitus globosus* for 80 min and cellular materials were discharged from the *Brachionus*; these showed that hemolytic cytotoxins are thought to change the permeability of the cell membrane to a range of ions, leading to cell death through swelling and eventually osmotic lysis of cells.[87]

3.3.3 Sister Chromatid Exchange

The measurement of sister chromatid exchanges (SCEs) is perhaps the most commonly used. SCEs are regarded as interchanges between DNA molecules at homologous loci within a replicating chromosome.[88] It can be performed on a diversity of cells, including cells from sentinel species like mussels and fish, which makes this assay extremely useful for environmental monitoring.[64] Around 20 species of fish have been used to assess SCEs as end-points with metals and organic compounds.[82] The practicality of SCE induction as a measure of low-level radiation effect was examined in a benthic marine worm, *Neanthes arenaceodentata*.[89] Larvae were exposed to ^{60}Co radiation;[90] the length of chromosome pair number one differed among metaphase spreads and was used as an index of chromosome condensation in a given metaphase.[90]

An *in vivo* SCE assay was applied to the larvae of the mussel *Mytilus edulis* exposed to mutagens bromodeoxyuridine, mitomycin C and methyl methanesulfonate. The sensitivity of the SCE response of mussel larvae to these mutagens compares favorably with that of other systems and indicates this assay to be valuable in marine genetic toxicological monitoring.[91] A special case is rotifers of the genus Bdelloidea, which have the ability to resist large amounts of ionizing radiation; the researchers propose that this attribute is because bdelloid primary oocytes are in G1 and therefore lack sister chromatids and collinear chromosome pairs are maintained as templates for the repair of DNA double-strand breaks caused by the frequent desiccation and rehydration characteristics of bdelloid habitats.[64,65]

3.4 DNA Damage

DNA damage includes physical or chemical changes in its structure, which can have many immediate effects depending on the type of damage and affected genomic region,[92] and may take the form of DNA adducts.[93] These alterations can disrupt the gene expression and cell functioning, uncoupling the transcription, and delaying the cell cycle and cell death (apoptosis). Damage to DNA can be from extrinsic sources, such as chemicals, UV radiation and viruses,[92,93] and intrinsic sources, such as spontaneous chemical reactions and reactive oxygen species (ROS, *e.g.* superoxide, hydrogen peroxide, radical hydroxyl and singlet oxygen).[94] The imbalance of ROS by antioxidant mechanisms in an organism is called oxidative stress.[95] Although ROS also participate in beneficial physiological processes, they cause damage to the interior of cells, including oxidative lesions, such as 8-hydroxy-2′-deoxyguanosine (8-OHdG u 8-oxodG), thymine glycols, cyclopurines, and different types of breaks of single or double DNA strands (SSBs, DSBs). In addition, alkylating agents, estrogen, cholesterol metabolites and reactive carbonyl species damage DNA. DNA lesions are also formed naturally by hydrolysis or changes in nucleotide-releasing non-informative abasic sites, or altered nucleotides not coding (miscoding).[96]

Transition metals, such as copper, chromium, nickel and cadmium, as well as quinones, dyes, herbicides, bipyridyls and nitro-aromatic compounds have cycling redox properties and the potential for causing oxidative stress, particularly concerning the peroxidation of lipids and carcinogenesis. The polyanionic nature of DNA provides a surface for adhesion of cationic metals, generating free radicals and leading to oxidative damage. Other HO attacks can be directly to the skeleton of the sugar–phosphate backbone, causing injuries that removed the base, fragmenting the deoxyribose and oxidizing the sugar.[97] Ionizing radiation can break one or two strands of DNA and produce several oxidative lesions.[98] Oxidized DNA damage includes: (a) oxidation of bases; (b) abasic sites (AP); (c) deamination; (d) SSBs or DSBs; and (e) cyclobutane pyrimidine dimers (CPDs).[98] It is possible to form adducts, which is a type of structural change that involves covalent bonds of a chemical or its metabolites with DNA. It has been assumed that most DNA adducts are harmful; however some adducts can be repaired by excision without consequences. Some adducts of nucleotides in inert DNA regions also do not produce adverse effects, noting that the DNA adducts are evidence of the specific exhibition that goes through all toxicokinetic barriers.[98]

3.4.1 DNA Damage and Rotifers

During their evolution, biological systems have developed enzymatic and nonenzymatic antioxidant mechanisms to protect their cellular components from oxidative damage[41,94] or to decrease its damage, although sometimes it is not enough, causing oxidative stress.[94] From green algae to marsupials, photoenzymatic repair (PER) occurs through the photolyase enzyme, with the rate of repair varying depending on the species and stage of development.[95] There is also a multiprotein Nucleotide Excision Repair System (NER) that removes DNA helix-destabilizing lesions; this recognizes changes in the structure of the DNA as rigidity of the distortion of the propeller and injuries in genes that block the active transcription of RNA polymerase II (RNAPIIo).[96] Organisms in clear aquatic systems receive harmful levels of UV radiation; wavelengths lower than 305 nm are the most harmful to DNA, managing to reduce the survival of aquatic invertebrates during at least one stage of life.[97] The molecular response in rotifer has been little studied.[98] There is no evidence of PER in the rotifer *Asplachna brightwelli* exposed to UV radiation from 50 to 4800 $J\,m^{-2}$, which reduced their longevity and pre-reproductive stage.[99] The response of *Bk-Dmrt* genes (*Dmrt11E, Dmrt93B*, and *Dmrt99B*) in *B. koreanus* exposed to UV-B radiation and benzo[*a*]pyrene (B[*a*]P) was evaluated,[98] and they were found to decrease in response to high doses of UV-B and B[*a*]P, causing delays in growth. Genes associated with DNA repair showed patterns of regulation in response to sublethal doses of gamma radiation in *B. koreanus*, decreasing population parameters.[41] Moreover, it was reported that sublethal exposure of UV-B increase the sensitivity of *B. calyciflorus* to pentachlorophenol (PCP) and mercury.[100]

3.4.2 Rotifer Strategies

Rotifers from the class Bdelloidea are known for their extreme desiccation tolerance. At any stage of their life cycle they can enter a metabolically quiescent state of anhydrobiosis for a prolonged period of time. At the cellular level, desiccation survival requires adaptations that retain the functioning of macromolecules (proteins and DNA) and membranes despite dehydration. This can be through the preservation of the integrity of these molecules or by repairing them after suffering damage during the drying process.[66] Bdelloids such as *Adineta vaga* and *Philodina roseola* are extremely resistant to ionizing radiation in comparison to *Euchlanis dilatata*, a rotifer that is not tolerant to desiccation and with optional sexually, as they are able to assemble long blocks of a crushed genome DNA.[13] The desiccation process in bdelloids allows them to reorganize themselves over time.[85] Therefore, the dehydration induces DSBs in DNA and damage is repaired after rehydration in *A. vaga*.[66] Some authors suggest that this rotifer species contains powerful antioxidant machinery that protects their cellular components from oxidative damage from high-dose ionizing radiation.[66,85] Thus, there is protection against oxidative damage to proteins involved in DNA DSB repair.[66] Consequently, *A. vaga* is very resistant to protein carbonylation induced by ionizing radiation.[85]

3.4.3 DNA Biomarkers

Aquatic organisms can provide experimental data for the evaluation of sublethal effects of oxidative stress, mutagenicity and other adverse effects of pollutants.[97] Structural alterations caused by xenobiotics can irreversibly change the DNA molecule, resulting in the expression of another cell chromosomal aberration response and activation of oncogenes.[95] The sequence of a number of genes has been studied in rotifers *B. plicatilis*, *B. ibericus* and *B. calyciflorus*.[98] DNA damage has been proposed as a useful variable to evaluate genotoxicity of pollutants, and can be used as a biomarker of exposure and effect based on the evaluation of structural changes in DNA by measuring DNA repair, directly or indirectly, and mutations produced in the genome.[99] A biomarker of cellular oxidative stress and genotoxicity in living organisms is 8-hydroxy-2′-deoxyguanosine (8-OHdG u 8-oxodG), which is considered the most important DNA damage, after the attack of ROS.[101]

3.4.4 Mutations in Rotifers

An important aspect of genotoxic exposure is the creation of mutations in the DNA, which lead to alterations in the functioning of genes. If such genetic aberrations are correlated with abnormal cellular morphologies, these mutations can lead to somatic events occurring during the formation of abnormal cells.[65] Damage to DNA can also lead to mutations during repair or replication. These mutations and other changes in the DNA, such as imbalances, deletions, insertions, substitutions or imbalances of the base pairs producing dysfunctional proteins, loss of heterozygosity and direct chromosome aberrations,[33,101] are caused by breakdown of individual strands or bad segregation and can cause tumors or congenital diseases if they occur in germ cells.[83] The creation of mutations in DNA leads to alterations in the functioning of genes. If these genetic aberrations are correlated with abnormal cellular morphologies, then somatic events may occur during the formation of abnormal cells. Specific nucleotide changes are considered an exchange in chemical adducts of DNA profile and the repair capacity of the cell on particular forms of adducts.[102] Alvarado-Flores *et al.*[74] recorded morphological abnormalities in *B. calyciflorus* exposed to the herbicide vinclozolin: in spite of these abnormalities, rotifers were still able to reproduce. There is the possibility of implementing the technique of PCR with *B. plicatilis* obtaining specific sequences of DNA to assess the changes in the abundance of the mRNA counterpart in rotifers subjected to stress, suggesting their relevance for developing biomarkers of environmental pollution.[103] Mitochondrial DNA should be particularly sensitive to these effects because of maternal inheritance and not being subject to recombination. However, this sensitivity may be owing to high levels of adduct formation in mtDNA in comparison with in nuclear DNA.[95] The complete mitochondrial genome of *B. koreanus* is very similar to what was reported for the mitochondrial genome of *B. plicatilis*, however with different arrangements.[35] Hwang *et al.*[35] suggest the need to assess changes in the expression of genes for potential use as xenobiotic-induced early environmental indicators.

References

1. R. Rico-Martínez, I. A. Pérez-Legaspi, J. C. Arias-Almeida and G. E. Santos-Medrano, Rotifers in Ecotoxicology, in: *Encyclopedia of Aquatic Ecotoxicology*, ed. J. F. Férard and C. Blaise, Springer, Berlin, Germany, 2013, pp. 973–996.
2. G. E. Santos-Medrano, E. M. Ramírez-López, S. Hernández-Flores, P. M. Azuara-Medina and R. Rico-Martínez, Determination of toxicity levels in the San Pedro River watershed, Aguascalientes, Mexico, *J. Environ. Sci. Health, Part A: Toxic/Hazard. Subst. Environ. Eng.*, 2007, **42**, 1403–1410.
3. S. S. S. Sarma, H. E. Trujillo-Hernández and S. Nandini, Population growth of herbivorous rotifers and their predator (*Asplanchna*) on urban wastewaters, *Aquat. Ecol.*, 2003, **37**, 243–250.
4. M. Isidori, M. Lavorgna, A. Nardelli and A. Parrella, Toxicity identification evaluation of leachates from municipal solid waste landfills: a multispecies approach, *Chemosphere*, 2003, **52**, 85–94.
5. G. S. Park, C. S. Chung, S. H. Lee, G. Hong, S. H. Kim, S. Y. Park, S. J. Yoon and S. M. Lee, Ecotoxicological evaluation of sewage sludge

using bioluminescent marine bacteria and rotifer, *Ocean Sci. J.*, 2005, **40**, 91–100.
6. S. B. José de Paggi and M. Devercelli, Land use and basin characteristics determine the composition and abundance of the microzooplankton, *Water, Air, Soil Pollut.*, 2010, **218**, 93–108.
7. R. L. Wallace, T. W. Snell, C. Ricci and T. Nogrady, Rotifera. Biology, Ecology and Systematics, in: *Guides to the Identification of the Microinvertebrates of the Continental Waters of the World*, ed. H. J. F. Dumont, Kenobi Productions, Ghent, Belgium, 2006, p. 299.
8. J. Cairns Jr., A. L. Buikema Jr., A. G. Heath and B. C. Parker, Effects of temperature on aquatic organism sensitivity to selected chemicals, *Virginia Water Resources Research Center Bulletin 106*, Virginia, USA, 1978, 97.
9. T. W. Snell and C. R. Janssen, Rotifers in ecotoxicology: a review, *Hydrobiologia*, 1995, **313/314**, 231–247.
10. J. Alvarado-Flores, M. R. Montoya-Garcia, J. Ventura-Juárez and R. Rico-Martínez, Immunodetection of Luteinizing Hormone (LH), Follicle-Stimulating Hormone (FSH), Thyroid Stimulating Hormone (TSH) and Prolactin (PRL) in *Brachionus calyciflorus* (Rotifera: Monogononta), *Rev. Biol. Trop.*, 2009, **57**, 1049–1058.
11. R. Rico-Martínez, I. A. Pérez-Legaspi, G. E. Quintero-Díaz, M. A. Hernández-Rodríguez, M. G. Rodríguez-Martínez and J. Zaragoza-Almaráz, Effect of copper addition to a laboratory maintained microcosm of Presidente Calles Reservoir, Aguascalientes, México, *Aquat. Ecosyst. Health Manage.*, 1998, **1**, 323–332.
12. M. J. González and T. M. Frost, Comparisons of laboratory bioassays and a whole-lake experiment: rotifer responses to experimental acidification, *Ecol. Appl.*, 1994, **4**, 69–80.
13. D. J. Versteeg, S. E. Belanger and G. J. Carr, Understanding single-species and model ecosystem sensitivity: data-based comparison, *Environ. Toxicol. Chem.*, 1999, **18**, 1329–1346.
14. M. McDaniel and T. W. Snell, Probability distributions of toxicant sensitivity for freshwater rotifer species, *Environ. Toxicol.*, 1999, **14**, 361–366.
15. C. Joaquim-Justo, V. Gosselain, J. P. Descy and J. P. Thomé, Relative importance of the trophic and direct pathways on PCB contamination in the rotifer species *Brachionus calyciflorus* (Pallas), *Hydrobiologia*, 1995, **313/314**, 249–257.
16. M. G. Dobbs, D. S. Cherry and J. Cairns Jr., Toxicity and bioaccumulation of selenium to a three-trophic level food chain, *Environ. Toxicol. Chem.*, 1996, **15**, 340–347.
17. S. Hong-Wen, D. Shu-Gui and H. Guo-Lan, Bioaccumulation of butyltins via an estuarine food chain, *Water, Air, Soil Pollut.*, 2001, **125**, 5–58.
18. I. Rubio-Franchini and R. Rico-Martínez, Evidence of lead biomagnification in invertebrate predators from laboratory and field experiments, *Environ. Pollut.*, 2011, **159**, 1831–1835.

19. E. J. Walsh, Rotifer genetics: integration of classic and modern techniques, *Hydrobiologia*, 1993, **256/257**, 193–204.
20. T. W. Snell, M. J. Childress and B. C. Winkler, Characteristics of the mate recognition factor in the rotifer *Brachionus plicatilis*, *Comp. Biochem. Physiol., Part A: Mol. Integr. Physiol.*, 1988, **89**, 481–485.
21. T. W. Snell and M. A. Nacionales, Sex pheromone communication in *Brachionus plicatilis* (Rotifera), *Comp. Biochem. Physiol., Part A: Mol. Integr. Physiol.*, 1990, **97**, 211–216.
22. K. K. Rishi, A. K. Datta-Gupta and Kamaljeet, A report on the chromosome numbers of some rotifer species, *CIS, Chromosome Inf. Serv.*, 1989, **46**, 8–9.
23. E. J. Walsh and L. Zhang, Polyploidy in a natural population of the rotifer *Euchlanis dilatata*, *J. Evol. Biol.*, 1992, **5**, 345–353.
24. J. L. M. Welch and M. Meselson, Karyotypes of bdelloid rotifers from three families, *Hydrobiologia*, 1998, **387/388**, 403–407.
25. R. K. Johnston and T. W. Snell, Moderately lower temperatures greatly extend the lifespan of *Brachionus manjavacas* (Rotifera): Thermodynamics or gene regulation?, *Exp. Gerontol.*, 2016, **78**, 12–22.
26. M. V. Sørensen and G. Giribet, A modern approach to rotifer an phylogeny: Combining morphological and molecular data, *Mol. Phylogenet. Evol.*, 2006, **40**, 585–608.
27. T. W. Snell, T. L. Shearer, H. A. Smith, J. Kubanek, K. E. Gribble and D. B. M. Welch, Genetic determinants of mate recognition in *Brachionus manjavacas* (Rotifera), *BMC Biol.*, 2009, **7**, 60.
28. ASTM E1440-91. Standard Guide for Acute Toxicity Test with the Rotifer *Brachionus*, ASTM International, West Conshohocken, PA. 2012.
29. ISO 20666. 2008. Water quality. Determination of the chronic toxicity to *Brachionus calyciflorus* in 48 h. International Organization for Standardization. Geneva, Switzerland.
30. J. S. Lee, R. O. Kim, J. S. Rhee, J. Han, D. S. Hwang, B. S. Choi, C. J. Lee, Y. D. Yoon, J. S. Lim, Y. M. Lee, G. S. Park, A. Hagiwara and I. Y. Choi, Sequence analysis of genomic DNA (680 Mb) by GS-FLX-Titanium sequencer in the monogonont rotifer, *Brachionus ibericus*, *Hydrobiologia*, 2011, **662**, 65–75.
31. B. Y. Lee, H. S. Kim, D. S. Hwang, E. J. Won, B. S. Choi, I. Y. Choi, H. G. Park, J. S. Rhee and J. S. Lee, Whole transcriptome analysis of the monogonont rotifer *Brachionus koreanus* provides molecular resources for developing biomarkers of carbohydrate metabolism, *Comp. Biochem. Physiol., Part D: Genomics Proteomics*, 2015, **14**, 33–41.
32. M. Caprioli, A. K. Katholm, G. Melone, H. Ramlbv, C. Ricci and N. Santo, Trehalose in desiccated rotifers: a comparison between a bdelloid and a monogonont species, *Comp. Biochem Physiol.*, 2004, **139**, 527–532.
33. J. F. Flot, B. Hespeels, X. Li, B. Noel, I. Arkhipova, E. G. Danchin, A. Hejnol, B. Henrissat, R. Koszul, J. M. Aury, V. Barbe, R. M. Barthélémy, J. Bast, G. A. Bazykin, O. Chabrol, A. Couloux, M.

Da Rocha, C. Da Silva, E. Gladyshev, P. Gouret, O. Hallatschek, B. Hecox-Lea, K. Labadie, B. Lejeune, O. Piskurek, J. Poulain, F. Rodriguez, J. F. Ryan, O. A. Vakhrusheva, E. Wajnberg, B. Wirth, I. Yushenova, M. Kellis, A. S. Kondrashov, D. B. Mark Welch, P. Pontarotti, J. Weissenbach, P. Wincker, O. Jaillon and K. Van Doninck, Genomic evidence for ameiotic evolution in the bdelloid rotifer *Adineta vaga*, *Nature*, 2013, **500**, 453–457.
34. G. S. Min and J. K. Park, Eurotatorian paraphyly: Revisiting phylogenetic relationships based on the complete mitochondrial genome sequence of *Rotaria rotatoria* (Bdelloidea: Rotifera: Syndermata), *BMC Genomics*, 2009, **10**, 533.
35. D. S. Hwang, K. Suga, Y. Sakakura, H. G. Park, A. Hagiwara, J. S. Rhee and J. S. Lee, Complete mitochondrial genome of the monogonont rotifer, *Brachionus koreanus* (Rotifera, Brachionidae), *Mitochondrial DNA*, 2014, **25**, 29–30.
36. G. Kaneko, T. Yoshinaga, Y. Yanagawa, S. Kinoshita, K. Tsukamoto and S. Watabe, Molecular characterization of Mn-superoxide dismutase and gene expression studies in dietary restricted *Brachionus plicatilis* rotifers, *Hydrobiologia*, 2005, **546**, 117–123.
37. A. K. S. Oo, G. Kaneko, M. Hirayama, S. Kinoshita and S. Watabe, Identification of genes differentially expressed by calorie restriction in the rotifer (*Brachionus plicatilis*), *Comp. Biochem. Physiol., Part B: Biochem. Mol. Biol.*, 2010, **180**, 105–116.
38. Y. Ozaki, G. Kaneko, Y. Yanagawa and S. Watabe, Calorie restriction in the rotifer *Brachionus plicatilis* enhances hypoxia tolerance in association with the increased mRNA levels of glycolytic enzymes, *Hydrobiologia*, 2010, **649**, 267–277.
39. M. Kailasam, G. Kaneko, A. K. S. Oo, Y. Ozaki, A. R. Thirunavukkarasu and S. Watabe, Effects of calorie restriction on the expression of manganese superoxide dismutase and catalase under oxidative stress conditions in the rotifer *Brachionus plicatilis*, *Fish. Sci.*, 2011, **77**, 403–409.
40. M. Y. Jung and Y. M. Lee, Expression profiles of heat shock protein gene families in the monogonont rotifer *Brachionus koreanus* - Exposed to copper and cadmium, *J. Toxicol. Environ. Health Sci.*, 2012, **4**(4), 235–242.
41. J. Han, E. J. Won, D. S. Hwang, J. S. Rhee, I. C. Kim and J. S. Lee, Effect of copper exposure on GST activity and on the expression of four GSTs under oxidative stress condition in the monogonont rotifer, *Brachionus koreanus*, *Comp. Biochem. Physiol., Part C: Toxicol. Pharmacol.*, 2013, **158**(2), 91–100.
42. B. Yim, H. Kim, M. Y. Jung and M. Y. Lee, Cadmium modulates the mRNA expression and activity of glutathione S-transferase in the monogonont rotifer *Brachionus koreanus*, *Toxicol. Environ. Health Sci.*, 2015, **7**, 217–223.
43. B. M. Kim, J. W. Lee, J. S. Seo, K. H. J. S. Rhee and J. S. Lee, Modulated expression and enzymatic activity of the monogonont rotifer *Brachionus*

koreanus Cu/Zn- and Mn-superoxide dismutase (SOD) in response to environmental biocides, *Chemosphere*, 2015, **120**, 470–478.

44. A. X. Yi, J. Han, J. S. Lee and K. M. Y. Leung, Toxicity of triphenyltin chloride to the rotifer *Brachionus koreanus* across different levels of biological organization, *Environ. Toxicol.*, 2016, **31**, 13–23.
45. R. O. Kim, J. S. Rheeb, E. J. Wonc, K. W. Leea, C. M. Kangd, Y. M. Leee and J. S. Lee, Ultraviolet B retards growth, induces oxidative stress, and modulates DNA repair-related gene and heat shock protein gene expression in the monogonont rotifer, *Brachionus* sp, *Aquat. Toxicol.*, 2011, **101**, 529–539.
46. J. Han, E. J. Wona, I. L. Kimb, J. H. Yimb, S. U. Leec and J. S. Lee, Sublethal gamma irradiation affects reproductive impairment and elevates antioxidant enzyme and DNA repair activities in the monogonont rotifer *Brachionus koreanus*, *Aquat. Toxicol.*, 2014, **155**, 101–109.
47. J. S. Rhee, R. O. Kim, B. M. Kim, H. U. Dahms and J. S. Lee, Genomic organization of selected genes in the small monogonont rotifer, *Brachionus koreanus. Gene*, 2012, **505**, 108–113.
48. R. van der Oost, J. Beyer and N. P. Vermeulen, Fish bioaccumulation and biomarkers in environmental risk assessment: a review, *Environ. Toxicol. Pharmacol.*, 2003, **13**, 57–149.
49. R. F. Lee and S. Steinert, Use of the single cell gel electrophoresis/comet assay for detecting DNA damage in aquatic (marine and freshwater) animals, *Mutat. Res.*, 2003, **544**, 43–63.
50. T. H. Ward and B. Marples, Technical report: SYBR Green I and the improved sensitivity of the single-cell electrophoresis assay, *Int. J. Radiat. Biol.*, 2000, **76**, 61–65.
51. G. Frenzilli, M. Nigro and B. P. Lyons, The Comet assay for the evaluation of genotoxic impact in aquatic environments, *Mutat. Res.*, 2009, **681**, 80–92.
52. V. Pellegri, G. Gorbi and A. Buschini, Comet Assay on *Daphnia magna* in eco-genotoxicity testing, *Aquat. Toxicol.*, 2014, **155**, 261–268.
53. R. Rico-Martínez, M. A. Arzate-Cárdenas, D. Robles-Vargas, I. A. Pérez-Legaspi, J. Alvarado-Flores and G. E. Santos-Medrano, Rotifers as models in toxicity screening of chemicals and environmental samples, Invertebrates, in: *Invertebrates. Experimental Models in Toxicity Screening*, ed. M. Larramendy and S. Soloneski, InTech, 2016, pp. 57–59.
54. E. J. Walsh, L. May and R. L. Wallace, A metadata approach to documenting sex in phylym Rotifera: diapausing embryos, males, and hatchlings from sediments. *Hydrobiologia*, 2016, DOI 10.1007/s10750-016-2712-z. Published on line May 9th, 2016.
55. F. A. Pray, Studies of the early development of the rotifer *Monostyla cornuta* Müller, *Trans. Am. Microsc. Soc.*, 1965, **84**, 210–216.
56. J. M. Smith, A. G. Cridge and P. K. Dearden, Germ cell specification and ovary structure in the rotifer *Brachionus plicatilis*, *EvoDevo*, 2010, **1**, 5–10.
57. N. Y. Denekamp, R. Reinhardt, M. W. Albrecht, M. Drungowski, M. Kube and E. Lubzens, The expression pattern of dormancy-associated genes in

multiple life-history stages in the rotifer *Brachionus plicatilis*, *Hydrobiologia*, 2011, **662**, 51–63.
58. E. A. Gladyshev and I. R. Arkhipova, Genome structure of Bdelloid rotifers: shaped by asexuality or desiccation?, *J. Hered.*, 2010, **101**, S85–S93.
59. K. Suga, D. M. Welch, Y. Tanaka, Y. Sakakura and A. Hagiwara, Analysis of expressed sequence tags of the cyclically parthenogenetic rotifer *Brachionus plicatilis*, *Plos One*, 2007, **8**, 1–7.
60. L. Bo-Young, K. Hui-Su, H. Dae-Sik, W. Eun-Ji, C. Beom-Soon, C. Ik-Young, P. Huem-Gi, R. Jae-Sung and L. Jae-Seong, Whole transcriptome analysis of the monogonont rotifer *Brachionus koreanus* provides molecular resources for developing biomarkers of carbohydrate metabolism, *Comp. Biochem. Physiol., Part D: Genomics Proteomics*, 2015, **14**, 33–41.
61. M. Fenech, The *in vitro* micronucleus technique, *Mutat. Res.*, 2000, **455**, 81–95.
62. J. A. Heddle, M. C. Cimino, M. Hayashi, F. Romagna, M. D. Shelby, J. D. Tucker, P. Vanparys and J. T. MacGregor, Micronuclei as an index of cytogenetic damage: past, present, and future, *Environ. Mol. Mutagen.*, 1991, **18**, 277–291.
63. C. Bolognesi, E. Perrone, P. Roggieri, D. M. Pampanin and A. Sciutto, Assessment of micronuclei induction in peripheral erythrocytes of fish exposed to xenobiotics under controlled conditions, *Aquat. Toxicol.*, 2006, **78**, 93–98.
64. D. B. Mark Welch, J. L. Mark Welch and M. Meselson, Evidence for degenerate tetraploidy in bdelloid rotifers, *Proc. Natl. Acad. Sci.*, 2008, **105**, 5145–5149.
65. E. Gladyshev and M. Meselson, Extreme resistance of bdelloid rotifers to ionizing radiaton, *Proc. Natl. Acad. Sci. U. S. A.*, 2008, **105**, 5139–5149.
66. E. J. Won, J. Han, A. Hagiwara, S. Oda, H. Mitani and J. S. Lee, Acute toxicity of gamma radiation to the monogont rotifer *Brachionus koreanus*, *Bull. Environ. Contam. Toxicol.*, 2016, **97**, 387–391.
67. B. L. Jones, J. VanLoozen, M. H. Kim, S. J. Miles, C. M. Dunham, L. D. Williams and T. W. Snell, Stres granules form in *Brachionus manjavacas* (Rotifera) in response to a variety of stressors, *Comp. Biochem. Physiol., Part A: Mol. Integr. Physiol.*, 2013, **166**, 375–384.
68. R. Herold and N. D. Meadow, Age-related changes in ultrastructure and histochemistry or rotiferan organs, *J. Ultrastruct. Res.*, 1970, **33**, 203–218.
69. C. Verdone-Smith and H. E. Enesco, The effect of dietary restriction on cell division potential DNA content and enzyme levels in the rotifer *Asplanchna brightwwelli*, *Exp. Gerontol.*, 1982, **17**, 463–471.
70. A. Colato and J. F. Fontanari, Soluble model for the acumulation of mutations in asexual populations, *Phys. Rev. Lett.*, 2001, **87**, 1–4.
71. W. S. Hsu, Oogenesis in the Bdelloidea rotifer *Philodina roseola* Ehrenberg, *Cellule*, 1956, **57**, 283–296.

72. F. Jean-Fracois, B. Hespeels, X. Li, B. Noel, I. Arkhipova, E. G. J. Danchin, A. Hejnol, B. Henrissat, O. Chabrol, A. Coulox, M. Da Rocha, C. Da Silva, E. Gladyshev, P. Gouret, O. Hallatschek, B. Hecox-Lea, K. Labadie, B. Lejeune, O. Piskurek, J. Poulian, F. Rodríguez, J. F. Ryan, O. A. Vakhrusheva, E. Wajnberg, B. Wirth, I. Yushenova, M. Kellis, A. S. Kondrashov, D. B. Mark Welch, P. Pontarotti, J. Weissenbanch, P. Wincker, O. Jaillon and K. Van Doninck, Genomic evidence for ameiotic evolution in the bdelloid rotifer *Adineta vaga*, *Nat. Lett.*, 2013, **500**, 453–457.
73. R. Ortells, T. W. Snell, A. Gómez and M. Serra, Patters of genetic differentiation in resting eggs banks of a rotifer species complex in Spain, *Arch. Hydrobiol.*, 2000, **149**, 52–551.
74. J. Alvarado-Flores, R. Rico-Martinez, A. Adabache-Ortiz and M. Silva-Briano, Morphological alterations in the freshwater rotifer *Brachionus calyciflorus* Pallas 1766 (Rotifera:Monogononta) caused by vinclozolin chronic exposure, *Ecotoxicology*, 2015, **24**, 915–925.
75. G. F. Fussman, Rotifers: excellent subjects for the study of macro- and microevolutionary change, *Hydrobiologia*, 2011, **662**, 11–18.
76. A. Aránguiz-Acuña and M. Serra, Diapause as escape strategy to exposure to toxicants: response of *Brachionus calyciflorus* to arsenic, *Ecotoxicology*, 2016, **4**, 708–719.
77. P. D. Hansen, Biosensors and Ecotoxicology, *Eng. Life Sci.*, 2008, **8**, 26–31.
78. H. Seibert, M. Gülden, M. Kolossa and G. Schepers, Evaluation of the relevance of selected in vitro toxicity test systems for acute systemic toxicity, *Altern. Lab. Anim.*, 1992, **20**, 240–245.
79. S. S. Umang, Importance of Genotoxicity & SA2 guidelines for genotoxicicty testing for pharmaceuticals, *IOSRJPBS.*, 2012, **1**, 43–54.
80. C. Clemedson, E. McFarlane-Abdulla, M. Andersson, F. A. Barile, M. C. Calleja, C. Chesné, R. Clothier, M. Cottin, R. Curren, E. Daniel-Szolgay, P. Dierickx, M. Ferro, G. Fiskesjö, L. Garza-Ocanas, M. J. Gómez-Lechón, M. Gülden, B. Isomaa, J. Janus, P. Judge, A. Kahru, R. B. Kemp, G. Kerszman, U. Kristen, M. Kunimoto, S. Kärenlampi, K. Lavrijsen, L. Lewan, H. Lilius, T. Ohno, G. Persoone, R. Roguet, L. Romert, T. Sawyer, H. Seibert, R. Shrivastava, A. Stammati, N. Tanaka, A. O. Torres, J.-U. Voss, S. Wakuri, E. Walum, X. Wang, F. Zucco and B. Ekwall, MEIC evaluation of acute systemic toxicity. Part I. Methodology of 68 *in vitro* toxicity assays used to test the first 30 reference chemicals, *ATLA, Altern. Lab. Anim.*, 1996, **24**, 249–272.
81. M. C. Calleja and G. Persoone, Cyst-based toxicity tests: 4.The potential of ecotoxicological tests for the prediction of acute toxicity in man as evaluated on the first ten chemicals of the MEIC programme, *Altern. Lab. Anim.*, 1992, **20**, 396–405.
82. M. Uribe-Alcocer and P. Díaz-Jaimes, Fish chromosomes as biomarkers of genotoxic damage and proposal for the use of tropical catfish species for short-term screening of genotoxic agents, *Biomonitors and Biomarkers as Indicators of Environmental Change 2*, 2001, vol. 56, pp. 361–390.

83. A. Krisko, M. Leroy, M. Radman and M. Meselson, *Proc. Natl. Acad. Sci.*, 2012, **109**, 2354–2357.
84. N. A. Rao, Acute toxicity and cytogenetic effects of monocrotophos in *Paramecium Caudatum* and *Oxytricha Fallax*, *Ind. J. Fund. Appl. Life Sci.*, 2011, **1**, 65–70.
85. R. Lal and D. M. Saxena, Effect of DDT on cell population growth, cell division, and DNA synthesis in *Stylonychia notophora* (Stokes), *Arch. Environm. Contam. Toxicol.*, 1980, **9**, 163–170.
86. X. Yl, Z. X. Chu and X. Xu, Effect of four organochlorine pesticides on the reproduction of freshwater rotifer *Brachionus calyciflorus* pallas, *Environ. Toxicol. Chem.*, 2007, **26**, 1695–1699.
87. F. H. Chang, Cytotoxic effects of *Vicicitus globosus* (Class Dictyochophyceae) and *Chattonella marina* (Class Raphidophyceae) on rotifers and other microalgae, *J. Mar. Sci. Eng.*, 2015, **3**, 401–411.
88. G. M. Williams, Methods for evaluating chemical genotoxicity, *Annu. Rev. Pharmacol. Toxicol.*, 1989, **29**, 189–211.
89. J. Brockmeyer, S. Spelten, T. Kuczius, M. Bielaszewska and H. Karch, Structure and function relationship of the autotransport and proteolytic activity of EspP from Shiga toxin-producing *Escherichia coli*, *PloS One*, 2009, **4**, 1–11.
90. F. L. Harrison and D. W. Rice, Effect of low ^{60}Co dose rates on sister chromatid exchange incidence in the benthic worm *Neanthes arenaceodentata*, Report Lawrence Livermore Natl. Lab., 1981, NTIS UCRL-53205, Livermore, 1–26.
91. F. L. Harrison and I. M. Jones, An *in vivo* sister-chromatid exchange assay in the larvae of the mussel *Mytilus edulis*: response to 3 mutagens, *Mutat. Res.*, 1982, **105**, 235–242.
92. A. A. Freitas and J. P. de Magalhães, A review and appraisal of the DNA damage theory of ageing, *Mutat. Res.*, 2011, **728**, 12–22.
93. L. Shugart, J. Bickham, G. Jackim, G. McMahon, W. Ridley, J. Stein and S. Steinert in *Biomarkers: Biochemical, Physiological, and Histological Marker of Anthropogenic Stress*, ed. R. Huggett, R. A. Kimerle, P. M. Mehrle and H. L. Bergman, Lewis Publishers, 1992, p. 347.
94. S. Lagerwerf, M. G. Vrouwe, R. M. Overmeer, M. I. Fousteri and L. H. F. Mullenders, DNA damage response and transcription, *DNA Repair*, 2011, **10**, 743–750.
95. A. Valavanidis, T. Vlahogianni, M. Dassenakis and M. Scoullos, Molecular biomarkers of oxidative stress in aquatic organisms in relation to toxic environmental pollutants, *Ecotoxicol. Environ. Saf.*, 2006, **64**, 178–189.
96. K. Diderich, M. Alanazi and J. H. J. Hoeijmakers, Premature aging and cancer in nucleotide excision repair-disorders, *DNA Repair*, 2011, **10**, 772–780.
97. G. Grad, C. E. Williamson and D. M. Karapelou, Zooplankton survival and reproduction responses to damaging UV radiation: A test of reciprocity and photoenzymatic repair, *Limnol. Oceanogr.*, 2001, **46**, 584–591.

98. B. M. Kim, C. B. Jeong, I. C. Kim, J. H. Yim, Y. S. Lee, J. S. Rhee and J. S. Lee, Identification of three doublesex genes in the monogonont rotifer *Brachionus koreanus* and their transcriptional responses to environmental stressor-triggered population growth retardation, *Comp. Biochem. Physiol., Part B: Biochem. Mol. Biol.*, 2014, **174**, 36–44.
99. M. Sawada and H. E. Enesco, Effects of UV radiation on the lifespan of the rotifer Asplachna brightwelli, *Exp. Gerontol.*, 1984, **19**, 289–296.
100. B. L. Preston, T. W. Snell and R. R. Kneisel, Uv-B exposure increases acute toxicity of pentachlorophenol and mercury to the rotifer *Brachionus calyciflorus*, *Environ. Pollut.*, 1999, **106**, 23–31.
101. B. Hespeels, M. Knapen, D. Hanot-Mambres, A. C. Heuskin, F. Pineux, S. Lucas, R. Koszul and K. Van Doninck, Gateway to genetic exchange? DNA double-strand breaks in the bdelloid rotifer *Adineta vaga* submitted to desiccation, *J. Evol. Biol.*, 2014, 1–12; J. F. Flot, B. Hespeels, X. Li, B. Noel, I. Arkhipova, E. G. J. Danchin, A. Hejnol, B. Henrissat, R. Koszul, J. M. Aury, V. Barbe, R. M. Barthélémy, J. Bast, G. A. Bazykin, O. Chabrol, A. Couloux, M. D. Rocha, C. D. Silva, E. Gladyshev, P. Gouret, Hallatschek, B. Hecox-Lea, K. Labadie, B. Lejeune, O. Piskurek, J. Poulain, F. Rodriguez, J. F. Ryan, O. A. Vakhrusheva, E. Wajnberg, B. Wirth, I. Yushenova, M. Kellis, A. S. Kondrashov, D. B. M. Welch, P. Pontarotti, J. Weissenbach, P. Wincker, O. Jaillon and K. V. Doninck, Genomic evidence for ameiotic evolution in the bdelloid rotifer *Adineta vaga*, *Nature*, 2013, **500**, 453–457.
102. H. U. Dahms, A. Hagiwara and J. S. Lee, Ecotoxicology, ecophysiology, and mechanistic studies with rotifers, *Aquat. Toxicol.*, 2011, **101**, 1–12.
103. B. J. Cochrane, Y. V. Mattley and T. W. Snell, Polymerase chain reaction as tool for developing stress protein probes. Environmental Toxicology and Chemistry, *Environ. Toxicol. Chem.*, 1994, **13**, 1221–1229.

CHAPTER 4

Prospects of Molluscan Immunomarkers in Monitoring Aquatic Toxicity

SAJAL RAY,* MITALI RAY, ANINDYA SUNDAR BHUNIA, PALLAB BANERJEE AND SOUMALYA MUKHERJEE

Aquatic Toxicology Laboratory, Department of Zoology, University of Calcutta, 35 Ballygunge Circular Road, Kolkata 700019, West Bengal, India
*Email: raysnailmail@rediffmail.com

4.1 Introduction

Mollusca is the second largest invertebrate phylum after Arthropoda. They are represented by clams, snails, mussels, slugs, squids, abalones and octopi, among others. They exhibit a wide range of variation in food habit, body plan, habitat preference and immunophysiological status. Their adaptational success is assumed to be related to the evolution of an advanced innate immunological machinery, which is less researched in many aquatic species. Freshwater molluscs belonging to the Classes Gastropoda and Bivalvia are generally stenohaline in nature and present immunological sensitivity against environmental toxins. Current understanding of molluscan immunity is based on studies carried out on a few species mostly belonging to the Classes Gastropoda and Bivalvia. Many of these investigated species bear medicinal, aquacultural and ecological importance. Freshwater molluscs are generally benthic in nature with slow mobility. Hemocytes, the blood cells of molluscs, are a major functional component of their immunity.

Issues in Toxicology No. 33
Ecotoxicology and Genotoxicology: Non-traditional Aquatic Models
Edited by Marcelo L. Larramendy

Published by the Royal Society of Chemistry, www.rsc.org

The gill, on the other hand, is the primary respiratory organ of aquatic molluscs and is actively involved in the processes of filter feeding and immunosurveillance.[1] Freshwater ecosystems of ponds, lakes and rivers support a wide range of molluscan species, mostly belonging to the Classes Gastropoda and Bivalvia. These waterbodies of different dimensions may be perennial or seasonal in nature, harbouring multiple species of flora and fauna. Thus, the freshwater ecosystem supports a wide range of global bioresources that bear ecological, economical, biotechnological and medicinal significance.[2] In recent years, contamination of freshwater aquifers by chemical toxins has been identified as a major environmental concern. Development of effective markers of aquatic toxicity using indicator species appears to be a challenge for present day ecotoxicologists. Available report suggests that molluscs can act as suitable monitoring invertebrates for aquatic toxicity. Several cellular and humoral parameters of the molluscan immunological system have been proposed as effective markers of toxicity of environmental chemicals. Many of these markers bear prospects as monitoring tools for toxicities of pesticides, manures, heavy metals, nanoparticles, organic effluents, emerging pollutants and industrial xenobiotics of anthropogenic origin. The density shift of immunocytes, phagocytic response of blood cells, dye retention efficacy and cytotoxicity have already been reported or suggested as toxicity immunomarkers of molluscan origin. Molluscs, in general, appear to function as suitable indicators of environmental perturbations including ecotoxicological stress. In this chapter, we review the type, efficacy and prospects of the immunological markers of aquatic toxicity reported in different species of molluscs.

4.2 Aquatic Toxins

In recent times, the freshwater ecosystem, the preferred habitat of molluscs, has been encountering a severe ecotoxicological threat of contamination by various xenobiotics.[3] Wastes liberated from diverse industries find their way into the waterbodies and adversely affect the aquatic ecosystem. Toxic metals, cyanides, metalloids, and inorganic and organic conjugates of diverse chemistry pose serious environmental risks in many regions of planet Earth.[4] Heavy metals like cadmium, chromium, lead, mercury and arsenic have been reported to be the major xenobiotics of industrial and geogenic nature. Commercial detergents and their components, like washing soda, have been established as major immunomodulators in many aquatic invertebrates.[2,5,6] Sulfuric, hydrochloric and nitric acids appear to be highly toxic to many aquatic invertebrates and fish. Agrotoxins, including pesticides and manures, have long been established as principal contaminants of the top soil and natural water. Pesticides, such as organochlorides, organophosphates, carbonates, and pyrethroids are widely used in many countries, resulting in an ecotoxicological crisis. A new generation of environmental contaminants termed "emerging pollutants" has been creating environmental hazards both in aquatic and terrestrial ecosystems.[7] Characterisation and quantitation of this group of environmental toxins have been a serious challenge to the toxicologists and chemists. Endocrine disrupting agents, antibiotics, analgesics and anti-inflammatory agents, antiepileptics, beta blockers, blood lipid regulators, disinfectants, plasticisers and others have been identified as major classes of emerging pollutants.[7] According to the authors, effective biomonitoring of the toxicity of these environmental toxins is needed through an organised research initiative. Innovative development of biomonitoring tools and regulatory strategies are required to reduce the pollution-related crisis of social and public health.

Industrial nanoparticles and nanotubes are another new generation of contaminants of the freshwater ecosystem. Nanoparticles of copper, titanium, silver, gold, selenium and aluminum are reported to generate high levels of toxicity at different trophic levels.[8] Bivalve molluscs act as a unique target group for monitoring the toxicity of nanoparticles.[9] Industrial nanowaste often enters the natural aquifers and generates toxicity in many aquatic organisms, including molluscs. Authors reported the immunomodulatory roles of nanotoxins in the hemocytes of the mollusc *Mytilus galloprovincialis* with reference to phagocytosis, reactive oxygen species generation and bacteriolysis. According to a recent report, organisms like freshwater molluscs have been identified as a suitable model for monitoring the health of aquatic ecosystems.[10] Many molluscan species are slow-moving and toxicologically sensitive groups of invertebrates distributed both in water and land, and occupy almost all levels of aquatic niches, including the benthic, littoral and mud–water interface of wetlands. Toxins, in general, display dual fates in the dissimilar milieu of the external environment and cells. The fate of a chemical toxin present in the open environment is influenced by various physicochemical factors, such as humidity, temperature, sunlight, wind, and adsorptive force and hydrological parameters, such as acidity, alkalinity and hardness. The rate of degradability or half life of the toxin is largely influenced by variable environmental factors. Availability, environmental stability, effective concentration and relative span of exposure of toxins have been established as major issues of ecotoxicological monitoring. Chemical affinity of toxin towards the solvents of various polarities may also determine the relative fate of a toxin. Upon their entry into the cells, toxins encounter xenometabolic enzymes, which in turn may transform the toxin into a metabolite of differential toxicity. Enzyme-catalysed oxidation and conjugation reactions play a vital role in the fate and excretion of toxins from biological system. The effective toxicity of a chemical substance is thus assumed to be governed by its fate in the open environment.

4.3 Routes of Toxin Entry in Molluscs: An Important Aspect of the Biomarker Approach to Toxicity Monitoring

Organ toxicity is a general proposition and is relevant to various factors like route of entry, concentration and span of toxin exposure.[11] The bodies of many aquatic molluscan species may be externally covered by a thick or thin layered shell secreted by the mantle. These shells are calcareous in nature and function as innate immunological barriers to environmental toxins and pathogens. Gastropod and bivalve molluscs differ grossly in relation to anatomy, body plan, immune strategy and behaviour.[12] Gastropods, bearing a coiled shell, evolved a large operculum and mouth. They are capable of closing the opercular cavity during stress, inactivity and predatory attack. Being obligatory filter feeders, molluscs are continuously and intimately exposed to toxins of ambient water and thus are considered as effective monitoring species for chemical pollution. However, during opening of the operculum, ambient water laden with toxins and pathogens enters into the body through the mouth, gill or mantle. Their mode of circulatory functioning and hemocoelic body organisation facilitate the rapid entry of toxins into the interior. Entry of toxins through the hard calcareous shell is not uncommon owing to its structural porosity and regenerative origin.

Bivalves, on the other hand, developed a paired valve bearing adductor muscles. These strong muscles occasionally close and hold the valves together, preventing entry of ambient water into the viscera. During favorable conditions and filter feeding, the bivalves selectively open the valves, allowing the water to flush the gill for the purpose of gaseous exchange and feeding. Toxins and parasites of ambient water enter the internal organs and hemocoel during filter feeding, locomotion, egg laying and other activities. Thus, the mouth, gill, opercular cavity, shell and mantle are assumed to be the organs through which aquatic toxins usually enter the viscera of molluscs at different rates. Through the oral route, the environmental toxins may also enter the gastrointestinal tract during feeding.

4.4 Molluscs as Toxicity-monitoring Species

Molluscs have been gaining special importance as candidate species for monitoring aquatic toxicity.[13] Their wide range of distribution, immunological sensitivity, and cellular and molecular responsiveness qualify them to act as effective monitoring species for environmental toxicity. Aquatic molluscs are reported to exhibit immunological resilience or plasticity in different physiological conditions. Bhunia *et al.* reported immunological resilience in an amphibian mollusc, *Pila globosa*, during experimental starvation and estivation.[14] They reported hemocyte density, phagocytosis, cytotoxicity and anti-oxidative defense as resilient immunological parameters during estivation and starvation. In the USA, a specific and well-organised "mussel watch programme" was launched to monitor the overall health of the aquatic ecosystem.[15] The success of this environmental monitoring programme depended on the relative responsiveness of mussels at the immunological, biochemical and cellular levels. Molluscs are also reported to be efficient bioaccumulators of environmental toxins. The higher life span of molluscs in nature appears to be an advantageous factor in the estimation of the chronic toxicity of a xenobiotic distributed in aquatic ecosystem.

These important issues prompted scientists and monitoring agencies to search for effective immunomarkers in many aquatic molluscs distributed in the contaminated environment.

In a review, the efficacies of multiple biomarkers were highlighted, which enabled ecotoxicologists to monitor environmental pollution.[16] The biomarker approach of monitoring pollution involves assessment of the biological response of an organism to exposure to a definite concentration of an environmental toxin. The authors reported the analysis of genomic DNA and mRNA, proteomics and target enzyme analysis as effective biomarkers of heavy metal, cyanides, pyrethroid and organophosphate toxicities. Cell culture technology has been identified as another process for monitoring pollution *in vitro*. They highlighted the importance of several invertebrate models as sources of biomarkers of environmental pollution. In recent years, different immunological parameters of invertebrate species have been reported as effective markers of exposure and toxicity of environmental chemicals. Information on the molluscan immunological system appears to be limited owing to inadequate data being available in many species distributed in different geographical areas.

4.5 Immunological Characteristics of Molluscs

Molluscs evolved around 500–550 million years ago and left rich paleontological evidence. Their innate mode of immunity is considered as a primitive strategy of biological defense and is highly effective in evading the toxic insult of environmental pathogens, parasites and toxins. The fundamental characteristics of the innate immunity of invertebrates include its effectiveness and instancy of action. One of the fundamental characteristics of molluscan immunity is the absence of antibodies and complement proteins as reported in mammals. Molluscs are assumed to lack an effective immunological memory mechanism as reported in higher invertebrates. However, these are not considered as "evolutionary handicaps" since these organisms are well capable of eliciting immunological defense reactions against environmental toxins, pathogens and parasites.

External physicochemical barriers, which comprise the calcareous shell, mantle and mucus, provide the first and most important line of immunological defense in molluscs. Once this barrier is breached, toxins and pathogens opportunistically invade the hemocoel. The molluscan immune system is characterised by the presence of hemocyte-mediated cellular immunity and humoral components comprised of lectins, antimicrobial peptides, lysozymes and non-specific agglutinins. Hemocytes, the blood cells of the hemolymph, are the chief immunoeffective cells of molluscs. The hemocyte is established as the major component of the molluscan immunological system and is capable of performing diverse physiological functions, such as nutrient transport, phagocytosis, encapsulation, cytotoxicity, surface adhesion and aggregation. Upon entry into the hemocoel, pathogens encounter a highly effective and interactive mode of immunoreactions participated in by the cellular and humoral components of immunity. During these cell–humoral and cell–cell immune cooperative reactions, pathogens and toxins are teleologically deactivated or eliminated depending on the magnitude of the reactions. Phagocytosis has been established as a highly effective mechanism of molluscan defense that is capable of eliminating toxic pathogens from body. Recognition, engulfment and cytotoxic destruction of environmental pathogens by hemocytes have well been researched in many species of molluscs.

Hemocytes, in general, exhibit a wide range of morphofunctional variation that is yet to be explored in many molluscan species. The absolute and relative densities of mobile and sessile hemocytes have been reported as important immunological parameters of molluscs. Grossly, hemocytes are recognisable as granulocytes and agranulocytes, which resemble their mammalian counterparts. Intrahemocytic lysosomes are reported to participate in the process of destroying engulfed pathogens within phagolysosomes. Upon exposure to environmental toxins, lysosomes undergo structural damage and release hydrolytic enzymes into the cell and adjacent tissues. Lysosomal hydrolases and proteases initiate substantial cellular damage of the host and appear to be autolytic in function. Enzymatically, damaged cells may often be eliminated by cellular apoptosis, a reported immunological response that is less studied in molluscs. Hemocyte-mediated particle encapsulation has also been reported as an effective immune reaction that is established in many invertebrates, including molluscs.[17] Production of specific and broad-spectrum antimicrobial peptides in molluscs has been reported. However, detailed molecular characterisation of these proteins has been documented in only a few species.

4.6 Molluscan Immunomarker Approach for Estimation of the Toxicity of Natural Water

Pollution is an undesirable alteration in the physical, chemical and biological qualities of the environment that may adversely affect the health of different organisms, including humans. Chemical and biological monitoring of pollutants and their effects is an important branch of research that involves the participation of scientists from diverse disciplines. In recent years, invertebrate immunologists have identified many species of aquatic molluscs as potential sources of immunomarkers of environmental toxicity. The effects of metals, polychlorinated biphenyls, pesticides, polycyclic aromatic hydrocarbons and different environmental pollutants on the immune system of molluscs were examined.[18] According to Mydlarz *et al.*, aquatic invertebrates permit prediction of the potential influence of ecological factors on their immunological responses.[19] The normal profile and general attributes of molluscan immune system are under the functional influence of diverse environmental factors, including chemical toxicity and related stress. The immunomodulatory roles of cadmium, copper, ozone, nonylphenol, estradiol, novobiocin, benzo[*a*]pyrene, paraoxon, atrazine, alachlor, carbaryl, and paraquat-like toxins have been reported in many molluscan species in aquatic environments.[18] Chemical toxin-mediated modulation of molluscan immune parameters involves interference with the density and other functional attributes of hemocytes. Phagocytic activity, metallothionein concentration, generation of reactive oxygen species, superoxide anion and nitrogen dioxides, cell adherence, lipoperoxidation, lysosomal enzyme release, cell mortality, lysozyme concentration, apoptosis, granulocyte percentage and glutathione-*S*-transferase activity were reported to be effective immunological parameters to estimate environmental contamination owing to anthropogenic activities.

In recent years, immunomarker and biomarker approaches for monitoring the environmental quality have been gaining special significance in environmental toxicology.[20,21] Immunomarking of ecological toxicity appears to be a highly effective method owing to its precision, accuracy and relative inexpensiveness. Environmental toxins, in general, may act as biological and immunological response modifiers for a defined concentration and span of exposure. The majority of environmental contaminants are reported as immunotoxins to many aquatic invertebrates, including molluscs.[4] However, the toxic effects of these chemical compounds on specific humoral and cell-mediated immunological parameters may exhibit species specificity. Moreover, the effects of age, sex and nutritional status of a monitoring species on toxicity cannot be ruled out. Xenobiotics, in general, often exhibit target specificity at the cellular, subcellular, enzymatic and macromolecular levels. Criteria for selection of suitable immunological parameters for toxicity assessment is an essential aspect of the invertebrate immunomarker approach. Molluscan species, in general, are relatively convenient for culture and maintenance in controlled laboratory conditions. Acclimation of the candidate mollusc prior to immunomarking is a critical issue for the toxicity biomonitoring of water. Moreover, molluscs, in general, are a widely distributed group of invertebrates that are optimally sensitive to diverse types of physical and chemical stressors.

A defined magnitude of immunological resilience has been reported in *P. globosa*, a widely distributed freshwater gastropod of India.[14] An optimal level of resilience of an experimental parameter against chemical stress indicates the suitability of the estimated response to function as an immunity biomarker in a monitoring species. The molluscan immunomarker approach involves an analysis of the toxin-induced shift in the cellular and humoral immune parameters. These generally include the immunological characterisation and estimation of a toxin-mediated shift in the immunological response of hemocytes and other immunocytes. The immunological characterisation of hemocytes in different molluscan taxa is awaited. Various authors have described the effectiveness of selected hemocyte-related immunological parameters as biomarkers of environmental pollution and pathogenesis. A novel immunomarker approach of evaluation of the degree of environmental contamination was applied in a mollusc, *Venerupis philippinarum*, distributed in two aquatic regions.[22] A new approach of ecological immunology is proposed to be an effective method of toxicity screening by utilising molluscan immunomarkers. The principal immunological markers considered in molluscs include total hemocyte count, hemocyte diameter and volume, lysozyme activity in hemocyte lysate and cell-free hemolymph, and the activities of superoxide dismutase and catalase of the gill and digestive gland. Experimental specimens with dissimilar ecotoxicological histories exhibited different levels of immunological responses, as reported by the authors. The immunomodulatory effect of copper in molluscs was considered as an effective immunomarker of toxicity in the same species. Following the immunomarker strategy principle, Matozzo *et al.* also examined the cellular response of the Manila clam *Ruditapes philippinarum* collected from two different sites of a lagoon in Venice.[22]

Several immunological and non-immunological parameters of molluscs have been tested, analysed and proposed as potential markers of ecotoxicity. While monitoring the pollution in the Antarctic environment, the esterase activity of the bivalve mollusc *Adamussium colbecki* was reported as a biomarker of toxicity.[23] Esterase is established as a key enzyme that is related to metabolism and other physiological activities. The immunomodulatory effects of cadmium are reported in marine bivalves and the result is indicative of the studied parameters as immunomarkers of aquatic toxicity.[24] Environmental antibiotics like ciprofloxacin, erythromycin and novobiocin generated immunological toxicity in a freshwater mollusc, *Elliptio complanata*.[25] The authors presented a detailed evaluation of the immunological status of the mussel with reference to immunocyte viability, reactive oxygen species generation, lysozyme activity and nitric oxide generation. Many of these immunological parameters have good prospects to function as immunomarkers of environmental antibiotic toxicity. Enzyme catalase plays a pivotal role in eliciting the immunological response of the clam *Meretrix meretrix* against oxidative stress and pathogen challenge.[26] The mussel *Mytilus trossulus* exhibited a biomarker response along a pollution gradient near an oil terminal in the Gulf of Finland.[27] This integrated level of biomarker approach involves consideration of multiple immunological parameters, such as activities of superoxide dismutase, catalase and glutathione-*S*-transferase, and lysosomal membrane stability, along with chemical analyses of water. Selected immune-associated parameters of a gastropod mollusc, *Haliotis diversicolor*, had been reported to be altered by tributyltin and benzo[*a*]pyrene.[28] According to this report, immunological parameters like total hemocyte count, superoxide anion, nitric oxide, and inducible nitric oxide synthase and myeloperoxidase activities exhibited a significant shift under the exposure of these toxins. Subsequent responses of recovery of these parameters in the same species were indicative of the efficacy of these parameters to act as effective immunomarkers of aquatic toxicity.

The tropical green mussel, *Perna viridis*, is reported as a test and source species of immunity biomarker of pollution.[29] According to the workers, the phagocytic response along with the acetylcholinesterase activity of the mussel act as significant tools for pollution monitoring in Indonesian water. The freshwater snail, *Lymnaea stagnalis*, is reported to exhibit a high degree of immunological sensitivity to exposure to municipal effluents.[25] Parameters like total hemocyte count, hemocyte viability, reactive oxygen species generation, phagocytic activity, activities of superoxide dismutase, glutathione reductase and nitric oxide synthetase, and toll-like receptor expression of *L. stagnalis* exhibited a significant alteration in response to municipal effluent. Municipal waste generated from human habitation points often contaminates freshwater ponds and lakes, the natural habitat of many invertebrate species, including molluscs. In this paper, the authors reported the immunotoxicological adversities of municipal waste in molluscs and indicated the potential of the studied responses as immunomarkers of aquatic toxicity. A battery of potential biomarkers was experimentally screened in molluscs exposed to copper, a serious environmental contaminant.[30] Luengen *et al.* evaluated the immunological responses of mussels as indicators of contamination in San Francisco bay.[31] They evaluated immunological reactivity in two species of mussels, *Mytilus californianus* and *M. galloprovincialis*, as sources of biological markers of contamination. In their study, two important immune parameters, *i.e.* hemocyte density and phagocytic response, were evaluated in depth along with hydrological analyses. They concluded that phagocytosis and the hemocytic index act as the most responsive assay methods of immunomarking. Particle engulfment or phagocytosis, a classical immunological mechanism of molluscs, was suggested as a sensitive response to evaluate the degree of environmental contamination.

4.7 Potential Immunomarkers of Aquatic Toxicity

Establishment of immunomarkers in molluscs requires a thorough screening of immunological responses in indicator species exposed to toxins both in field and experimental conditions. Dose response analysis of a toxin is an important issue for standardisation of an immunomarker in molluscs. Selected parameters of cellular and humoral immunity of molluscs have been reported as potential immunomarkers (Figure 4.1) of toxicity and exposure of aquatic xenobiotics. However, a thorough and in-depth investigation of the immunomarker potential of these parameters is still awaited in many species. The functional efficacy of cellular and humoral immunomarkers of toxicity largely depends on the optimal sensitivity, recovery potential and immunological resilience of the candidate parameters against the exposure of a discrete concentration of toxin for a unit span of exposure. The roles of static and flowing water environments are considered as important issues in immunotoxicity screening in aquatic molluscs. Malacologists and ecotoxicologists have identified the components of the external physicochemical barrier, and cellular and humoral immunity parameters as potential sources of immunomarkers in molluscs.

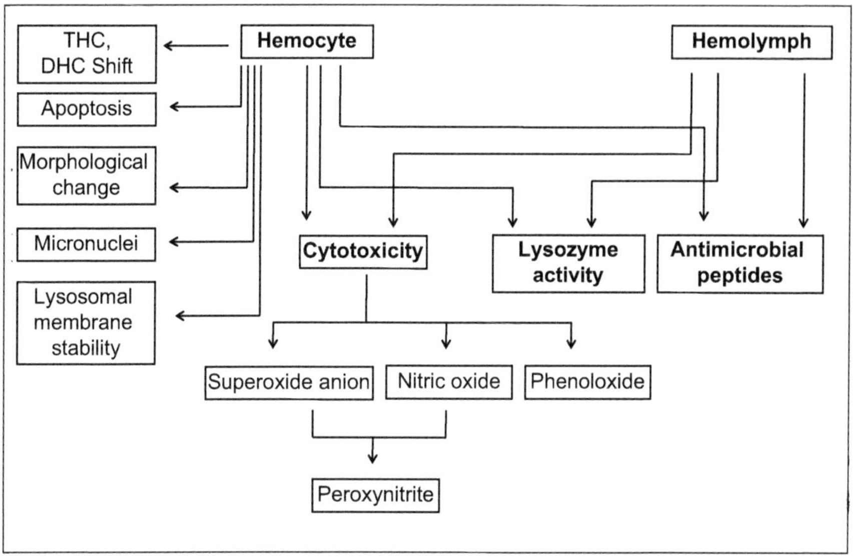

Figure 4.1 Mono- or poly-marker monitoring approaches involve consideration of a single or multiple immune-associated parameters, respectively, for estimation of environmental toxicity. Many such parameters are of hemocyte or hemolymph in origin.

4.7.1 Mucus Exudation Response of Molluscs as a Marker of Toxicity of Suspended Particulates of Water Column

The molluscan body plan is characterised by the presence or absence of an external calcareous shell, an effective immunological barrier to environmental toxins or pathogens. The innate immunological status of molluscs is assumed to be influenced by the structural and functional attributes of the indicator species occupying a specific habitat. Mucus is a slimy and viscous secretory product that is involved in multiple physiological and behavioral activities. In many molluscan species, mucus is secreted from the specially developed mucus cells of the mantle, a thin muscular flap covering the body cavity. Chemically, the mucus bears a high content of water, carbohydrate, protein, minerals and other nutrients as minor components. It is physiologically involved in the maintenance of internal water homeostasis, nutrition, immune defense and lubrication-related activities in molluscs.

L. marginalis is a pearl-forming bivalve distributed widely in the freshwater ponds and lakes of Indian subcontinent. This commercially important mussel is reported to exude high concentrations of mucus under exposure to azadirachtin- and pyrethroid-based pesticides, established contaminants of the freshwater ecosystem.[32,33] The kinetics of mucus release from the mantle were investigated under toxin exposure against a standard control. Continuous and unrestricted release of mucus by toxin-treated molluscs appears to be a typical stress response in monitoring species. Overexudation of mucus may often lead to a physiological state of desiccation and immune impairment owing to the substantial thinning of the internal mucosal sheath, a component of the pathogen and toxin barrier. The phenomenon of rapid expulsion of injected yeast particulates through mucus was reported in a freshwater snail, *P. globosa*.[4] It is an amphibian mollusc that spends a major part of its life cycle in freshwater aquifers. According to the authors, within 30 minutes of injection of unit density of cultured yeast particulates, the experimental specimens exhibited a characteristic immunological reactivity by copious release of mucus. More than 80% of non-self yeast particles were expelled from the body through mucus within a limited period of time.

Upon their natural and opportunistic invasion into the hemolymph and other tissues, environmental pathogens and xenobiotics may cause similar mode of mucus expulsion as the innate immune response. This response of mucus release presents a unique degree of immunological resilience by exhibiting the states of recovery and restoration after the phase of hypersecretion. The promptness of the secretory reactivity and the resilience of this response indicate a high degree of efficacy of this marker in assessing the aquatic toxicity of particulate xenobiotics. In recent years, the effective immunological monitoring of nanotoxins has been a serious ecotoxicological challenge. Industrial nanotoxins are being deposited and concentrated in aquatic reservoirs at an alarming rate. Being filter feeders, aquatic gastropods and bivalves become easy targets for nanotoxins of diverse chemistry and toxicity. Kinetic analysis of mucus release is thus assumed to function as an effective immunomarker of different xenobiotics, including nanotoxins.

4.7.2 Hemocytes: A Major Source of Cellular and Humoral Immunomarkers of Freshwater Toxicity

Hemocytes or circulating blood cells have long been established as immunocytes of molluscs. Current scientific understanding of molluscan immunology centers around the information generated on the structure and function of these cells. Hemocytes are distributed in various organs and tissues at different densities. The hemolymph or blood of molluscs has been identified as a tissue with a high density of hemocytes. The presence of immunoactive sessile hemocytes in different organs has also been reported. Several workers have attempted to classify the hemocytes of molluscs from different viewpoints. Morphological variation, species specificity and non-uniform nomenclature have been identified as the factors hampering the formulation of a uniform scheme of hemocyte classification. George and Ferguson proposed a scheme of classification for gastropod hemocytes with a report on their roles in eliciting non-specific immunoreactivity.[34] A detailed scheme of classification for hemocytes of Gastropoda and Bivalvia was proposed employing microscopy and flow cytometry.[35] While discussing the comparative analyses of hemocytes, the authors reported the principal immunological activities of various subtypes of hemocyte. The phagocytic and cytotoxic responses of hemocyte variants were discussed with reference to the cellular immunological status of three freshwater species *B. bengalensis, P. globosa*, and *L. marginalis*. Fluorescence activated cell sorting (FACS) analysis allowed the scientists to study the relative granularity of cells in terms of structural complexity and size (Figure 4.2) in invertebrates.[36] FACS is a novel technology that has been applied widely in the field of molluscan immunology for the last two decades. The importance of this technology in molluscan immunological research has been highlighted.[37] Employing FACS technology, a large number of cellular and subcellular analyses of molluscan immunocytes can be carried out with millions of cells within a short span of time. The molluscan hemocytes were classified as agranulocytes, semigranulocytes and granulocytes.[35] Microscopic analyses of these major types of cells revealed the existence of several subtypes (Figure 4.3) of hemocytes with discrete immunological functions. Major subtypes of hemocytes include blast-like cells, round hyalinocytes, spindle hyalinotcytes, semigranular asterocytes, round semigranulocytes, round granulocytes, spindle granulocytes and granular asterocytes.

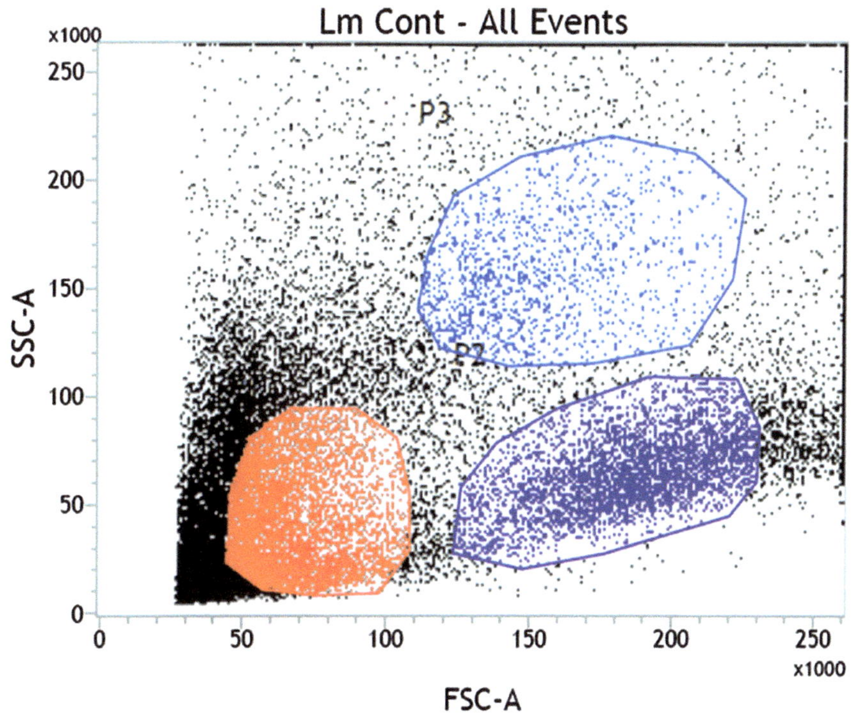

Figure 4.2 Forward and side scattering by hemocyte variants of molluscs are the key components of gatewise FACS analysis of multiple immunomarker parameters.

4.7.2.1 Total and Differential Counts of Hemocytes as Immunomarkers of Hydrological Toxicity

Most of the immunological reactivities of molluscs are mediated by these immunocytes of varied morphology and function. Hemocytes, after being generated in different hematopoietic organs, are released in the hemolymph as circulating blood cells. A number of specific populations are assumed to be sessile in nature and may be located in different tissues and organs. Both of these mobile and sessile hemocytes may exhibit a characteristic mode of trafficking from organ to hemolymph and *vice versa*.[38] However, the overall movement of cells from their stationary to mobile state is reported to control the blood cell homeostasis of molluscs. The absolute density of circulating hemocytes is expressed as the total number of cells present in a unit volume of hemolymph. This can be determined by microscopic hemocytometry,

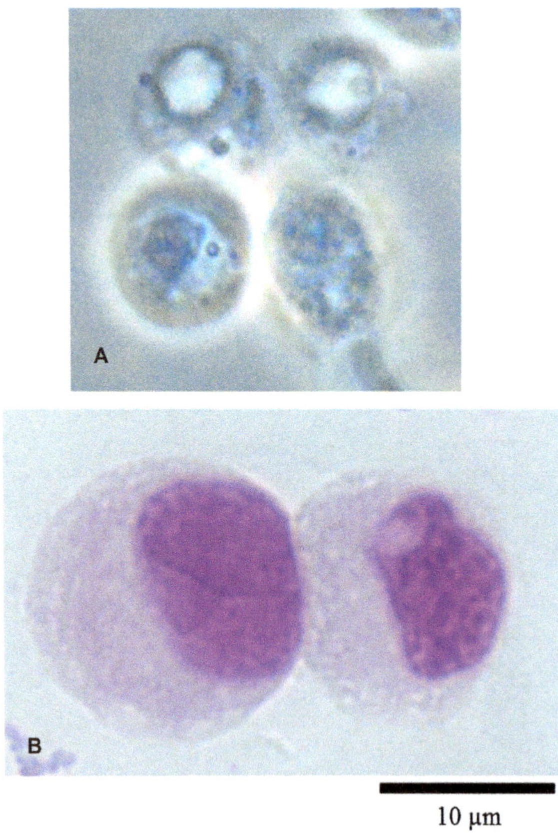

Figure 4.3 Hemocyte, the chief immunoeffector cell of molluscs, is the major source of immunomarkers of aquatic toxicity. Photomicrographs of hemocytes of *B. bengalensis* as obtained under phase contrast (A) and bright field optics (B). Light microscope images, magnification: ×1000; scale bar: 10 μm.

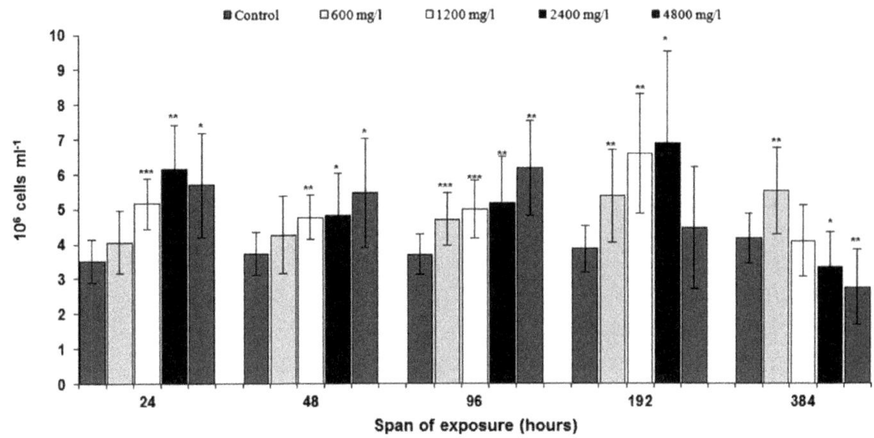

Figure 4.4 Modulation of total hemocytes count of the gastropod *B. bengalensis* exposed to azadirachtin. Data are presented as the mean ± standard deviation ($n = 10$). The asterisks (*) indicate values that are significantly different (*: $p < 0.05$, **: $p < 0.01$, ***: $p < 0.001$) from the control.

FACS and other methods. Relative or differential densities of hemocytes refer to the absolute density of each cell variant present in the fluid phase of hemolymph. Absolute and differential densities of hemocytes act as sensitive markers of toxicity, pathogenesis and chemical stressors. Many environmental toxins, such as azadirachtin, have been reported to alter the total and differential densities of hemocytes of molluscs under stressed conditions (Figure 4.4).

Ray *et al.* determined the differential densities of agranulocytes, semi-granulocytes and granulocytes in three species of freshwater mollusc (*P. globosa, B. bengalensis* and *L. marginalis*) employing FACS and microscopy.[35] Since the blood cell density of molluscs is a sensitive physiological parameter, a careful acclimation of the test species is mandatory prior to such cellular analyses. A detailed characterisation of the immune-related activities of hemocytes of gastropod molluscs was carried out.[39] The authors reported the total count, viability, lysosomal stability, phagocytosis and oxidative activities of the hemocyte as immune-related parameters in molluscs. High hemocyte load has been reported as a causative factor of increased resistance in the insect *Drosophila suzukii* exposed to a parasitoid wasp.[40] The data indicated a direct relationship between total hemocyte count and relative immunity in invertebrates. Hemocyte density is thus feasible for use as a functional agent determinant of effective immunological response in invertebrates.

Chakraborty *et al.* studied the dynamics of hemocyte density in a freshwater mollusc (*L. marginalis*) treated with environmentally realistic concentrations of arsenic, a precarious contaminant of the aquatic ecosystem of India.[13] Arsenic, a carcinogenic metalloid, is released from the Earth's crust and contaminates the water of ponds and lakes. Reports of the immunotoxicity of arsenic in aquatic invertebrates are limited. Total and differential counts of hemocytes of *L. marginalis* were estimated following exposure to 1, 2, 3, 4 and 5 mg L^{-1} of sodium arsenite for 96 hours. Arsenic treatment yielded a marked suppression of the total hemocyte count in a dose-dependent manner against the control. Relative densities in terms of the differential count of each subpopulation of hemocyte were reported to be affected grossly by arsenic. A recovery assay was conducted by determining the total hemocyte count in the post-treated specimens after their maintenance in arsenic-free water for 15 and 30 days. However, the percentage of restoration of normal density was not 100%, indicating the organism's moderate range of immunological resilience under arsenic exposure. While discussing the experimental data, the authors highlighted the hemocyte density of *L. marginalis* as an effective immunological marker of arsenic toxicity. According to them, both the total and differential hemocyte count appear to be physiologically sensitive to exposure to arsenic. Scientists claimed these cellular parameters as markers of immunotoxicity, which supported the view of Oliver and Fisher.[41]

Azadirachtin was reported to modulate the total count of hemocytes of *L. marginalis*.[42] Azadirachtin is a neem plant (*Azadirachta indica*)-derived steroid like tetranorterpinoid that is applied in agricultural fields for crop protection. It is a biopesticide that often contaminates freshwater ponds and lakes during monsoons and floods. After proper acclimation, experimental specimens of *L. marginalis* were exposed to sublethal concentrations of azadirachtin for 24, 48, 72 and 96 hours. The azadirachtin-mediated shift in this parameter is thus suggested to be applied in toxicity monitoring of the pesticide and allied immunotoxins. Industrial-, agricultural- and urbanization-related projects liberate different types of chemical toxins that are often deposited in freshwater ponds and lakes through multiple routes. Effective biomonitoring of these chemicals appears to be a challenge to environmental toxicologists.

The pyrethroid group of pesticides is widely used in agricultural fields for effective pest control. The immunotoxicity of pyrethroid pesticides in freshwater aquatic molluscs has been reported. The dynamics of hemocyte density in *L. marginalis* were reported in response to exposure to cypermethrin, a pyrethroid pesticide.[43] The authors examined the shift of patterns in the total and differential hemocyte counts in the mentioned specimen. A dose-dependent alteration in the densities of agranulocytes and asterocytes indicated their potential to function as effective immunomarkers of cypermethrin toxicity in the freshwater environment. Exposure of *L. marginalis* to 0.03, 0.05, 0.07 and 0.09 mg L^{-1} of cypermethrin for 15 days yielded a maximum inhibition in the densities of agranulocytes and asterocytes. Heart rate and hemocyte count were established as stress indicators in the snail *Helix pomatia*.[44] Scientists considered disturbed hibernation as a stress and enumerated hemocyte densities at different heart rate conditions. An experimental damage made to the shell resulted in a sharp increase in the total hemocyte count. This damage-induced increase in the cell density was indicative of a characteristic stress response of molluscs. Total hemocyte count is established as a marker of environmental pollution in the mollusc *P. globosa*.[45] The authors reported a dose-dependent alteration in the total hemocyte count in the hemocytes of an amphibian snail exposed to toxic biofuel smoke. *P. globosa*, an inhabitant of the freshwater ecosystem, exhibited a variation in hemocyte density under different levels of smoke exposure. The immunomodulatory effect of fenvalerate, a pyrethroid pesticide, was reported in the freshwater snail *B. bengalensis* with reference to a shift in hemocyte densities and surface adhesion.[46] Sublethal exposure to fenvalerate yielded a substantial shift in these immunological parameters in this viviparous snail. Total and differential densities of circulatory hemocytes have long been established as important immunological parameters of many organisms, including molluscs. A simple method of collection and laboratory culture of hemocytes renders these cells effective for use as an easy source of immunomarker of aquatic toxicity.

4.7.2.2 *Phagocytic Response is an Established Immunomarker of Water Toxicity*

Phagocytosis is the process by which a cell recognises, attaches and engulfs foreign particles and pathogens.[47] Phagocytosis is established as the principal immunological response of molluscs against invading pathogens, parasites and toxins. It has been reported in all major animal phyla and is well investigated in mammals. Phagocytosis involves several functional phases characterised by recognition, chemotaxis, attachment (Figure 4.5B), engulfment and ingestion or destruction of pathogens. Post-engulfment lysis of bacteria and other pathogens may involve the oxygen-dependent or oxygen-independent biochemical reactions, which have been investigated in a few invertebrates.

In molluscs, hemocytes have been identified as the major cell type that exhibits high potential for phagocytosis. The phagocytic activity of hemocytes of the freshwater bivalve *L. marginalis* was established as a biomarker of arsenic toxicity, a prominent geogenic toxin of ponds and lakes.[10] Phagocytic responses of hemocytes were quantitated both in the presence and absence of arsenic. In both the experiments, arsenic exhibited concentration-dependent inhibition in the phagocytic response of hemocytes. The maximum inhibition was reported with 5 $mg\,L^{-1}$ of arsenic for 30 days. A marked restoration of the inhibited response of phagocytosis indicated a high level of effectiveness of this parameter to act as an immunomarker of arsenic toxicity. According to Fournier *et al.*, utilisation of phagocytosis as an *in situ* tool in toxicity monitoring appears to be ideal owing to no requirement for the sensitisation of the bioindicator species collected from the natural environment.[48] Effective monitoring of the toxicity of diverse xenobiotics is an important issue related to biodiversity conservation and environmental toxicity analysis. Dose-dependent immunomodulation of phagocytic response and the

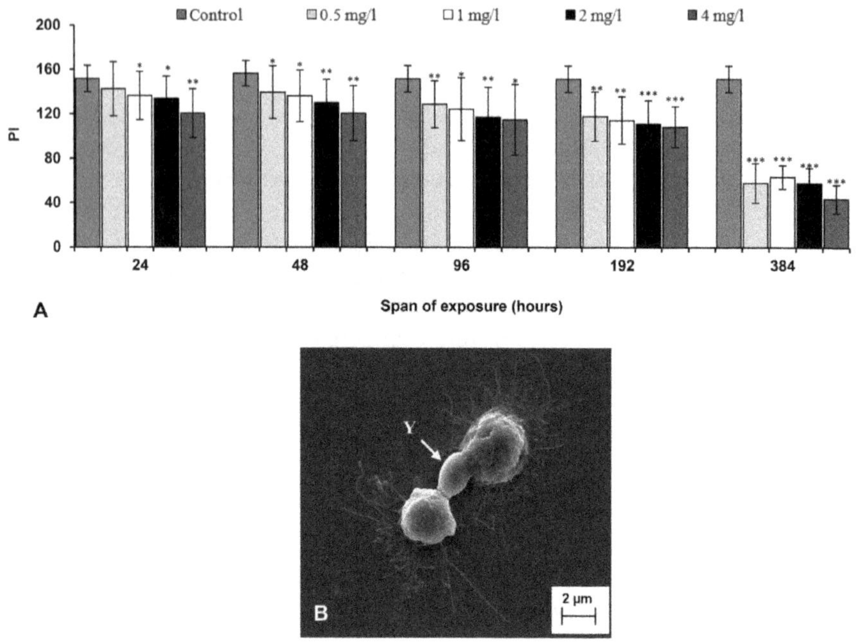

Figure 4.5 Toxin-dependent reduction of phagocytic response might act as an immunomarker of ecotoxicity. Fenvalerate-induced alteration in the phagocytic indices of hemocytes of *B. bengalensis* (A). Data are presented as the mean ± standard deviation ($n = 10$). The asterisks (*) indicate the values that are significantly different (*: $p < 0.05$, **: $p < 0.01$, ***: $p < 0.001$) from the control. A cultured yeast particle (Y) is being phagocytosed by hemocytes of a gastropod *Telescopium telescopium* exposed to 0.025% diesel for 96 hours (B). Pseudopodial extensions of the cells in the form of an aster, as observed under scanning electron microscope, was considered as a characteristic response of many molluscan immunocytes.

subsequent response of recovery from arsenic toxicity is strongly suggestive of the utilisation of this parameter as an immunomarker of aquatic toxicity. Fenvalerate exposure yielded a decrease in the phagocytic response of hemocytes of the snail *B. bengalensis* (Figure 4.5A).

Mukherjee *et al.* quantitated the phagocytic response of hemocytes of *L. marginalis* treated with an azadirachtin-based pesticide.[49] Hemocytes of the indicator species were exposed to sublethal concentrations of pesticides for different times. Experimental specimens received *in vivo* and *in vitro* exposure of azadirachtin for 1, 2, 3, 4, and 7 days and one hour, respectively. For both the treatment sets, azadirachtin exhibited a concentration-dependent inhibition in the phagocytic response against charcoal. Charcoal is a component of the industrially polluted environment and was used in this experiment as a particulate xenobiotic. The authors established phagocytosis of charcoal particulates by hemocytes as an immunological marker of azadirachtin exposure and toxicity. While examining the toxicity of biofuel smoke in an amphibian mollusc *P. globosa*, they reported toxin-induced inhibition in the phagocytic response of the hemocytes in the experimental species. The pattern of inhibition of phagocytosis appeared to be dependent on the duration of smoke exposure. The experimental results suggested phagocytosis as a marker of smoke toxicity in *P. globosa*.

A detailed comparative analysis was presented on the morphological variation, phagocytic activity and cytotoxic status of hemocytes of three freshwater molluscs in India.[35] Workers reported a non-specific phagocytic response in the isolated hemocytes of *B. bengalensis, P. globosa* and *L. marginalis*. In *P. globosa,* the semigranulocytes and granulocytes were identified as the major phagocytes, whereas the agranulocytes and granulocytes of *B. bengalensis* and *L. marginalis* had been reported as their principal phagocytic cells. Four distinct subpopulations of hemocytes were identified in the bivalve *M. galloprovincialis* as professional phagocytes.[50] Hemocytes were identified as classical immune active cells capable of performing phagocytosis of non-self particulates in many species.[51–54] A decrease in phagocytosis by the hemocytes was reported in *V. philippinarum* exposed to fluoride.[55] In *Mercenaria mercenaria*, phenol exposure resulted in inhibition of the phagocytic activity.[56] Diesel, released from the motorised vessels of the Sunderbans estuary of India and Bangladesh, contaminates the water and silt of the mangrove forests of this region. This organic toxin was reported to adversely affect the phagocytic potential of the hemocytes of the mudwhelk *Telescopium telescopium* (Figure 4.5B).

The establishment of a physiological marker of toxicity or pathogenesis is influenced by two important factors *i.e.* easy availability of biomonitoring species and relative simplicity of monitoring protocol. In a few species of molluscs, phagocytosis has already been established or indicated as an effective immunomarker of ecotoxicity. The major advantages of estimation of phagocytic response include a non-complex protocol that requires optimal consumption of time. At the cellular level, advanced microscopy is the preferred method of estimation of phagocytosis for many cell biologists. Currently, flow cytometry has become a more popular and accepted method of estimation of phagocytic response in a large number of hemocytes within a short period of time. The determination and quantitation of phagocytic activity in the hemocytes of differential granularities, sizes and metabolic states is a major advantage of this technology.

4.7.2.3 Immunocytotoxic Potential of Molluscs as a Marker of Aquatic Contamination

The cytotoxic potential of the immunocytes of molluscs has been established as a major strategy of immune defense in many aquatic species. Pathogenic microorganisms and environmental toxins are reported to induce the cytotoxic potential of the host upon their entry into the tissue and cells. Cytotoxic molecules in most cases have been identified as reactive oxygen intermediates exhibiting the toxic potential of cells to destroy foreign agents after their engulfment by the professional phagocytes. These molecules are short-lived cytotoxic agents generated inside the immunocytes and hemolymph following a biochemical pathway. The cytotoxic potential of immunocytes is physiologically dependent on the immunological potential of generation of different reactive oxygen species, such as superoxide anion, hydrogen peroxide, hydrogen radicals and hypochlorous acid, and reactive nitrogen species.[57] Reactive nitrogen intermediates include nitric oxide, nitrogen dioxide, nitrous acid and peroxynitrile ions, which have been identified as potential intracellular killing agents of pathogens. Collective generation of reactive oxygen species is functionally associated with destruction and elimination of environmental bacteria, viruses and protozoa by itself or in association with lysosomal hydrolases. In this mechanism of pathogen destruction and elimination, the role of apoptosis is also considered to be significant. However, reactive oxygen and nitrogen species in the hemocytes of molluscs have been proposed as immunomarkers of toxicity in the freshwater ecosystem.

4.7.2.4 Cellular Oxidative Stress: Roles of Superoxide Anion and Reactive Oxygen Species in Monitoring the Health of Aquatic Ecosystems

The superoxide anion is a cytotoxic molecule that mediates oxygen-dependent killing of pathogens by respiratory burst activity. Chemically, superoxide anions are generated from oxygen by the catalytic activity of nicotinamide adenine dinucleotide phosphate hydrogenase oxidase. The superoxide anion is a cytologically toxic and short-lived molecule that is effective in the innate immunity of invertebrates and mammals. Generation of superoxide anions is largely controlled by the pro- and anti-oxidative potentials of cells to maintain a state of cellular homeostasis. In this cellular process, effective biochemical scavenging of superoxide anions by the enzyme superoxide dismutase is considered as a major immunological reaction that is examined in selected organisms. Enzymatic dismutation transforms the superoxide anion into hydrogen peroxide, which is subsequently converted to water by action of catalase. Granulocytes, semi-granulocytes and agranulocytes of different freshwater molluscs have been reported as major generators of superoxide anions.[35] Flow cytometric isolation of cells and spectrophotometric estimation of superoxide anions indicated that gastropods and bivalves non-specifically generate this cytotoxic molecule in their blood cells as an innate immunological response. The assay methodology involved the principle of reduction of nitroblue tetrazolium into a coloured precipitated form of insoluble formazan whose absorption is spectrophotometrically estimated at 630 nm. The activity of phenoloxidase and generation of superoxide anion and nitric oxide in the hemocytes and hemolymph of the freshwater snail *P. globosa* exhibited a unique mode of resilience during estivation and experimental starvation.[14]

The natural freshwater reservoirs of the Indo-Gangetic floodplains are reported to be under ecotoxicological threat from arsenic. Immunological and oxidative stress from arsenic were examined in *L. marginalis*.[58] The freshwater bivalve was exposed to different environmentally realistic concentrations of arsenic and generation of superoxide anion and nitric oxide as well as the activities of transaminases, phosphatase, acetylcholinesterase, phenoloxidase, catalase and glutathione-*S*-transferase were estimated in different tissues and hemocytes. A detailed analysis of control and arsenic-treated hemocytes was also reported along with the histopathological status of the digestive gland of the same specimen. Superoxide anion generation exhibited a concentration-dependent increase up to 96 hours of exposure. However, prolonged treatment with arsenic for 15 days resulted in inhibition in the generation of superoxide anions in the hemocytes of test specimens. According to them, the dose-responsive generation of superoxide anions and activities of the studied enzyme validated their immunomarker potential in *L. marginalis* distributed in contaminated environments. Modulation of generation of cytotoxic molecules like superoxide anion and nitric oxide as well as the activity of phenoloxidase in the face of variations of environmental temperatures and pH have been reported.[59] Mukherjee *et al.* quantitated the oxidative stress from azadirachtin in the hemocytes of *L. marginalis*.[60] They carried out cytochemical localisation along with spectrophotometric analyses of superoxide anion in the hemocytes of *L. marginalis* exposed to sublethal concentrations of azadirachtin. Exposure to 0.006, 0.03, 0.06 and 0.09 $mg L^{-1}$ of azadirachtin resulted in a significant shift in the generation of superoxide anion in the hemocytes of molluscs. A dose-dependent increase of generation of superoxide anion was reported up to 7 days of treatment in response to the experimental concentrations of the toxin. However, prolonged exposures of 15 and 30 days yielded inhibition in the generation of superoxide anion in the hemocytes. A low concentration of azadirachtin was reported to generate a substantial level of cellular stress in the blood cells of molluscs. This detectable shift in the generation of superoxide anion is thus considered as an effective marker of toxicity of this pesticide present in the contaminated ecosystem.

The toxic effects of tributyltin and benzo[*a*]pyrene were screened in the gastropod *H. diversicolor*.[28] The authors reported the toxicity of these aquatic chemicals on the immune-associated activities of hemocytes and subsequent responses of stress recovery. Parameters like total hemocyte count, phagocytosis, lysosomal membrane activity, generation of superoxide anion and nitric oxide, and activity of nitric oxide synthase were estimated in abalone exposed to tributyltin and benzo[*a*]pyrene. Both of these environmental contaminants affected the generation of extracellular superoxide anion in the hemocytes subjected 3, 7, 14 and 21 days of exposure. A 7 and 14 days restoration assay revealed that the toxicity recovery response was not 100% in comparison to the control. Spectrophotometric estimation of generation of superoxide anion in the intracellular compartment of hemocytes showed an induction in the generation of the anion in molluscs exposed to these toxins for up to 21 days of treatment. For tributyltin, the recovery response was reported to be less than 100%. Toxin-induced shift in the generation of superoxide anion in the extra- and intra-cellular compartments of hemocytes and the toxicity recovery response indicated the efficacy of this parameter to act as an effective immunomarker of tributyltin and benzo[*a*]pyrene toxicity in aquatic molluscs.

Oxidative stress of aquatic invertebrates has been reported to be functionally related to chemical contamination of water.[61] Flow cytometric estimation of reactive oxygen species has been considered as an effective method of assessment of cellular oxidative stress in many organisms.[6] This novel method permits estimation of the generation of superoxide anion, hydroxyl radical, and nitric oxide in the selected immunocytes collectively. The principle of this assay involves the chemical reaction of 2′7′-dicholorofluorescein diacetate (DCFH-DA), a membrane-permeable probe of non-fluorescent nature. After its entry into hemocytes, diacetate is hydrolysed by cellular esterase. Cellular superoxide anion and hydrogen peroxide, in turn, oxidise DCFH to DCF, a fluorescent product. DCFH can also be oxidised by nitrate radicals, resulting in generation of detectable green fluorescence. Detection of the resulting green fluorescence of DCF is carried out by flow cytometer. DCF-mediated green fluorescence appears to be proportional to generation of reactive oxygen species in hemocytes. Mitochondrial involvement of reactive oxygen species generation is established in the hemocytes of the oyster *Crassostrea gigas* under unstimulated conditions.[62] The authors employed FACS technology in determination of reactive oxygen species in the hemocytes of molluscs. Xenobiotic accumulation in the lysosome was reported to enhance generation of reactive oxygen species in the hemocytes of *M. edulis*, indicating its potential to act as immunomarker of water pollution. Treatment with cadmium in the cultured hemocytes of molluscs resulted in an alteration in the activity of phenoloxidase and reactive oxygen species generation. An increase of reactive oxygen species by 130% was reported in the hemocytes of *H. tuberculata* exposed to 50 000 $\mu g L^{-1}$ of cadmium chloride for 10 days.[63] The assay protocol was based on the principle of DCFH-DA-mediated fluorescence of cells.

4.7.2.5 *Pro-oxidative, Anti-oxidative and Detoxification Responses of Molluscs as Indicators of Environmental Toxicity*

Oxidative stress is considered to result from the effects of pro- and anti-oxidative agents generated inside the cell exposed to toxins. The oxidative status of an organism is thus maintained by the cumulative reactivities of oppositely functioning pro- and anti-oxidative molecules generated in an organism during health and diseased conditions. Several scientific reports indicate that the generation of pro- and anti-oxidative agents is influenced by different chemical contaminants in the freshwater ecosystem. Biochemical sensitivity and dose-dependent generation of these agents can be considered as immunological markers of toxin-induced stress in molluscs. Untreated municipal sewage water was reported to increase hydrogen peroxide and nitric oxide production and phagocytosis in the blue mussel *M. edulis*. Induction in generation of hydrogen peroxide continued up to 21 days of exposure.[64] Superoxide dismutase and catalase, the reported antioxidant enzymes, exhibited seasonal variations in activity in bivalves collected from water with a history of contamination by polychlorinated biphenyls. Exposure to 60 and 110 μg copper L^{-1} resulted in inhibition of the activity of superoxide dismutase of hemocytes of the clam *Tapes philippinarum*.[65] A low dose of a mixture of the pesticides atachlor, metolachlor, atrazine, terbuthylazine, diurin, fosetyl aluminum, carbaryl and glyphosate resulted in variation in the activities of catalase and glutathione-*S*-transferase, a detoxifying enzyme, in the oyster *C. gigas*.[66] Catalase activity was claimed as a potential marker of oxidative stress in the mussel *M. galloprovincialis*.[67] According to this report, acute exposure to benzo[*a*]pyrene resulted in a typical bell-shaped response with a maximal peak after 48 hours of exposure to the toxin. Catalase and superoxide dismutase activities of molluscs were proposed as biomarkers of pollution in the Hilla river of Iraq.[68] This suite of biomarkers of catalase, superoxide dismutase, glutathione-*S*-transferase activities is suitable for monitoring water toxicity along a gradient near an oil terminal at Finland.[27] However, many of these reported parameters of molluscs exhibit a high level of sensitivity towards different chemical species of aquatic toxins. Thus, these immunological parameters may be assumed to function as immunological markers of pollution either alone or in association with others. An integrative approach to environmental monitoring involving multiple markers and chemical analyses is presumed to be highly effective in monitoring ecosystem health.

4.7.2.6 Reactive Nitrogen Species: A Sensitive Immunomarker of Freshwater Toxicity

The cytotoxic response generates reactive nitrogen intermediates in the immunocytes and other tissues of the host for the purpose of inactivation of internalised environmental pathogens. Information regarding nitrogen intermediates is more abundant in mammals in comparison to that in invertebrates. Principal intermediates formed in the cells during immune elicitation include nitrogen dioxide, nitrous acid and toxic peroxynitrite anions. During the conversion of L-arginine to L-citrulline involving inducible nitric oxide synthase, nitric oxide is generated in hemocytes as a major cytotoxic agent. Nitric oxide may also conjugate with superoxide anion to generate a highly toxic pathogen-killing agent known as peroxynitrite. Nitric oxide, superoxide anion and peroxynitrite can act as potential cytotoxic agents under adequate immune elicitation.

Scientific reports that are available for molluscs and other organisms have suggested the role of these cytotoxic molecules in monitoring the quality and health of aquatic ecosystem. Generation of nitric oxide was reported in different subpopulations of hemocytes of three freshwater molluscs as non-specific cytotoxic agents.[35] Flow cytometric and spectrophotometric quantification of nitric oxide revealed its differential generation among the agranulocytes, semigranulocytes and granulocytes of *P. globosa, B. bengalensis* and *L. marginalis*. The level of nitric oxide generation was reported to be higher in the granulocytic hemocytes of three experimental species, indicating their high cytotoxic potential. *P. globosa*, an estivating gastropod, exhibited a characteristic pattern of immunological resilience with reference to their cytotoxic potential.[14]

Generation of nitric oxide in the hemocytes of the bivalve *L. marginalis* as a marker of arsenic toxicity was established.[10] The authors examined the dynamics of nitric oxide generation in the hemocytes of molluscs exposed to environmentally realistic concentrations of arsenic. They also reported a dose-dependent decrease of nitric oxide generation in the hemocytes both in the presence and absence of yeast phagocytosis. Sublethal concentrations of tributyltin and benzo[*a*]pyrene significantly induced multiple immune-related parameters in the gastropod *H. diversicolor*, including generation of nitric oxide, nitric oxide synthase and myeloperoxidase activities.[28] According to the authors, tributyltin exhibited a high level of immunotoxicity in *H. diversicolor* in comparison to that of benzo[*a*]pyrene, which presented a better recovery response. Reports of the immunotoxicity of environmental chemicals on nitric oxide generation and nitric oxide synthase activity are limited in molluscs. A comparative report indicated a higher level of toxicity of tributyltin on the activity of nitric oxide synthase in the experimental specimen.[28] However, the recovery of nitric oxide synthase was reported to be almost 100%.

4.7.2.7 Phenoloxidase Activity and Ecotoxicological Monitoring of Water

The phenoloxidase cascade has been established as a major component of the innate immunity of many invertebrates, including molluscs. It involves a unique coordination of cellular and biochemical reactions, resulting in protease-dependent conversion of the enzyme prophenoloxidase into active phenoloxidase, a reported mediator of molluscan immunity. Phenoloxidase is assumed to be involved in eliciting phagocytosis, non-self recognition, melanisation, cytotoxicity and wound-healing processes. Gelation of body fluid, a characteristic immunological response of invertebrates, may functionally be associated with conversion of phenoloxidase from its inactive proenzyme prophenoloxidase by serine protease. Activation of the phenoloxidase cascade is influenced by external nutritional factors, plasma gelation, calcium ion concentration, pH and other environmental toxins.

The activity of phenoloxidase has been reported in different subpopulations of hemocytes of freshwater molluscs *P. globosa, B. bengalensis* and *L. marginalis*.[35] The authors reported elevated activity of phenoloxidase in the agranulocytes of *P. globosa*. Differential levels of activity of this enzyme in three experimental species appear to be owing to a taxon-specific immunological strategy of molluscs.[12] The dynamics of phenoloxidase activity in the snail *P. globosa* were reported during experimental estivation and starvation.[14] The resilience of activity of phenoloxidase in different environmental conditions is suggestive of its prospects as a marker of stress and environmental perturbation.

The toxic effect of cadmium chloride was examined on the activity of phenoloxidase of hemocytes of the gastropod *H. tuberculata in vitro*.[63] A high concentration of 10 000 $\mu g L^{-1}$ of cadmium chloride resulted in a dose-dependent increase in phenoloxidase activity in the cultured hemocytes of the mentioned species. A maximum level of 9% induction was reported against the highest experimental concentration of cadmium. Published data strongly suggest this parameter as a marker of cadmium toxicity in water. Zinc, an aquatic toxin, affected several immunological parameters, including phenoloxidase activity, phagocytosis and reactive oxygen species, of the gastropod *H. tuberculata*.[69] Zinc treatment *in vitro* yielded a 26-fold increase in the phenoloxidase activity of hemocytes of this specimen. The stimulatory roles of lipopolysaccharide and propargylglycine on the phenoloxidase activity of the oyster *C. gigas* have been reported.[70] Phenoloxidase grossly influences the biochemical steps of melanisation following a cascade. Phenoloxidase acts as a pivotal enzyme in the phenoloxidase cascade under different immunological challenges. Available information indicates a unique level of sensitivity of this enzyme to exposure to different xenobiotics. Considering the high level of molecular and immunological sensitivity of phenoloxidase, this parameter has been proposed as a characteristic marker of exposure to and toxicity of selected aquatic toxins.

4.7.2.8 Toxin-induced Lysosomal Membrane Destability of Molluscs is an Effective Immunomarker of Toxicity of Water Contaminants

Lysosomes contain a cluster of hydrolytic and other enzymes involved in the processes of autolysis and phagolysosomal destruction of pathogens and others. They play a significant role in the physiology and immunology of the host in the face of pathogen and toxin challenge. The functional attributes of lysosome largely depend on the structural integrity of the membrane of this organelle. The physical integrity of the lysosomal membrane appears to be sensitive to xenobiotic insults originating from the ambient environment of the aquatic organism. Toxin-induced damage to the membrane may lead to leaching of lysosomal content into the cytoplasm and extracellular compartment, resulting in enzyme-mediated damage of the host tissue.

The structural integrity or stability of lysosomal membranes of hemocyte and other cells can be estimated following the principle of vital dye retention, an accepted method of screening the structural and functional status of hemocytes. Several workers have proposed the kinetics of dye release as an effective marker of environmental toxicity. The simple and inexpensive protocol of dye retention analysis involves staining of the isolated hemocytes with neutral red, a cationic probe of red color. The rate of diffusion of the probe from the lysosomal compartment into the cytoplasm per unit time is examined microscopically to determine the neutral red retention time (NRRT). The time gap between probe application and the first appearance of the diffused dye into the cytosol of the examined hemocytes is considered as the NRRT in bivalves. Healthy and normal hemocytes theoretically exhibit a longer NRRT in comparison to cells damaged by toxic exposure to an aquatic contaminant. Normally, more than 50% of the hemocytes of the control and exposed sets were quantified to determine NRRT in hemocytes. Spectrophotometric estimation of neutral red trapped in the hemocytes is considered as another method to determine lysosomal membrane stability.[6,28] Both of these methods of estimation appear to be effective in the determination of lysosomal membrane stability.

Lysosomal stability of the hemocytes of *L. marginalis* was claimed as a biomarker of arsenic toxicity in a freshwater ecosystem.[71] The authors reported that a vast stretch of land on both banks of the river Hooghly of West Bengal, India has been identified as an arsenic-affected zone and they collected the experimental bivalves from these regions for different analyses. Unexposed specimens of *L. marginalis* were treated with environmentally realistic concentrations of 1, 2, 3, 4 and 5 $mg\,L^{-1}$ of sodium arsenite for 24, 48, 72 and 96 hours and 15 and 30 days. Arsenic exposure yielded a concentration-dependent decrease in NRRT in comparison to that of the control. Photodocumentation of the release of lysosomal dye into the cytosol indicated substantial membrane damage owing to arsenic treatment. According to the authors, the neutral red retention assay for estimation of the membrane stability can thus be applied as an immunomarker of arsenic toxicity both in the laboratory and field conditions. Molnar and Fong reported copper- and cadmium-induced destabilisation of the lysosomes in the digestive gland of two species of freshwater gastropod, whereas no alteration of lysosomal integrity was reported after exposure to pesticide methoxychlor.[72] Lysosomal membrane stability by neutral red retention assay of molluscan cells was confirmed as a marker for the assessment of water quality.[73] Fluoride, a contaminant of waterbodies, presented immunotoxicity in the hemocytes of the manila clam *V. philippinarum*.[55] A low level of lysosomal membrane stability had been reported in the mussel *M. trossulus* distributed along a pollution gradient near an oil terminal in the Baltic Sea.[27] The authors highlighted the biomarker response of this parameter in monitoring the water quality of the Gulf of Finland. Fenvalerate treatment yielded a substantial destability in the lysosomal membrane of hemocytes of *B. bengalensis* (Figure 4.6).

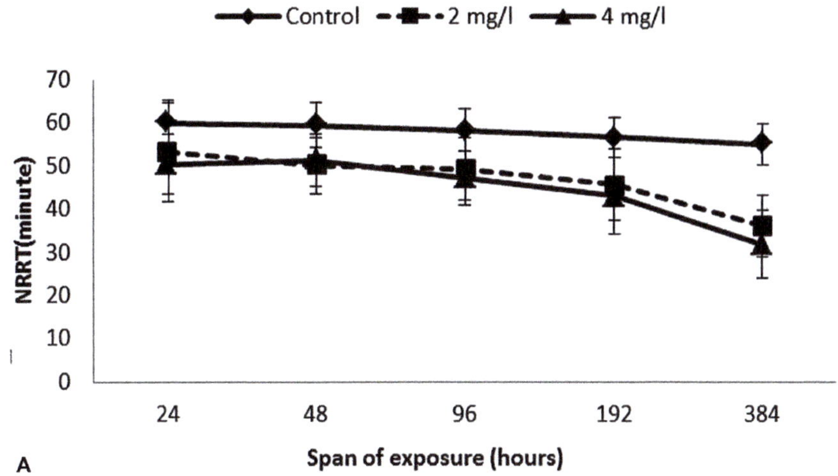

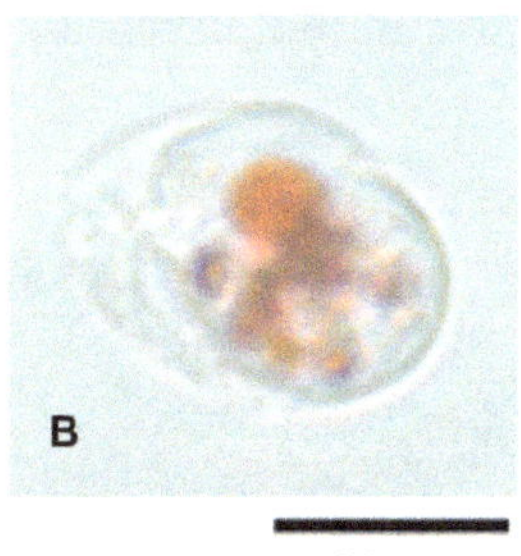

Figure 4.6 A dose- and time-dependent decrease in the neutral red retention time of hemocyte lysosome of *B. bengalensis* treated with 2 and 4 mg L^{-1} fenvalerate for different spans of exposure (A). Data are presented as the mean $\pm$ standard deviation ($n = 10$). The asterisks (*) indicate the values that are significantly different (*: $p < 0.05$, **: $p < 0.01$, ***: $p < 0.001$) from the control. Retention of neutral red probe within the lysosomes of hemocytes of *B. bengalensis* treated with 3 mg L^{-1} of fenvalerate for 48 hours at the start point of the membrane stability assay (B). Light microscopic images, magnification: ×1000; scale bar: 10 μm.

Gopalakrishnan *et al.* screened the toxicity of tributyltin and benzo[*a*]-pyrene on several immune associated parameters including lysosomal stability in *H. diversicolor*, a gastropod.[28] The efficacy of neutral red retention was evaluated in cultivated mussels and it was established as an indicator of stress in relation to their handling and processing for commercial purposes.[74] The sensitivity of the neutral red retention assay against variations in environmental temperatures and salinity has been reported.[75] Several reports suggest the neutral red retention efficacy of lysosome as an effective immunological marker of environmental toxicity. Its relative inexpensiveness, non-involvement of major instrumentation in estimation procedure and dose-dependent toxic response indicate the suitability of this immunological marker for monitoring the quality of water.

4.7.2.9 *Apoptotic Response as a Marker of Pollutant Exposure*

Apoptosis or programmed cell death bears broad-ranging physiological implications in both invertebrates and vertebrates. It is recognised as an important cellular process that largely influences the developmental and post-developmental phases of an organism. Apoptosis bears immense immunotoxicological importance, but it is insufficiently investigated in many species of aquatic molluscs. Apoptosis is characterised by membrane blebbing, DNA fragmentation and shrinkage in cell volume. Assay methodologies like terminal deoxynucleotidyl transferase dUTP nick end labeling (TUNEL) assay and cellular staining by annexin V and propidium iodide enabled the researchers to estimate apoptosis following the principle of immunochemistry (Figure 4.7). Physiological and toxin-induced apoptosis in molluscs is a less studied area of research in comparison to other organisms. Translocation of phosphatidylserine from the inner leaflet of the cell membrane to its outer one has been identified as a signature event of apoptosis and is the basis of immunofluorescent detection by FACS.[76]

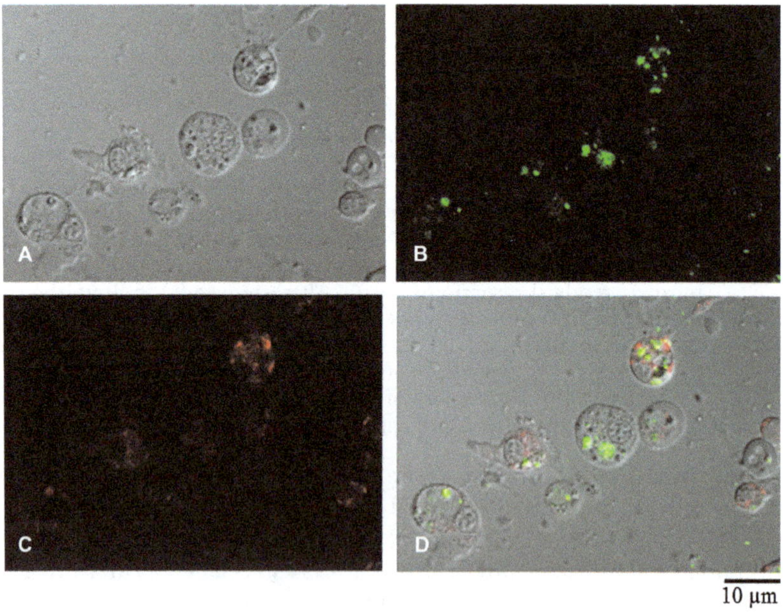

Figure 4.7 Immunofluorescent detection of apoptotic (B) and necrotic (C) responses of the hemocytes of *T. telescopium* treated with 0.025% diesel for 96 hours; the cells after staining with annexin V-FITC and propidium iodide yielded green and red fluorescence, respectively, as observed under confocal optics. Plate (A) depicts the image under differential interference contrast (DIC) optics, whereas, plate (D) represents the superimposition of two fluorescent signals. Diesel is an aquatic toxin of vehicular origin in the estuarine water of India.

Apoptosis is assumed to be a phylogenetically conserved response of organisms that is often modulated by environmental stressors, including chemical toxins.[76,77] Apoptosis was suggested as an immunological response of molluscs that was modulated by treatment with fomesafen, a herbicide.[78] Cypermethrin and fenvalerate treatment inhibited hemocyte apoptosis in two freshwater molluscs, *B. bengalensis* and *L. marginalis*, as revealed by phase contrast, fluorescence microscopy and flow cytometry.[79] The authors quantitated apoptosis by coupled reaction with fluorescein isothiocyanate (FITC) conjugated annexin V and propidium iodide. Annexin V exhibits strong affinity to phosphatidylserine, whereas propidium iodide intercalates DNA of the nucleus of late apoptotic or necrotic hemocytes. Thus, this method allows researchers to quantitate both the apoptotic and necrotic responses of hemocytes of molluscs exposed to toxins. Perry and Lynn detected physiological and pesticide-induced apoptosis in the early developmental stages of bivalves.[80] Toxin-dependent induction of apoptosis was observed by TUNEL assay in molluscs.[55] Workers pointed out a state of apoptosis-associated oxidative stress in the hemocytes of *V. philippinarum* upon exposure to toxic fluoride.

4.7.2.10 *Lysozyme Activity is an Innate Immunological Marker of Water Toxicity*

Lysozyme is a bactericidal enzyme that is functionally involved in deactivation of invading microorganisms in molluscs. Lysozyme is found both in the cellular and extracellular compartments. Reports regarding lysozyme in molluscs are limited. Inhibition of the lysozyme activity was observed in abalone treated with tributyltin and benzo[*a*]pyrene.[28] The authors reported a 7 day recovery of lysozyme activity in the molluscan specimens pretreated with tributyltin. Their report supported the observations of Zhou *et al.*[81] According to them, toxin-induced reduction in lysozyme activity is a physiological indication of immune suppression in molluscs. In an another study, scientists reported that cadmium exposure inhibited the lysozyme activity of hemocytes and hemolymph in marine bivalves.[24] The effects of different environmental parameters on the immunity of bivalves have been examined.[22] Lysozyme activity in the hemocyte lysate and the cell-free hemolymph of molluscs collected from two differentially contaminated field spots presented a non-uniform result. A persistent difference in the immunological parameters, including the lysozyme activity, indicated the immunomodulatory role of copper in molluscs. Washing soda was reported to induce a shift in the activities of lysozyme and acetylcholinesterase, reactive oxygen species generation and cell aggregation responses in a non-molluscan invertebrate of a freshwater ecosystem.[6] Lysozyme is assumed to play a significant role in maintaining the innate immunological status of aquatic molluscs. Toxin-induced suppression of the activity may render a species vulnerable to pathogen invasion and diseases. However, the sensitivity of the activity of this enzyme to exposure to xenobiotics indicates its effectiveness as an immunological marker of environmental toxicity.

4.7.2.11 Hemocyte Aggregation Response as an Immunological Tool for Ecotoxicological Assessment

Circulating hemocytes of aquatic molluscs undergo cell–cell aggregation and/or cell–surface adhesion both in healthy and stressed conditions. Hemocyte aggregation, a less researched area of malacology, is assumed to be functionally involved during the processes of the bacterial encapsulation response and formation of a hemocyte plug at the site of wound formation to arrest hemolymph loss. The aggregation response has been identified as an important cell-mediated immunological reaction in molluscs. Damage to external tissue by injury or parasitic invasion may lead to a fatal loss of body fluid from the damaged site. Many invertebrates evolved a unique mechanism of arrestation of blood loss through the process of fluid gelation and hemocyte plug formation by exhibiting a typical cell aggregation response. At the site of coagulation or gelation, specific immunocytes aggregate and release their contents at the damaged site. Cellular discharge stimulates the aggregation response of hemocytes to form a cellular plug at the wound site, which in turn arrests the loss of blood from the damaged site. A method of determination of hemocyte aggregation was established by experimental induction of 'no aggregation response' using a chemical fixative.[82] Auffret and Oubella reported that pesticides induced a shift in the aggregation response of hemocytes of *C. gigas*.[83] The authors examined the effects of tributyltin, copper and cadmium on the hemocyte aggregation response. Both copper and cadmium exhibited concentration-dependent inhibition of hemocyte aggregation with respect to the control. The protocol reported in this paper deals with the determination of indirect numerical values indicating hemocyte aggregation. An alternative method of aggregation assay involved estimation of turbidity changes of a hemocyte suspension using a platelet aggregometer. Aufferet and Oubella estimated hemocyte aggregation *in vitro* by Coulter counter technology. From this report, it was evident that low levels of copper and cadmium might result in a significant change in the aggregation index of hemocytes.[84] A microquantity of tributyltin and a mixture of pesticides atrazine, diuron and isoproturon yielded a marked shift in hemocyte aggregation response. The high level of sensitivity of the cell aggregation response to xenobiotics can thus be applied in the process of monitoring toxicity in water.

Molluscs have been identified as the best non-vertebrate organisms for monitoring environmental toxicity.[85] Several immunological and related parameters of molluscs are claimed to be effective markers of aquatic pollution. A number of aquatic species mostly belonging to classes Gastropoda and Bivalvia were established as indicator species of water toxicity. An integrative approach of chemical monitoring and biomarker analysis was suggested to be effective in pollution monitoring in a lagoon using the clam *R. philippinarum.*[20] The authors considered lysozyme activity and total hemocyte count as immunological markers of toxicity of several water contaminants, such as polychlorinated biphenyls, polycyclic aromatic hydrocarbons, trichloro-bis(*p*-chlorophenyl) ethane and its breakdown products, and hexachlorobenzene. They strongly encouraged this integrated approach of biomonitoring for the purpose of environmental risk assessment. Consideration of a set of immunomarkers may provide a better scope and results when monitoring the toxicity in a contaminated ecosystem. The usefulness of bioindicators and biomarkers in monitoring pollution was highlighted by Hamza-Chaffai.[86] The authors mentioned the effectiveness of filter-feeding bivalves as ideal indicator species of aquatic pollution. The objective of ecotoxicological analysis was stated to be the accurate prediction of toxic effects of pollutants to avoid detrimental consequences. In a review, Gupta and Singh specified molluscs as sensitive bioindicators of metal pollution in aquatic systems.[87] Molecules, cells, and organelles are suggested as the important functional components of a monitor species, which are capable of generating early biomarker signals. Expression of interferon gamma, tumor necrosis factor alpha and inducible nitric oxide synthase has been reported in the hemocytes of gastropod and bivalves.[88] These molecules might also bear immunomarker potential for toxicity of pollutants distributed in the freshwater environment. However, in-depth research is still awaited in many species of aquatic molluscs. In another review, Zhou *et al.* discussed the role of bivalve and gastropod molluscs in monitoring metal toxicity in aquatic ecosystems.[21] Emerging pollutants, a new threat to the aquatic ecosystem, pose a serious ecotoxicological risk in many countries. A major challenge in biomonitoring the toxicity of emerging pollutants involves establishment of effective immunomarkers in the common varieties of filter-feeding molluscs. Ecological risk, bioremediation and monitoring of toxicity of emerging pollutants were discussed in a review article.[7] Bivalve molluscs have been identified as a unique target group for nanotoxins in freshwater ecosystems.[9] Many nanoparticles are reported to affect phagocytosis, generation of reactive oxygen species, enzyme activity and bacterial killing. In this research article, the immunotoxicity of diverse nanoparticles was discussed in multiple molluscan species in aquatic habitats. Development of cellular and molecular immunomarkers in molluscan models for screening the toxicity of diverse chemical toxins appears to be an emerging field in ecotoxicology. From the available data, it is evident that the filter-feeding molluscs in freshwater ecosystems have bright prospects as potential sources of immunomarkers of exposure and toxicity of xenobiotics in the current aquatic environment.

Acknowledgements

Authors thankfully acknowledge Abhishek Ray of the Aquatic Toxicology Laboratory for his efforts in preparing the manuscript. The scientific contributions of the former and current research students of the laboratory are also acknowledged.

References

1. S. Chakraborty, M. Ray and S. Ray, Toxicity of sodium arsenite in the gill of an economically important mollusc of India, *Fish Shellfish Immunol.*, 2010, **29**, 136–148.
2. S. Mukherjee, M. Ray and S. Ray, Immunotoxicity of washing soda in a freshwater sponge of India, *Ecotoxicol. Environ. Saf.*, 2015, **113**, 112–123.
3. S. Ray, M. Ray, S. Chakraborty and S. Mukherjee, Immunotoxicity of environmental chemicals in the pearl forming mussels of India- a review, in *Mussels: Anatomy, Habitat and Environmental Impact*, ed. L. E. McGevin, Nova Science Publishers Inc., USA, 2011, pp. 429–440.
4. S. Ray, S. Mukherjee, N. S. Bhunia, A. S. Bhunia and M. Ray, Immunotoxicological threats of pollutants in aquatic invertebrates, in *Emerging Pollutants in the Environment- Current and Further Implications*, ed. M. L. Larramendy and S. Soloneski, InTech, Croatia, 2015, ch. 6, pp. 147–165.
5. S. Mukherjee, M. Ray, M. K. Dutta, A. Acharya, S. K. Mukhopadhyay and S. Ray, Morphological alteration, lysosomal membrane fragility and apoptosis of the cells of Indian freshwater sponge exposed to washing soda (sodium carbonate), *Ecotoxicol. Environ. Saf.*, 2015, **122**, 331–342.
6. S. Mukherjee, M. Ray and S. Ray, Shift in aggregation, ROS generation, antioxidative defense, lysozyme and acetylcholinesterase activities in the cells of an Indian freshwater sponge exposed to washing soda (sodium carbonate), *Comp. Biochem. Physiol., Part C: Toxicol. Pharmacol.*, 2016, **187**, 19–31.
7. M. Gavrilescu, K. Demnerová, J. Aamand, S. Agathos and F. Fava, Emerging pollutants in the environment: present and future challenges in biomonitoring, ecological risks and bioremediation, *New Biotechnol.*, 2015, **32**(1), 147–156.
8. S. K. Rajkishore, K. S. Subramanian, N. Natarajan and K. Gunasekaran, Nanotoxicity at various trophic levels: a review, *Bioscan*, 2013, **8**(3), 975–982.
9. L. Canesi, C. Ciacci, R. Fabbri, A. Marcomini, G. Pojana and G. Gallo, Bivalve molluscs as a unique target group for nanoparticle toxicity, *Mar. Environ. Res.*, 2012, **76**, 16–21.
10. S. Chakraborty, M. Ray and S. Ray, Evaluation of phagocytic activity and nitric oxide generation by molluscan haemocytes as biomarkers of inorganic arsenic exposure, *Biomarkers*, 2009, **14**, 539–546.
11. S. Ray, Levels of toxicity screening of environmental chemicals using aquatic invertebrates - a review, in *Invertebrates- Experimental Models in Toxicity Screening*, ed. M. L. Larramendy and S. Soloneski, InTech, Croatia, 2016, ch. 1, pp. 1–12.

12. N. V. Yakovleva, M. P. Samoilovic and A. M. Gorbushin, The diversity of strategies of defense from pathogens in molluscs, *J. Evol. Biochem. Physiol.*, 2001, **37**(4), 358–367.
13. S. Chakraborty, M. Ray and S. Ray, Sodium arsenite induced alteration of hemocyte density of *Lamellidens marginalis* – an edible mollusc from India, *Clean: Soil, Air, Water*, 2008, **36**, 195–200.
14. A. S. Bhunia, S. Mukherjee, N. S. Bhunia, M. Ray and S. Ray, Immunological resilience of a freshwater Indian mollusc during aestivation and starvation, *Aquacult. Rep.*, 2016, **3**, 1–11.
15. I. V. Hansen, J. M. Weeks and M. H. Depledge, Accumulation of copper, zinc, cadmium and chromium by the marine sponge *Halichondria panicea* Pallas and the implications for biomonitoring, *Mar. Pollut. Bull.*, 1995, **31**, 133–138.
16. S. Kotelevstev, V. Tonkopii and O. Hänninen, Biomonitoring of environmental pollution, *Encyclopedia Life Support Systems*, 2009, pp. 139–160.
17. B. Meena, S. Rose, S. S. Jayaraj and S. Vincent, Encapsulation response from the hemocytes of estuarine clam, *Meretrix casta*, *Int. J. Pharma Bio Sci.,* **1**, 2010, 1–8.
18. M. I. Girón-pérez, Relationships between innate immunity in bivalve molluscs and environmental pollution, *Invertebr. Surv. J.*, 2010, **7**, 149–156.
19. L. D. Mydlarz, L. E. Jones and D. Harvell, Innate Immunity, environmental drivers, and disease ecology of marine and freshwater invertebrates, *Annu. Rev. Ecol. Evol. Syst.*, 2006, **37**, 251–288.
20. V. Matozzo, A. Binelli, M. Parolini, L. Locatello and M. G. Marin, Biomarker responses and contamination levels in the clam *Ruditapes philippinarum* for biomonitoring the Lagoon of Venice (Italy), *J. Environ. Monit.*, 2010, **12**(3), 776–786.
21. Q. Zhou, J. Zhang, J. Fu, J. Shi and G. Jiang, Biomonitoring: an appealing tool for assessment of metal pollution in the aquatic ecosystem, *Anal. Chim. Acta*, 2008, **606**, 135–150.
22. V. Matozzo, M. Giacomazzo, L. Finos, M. G. Marin and M. Milan, Can ecological history influence immunomarker responses and antioxidant enzyme activities in bivalves that have been experimentally exposed to contaminants? A new subject for discussion in 'eco-immunology' studies, *Fish Shellfish Immunol.*, 2013, **35**(1), 126–135.
23. S. Bonacci, M. A. Browne, A. Dissanayake, J. A. Hagger, I. Corsi, S. Focardi and T. S. Galloway, Esterase activities in the bivalve mollusc *Adamussium colbecki* as a biomarker for pollution monitoring in the Antarctic marine environment, *Mar. Pollut. Bull.*, 2004, **49**, 445–455.
24. V. Ivanina, C. Hawkins and I. M. Sokolova, Immunomodulation by the interactive effects of cadmium and hypercapnia in marine bivalves *Crassostrea virginica* and *Mercenaria mercenaria*, *Fish Shellfish Immunol.*, 2014, **37**, 299–312.

25. M. Gust, M. Fortier, J. Garric, M. Fournier and F. Gagné, Immunotoxicity of surface waters contaminated by municipal effluents to the snail *Lymnaea stagnalis*, *Aquat. Toxicol.*, 2013, **126**, 1–11.
26. C. Wang, X. Yue, X. Lu and B. Liu, The role of catalase in the immune response to oxidative stress and pathogen challenge in the clam *Meretrix meretrix*, *Fish Shellfish Immunol.*, 2013, **34**, 91–99.
27. R. Turja, A. Soirinsuo, H. Budzinski, M. H. Devier and K. K. Lehtonen, Biomarker responses and accumulation of hazardous substances in mussels (*Mytilus trossulus*) transplanted along a pollution gradient close to an oil terminal in the Gulf of Finland (Baltic Sea), *Comp. Biochem. Physiol., Part C: Toxicol. Pharmacol.*, 2012, **157**(1), 80–92.
28. S. Gopalakrishnan, W. Huang, Q. Wang, M. Wu, J. Liu and K. Wang, Effects of tri-butyltin and benzo[a]pyrene on the immune-associated activities of hemocytes and recovery responses in the gastropod abalone, Haliotis diversicolor, *Comp. Biochem. Physiol., Part C: Toxicol. Pharmacol.*, 2011, **154**, 120–128.
29. K. Yaqin, B. W. Lay, E. Riani, Z. A. Masud and P. Hansen, Hot spot biomonitoring of marine pollution effects using cholinergic and immunity biomarkers of tropical green mussel (*Perna viridis*) of the Indonesian waters, *J. Toxicol. Environ. Health Sci.*, 2011, **3**, 356–366.
30. R. J. Brown, T. S. Galloway, D. Lowe, M. A. Browne, A. Dissanayake, M. B. Jones and M. H. Depledge, Differential sensitivity of three marine invertebrates to copper assessed using multiple biomarkers, *Aquat. Toxicol.*, 2004, **66**, 267–278.
31. C. Luengen, C. S. Friedman, P. T. Raimondi and A. R. Flegal, Evaluation of mussel immune responses as indicators of contamination in San Francisco Bay, *Mar. Environ. Res.*, 2004, **57**(3), 197–212.
32. S. Mukherjee, Studies on toxicological response of *Lamellidens marginalis* to azadirachtin based pesticide, PhD thesis, University of Calcutta, 2011.
33. C. Mandal, Studies on toxicity of fenvalerate in *Bellamya bengalensis* (Lamarck), PhD thesis, University of Calcutta, 2015.
34. W. C. George and J. H. Ferguson, The blood of gastropod molluscs, *J. Morphol.*, 1950, **86**, 315–324.
35. M. Ray, N. S. Bhunia, A. S. Bhunia and S. Ray, A comparative analyses of morphological variations, phagocytosis and generation of cytotoxic agents in flow cytometrically isolated hemocytes of Indian molluscs, *Fish Shellfish Immunol.*, 2013, **34**, 244–253.
36. S. Mukherjee, M. Ray and S. Ray, Phagocytic efficiency and cytotoxic responses of Indian freshwater sponge (*Eunapius carteri*) cells isolated by density gradient centrifugation and flow cytometry: a morphofunctional analysis, *Zoology*, 2015, **118**, 8–18.
37. M. Cossarizza, L. Pinti, Troiano and E. L. Cooper, Flow cytometry as a tool for analysing invertebrate cells, *Invertebr. Surv. J.*, 2005, **2**, 32–40.
38. R. Qubella, P. Maes, C. Paillard and M. Auffret, Experimentally induced variation in hemocyte density for *Ruditapes philippinarum* and *R. decussatus* (Mollusca, Bivalvia), *Dis. Aquat. Org.*, 1993, **15**, 193–197.

39. L. Donaghy, H. Honga, C. Lambert, H. Parkc, W. J. Shimd and K. Choi, First characterisation of the populations and immune-related activities of hemocytes from two edible gastropod species, the disk abalone, *Haliotis discus discus* and the spiny top shell, *Turbo cornutus*, *Fish Shellfish Immunol.*, 2010, **28**, 87–97.
40. B. Z. Kacsoh and T. A. Schlenke, High hemocyte load is associated with increased resistance against parasitoids in *Drosophila suzukii*, a relative of *D. melanogaster*, *PLoS One*, 2012, **7**(4), 1–16.
41. L. M. Oliver and W. S. Fisher, Appraisal of prospective bivalve immunomarkers, *Biomarkers*, 1999, **4**, 510–530.
42. S. Mukherjee, M. Ray and S. Ray, Azadirachtin induced modulation of total count of hemocytes of an edible bivalve *Lamellidens marginalis*, *Proc. Zool. Soc.*, 2006, **59**(2), 203–207.
43. K. Das, M. Ray and S. Ray, Cypermethrin induced dynamics of hemocyte density of Indian mollusc *Lamellidens marginalis*, *Anim. Biol. J.*, 2012, **3**(1), 1–11.
44. L. Renwrantz and F. Spielvogel, Heart rate and hemocyte number as stress indicators in disturbed hibernating vineyard snails, Helix pomatia, *Comp. Biochem. Physiol., Part A: Mol. Integr. Physiol.*, 2011, **160**, 467–473.
45. S. Ray, M. Ray and N. S. Bhunia, Toxicity of smoke in an amphibian mollusk from India, in *Environmental and Regional Air Pollution*, ed. D. Gallo and R. Mancini, Nova Science Publishers Inc. NY, USA, 2009, ch. 14, pp. 351–366.
46. K. Guria and S. Ray, Modulation of glass surface adhesion characteristics and sub population shift of hemocytes of *Bellamya bengalensis* by synthetic pyrethroid, *Indian J. Environ. Ecoplann.*, 2002, **6**(1), 175–178.
47. L. M. Stuart and R. Alan Ezekowitz, Phagocytosis and comparative innate immunity: learning on the fly, *Nat. Rev. Immunol.*, 2008, **8**, 131–141.
48. M. Fournier, D. Cyr, B. Blakley, H. Boermans and P. Brousseau, Phagocytosis as a biomarker of immunotoxicity in wildlife species exposed to environmental xenobiotics, *Am. Zool.*, 2000, **40**, 412–420.
49. S. Mukherjee, M. Ray and S. Ray, Phagocytosis of charcoal particulate as immunological marker of azadirachtin exposure, in *Perspective of Animal Ecology and Reproduction*, ed. V. K. Gupta, Daya Pub., New Delhi, India, 2011. pp. 55–67.
50. E. García-garcía, M. Prado-álvarez, B. Novoa, A. Figueras and C. Rosales, Immune responses of mussel hemocyte sub-populations are differentially regulated by enzymes of the P13-K, PKC and ERK Kinase families, *Dev. Comp. Immunol.*, 2008, **32**, 52–55.
51. C. J. Bayne, M. N. Moore, T. H. Carefoot and R. J. Thompson, Haemolymph functions in *Mytilus californianus*: the cytochemistry of hemocytes and their responses to foreign implants and haemolymph factors in phagocytosis, *J. Invertebr. Pathol.*, 1979, **34**, 1–20.
52. L. Renwrantz, T. Yoshino, T. Cheng and K. Auld, Size determination of hemocytes from the American oyster, *Crassostrea virginica*, and the

description of a phagocytosis mechanism, *Zool. Jb. Physiol.*, 1979, **83**, 1–12.
53. C. Lopez, M. J. Carballal, C. Azevedoc and A. Villalba, Morphological characterization of the hemocytes of the clam: *Ruditapes decussates* (Mollusca: Bivalvia), *J. Invertebr. Pathol.*, 1997, **69**, 51–57.
54. J. Sun, X. Wu and W. Zhang, Morphological, structural and functional characteristics of the hemocytes of the oyster, *Crassostrea ariakensis*, *J. Shellfish Res.*, 2006, **25**, 55–64.
55. L. Ballarin, V. Covre, L. Masiero and S. Casellato, Immunotoxic effects of fluoride on the hemocytes of *Venerupis philippinarum*, *Invertebr. Surv. J.*, 2014, **11**, 22–29.
56. C. R. Fries and M. R. Tripp, Depression of phagocytosis in *Mercenaria* following chemical stress, *Dev. Comp. Immunol.*, 1980, **4**, 233–244.
57. Fridovich, Superoxide radical and superoxide dismutases, *Annu. Rev. Biochem.*, 1995, **64**, 97–112.
58. S. Chakraborty, M. Ray and S. Ray, Cell to organ: physiological, immunotoxic and oxidative stress responses of *Lamellidens marginalis* to inorganic arsenite, *Ecotoxicol. Environ. Saf.*, 2013, **94**, 153–163.
59. S. Mukherjee, A. S. Bhunia, N. S. Bhunia, M. Ray and S. Ray, Immunomodulatory effects of temperature and pH of water in an Indian freshwater sponge, *J. Therm. Biol.*, 2016, **59**, 1–12.
60. S. Mukherjee, M. Ray and S. Ray, Oxidative stress in freshwater bivalve of India exposed to azadirachtin based pesticide, *Roman. J. Biol. Zool.*, 2012, **57**(1), 79–88.
61. G. W. Winston, M. N. Moore, A. I. Miles and C. Soverchia, Production of reactive oxygen species by hemocytes from the marine mussel, *Mytilus edulis*: lysosomal localization and effect of xenobiotics, *Comp. Biochem. Physiol., Part C: Pharmacol., Toxicol. Endocrinol.*, 1996, **113**, 221–229.
62. L. Donaghy, E. Kraffe, N. L. Goïc, C. Lambert, A. K. Volety and P. Soudant, Reactive oxygen species in unstimulated hemocytes of the pacific oyster *Crassostrea gigas*: a mitochondrial involvement, *PLoS One*, 2012, **7**(10), e46594.
63. T. Latire, C. Le Pabic, E. Mottin, A. Mottier, K. Costil, N. Koueta, J. M. Lebel and A. Serpentini, Responses of primary cultured haemocytes from the marine gastropod *Haliotis tuberculata* under 10-day exposure to cadmium chloride, *Aquat. Toxicol.*, 2012, **109**, 213–221.
64. F. M. Akaishi, S. D. St-jean, F. Bishay, J. Clarke, I. S. Rabitto and C. O. Ribeiro, Immunological responses, histopathological finding and disease resistance of blue mussel (*Mytilus edulis*) exposed to treated and untreated municipal wastewater, *Aquat. Toxicol.*, 2007, **82**, 1–14.
65. V. Matozzo, L. Ballarin, D. M. Pampanin and M. G. Marin, Effects of copper and cadmium exposure on functional responses of hemocytes in the clam, *Tapes philippinarum*, *Arch. Environ. Contam. Toxicol.*, 2001, **170**, 163–170.
66. F. Geret, T. Burgeot, J. Haure, B. Gagnaire, T. Renault, P. Y. Communal and J. F. Samain, Effects of low-dose exposure to pesticide mixture on

physiological responses of the Pacific oyster, *Crassostrea gigas*, *Environ. Toxicol.*, 2013, **28**(12), 689–699.
67. M. Banni, A. Negri, A. Dagnino, J. Jebali, S. Ameur and H. Boussetta, Acute effects of benzo[a]pyrene on digestive gland enzymatic biomarkers and DNA damage on mussel *Mytilus galloprovincialis*, *Ecotoxicol. Environ. Saf.*, 2010, **73**, 842–848.
68. M. J. Almamoori, J. M. Salman, R. Hughes and A. H. Al-saadi, Biochemical changes in two species of molluscs as environmental biomarkers of pollution in hilla river, Iraq, *Int. J. Sci. Nat.*, 2013, **4**, 40–43.
69. E. Mottin, C. Caplat, M. Mahaut, K. Costil, D. Barillier, J. Lebel and A. Serpentini, Effect of *in vitro* exposure to zinc on immunological parameters of haemocytes from the marine gastropod *Haliotis tuberculata*, *Fish Shellfish Immunol.*, 2010, **29**, 846–853.
70. Z. Sun, L. Wang, T. Zhang, Z. Zhou, Q. Jiang, Q. Yi and C. Yang, The immunomodulation of inducible hydrogen sulfide in Pacific oyster *Crassostrea gigas*, *Dev. Comp. Immunol.*, 2014, **46**, 530–536.
71. S. Chakraborty and S. Ray, Nuclear morphology and lysosomal stability of molluscan hemocytes as possible biomarkers of arsenic toxicity, *Clean: Soil, Air, Water*, 2009, **37**(10), 769–775.
72. N. Molnar and P. P. Fong, Toxic effects of copper, cadmium, and methoxychlor shown by neutral red retention assay in two species of freshwater molluscs, *Open Environ. Pollut. Toxicol. J.*, 2012, **3**, 65–71.
73. N. Koukouzika and V. K. Dimitriadis, Multiple biomarker comparison in *Mytilus galloprovincialis* from the Greece coast: "lysosomal membrane stability, neutral red retention, micronucleus frequency and stress on stress", *Ecotoxicology*, 2005, **14**, 449–463.
74. J. M. Harding, C. Couturier, G. J. Parsons and N. W. Ross, Evaluation of the neutral red assay as a stress response indicator in cultivated mussels (*Mytilus spp.*) in relation to post-harvest processing activities and storage conditions, *Aquaculture*, 2004, **231**, 315–326.
75. C. Hauton, L. E. Hawkins and S. Hutchinson, The use of the neutral red retention assay to examine the effects of temperature and salinity on haemocytes of the European flat oyster *Ostrea edulis* (L), *Comp. Biochem. Physiol., Part B: Biochem. Mol. Biol.*, 1998, **119**, 619–623.
76. T. Kiss, Apoptosis and its functional significance in molluscs, *Apoptosis*, 2010, **15**, 313–321.
77. L. I. Sweet, D. R. Passino-Reader, P. G. Meier and G. M. Omann, Xenobiotic-induced apoptosis: significance and potential application as a general biomarker of response, *Biomarkers*, 1999, **4**, 237–253.
78. J. Russo and L. Madec, Haemocyte apoptosis as a general cellular immune response of the snail, *Lymnaea stagnalis*, to a toxicant, *Cell Tissue Res.*, 2007, **328**, 431–441.
79. M. Ray, A. S. Bhunia, N. S. Bhunia and S. Ray, Density shift, morphological damage, lysosomal fragility and apoptosis of hemocytes of Indian molluscs exposed to pyrethroid pesticides, *Fish Shellfish Immunol.*, 2013, **35**, 499–512.

80. K. Perry and J. Lynn, Detecting physiological and pesticide-induced apoptosis in early developmental stages of invasive bivalves, *Hydrobiologia*, 2009, **628**, 153–164.
81. J. Zhou, Z. H. Cai, X. S. Zhu, L. Li and Y. F. Gao, Innate immune parameters and haemolymph protein expression profile to evaluate the immunotoxicity of tributyltin on abalone (*Haliotis diversicolor supertexta*), *Dev. Comp. Immunol.*, 2010, **34**, 1059–1067.
82. J. H. Chen and C. J. Bayne, Bivalve mollusc hemocyte behaviors: characterization of hemocyte aggregation and adhesion and their inhibition in the California mussel (*Mytilus califurnianus*), *Biol. Bull.*, 1995, **188**, 255–266.
83. M. Auffret and R. Oubella, Hemocyte aggregation in the oyster *Crassostrea gigas: in vitro* measurement and experimental modulation by xenobiotics, *Comp. Biochem. Physiol., Part A: Mol. Integr. Physiol.*, 1997, **118**, 705–712.
84. M. Auffret and R. Oubella, Cytology and cytometric analysis of bivalve mollusc hemocytes, in *Techniques in Fish Immunology, Vol. 4: Immunology and Pathology of Aquatic Invertebrates*, ed. J. S. Stolen, SOS Publication, Fair Haven, 1995, pp. 55–64.
85. Why Mollusc Toxicity Tests for Endocrine Disruptors and Other Chemicals Are Needed – CHEM Trust briefing November 2009.
86. A. Hamza-Chaffai, Usefulness of bioindicators and biomarkers in pollution biomonitoring, *Int. J. Biotechnol. Wellness Ind.*, 2014, **3**, 19–26.
87. S. K. Gupta and J. Singh, Evaluation of mollusc as sensitive indicator of heavy metal pollution in aquatic ecosystem, *IIOAB J.*, 2011, **2**, 49–57.
88. M. Ray, N. S. Bhunia, A. S. Bhunia and S. Ray, Expression analyses of interferon gamma, tumor necrosis factor alpha and inducible nitric oxide synthase in the hemocyte morphotypes of two commercially important Indian molluscs, *Aquacult. Rep.*, 2016, **4**, 30–35.

CHAPTER 5

*Application of the Zebra Mussel (*Dreissena polymorpha*) in the Toxicity Evaluation of Emerging Aquatic Pollutants*

MARCO PAROLINI* AND ANDREA BINELLI

Department of Biosciences, University of Milan, via Celoria 26, I-20133 Milan, Italy
*Email: marco.parolini@unimi.it

5.1 Introduction

Over the last two decades, several monitoring studies have revealed the presence of new compounds, called 'emerging pollutants', in aquatic environments.[1] Emerging pollutants are new products or chemicals without regulatory status and whose effects on the environment and human health are unknown.[1] Among them, Pharmaceutical and Personal Care Products (PPCPs) have received particular attention because of the increasing knowledge on their widespread occurrence in the environment and their potential hazard towards aquatic organisms. These molecules are mainly used for domestic, veterinary and hospital purposes. After administration, they are excreted unchanged and/or as metabolites reaching the sewage. Measurable

Issues in Toxicology No. 33
Ecotoxicology and Genotoxicology: Non-traditional Aquatic Models
Edited by Marcelo L. Larramendy

Published by the Royal Society of Chemistry, www.rsc.org

levels of hundreds of PPCPs belonging to diverse therapeutic classes, including analgesics, antibiotics, β-blockers, anticonvulsants, lipid-regulators, contrast and anti-cancer agents, hormones, fragrances and disinfectants, have been found in the inlets and outlets of wastewater treatment plants (WWTPs) in the $\mu g\,L^{-1}$ range. Since WWTPs were not originally designed for the elimination of xenobiotics, they have a low capacity for removing most PPCPs from wastewaters.[1,2] Thus, a number of these molecules enter the surface waters, contributing to the environmental contamination. Several monitoring programs have revealed the presence of different PPCPs at concentrations ranging between tens and hundreds of $ng\,L^{-1}$ in both river and lake waters.[2] Despite the growing information on the occurrence of PPCPs in the aquatic environment, the main problem regarding these emerging pollutants is the dearth of knowledge of their impact in the middle- or long-term on ecosystems. Since drugs have an intended biological activity, and are often persistent and bioavailable, they are potentially dangerous to non-target aquatic organisms, even at low environmental concentrations.[3] Although many studies have shown that acute effects occur only after exposure to diverse PPCPs at concentrations in the $mg\,L^{-1}$ range,[3–5] it is important to bear in mind that aquatic organisms are exposed to low PPCP concentrations over their whole life span, so that chronic effects are much more probable than acute ones.[6] Even though a growing number of laboratory studies have investigated the chronic toxicity of some PPCPs in diverse biological models,[7–9] the attention has been focused only on a few PPCPs, mainly non-steroidal anti-inflammatory drugs (NSAIDs), blood lipid lowering agents, antibiotics and sex hormones.[1] In addition, most of these investigations have been performed on classical model organisms of ecotoxicology, including crustaceans, marine mussels and fish.[1] Thus, there is a dearth of information regarding the toxicity of PPCPs on several aquatic species, including both invertebrates and vertebrates.

In freshwater ecotoxicology, the zebra mussel (*Dreissena polymorpha*) has been considered an excellent model species to monitor the presence and to assess the toxicity of environmental pollutants.[10] The zebra mussel (Figure 5.1) is a filter-feeding bivalve mollusk native to the Ponto-Caspian region that became one of the most invasive freshwater species. This bivalve is used as a bioindicator in both field and laboratory ecotoxicological investigations because of its peculiar physiological and ecological features:

1. Wide distribution owing to its planktonic larval stage, which allowed the colonization of both European and North American lentic and lotic freshwaters.
2. It is a sessile organism that colonizes hard substrates (both natural and anthropic). This feature suggests its use in biomonitoring programs because it can point out local sources of contamination.
3. Ease of sampling.
4. Longevity (3–5 years).
5. Easily maintained under laboratory conditions.

Figure 5.1 The zebra mussel (*Dreissena polymorpha*).
Reproduced from *Arch. Environ. Contam. Toxicol.*, Application of a biomarker response index for ranking the toxicity of five pharmaceutical and personal care products (PPCPs) to the bivalve *Dreissena polymorpha*, **64**, 2013, 439–447, M. Parolini, A. Pedriali and A. Binelli, (© Springer Science + Business Media New York 2012) With permission of Springer.

6. High filtration activity (ranging from 5 to 400 mL $bivalve^{-1} h^{-1}$)[11] allowing fast intake of environmental pollutants and the subsequent quick identification of their negative effects.

Despite its invasive status, this apparent disadvantage could represent one of the pivotal reasons that suggest the use of the zebra mussel in both field biomonitoring and laboratory studies aimed at assessing the toxicity of diverse environmental pollutants. In fact, the sampling of this species does not have ethical consequences and its removal from freshwater can safeguard the populations of native bivalve species (*e.g.*, Unionids), whose health status is threatened because their valves are used by zebra mussel specimens as growing substrates.

A number of studies have focused on the evaluation of the toxicity of different environmental pollutants by using the zebra mussel as a model organism, highlighting its usefulness in this kind of research. The attention has been mainly addressed to the toxicity of the so-called legacy pollutants (*e.g.*, *p,p′*-DDT and relative homologues, polycyclic aromatic hydrocarbons, and polychlorinated biphenyls)[12–15] because of their widespread diffusion in aquatic ecosystems and their well-known hazards towards organisms. However, in the last two decades, a growing number of investigations have used the zebra mussel as model organisms to assess the potential adverse effects induced by emerging aquatic pollutants,[16–18] showing the sensitivity of this bivalve species to diverse chemicals belonging to this wide class of contaminants.

The present chapter presents the suitability of the zebra mussel as a sensitive non-conventional model organism to assess the chronic toxicity of emerging pollutants, such as PPCPs. Here we show the results from our recent investigations, highlighting sub-lethal effects to the zebra mussel induced by short-term exposure (96 hours) to five PPCPs commonly found in freshwater environments, namely triclosan (TCS),[19] trimethoprim (TMP),[20] paracetamol (PCM),[21] diclofenac (DCF)[22] and ibuprofen (IBU).[23] TCS is an antibacterial agent used in several healthcare products, while TMP is an antibiotic with potent microbicide activity against a wide variety of bacterial species. DCF and IBU are two of the most frequently used non-steroidal anti-inflammatory drugs (NSAIDs) worldwide, while PCM, also known as acetaminophen, is an over-the-counter antipyretic used to treat pain and fever. Our investigations have focused on these specific molecules because they are extensively used and are among the most detected PPCPs in the aquatic environment worldwide. In fact, NSAIDs and antibiotics are the most relevant therapeutic classes in terms of freshwater contamination, since they represent 16% and 15%, respectively, of the pharmaceuticals detected in monitoring surveys globally.[2] In addition, their increasing environmental levels are causing growing concern regarding their potential hazard towards non-target aquatic organisms. Although our previous experiments tested the toxicity of three different concentrations of each of the PPCPs mentioned above, in the present dissertation we just focus on the effects of selected PPCPs towards zebra mussel specimens exposed to a similar concentration (1 nM) for each drug, reflecting the mean of the concentrations of each PPCP measured in surface waters worldwide. In detail, we compared the toxicity on zebra mussel specimens separately exposed to 290 ng L^{-1} of TCS, 290 ng L^{-1} of TMP, 318 ng L^{-1} of DCF, 200 ng L^{-1} of IBU and 154 ng L^{-1} of PCM. We centered our attention on cyto-genetic effects and the potential occurrence of an oxidative stress situation. Briefly, genotoxicity was evaluated by the Single Cell Gel Electrophoresis (SCGE) assay, the DNA Diffusion assay (a measure of apoptotic frequency) and the micronucleus test (MN test). The Neutral Red Retention Assay (NRRA), by evaluating the lysosome membrane stability, was used to investigate if PPCPs promote a situation of general stress in bivalves. Lastly, the activities of three antioxidant phase I enzymes, namely catalase (CAT), superoxide dismutase (SOD) and glutathione peroxidase (GPx), as well as the phase II detoxifying enzyme glutathione *S*-transferase (GST), were measured to study the modulation of enzymatic defense in response to exposure to selected PPCPs.

5.2 Experimental Design

Exposure procedures and biomarker methods have been thoroughly described in previous studies[19–23] so they are just briefly reported here. Zebra mussel specimens were sampled in the Lake of Lugano, which is considered a reference site because of its low PPCP pollution,[24] transferred to the laboratory in bags filled with lake water, and then kept in depuration under controlled laboratory conditions with a natural photoperiod at constant temperature (20 ± 1 °C), pH (7.5) and oxygenation saturation. Bivalves were fed and water was renewed on a daily basis to gradually purify the mollusks from pollutants accumulated in their soft tissues in their native place. Zebra mussel health status was checked by measuring biomarker baseline levels weekly. Bivalves were exposed to PPCPs only when biomarker baseline levels were comparable with those obtained in our previous laboratory studies.[19,20] Exposure to single PPCPs was performed under semi-static conditions for 96 hours in 10 L aquaria.[19–23] Previous studies demonstrated that this period of time is enough to show sub-lethal effects on zebra mussel specimens exposed to different xenobiotics.[25,26] All experimental groups (control and 1 nM treatment per molecule) were processed at the same time. Three hundred zebra mussel specimens were put in each aquarium. We prepared a working solution of each PPCP (10 mg L^{-1}) by dissolving the powder standard in dimethyl sulfoxide (DMSO), which was added daily to the aquaria in order to reach the desired concentration. The entire water volume and PC were renewed daily in order to guarantee a constant solution concentration of PPCPs over each 24 h period of exposure, and to prevent loss of contaminant and the degradation of each single parental compound into its corresponding metabolites. Every 24 h for 5 days (up to 96 h of exposure), we collected 33 zebra mussel specimens from each aquarium. We collected hemolymph from 10 of these individuals to analyze cytogenotoxicity biomarkers, namely the Neutral Red Retention Assay (NRRA), the Single Cell Gel Electrophoresis (SCGE) assay, the DNA Diffusion assay and the Micronucleus test (MN test), on hemocytes. The whole soft tissue of the other individuals was quickly frozen in liquid nitrogen and stored at −80 °C until the enzymatic activity was measured.

5.2.1 Biomarker Methods

Since the methods and procedures for all the biomarkers considered in the present dissertation were described in detail elsewhere,[21] we report here only a brief description. The NRRA was applied on zebra mussel hemocytes according to a previously validated method.[25] The activities of SOD, CAT, GPx, and GST were measured in triplicate in the cytosolic fraction extracted from a pool of three whole mussels ($\approx$0.3 g fresh weight) homogenized in an opportune phosphate buffer using a Potter homogenizer. Dithiothreitol (DTT, 100 mM) and specific protease inhibitors (1 : 10 v/v), namely phenanthroline (10 mM) and trypsin inhibitor (10 mg mL^{-1}), were also added to the buffer. The homogenate was centrifuged at 15 000$\times g$ for 1 hour at 4 °C. The samples were immediately processed for the determination of protein according to the Bradford method and enzymatic activities were determined spectrophotometrically.[26] Briefly, the CAT activity was determined by measuring the consumption of hydrogen peroxide at 240 nm. The SOD activity was determined by evaluating the inhibition of cytochrome c (10 μM) reduction at 550 nm by the superoxide anion generated by the xanthine oxidase (1.87 mU mL^{-1})/hypoxanthine (50 μM) reaction. The GPx activity was measured by monitoring the consumption of NADPH at 340 nm using 0.2 mM hydrogen peroxide substrate in phosphate buffer containing glutathione (2 mM), sodium azide (NaN_3; 1 mM), glutathione reductase (2 U mL^{-1}), and NADPH (120 μM). The GST activity was measured by adding reduced glutathione (1 mM) and 1-chloro-2,4-dinitrobenzene in phosphate buffer to the cytosolic fraction and monitoring the resulting reaction for 1 min at 340 nm. Genetic biomarker analysis was performed on zebra mussel hemocytes. The alkaline (pH > 13) SCGE assay was performed following the method adapted for the zebra mussel by Buschini *et al.*:[27] 50 cells per slide were analyzed using an image analysis system (Comet Score®), for a total of 500 analyzed cells per individual. The ratio between migration length and comet head diameter (LDR) was considered as the endpoint to assess DNA fragmentation. The apoptotic cell frequency was evaluated through the protocol described by Singh.[28] 200 cells per slide were analyzed (1000 cells per sample). The MN test was performed according to the method of Pavlica *et al.*[29] 400 cells were counted per each slide for a total of 4000 cells/treatment and micronuclei were identified by criteria described in the literature.[30]

5.2.2 Biomarker Response Index (BRI)

To compare the toxicity of selected PPCPs and to accurately rank their toxic potential towards the zebra mussel, we integrated the single biomarker responses into a biomarker response index (BRI) according to a previously proposed approach,[31,32] with opportune modification.[33] Briefly, we calculated the percentage alteration level (AL) as the deviation of the mean value of each biomarker per treated sample and time of exposure from the corresponding control. Since no significant differences ($p > 0.05$) were observed between the baseline levels at the beginning ($t = 0$ h) of each single exposure per tested biomarker, we considered the $t = 0$ h value as the control. After calculating AF, it was assigned a specific score. The biomarker responses showing a ±20% difference compared to the corresponding control, although statistically significant, were not considered of biological relevance and were assigned a score of 1. Deviations from control ranging between ±20% and ±50%, indicating an early physiological response of the organism, were assigned a score of 2, while large differences from controls (±50–100%), owing to notable alterations induced by stressors, were assigned a score of 3. Deviations that largely overcome the controls (more than ±100%) were scored 4. Since it is well-known that enzyme activity follows a bell-shape trend and its decrease after the achievement of the maximum activity shows a situation of high stress, when an enzymatic activity showed a bell-shape trend we decided to calculate its AF compared to the highest obtained value, which was scored as described above. In addition, each biomarker was also weighted according to its level of biological organization because it is assumed that an alteration at the cellular level has a greater impact on the health status of the organisms than changes at the biochemical level;[33] cellular biomarkers were weighted as 2, while biochemical assays as 1. The overall 96 h BRI value for each tested PPCP was then calculated according to the formula reported by Parolini *et al.*[33]

5.2.3 Statistical Analysis

Data normality and homoscedasticity were verified using the Shapiro-Wilk and Levene's tests, respectively. One-way analysis of variance (ANOVA) was performed followed by the Bonferroni *post hoc* test to investigate the effects of each treatment on single considered endpoints over the 96 h exposure. In addition, factorial ANOVA followed by a Bonferroni *post hoc* test was performed to point out significant differences ($p<0.05$) among PPCP-induced effects for each measured biomarker. All statistical analyses were performed using the STATISTICA 7.0 software package.

5.3 Discussion of Obtained Results

The present dissertation showed that low environmental concentrations of TCS, TMP, IBU, DCF and PCM, similar to those currently found in surface water worldwide, may represent an unneglectable threat to zebra mussel health status. Data obtained from the NRRA (Figure 5.2) highlighted a significant decrease in the stability of lysosome membranes after TCS ($F=15.86$; $p<0.05$), DCF ($F=3.80$; $p<0.05$) and PCM treatment ($F=6.18$; $p<0.01$).

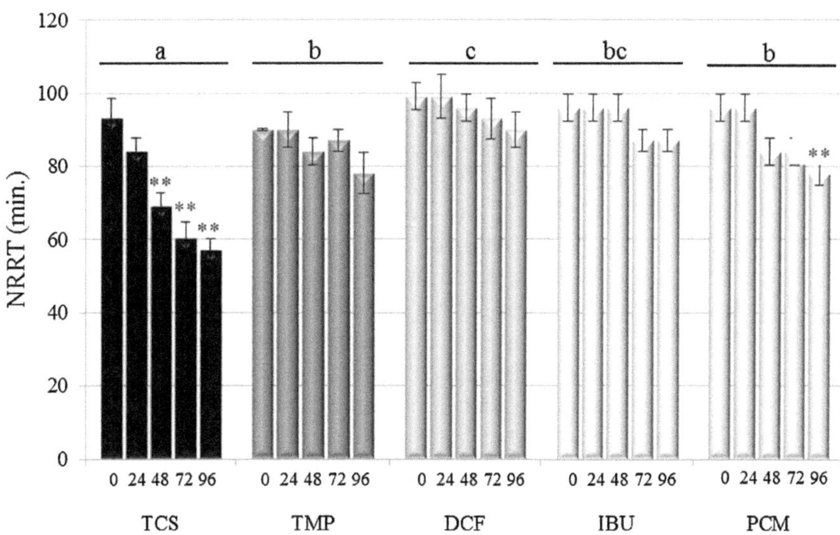

Figure 5.2 Mean of Neutral Red Retention Time (NRRT) ($\pm$SEM) in lysosomes from hemocytes of zebra mussels exposed to 1 nM of selected drugs. Significant differences (two-way ANOVA, Bonferroni *post hoc* test; ** $p<0.01$) between treated mussels and the corresponding control are shown. Different letters above histograms indicate significant differences among PPCP treatments ($p<0.01$).
Reproduced from *Arch. Environ. Contam. Toxicol.*, Application of a biomarker response index for ranking the toxicity of five pharmaceutical and personal care products (PPCPs) to the bivalve *Dreissena polymorpha*, **64**, 2013, 439–447, M. Parolini, A. Pedriali and A. Binelli, (© Springer Science + Business Media New York 2012) With permission of Springer.

Differently, although a significant effect of exposure time was noticed for TMP ($F = 2.76$; $p < 0.05$) and IBU ($F = 2.68$; $p < 0.05$), no significant NRRT decrease ($p > 0.05$) in treated bivalves compared to control was found.

These findings confirmed that TCS-, DCF- and PCM-treated zebra mussels suffered a cellular stress situation that is likely to be linked to the overproduction of reactive oxidative species (ROS), which could lead to an oxidative stress situation. In fact, the stability of lysosome membranes in mussels may be affected by the production of oxyradicals generated by exposure to environmental contaminants[34] and alterations to these organelles have been related to the increase of peroxidative processes.[35] Toxic effects of diverse xenobiotics often occur as a consequence of increased cellular levels of ROS owing, alternatively, to the straightforward activation of processes leading to their synthesis or indirectly to a direct action on antioxidant enzymes and scavengers, decreasing cell defenses.[36] The antioxidant defense system plays a pivotal role in processes aimed to neutralize the toxicity of ROS and the evaluation of the activity of SOD, CAT and GPx, as well as GST, provides useful information on responses of organisms to pollutant-induced levels of ROS. Slight imbalances of SOD and CAT activity were found after the exposure to all the tested therapeutics (Table 5.1), indicating that an environmentally relevant concentration of these drugs may promote the increase of both superoxide anion and hydrogen peroxide, respectively.

Similarly, the significant increase of GPx levels noticed for TCS ($F = 10.46$; $p < 0.05$), IBU ($F = 29.91$; $p < 0.05$) and PCM ($F = 12.83$; $p < 0.05$) exposures suggested that these PPCPs induced the overproduction of H_2O_2, which was actively counterbalanced by the enzymatic activity. Lastly, GST activity showed a wide range of response variation after the exposure to TCS, TMP, IBU and DCF, while PCM did not affect its activity over the whole experiment. The changes in the activities of defense enzymes should suggest that the exposure to tested PPCPs led to overproduction of ROS, whose toxicity was generally counterbalanced by enzymatic defense system. However, a small amount of ROS can escape the antioxidant shield and it could promote the occurrence of oxidative damage to cellular macromolecules, including DNA. Many studies demonstrated that the increase of pollutant-induced ROS caused remarkable changes in the integrity of DNA in different aquatic organisms.[37,38] TCS induced significant primary genetic lesions (pointed out by the LDR endpoint) in hemocytes from treated zebra mussels after as little as 24 hours of exposure according to a significant time-dependent relationship ($F = 292.17$; $p < 0.01$). In contrast, the other PPCPs did not cause a significant increase of DNA fragmentation (Figure 5.3).

Accordingly, TCS induced a significant time-dependent ($F = 105.90$; $p < 0.01$) increase of apoptotic and micronucleated cell frequency (Figures 5.4 and 5.5, respectively).

A significant time-dependent ($F = 18.02$; $p < 0.01$) increase of apoptotic frequency was also noticed after 72 hours of exposure to TMP, but this antibiotic did not promote the formation of micronuclei (Figure 5.5). Lastly,

Table 5.1 Mean SOD, CAT, GPx and GST activity (±SEM) measured in zebra mussel homogenates after exposure to 1 nM of each single PPCP. Different superscript letters next to the drug names refer to significant differences among PPCP treatments ($p < 0.01$). Reproduced from *Arch. Environ. Contam. Toxicol.*, Application of a biomarker response index for ranking the toxicity of five pharmaceutical and personal care products (PPCPs) to the bivalve *Dreissena polymorpha*, **64**, 2013, 439–447, M. Parolini, A. Pedriali and A. Binelli, (© Springer Science + Business Media New York 2012) With permission of Springer.

		SOD U mg prot.$^{-1}$	CAT mM min^{-1} mg prot.$^{-1}$	GPx μM min^{-1} mg prot.$^{-1}$	GST μM min^{-1} mg prot.$^{-1}$
TCS[a]	$t=0$	6.720 ± 0.370	0.092 ± 0.005	5.262 ± 0.484	0.175 ± 0.012
	$t=24$	8.002 ± 0.741	0.095 ± 0.001	5.559 ± 0.373	**0.241 ± 0.011[e]**
	$t=48$	7.703 ± 0.693	0.091 ± 0.003	**7.109 ± 0.250[e]**	**0.268 ± 0.009[e]**
	$t=72$	7.256 ± 0.412	**0.069 ± 0.002[e]**	**7.310 ± 0.182[e]**	**0.212 ± 0.014[d]**
	$t=96$	9.019 ± 0.328	0.078 ± 0.015	**7.431 ± 0.417[e]**	**0.213 ± 0.010[d]**
TMP[b]	$t=0$	21.050 ± 1.270	0.252 ± 0.014	16.816 ± 0.650	0.401 ± 0.017
	$t=24$	19.826 ± 1.394	**0.152 ± 0.006[e]**	**13.054 ± 0.653[e]**	**0.467 ± 0.010[d]**
	$t=48$	19.315 ± 1.199	0.238 ± 0.011	**14.127 ± 0.584[d]**	**0.540 ± 0.016[e]**
	$t=72$	**13.680 ± 0.705[d]**	0.223 ± 0.004	**12.907 ± 0.366[e]**	**0.486 ± 0.009[e]**
	$t=96$	**8.235 ± 0.553[e]**	0.269 ± 0.007	**12.275 ± 0.428[e]**	0.411 ± 0.007
IBU[b]	$t=0$	13.264 ± 0.554	0.070 ± 0.003	14.733 ± 0.275	0.208 ± 0.004
	$t=24$	14.130 ± 1.297	0.060 ± 0.011	**38.337 ± 0790[e]**	0.250 ± 0.010
	$t=48$	17.046 ± 2.247	0.088 ± 0.006	**34.729 ± 2.362[e]**	**0.359 ± 0.030[e]**
	$t=72$	13.922 ± 0.564	0.064 ± 0.008	**23.634 ± 0.640[d]**	0.239 ± 0.003
	$t=96$	**22.758 ± 2.022[e]**	**0.098 ± 0.010[e]**	**33.073 ± 1.453[e]**	**0.316 ± 0.011[e]**
DCF[b]	$t=0$	13.432 ± 1.559	0.134 ± 0.011	25.004 ± 0.401	0.565 ± 0.057
	$t=24$	14.941 ± 1.079	0.141 ± 0.014	23.276 ± 0.867	**0.725 ± 0.030[e]**
	$t=48$	17.058 ± 1.081	0.122 ± 0.013	**20.595 ± 0.444[e]**	0.569 ± 0.005
	$t=72$	16.358 ± 0.740	0.120 ± 0.006	23.817 ± 0.499	**0.429 ± 0.001[d]**
	$t=96$	15.199 ± 1.020	0.169 ± 0.002	**15.739 ± 0.470[e]**	**0.433 ± 0.003[d]**
PCM[c]	$t=0$	14.852 ± 1.615	0.177 ± 0.010	10.944 ± 1.001	0.484 ± 0.014
	$t=24$	**33.498 ± 1.991[e]**	0.192 ± 0.012	12.025 ± 0.552	0.432 ± 0.031
	$t=48$	23.780 ± 1.763	0.170 ± 0.012	14.232 ± 1.024	0.444 ± 0.028
	$t=72$	15.529 ± 1.182	0.185 ± 0.010	**17.737 ± 1.419[e]**	0.518 ± 0.031
	$t=96$	11.958 ± 0.944	0.173 ± 0.007	**21.079 ± 1.525[e]**	0.415 ± 0.027

the three tested NSAIDs did not trigger the apoptotic process (Figure 5.4) and did not cause an increase of MN frequency (Figure 5.5).

The slight variations in cellular and genetic biomarkers induced by the 96 hours exposure to 1 nM of TMP, DCF, IBU and PCM suggest a low hazard of these therapeutics towards the zebra mussels. In contrast, alarming adverse effects were found in response to TCS exposure, as shown by the significant increases of cytogenetic damage after only 24 hours of exposure, together with relevant imbalances of enzymatic activities. These findings suggest that TCS is the most toxic molecule towards the zebra mussel among

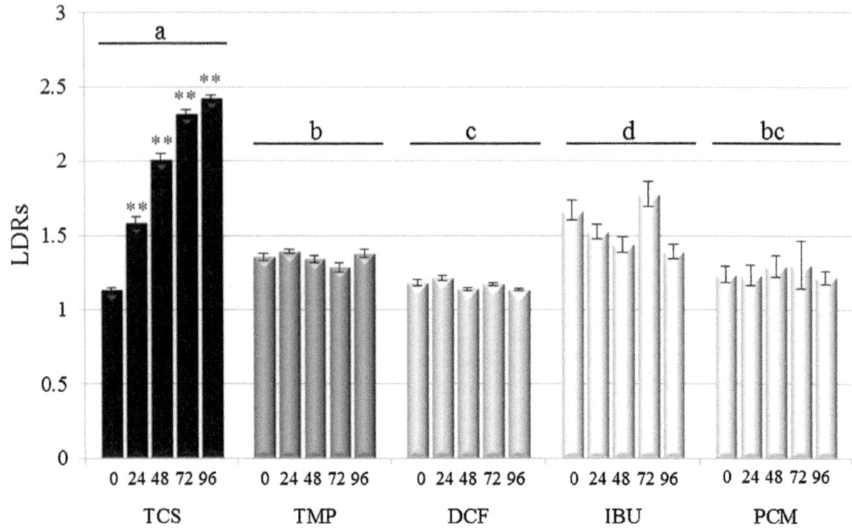

Figure 5.3 Primary DNA damage expressed by the length/diameter ratio (LDR, mean values ± SEM) measured in hemocytes of the zebra mussel exposed to 1 nM of selected drugs. Significant differences (two-way ANOVA, Bonferroni *post hoc* test; ** $p < 0.01$) between treated mussels and the correspondent control are shown. Different letters above histograms indicate significant differences among PPCP treatments ($p < 0.01$). Reproduced from *Arch. Environ. Contam. Toxicol.*, Application of a biomarker response index for ranking the toxicity of five pharmaceutical and personal care products (PPCPs) to the bivalve *Dreissena polymorpha*, **64**, 2013, 439–447, M. Parolini, A. Pedriali and A. Binelli, (© Springer Science + Business Media New York 2012) With permission of Springer.

the tested PPCPs. The statistical analysis (factorial ANOVA, Bonferroni *post hoc* test; $p < 0.05$) showed significant differences between the adverse effects induced by TCS and all the other drugs for all the considered endpoints, excluding SOD and CAT. According to these approaches, the second most toxic drug seemed to be the IBU, followed by TMP, DCF and PCM. However, because of the wide variability in biomarker responses, showing different trends depending on the considered drug (Figures 5.2–5.5 and Table 5.1), the mere interpretation of single biomarker responses, also when supported by the statistical analysis, could not be exhaustive to accurately rank the toxicity of PPCPs towards the model organism. Thus, although the assessment of several biomarkers is considered the best approach to understand the toxicity and the mechanism of action of environmental contaminants towards an organism,[38] the analysis of the changes of dissimilar biological endpoints is not sufficient to rank the toxicity of different molecules. To bypass this limitation, the application of procedures aimed at integrating the biomarker results into a synthetic index could help to minimize the variation of responses.[39] The biomarker response index we applied in the present study was computed using all the measured endpoints under our

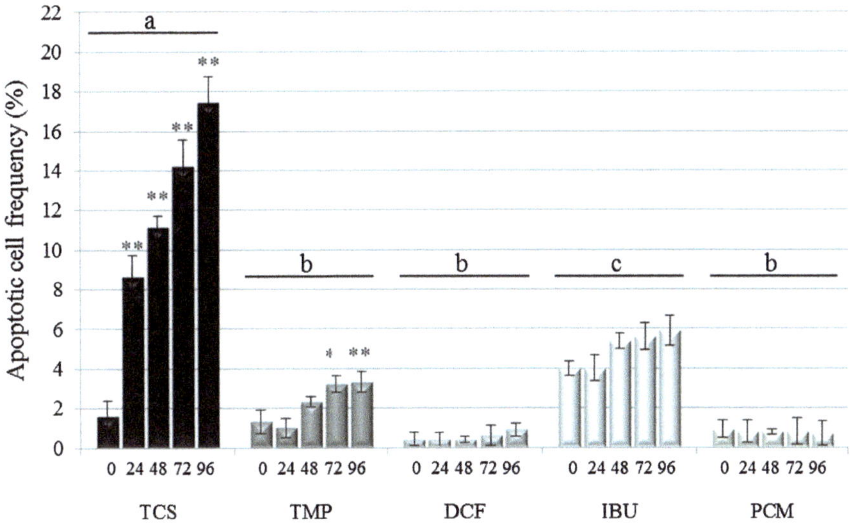

Figure 5.4 Percentages of apoptotic hemocytes (mean values ± SEM) measured by the DNA diffusion assay for zebra mussel specimens exposed to 1 nM of selected drugs. Significant differences (two-way ANOVA, Bonferroni *post hoc* test; ** $p<0.01$) between treated mussels and the correspondent control are shown. Different letters above histograms indicate significant differences among PPCP treatments ($p<0.01$).
Reproduced from *Arch. Environ. Contam. Toxicol.*, Application of a biomarker response index for ranking the toxicity of five pharmaceutical and personal care products (PPCPs) to the bivalve *Dreissena polymorpha*, **64**, 2013, 439–447, M. Parolini, A. Pedriali and A. Binelli, (© Springer Science + Business Media New York 2012) With permission of Springer.

experimental conditions (Figure 5.6) and efficiently reduced the wide degree of variability of biomarker responses. The BRI approach confirmed the TCS as the compound that most negatively affected the zebra mussel's health status. Moreover, it allowed the accurate ranking of the other tested PPCP toxicity towards the model organism as follows: TMP > IBU > DCF ≈ PCM. It is interesting to note that the BRI results overturned the toxicity ranking based on the interpretation of single biomarker responses and statistical analysis, highlighting that TMP was more toxic than IBU. In addition, the BRI-based ranking disentangled the toxicity of tested NSAIDs; whilst the statistical analysis did not show significant differences both between single endpoints and whole drug toxicity, the BRI showed that IBU can be considered more toxic than DCF and PCM towards the zebra mussel.

The highest toxicity of TCS compared to the other tested drugs is consistent with other comparative studies on the toxicity of PPCPs and could be owing to its lipophilic nature ($\log K_{ow} = 4.76$). However, despite of the TCS findings, our BRI-based toxicity ranking partially disagreed the theory according to which the toxicity of a molecule is related to its $\log K_{ow}$. In fact, although TMP is a hydrophilic drug ($\log K_{ow} = 0.91$) it resulted as the second

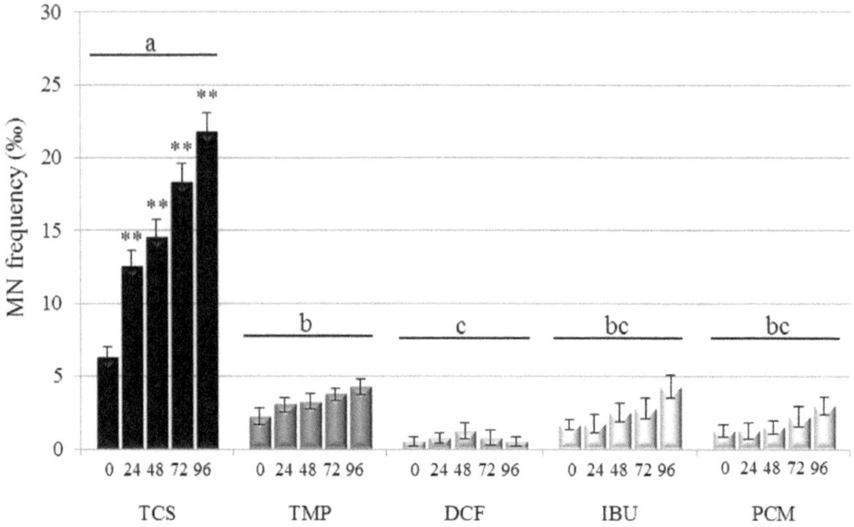

Figure 5.5 Frequency of micronucleated hemocytes (mean values ± SEM) measured by the MN test zebra mussel specimens exposed to 1 nM of selected drugs. Significant differences (two-way ANOVA, Bonferroni *post hoc* test; ** $p < 0.01$) between treated mussels and the correspondent control are shown. Different letters above histograms indicate significant differences among PPCP treatments ($p < 0.01$).
Reproduced from *Arch. Environ. Contam. Toxicol.*, Application of a biomarker response index for ranking the toxicity of five pharmaceutical and personal care products (PPCPs) to the bivalve *Dreissena polymorpha*, **64**, 2013, 439–447, M. Parolini, A. Pedriali and A. Binelli, (© Springer Science + Business Media New York 2012) With permission of Springer.

most toxic molecule towards the zebra mussel among those we tested, showing a higher toxicity than IBU ($\log K_{ow} = 3.97$) and, above all, than DCF ($\log K_{ow} = 4.4$). We unexpectedly found a very low toxicity of DCF, disagreeing with a number of findings showing that the exposure to this therapeutic can induce diverse adverse effects towards aquatic vertebrates,[40,41] so that it was included in the recent list of priority substances by the European Commission.[42] Moreover, the toxicity of the tested NSAIDs to the zebra mussel was also inconsistent with results obtained by Cleuvers,[43] who found that the acute toxicity of DCF was 1.5- and 5-fold higher compared to that of IBU towards *Daphnia magna* and *Desmodesmus subspicatus*, respectively. In addition, our BRI-based NSAIDs scale of toxicity was also in reverse order compared to a previous ranking performed on the basis of our *in vitro* results from four cytogenotoxicity biomarkers applied on zebra mussel hemocytes, where DCF was the most toxic therapeutic.[44] This discrepancy could be owing to differences in experimental approaches since *in vitro* drug bioavailability could be different than that *in vivo*, and single cells could present a different sensitivity to the tested NSAIDs compared to the sensitivity of the whole organism. Thus, the toxicity ranking based on the BRI approach,

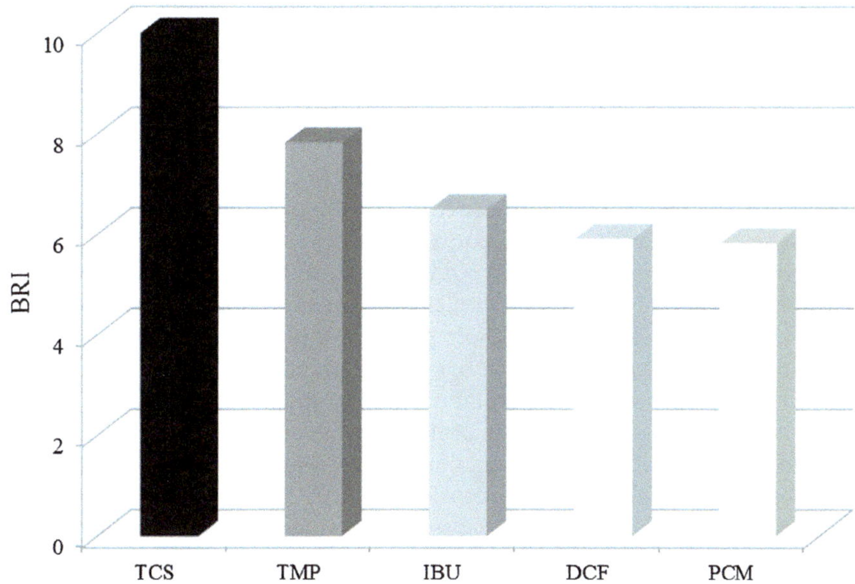

Figure 5.6 Biomarker response index (BRI) value calculated for exposure to 1 nM of each tested PPCP.
Reproduced from *Arch. Environ. Contam. Toxicol.*, Application of a biomarker response index for ranking the toxicity of five pharmaceutical and personal care products (PPCPs) to the bivalve *Dreissena polymorpha*, **64**, 2013, 439–447, M. Parolini, A. Pedriali and A. Binelli, (© Springer Science + Business Media New York 2012) With permission of Springer.

being built on *in vivo* data, can be considered more accurate and trustworthy than the *in vitro* one since it reflects the real effects that the organism could suffer under a natural scenario. Moreover, the BRI considered the responses of a suite of eight different assays belonging to different levels of biological organization, reporting a more complete picture on the alteration of zebra mussel health status after exposure to low environmental concentrations of PPCPs commonly found in the aquatic environment.

5.4 Conclusions

The present work shows the suitability and the usefulness of the zebra mussel in the toxicity evaluation of emerging aquatic pollutants, such as PPCPs. Moreover, our results show that this model organism is sensitive to exposure to these contaminants at low environmental concentrations, highlighting a number of adverse effects, from biochemical to cellular level. The analysis of single endpoints and, above all, the integration of biomarker data into a BRI, showed that TCS is the most hazardous molecule towards the zebra mussel among the tested PPCPs. However, although the toxicity of the other molecules was lower than that of TCS, it cannot be underestimated. This is particularly true considering that the effects we measured were from short-term exposure (96 hours), while in real ecosystems zebra mussel specimens undergo a more hazardous situation because they are exposed to low PPCP concentrations for their whole life-span, likely resulting in a more severe impairment of their health status. In addition, the hazard of all the tested PPCPs towards the zebra mussel and other aquatic non-target organisms should be considered with caution because the continuous and often growing consumption of these drugs could lead to notable increases in their water concentrations and, consequently, to more dangerous adverse effects compared to those pointed out in the present work. Lastly, the toxicity of the single PPCPs towards our model organism could also be enhanced in aquatic ecosystems because of their possible interactions with other contaminants to form complex mixtures whose hazard assessment represents one of the main challenge in aquatic ecotoxicology. In conclusion, our research confirmed that (1) because of its response sensitivity to contaminant exposure the zebra mussel can be considered a suitable non-conventional biological model for aquatic ecotoxicology, (2) PPCPs at environmental concentrations may represent a threat to the health status of non-target aquatic organisms, and (3) the biomarker data integrated into a synthetic response index may represent a useful tool to rank the toxicity of different molecules and to lay the foundations for an appropriate environmental risk assessment for PPCPs.

References

1. T. Deblonde, C. Cossu-Leguille and P. Hartemann, Emerging pollutants in wastewater: a review of the literature, *Int. J. Hyg. Environ. Health*, 2011, **214**, 442–448.
2. L. H. M. L. M. Santos, A. N. Araujo, A. Fachini, A. Pena, C. Deleure-Matos and M. C. B. S. M. Montenegro, Ecotoxicological aspects related to the presence of pharmaceuticals in the aquatic environment, *J. Hazard. Mater.*, 2010, **175**, 45–95.
3. K. Fent, A. A. Weston and D. Caminada, Ecotoxicology of human pharmaceuticals, *Aquat. Toxicol.*, 2006, **76**, 122–159.
4. T. Haap, R. Triebskorn and H. R. Hohler, Acute effects of diclofenac and DMSO to *Daphnia magna*: immobilisation and hsp70-induction, *Chemosphere*, 2008, **73**, 353–359.
5. B. Quinn, F. Gagné and C. Blaise, An investigation into the acute and chronic toxicity of eleven pharmaceuticals (and their solvents) found in wastewater effluent on the cnidarians, *Hidra attenuate*, *Sci. Total Environ.*, 2008, **389**, 306–314.
6. M. Crane, C. Watts and T. Boucard, Chronic aquatic environmental risks from exposure to human pharmaceuticals, *Sci. Total Environ.*, 2006, **367**, 23–41.
7. S. Franzellitti, S. Buratti, P. Valbonesi and E. Fabbri, The mode of action (MOA) approach reveals interactive effects of environmental

pharmaceuticals on *Mytilus galloprovincialis*, *Aquat. Toxicol.*, 2013, **140–141**, 249–256.

8. M. Gonzalez-Rey and M. J. Bebianno, Effects of non-steroidal anti-inflammatory drug (NSAID) diclofenac exposure in mussel *Mytilus galloprovincialis*, *Aquat. Toxicol.*, 2014, **148**, 221–230.
9. N. Laville, S. Aït-Aïssa, E. Gomez, C. Casellas and J. M. Porcher, Effects of human pharmaceuticals on cytotoxicity, EROD activity and ROS production in fish hepatocytes, *Toxicology*, 2004, **196**, 41–55.
10. A. Binelli, C. Della Torre, S. Magni and M. Parolini, Does zebra mussel (*Dreissena polymorpha*) represent the freshwater counterpart of *Mytilus* in ecotoxicological studies? A critical review, *Environ. Pollut.*, 2015, **196**, 386–403.
11. B. S. Baldwin, M. S. Mayer, J. Dayton, N. Pau, J. Mendilla, M. Sullivan, A. Moore, A. Ma and E. L. Mills, Comparative growth and feeding in zebra and quagga mussels (Dreissena polymorpha and Dreissena bugensis): implications for North American lakes, *Can. J. Fish. Aquat. Sci.*, 2002, **59**, 680–694.
12. A. Binelli, C. Riva, D. Cogni and A. Provini, Assessment of the genotoxic potential of benzo(a)pyrene and pp-dichlorodiphenyldichloroethylene in Zebra mussel (*Dreissena polymorpha*), *Mutat. Res.*, 2008, **649**, 135–145.
13. A. Binelli, C. Riva, D. Cogni and A. Provini, Genotoxic effects of p,p'-DDT (1,1,1-trichloro-2,2-bis-(chlorophenyl)ethane) and its metabolites in Zebra mussel (*D. polymorpha*) by SCGE assay and micronucleus test, *Environ. Mol. Mutagen.*, 2008, **49**, 406–415.
14. C. Riva, A. Binelli, D. Cogni and A. Provini, Evaluation of DNA damages induced by BDE-209 in haemocytes of Dreissena polymorpha using SCGE assay and MN test, *Environ. Mol. Mutagen.*, 2007, **48**, 735–743.
15. M. Faria, L. Carrasco, S. Diez, M. C. Riva, J. M. Bayona and C. Barata, Multibiomarker responses in the freshwater mussel *Dreissena polymorpha* exposed to polychlorobiphenyls and metals, *Comp. Biochem. Physiol., Part C: Toxicol. Pharmacol.*, 2009, **149**, 281–288.
16. V. Contardo-Jara, C. Lorenz, S. Pflugmacher, G. Nützmann, W. Kloas and C. Wiegand, Exposure to human pharmaceuticals Carbamazepine, Ibuprofen, and Bezafibrate causes molecular effects in *Dreissena polymorpha*, *Aquat. Toxicol.*, 2011, **105**, 428–437.
17. Y. De Lafontaine, F. Gagné, F. Blaise, G. Costan, P. Gagnon and H. M. Chan, *Aquat. Toxicol.*, 2000, **50**, 51.
18. V. Contardo-Jara, C. Lorenz, S. Pflugmacher, G. Nützmann, W. Kloas and C. Wiegand, Biomarkers in zebra mussels (*Dreissena polymorpha*) for the assessment and monitoring of water quality of the St Lawrence River (Canada), *Aquat. Toxicol.*, 2000, **50**, 51–71.
19. A. Binelli, D. Cogni, M. Parolini, C. Riva and A. Provini, *In vivo* experiments for the evaluation of genotoxic and cytotoxic effects of Triclosan in Zebra mussel hemocytes, *Aquat. Toxicol.*, 2009, **91**, 238–244.
20. A. Binelli, M. Parolini, D. Cogni, A. Pedriali and A. Provini, A multi-biomarker assessment of the impact of the antibacterial trimethoprim

on the non-target organism Zebra mussel (*Dreissena polymorpha*), *Comp. Biochem. Physiol., Part C: Toxicol. Pharmacol.*, 2009, **150**, 329–336.
21. M. Parolini, A. Binelli, D. Cogni and A. Provini, Multi-biomarker approach for the evaluation of the cyto-genotoxicity of paracetamol on the zebra mussel (*Dreissena polymorpha*), *Chemosphere*, 2010, **79**, 489–498.
22. M. Parolini, A. Binelli and A. Provini, Chronic effects induced by ibuprofen on the freshwater bivalve *Dreissena polymorpha*, *Ecotoxicol. Environ. Saf.*, 2011, **74**, 1586–1594.
23. M. Parolini, A. Binelli and A. Provini, Assessment of the potential cyto-genotoxicity of the nonsteroidal anti-inflammatory drug (NSAID) diclofenac on the zebra mussel (*Dreissena polymorpha*), *Water, Air, Soil Pollut.*, 2011, **217**, 589–601.
24. E. Zuccato, S. Castiglioni, R. Bagnati, C. Chiabrando, P. Grassi and R. Fanelli, Illicit drugs, a novel group of environmental contaminants, *Water. Res.*, 2008, **42**, 961–968.
25. D. M. Lowe and R. K. Pipe, Contaminant induced lysosomal membrane damage in marine mussel digestive cells—an *in vitro* study, *Aquat. Toxicol.*, 1994, **30**, 357–365.
26. A. Orbea, M. Ortiz-Zarragoitia, M. Solé, C. Porte and M. P. Cajaraville, Antioxidant enzymes and peroxisome proliferation in relation to contaminant body burdens of PAHs and PCBs in bivalve molluscs, crabs and fish from the Urdaibai and Plentzia estuaries (Bay of Biscay), *Aquat. Toxicol.*, 2002, **58**, 75–98.
27. A. Buschini, P. Carboni, A. Martino, P. Poli and C. Rossi, Effects of temperature on baseline and genotoxicant-induced DNA damage in haemocytes of *Dreissena polymorpha*, *Mutat. Res.*, 2003, **537**, 81–92.
28. N. P. Singh, simple method for accurate estimation of apoptotic cells, *Exp. Cell Res.*, 2000, **256**, 328–337.
29. M. Pavlica, G. I. V. Klobucar, N. Mojas, R. Erben and D. Papeš, Detection of micronuclei in haemocytes of zebra mussel and ramshorn snail exposed to pentachlorophenol, *Mutat. Res.*, 2000, **465**, 145–150.
30. M. Kirsch-Volders, T. Sofuni, M. Aaderma, S. Albertini, D. Eastmond, M. Fenech, M. Ishidate, E. Lorge, H. Norppa, J. Suralles, W. Von der Hude and A. Wakata, Report from the *in vitro* micronucleus assay working group, *Environ. Mol. Mutagen.*, 2000, **35**, 167–172.
31. J. A. Hagger, M. B. Jones, D. Lowe, D. P. R. Leonard, R. Owen and T. S. Galloway, Application of biomarkers for improving risk assessments of chemicals under the Water Framework Directive: a case study, *Mar. Pollut. Bull.*, 2008, **56**, 1111–1118.
32. J. A. Hagger, D. Lowe, A. Dissanayake, M. B. Jones and T. S. Galloway, The influence of seasonality on biomarker responses in *Mytilus edulis*, *Ecotoxicology*, 2010, **19**, 953–962.
33. M. Parolini, A. Pedriali and A. Binelli, Application of a Biomarker Response Index for Ranking the Toxicity of Five Pharmaceutical and Personal Care Products (PPCPs) to the Bivalve *Dreissena polymorpha*, *Arch. Environ. Contam. Toxicol.*, 2013, **64**, 439–447.

34. F. Regoli, M. Nigro and E. Orlando, Lysosomal and antioxidant responses to metals in the Antarctic scallop *Adamussium colbecki*, *Aquat. Toxicol.*, 1998, **40**, 375–392.
35. G. W. Winston, M. N. Moore, M. A. Kirchin and C. Soverchia, Production of reactive oxygen speciesc (ROS) by hemocytes from the marine mussel, *Mytilus edulis*, *Comp. Biochem. Physiol., Part C: Pharmacol., Toxicol. Endocrinol.*, 1996, **11**, 221–229.
36. A. Viarengo, D. Lowe, C. Bolognesi, E. Fabbri and A. Koehler, The use of biomarkers in biomonitoring: a 2-tier approach assessing the level of pollutant induced stress syndrome in sentinel organisms, *Comp. Biochem. Physiol., Part C: Toxicol. Pharmacol.*, 2007, **146**, 281–300.
37. F. Regoli, D. Pellegrini, G. W. Winston, S. Gorbi, S. Giuliani, C. Virno–Lamberti and S. Bompadre, Application of biomarkers for assessing the biological impact of dredged materials in the Mediterranean: the relationship between antioxidant responses and susceptibility to oxidative stress in the red mullet (*Mullus barbatus*), *Mar. Pollut. Bull.*, 2002, **44**, 912–922.
38. E. Mamaca, R. K. Bechmann, S. Torgrimsen, E. Aas, A. Bjornstad, T. Baussant and S. Le Floch, The neutral red lysosomal retention assay and Comet assay on haemolymph cells from mussels (*Mytilus edulis*) and fish (*Symphodus melops*) exposed to styrene, *Aquat. Toxicol.*, 2005, **75**, 191–201.
39. J. A. Hagger, M. B. Jones, D. R. P. Leonard, R. Owen and T. S. Galloway, Biomarkers and integrated environmental risk assessment: are there more questions than answers?, *Integr. Environ. Assess. Manage.*, 2006, **2**, 312–329.
40. L. H. Yang, G. G. Ying, H. C. Su, J. L. Stauber, M. S. Adams and M. T. Binet, Growth inhibiting effects of 12 antibacterial agents and their mixtures on the freshwater microalga *Pseudokirchneriella subcapitata*, *Environ. Toxicol. Chem.*, 2008, **27**, 1201–1208.
41. J. W. Kim, H. Ishibashi, R. Yamauchi, N. Ichikawa, Y. Takao, M. Hirano, M. Koga and K. Arizono, Acute toxicity of pharmaceutical and personal care products on freshwater crustacean (*Thamnocephalus platyurus*) and fish (*Oryzias latipes*), *J. Toxicol. Sci.*, 2009, **34**, 227–232.
42. European Commission, European Environment Agency, Brussels 2012, p. 6.
43. M. Cleuvers, Mixture toxicity of the anti-inflammatory drugs diclofenac, ibuprofen, naproxen, and acetylsalicylic acid, *Ecotoxicol. Environ. Saf.*, 2004, **59**, 309–315.
44. M. Parolini, A. Binelli, D. Cogni, C. Riva and A. Provini, An *in vitro* biomarker approach for the evaluation of the ecotoxicity of nonsteroidal anti-inflammatory drugs (NSAIDs), *Toxicol. In vitro*, 2009, **23**, 935–942.

CHAPTER 6

Crayfish: An Experimental Model for Examining Exposure to Environmental Contamination

RACHELLE M. BELANGER,*[a] SARA E. LAHMAN[b] AND PAUL A. MOORE[c]

[a] Biology Department, University of Detroit Mercy, Detroit, MI 48221, USA; [b] Department of Biological and Physical Sciences, University of Mount Olive, Mount Olive, NC 28365, USA; [c] Laboratory for Sensory Ecology, Department of Biological Sciences, J. P. Scott Center for Neuroscience, Mind, and Behavior, Bowling Green State University, Bowling Green, OH 43403, USA

*Email: belangra@udmercy.edu

6.1 Introduction

6.1.1 Background

The introductory material presented here on the ecology, behavior and phylogeny of the crayfish is very cursory. This material is written to give the reader a very basic introduction into the potential of the crayfish as a toxicology model. There are excellent books that give a far more detailed overview of the many different aspects of the crayfish. The reader, if interested, should investigate these ref. 1–3.

Issues in Toxicology No. 33
Ecotoxicology and Genotoxicology: Non-traditional Aquatic Models
Edited by Marcelo L. Larramendy

Published by the Royal Society of Chemistry, www.rsc.org

6.1.2 Phylogeny and Distribution

Crayfish are Malacostracan Crustaceans and are placed within one of the largest taxon of crustaceans: the decapods. Decapod translates into ten feet because the crustaceans in this group have five pairs of walking legs. Currently, there are more than over 14 000 species within this order, which also includes shrimps, lobsters and crabs.[4] Crustaceans within this order demonstrate bilateral symmetry, including a pair of compound eyes that are usually located on long stalks. In addition to eyes, these organisms possess numerous appendages dedicated to either mechanoreception and/or chemoreception, including pairs of antennae, antennules (lateral and medial), pereopods and three pairs of mouthparts (or maxillipeds).[5]

Within the decapods, crayfish fall across three phylogenetic families, namely Astacidae, Cambaridae, and Parastacidae.[6] Currently, there are over 600 recognized and named species of crayfish in the world.[7] Crayfish within the families of Astacidae and Cambaridae are almost found exclusively in the northern hemisphere, whereas the Parastacidae are found across the southern hemisphere. Although absent from the African and Antarctic continents, crayfish are found across the globe in a large variety of different habitats. An excellent description of the distribution of different genera of crayfish can be found elsewhere.[8] One of the most significant problems with regard to habitat and global distribution is the introduction of non-native species of crayfish. A number of species of crayfish have either been purposefully introduced for aquaculture reasons or accidentally introduced as a result of recreational fishing as crayfish are often used as bait for large predator fish.[9]

6.1.3 Habitat and Range

Crayfish inhabit a large variety of freshwater habitats, including lakes, rivers, wetlands, caves and swamps. Crayfish in the genus *Astacus* are located across Europe and can be found in lakes, estuaries, and even brackish waters.[7] The narrow-clawed crayfish (*A. leptodactylus*) can tolerate higher salinity waters as well as waters that can be eutrophic with lower oxygen content. In contrast, the noble crayfish (*A. astacus*) and the thick clawed crayfish (*A. pachypus*) tend to prefer colder climate fresh water rivers and streams that are rich in oxygen and free from pollutants.

Within the western part of the United States, the genus *Pacifastacus* is the dominant genus of crayfish. This genus originated west of the rocky mountains to the Pacific coast of the United States.[10] More recently, this genus has been introduced to Japan and some areas of Europe.[7] The signal crayfish (*P. leniusculus*) is one of the dominant invasive species of crayfish and one of the reasons is the large range of habitats where the crayfish can survive. The signal crayfish's original habitats are typically small streams and rivers. The transport and introduction by humans has resulted in an explosion of different habitats in which this crayfish has been found. The signal crayfish has been found in estuaries with salinities as high as 20 ppt and in lakes as deep as 100 m.[11] This crayfish creates simple burrows in the banks of rivers and lakes, and can reach considerable densities because of its burrowing capacity.[12]

In the northern to middle part of the United States, the genus *Procambarus* is a dominant genus. Among this group of crayfish is the commercially important red swamp crayfish (*P. clarkii*). As the name indicates, this species of crayfish can be found in swamps and other wetlands.[13] This species of crayfish has a very low tolerance for salinity ranges (<10 ppt), but is tolerant of lower oxygen levels that are sometimes present within periodically flooded wetlands. As with the signal crayfish, *P. clarkii* has been introduced across the globe and can now be found within lakes and rivers as well as wetlands. Two other *Procambarus* species, both referred to as the white river crayfish (*P. acutus acutus* and *P. zonangulus*), prefer faster flowing rivers and streams and are not typically found in the same habitats as *P. clarkii.*[14]

Another eastern North American genus is the *Orconectes*. Crayfish within this genus can be found from the Gulf of Mexico up to Central Canada.[15] Crayfish within *Orconectes* are found in lakes and rivers and rarely inhabit estuaries or flooded wetland habitats. A number of species within this genus can construct elaborate burrows in lakes, river banks, and even some semi-terrestrial habitats. *Orconectes* species are tolerant of large changes in the temperature and oxygen content of water. The rusty crayfish (*O. rusticus*) occurs in lakes and rivers and can often be found associated with large rocks and debris within streams (Figure 6.1A,B). The virile crayfish (*O. virilis*), like the rusty crayfish, inhabits rivers and lakes but has a higher tolerance for muddy environments when compared to the rusty crayfish (Figure 6.1C,D). A third species (*O. limosus*) prefers more sediment-rich habitats often with slower flowing rivers that may have clay or silt substrates.[8]

North America has a rich diversity of crayfish species and a third genus can be located east of the Rocky Mountains to the Atlantic seashore. The genus *Cambarus* consists of a wide diverse group of crayfish that can be found in habitats that range from cold, rocky, and fast-flowing rivers to stagnant drainage ditches.[16] In addition, this genus contains species of crayfish that can build elaborate burrows in soft sediments. The devil crayfish (*C. diogenes*) are primarily burrowing crayfish that may have the widest distribution of habitats of any crayfish species. The devil crayfish can be found in marsh or swamp like habitats as well as the banks of rivers and lakes. The Appalachian brook crayfish (*C. bartonii bartonii*) is typically found in small rivers and streams, but has a range that extends from Georgia in the United States to the Hudson Bay in Canada.[17]

The final genus covered in this chapter is located in Australia. This genus, *Cherax*, is another diverse group of crayfish that contain commercially important species in Australia.[18] The marron crayfish (*C. tenuimanus*) is found in heavily forested regions that have permanently flowing streams and high rainfall. As with many other crayfish species, the marron crayfish has been

Figure 6.1 (A) *O. rusticus* has a dark brown body and can be distinguished by rusty spots on the lateral side of the carapace. (B) The chelae of *O. rusticus* are described as being a similar color to the rest of the body and may even vary from green to a red color with black bands on the tips. (C) *O. virilis* are an overall brown or reddish brown color with yellow bumps on the medial side of the body, near the head. (D) The chelae of *O. virilis* can be either brown or blue in color and they also have bands on the tips, but they are typically orange or orange-red in color. Body and chelae colors may vary based on the diet of the animal.

introduce to regions of Africa, Japan, and China for aquaculture reasons. The yabby crayfish (*C. destructor* and *C. albidus*) include crayfish that can get as large as the Maine lobster (*Homarus americanus*). These crayfish are found in habitats ranging from cold, fast-flowing alpine streams to slow, warm and almost stagnant swamps.[18]

Although as a group crayfish inhabit a wide range of freshwater habitats, the central commonality, for the purposes of this chapter, is that all of these habitats are located near riparian zones that are often inhabited by human communities. Thus, these habitats have the potential to be altered or impacted by human activities that include the introduction of a wide variety of anthropogenic chemicals (*e.g.* pharmaceuticals, agricultural chemicals and metals).

6.1.4 Life History Strategies

Although crayfish life history strategies are as varied as the different habitats described in the previous section, some generalized concepts can be developed.[19] Female crayfish carry eggs on their swimmerets on the underside of the abdomen. Here the female oxygenates the eggs as well as provides support in the form of continual hygienic care. Eggs will typically incubate for two to four months depending on the species and environmental, conditions such as temperature and oxygen. After hatching, the juveniles of some species of crayfish will immediately leave the mother to live a solitary existence. In a few burrowing species, the juveniles will remain with the mother for years. Sexual maturity is obtained after a few weeks up to a year depending on the species.[1]

Life span also varies greatly across the different species of crayfish with *P. clarkii* found at the lower end of the spectrum (~1.5 years) and *Astacopsis-gouldi* at the upper range (over five decades). For most of the species of crayfish, continual and allometric growth has been measured across the entire live span. The classical concepts of *r*-selected and *K*-selected reproductive strategies can be applied to crayfish, but produce mixed results. As mentioned above, crayfish can be quite short lived, mate once, have numerous offspring, fast egg development, and high juvenile mortality, which would place these species within an *r*-selected range of strategies.[2] Still, other species are quite long lived, mate multiple times throughout their life, have lower fecundity, slow egg development, and some parental care.[2] In addition, some species have a single adult morphology that is sexually active, while other species cycle through a reproductive and non-reproductive form throughout the adult lifetime. Those species that cycle often have morphological and physiological differences between the non-reproductive and reproductive forms. Beyond hormonal and sexually-related morphological changes, levels of aggression and chelae to carapace morphological differences are also seen.[2]

6.1.5 General Anatomy and Physiology

Being decapod crustaceans, crayfish have five pairs of pereopods (including the major chelae and walking legs), bilateral symmetry and an exoskeleton. The anterior pereopods are often modified into major and minor chelae. Many of the appendages (walking legs, antennae, and antennules) demonstrate a segmented morphology. In general, the body plan contains a cephalothorax (often called the carapace) and the abdomen. The end of the abdomen contains the telson. Many of the appendages and other body parts of the crayfish are covered with sensory hairs (chemo and/or mechano) located on body parts specialized into sensory appendages. The majority of these appendages (*e.g.* antennae, antennules, maxillipeds and chelae) are located anteriorly on the cephalothorax. In addition to these appendages, a pair of stalked eyes is found at the head of the cephalothorax.[5]

The circulatory system of crayfish is an open system with a heart and several arteries and sinus.[5] While the heart serves as the major pumping organ, a minor role is played by the accessory pump, the cor frontale. The heart and the pericardial cavity are located dorsally with a major source of innervation travelling laterally to the gills on either side of the heart. The heart and gills are located in the cephalothorax region of the body. Also within this section is the digestive system, which is divided into three sections: the foregut (oesophagus and stomach), midgut (canal, caecum, and hepatopancreas), and the hind gut. Of special interest for ecotoxicology is the hepatopancreas (or digestive gland), which functions to detoxify compounds within the crayfish body, as well as to produce digestive enzymes. Given the central function of this organ in the digestion of food and protection of the body from toxins, the physiological state of this organ can provide some insight into the effects of ecotoxins. Finally, crayfish have a well-developed excretory system located anteriorly of the heart with a bladder system. The bladder contains a nephropore located just ventrally of the eyes, which is used to release the bladder contents.[5]

6.1.6 Crayfish Ecology

Crayfish are polytrophic and omnivorous. Essentially, these two distinctions mean that crayfish will consume a wide range of material, including detritus, live plant material, macroinvertebrates and small fish. In most environments, crayfish can play an important role in the structure of the food web because of their polytrophic status.[20] Crayfish will also consume benthic macroinvertebrates and crayfish, in turn, are consumed by fish populations. In some aquatic habitats, crayfish are keystone species based on their central role in food web dynamics and, as such, can play a key role as bioindicators of habitat quality.[20] Crayfish can function as shredding organisms in streams and lakes. "Shredders" are aquatic organisms that, through foraging, consume and break up terrestrial leaf material, which allows this material to be consumed by other aquatic macroinvertebrates. In addition, adult crayfish can have significant impacts on plant and algal communities in aquatic habitats. The nature of crayfish diets and the exact role that they play in aquatic food webs is dependent upon the size and age of the crayfish, the distribution of resources available, and the presence and distribution of predators.[20,21] The presence of crayfish in habitats is often an indicator of increased diversity of macroinvertebrates, macrophytes, and even fish populations. Aquatic habitats have a wide range of ecosystem services, such as biofiltering of sediments and contaminants, and crayfish populations play a key role in influencing a rich diverse biological community that produces these services.

As a group, crayfish inhabit rivers, lakes, swamps, wetlands, and a wide diversity of substrates, including gravel, sand, mud, clays, and macrophytes. Within these habitats, the distribution of crayfish is often randomly spread across a heterogeneous landscape. Regardless of the location or habitat, the key point for this chapter is that all of these habitats have strong linkage to the terrestrial habitat through the riparian zone (which is the terrestrial habitat that serves as an interface between aquatic and terrestrial habitats).[22] The importance of this point is that many different human activities (agricultural, industrial, wastewater treatment plants and industrial uses) often impact the riparian zone and subsequent habitat adjacent to the riparian zone.[23] These areas are often where crayfish are located as they tend to inhabit the shallower areas in aquatic habitats.

6.1.7 Human Activities That Impact Crayfish

Although crayfish can be found in a wide diversity of habitats, any particular species often has a limited range of abiotic and biotic factors that determine suitable habitats. A number of different human activities have contributed to alteration to or degradation of crayfish habitats.[24] Chief among these activities, and central to this chapter, is declining water quality through pollution.[25] Outflow from industrial facilities may contain organic compounds, such as polycyclic aromatic hydrocarbons (PAHs) and polychlorinated biphenyls (PCBs), or even heavy metals, such as copper, zinc, lead or nickel.[26] Runoff from large-scale agricultural fields can contain high concentrations of nitrogen or phosphorus as well as herbicides or insecticides. Wastewater treatment plants often contain untreated levels of pharmaceuticals that are contained within effluents or even find their way into streams through groundwater.

6.1.8 Using Crayfish as Bioindicators

Model systems for bioindicators should have a number of characteristics that can serve a broad and diverse field like ecotoxicology. There characteristics can be grouped based on their ecology, morphology and physiology.[27] First, ecological attributes should include a well-developed taxonomy, a global distribution, high densities, long life span and low mobility. Second, morphological attributes include large body size for identification and physiology studies as well as low genetic variability. Finally, the physiology of any indicator species should be well-studied and include sensitivity to toxic substances and accumulation of toxins within specific organs.[27] Crayfish have all of these attributes.

Crayfish can serve as an excellent model species to increase the knowledge base for toxicology. Overall, crayfish fulfill criteria described for bioindicators of environmental contamination.[2,28,29] One of the most important aspects of crayfish that makes them an ideal bioindicator is the important role that they play as a keystone species within aquatic systems.[21,30] Many different crayfish species have often been used as model organisms in toxicological research, which provides a solid research background for toxicological work.[27,31,32] Attributes such as their size, excellent fecundity and relatively short life-span create opportunities for crayfish to be laboratory models for toxicity testing. There is a rich and diverse research literature on their basic physiological functions, which provides a baseline for sublethal testing.[5] Given the large-sized and externally carried eggs, measurement and quantification for assays on fertility, reproduction success and mortality are readily done. The successful aquaculture of different crayfish species also opens doorways to examine how toxins affect the developmental processes.

Crayfish have often been considered bioindicator species of heavy metal pollution present in the aquatic environment as metals tend to accumulate in their tissues.[31] Traditionally, the toxicity of metals and agricultural chemicals has been investigated by examining the 24, 48, 72 or 96 hour (h) lethal concentration (LC) at which 50% of the population dies (*e.g.* 96 h LC_{50}). However, concentrations that cause high mortality rates may be well above environmentally relevant concentrations and thus may be of little biological significance.[32] Crayfish display other quantifiable responses to sublethal concentrations of contaminants, which include, but are not limited to, changes in behavior, chemoreception, mate success, morphology, tissue structure and toxicant accumulation, anti-oxidant activation, enzyme function, and cardiovascular system regulation. This chapter focuses on a variety of responses by crayfish to sublethal concentrations of contaminants. In addition to lethality, other physiological changes have been recorded when crayfish were exposed to pharmaceuticals, agricultural chemicals and metals, either under laboratory conditions or when collected from contaminated field sites. These responses make crayfish an invaluable bioindicator of contamination. Further, given their importance in the aquatic food web, they may be used as a sentinel species when examining contamination from pharmaceuticals, agricultural chemicals and metals as the uptake and toxic responses seen in crayfish are similar to those in other aquatic species.

6.2 Pharmaceuticals

As a class of compounds, pharmaceuticals are a relatively new group of exotoxins that has unfortunately grown over the past two decades.[33,34] Pharmaceuticals can enter aquatic habitats through their use in farming, disposal of unused medicines in landfills, and wastewater treatment plants.[33,35] The types of pharmaceuticals found in wastewater effluent are varied. Anti-inflammatory drugs, β-blockers, antibiotics, anxiolytics and endocrine compounds have all been detected in wastewater treatment plant effluent.[35,36] Some of these compounds are removed by wastewater treatment plants; however, the fact that so many remain at detectable levels after treatment is concerning. Pharmaceutical compounds that remain in treated effluent may affect aquatic organisms, especially in immediate areas of wastewater effluent discharge. Although a growing area of research, many of the toxicological studies on pharmaceutically active compounds have been performed on pelagic organisms, such as cnidarians and daphnia.[37] This work has shown that dosages as small as 1 $ng\,L^{-1}$ can have impacts on aquatic invertebrates.[37–39] Benthic macroinvertebrates, like crayfish, will likely be exposed to different concentrations of compounds than pelagic organisms.

Few studies have examined the effects of pharmacological agents on crayfish, but those that do are alarming. For example, Kulkarni *et al.*[40] showed that the presence of serotonin alters the reproductive cycle of crayfish. When female *P. clarkii* were treated with 15 $\mu g\,g^{-1}$ serotonin (5-HT), they demonstrated enhanced ovarian development owing to increased the release of ovary-stimulating hormone. This resulted in larger oocytes with enhanced amounts of vitelline, while 5-HT blockers have been shown to inhibit development. Exposure to serotonin reuptake inhibitors, such as fluoxetine, the active ingredient in Prozac, was shown to enhance crayfish (*O. rusticus*) growth in both form I and form II males (500 $\mu g L^{-1}$ for 10 days).[41] Additionally, Tierney *et al.*[41] showed that exposure to 2 and 500 $\mu g L^{-1}$ fluoxetine limits locomotory behavior, indicating that crayfish growth and locomotion can be manipulated by short-term exposure to fluoxetine. In addition, Goetz *et al.*[42] showed that long-term exposure to fluoxetine (100 $ng L^{-1}$) had significant effects on excitatory junction potential amplitudes of crayfish superficial extensor muscles, leading to long-term depression of potential amplitudes. BETAMAX VET® is a cypermethrin-based pharmaceutical used in the aquaculture industry to treat salmon louse (*Lepeophtheirus salmonis*) infestations in salmon farms. This compound is lethal to crayfish and has been used as a control agent for the spread of the invasive crayfish *P. leniusculus.*[43] Chloramine-T is a disinfectant used in medical and veterinary facilities that can enter aquatic habitats through wastewater treatment facilities. It was shown that 50 $mg L^{-1}$ was toxic to the narrow clawed crayfish (*A. leptodactylus*) and led to substantial loss of energy in these crayfish.[44] Overall, Kuklina *et al.*[44] found that crayfish tolerated short-term chloramine-T exposure, but had increased sensitivity to higher concentrations, making them an ideal model organism for examining environmental exposure to chloramine-T and other pharmaceuticals.

6.3 Agricultural Chemicals

In the Unites States, approximately 1 billion pounds of agricultural chemicals are used each year to control weeds, insects and other pests; however, the use of these chemicals may have adverse effects on non-target aquatic organisms.[45] Agricultural chemicals can enter the aquatic ecosystem typically through non-point sources. Non-point sources (*e.g.* runoff, seepage and deposition from the atmosphere) are the dominant sources of pesticides found in streams.[45] Typically aquatic organisms, like crayfish, are exposed to agricultural chemicals in the water, but they may also ingest them *via* their food source.[32] A key parameter for assessing the potential of a lipophilic chemical in the environment is the use of the octanol–water partition coefficient (K_{ow}). If a chemical has a high K_{ow}, it will partition appreciably into biota and organic matter, and this coefficient has been used successfully when estimating bioconcentration factors.[46] Substances with a high K_{ow}, (>4) are persistent in the environment and have the potential to accumulate and have long-term effects.[46] Most pesticides (herbicides, insecticides and fungicides) have a high K_{ow} and thus a strong hydrophobic character. Given this, these chemicals are mainly associated with particulate matter and sediments and may bioaccumulate in fatty tissues of organisms.[47]

6.3.1 Herbicides

Herbicides are used mainly in agricultural areas, but are also used in urban areas to increase product yield and to control unwanted urban vegetation. Some of the most commonly used pesticides are *s*-triazine herbicides (*e.g.* atrazine, hexazinone, metribuzine, prometryne, simazine and terbutryne). Because of their chemical properties, triazines are highly toxic and are commonly found in groundwater and surface water following heavy rainfall.[48,49] Triazines have been found to have toxic effects on various crayfish species and research supports the fact that crayfish may be used as an appropriate bioindicator when examining the toxicity of triazines. Velisek *et al.*[50] used the signal crayfish (*P. leniusculus*) to evaluate the 96 h LC_{50} toxicity values for atrazine, hexazinone, metribuzine, prometryne, simazine, and terbutryne. They reported that the LC_{50}s were 12.1 mg L^{-1} for atrazine, 13.9 mg L^{-1} for terbutryne, 14.4 mg L^{-1} for prometryne, 19.5 mg L^{-1} for hexazinone, 30.6 mg L^{-1} for metribuzine, and 77.9 mg L^{-1} for simazine. They also found that atrazine was the most toxic triazine examined.[50] When crayfish were treated with atrazine, they had difficulty navigating around a tank and often remained in the corners.[50] Belanger *et al.*[51,52] also showed that exposing crayfish (*O. rusticus* and *O. virilis*) to 80 μg L^{-1} atrazine affected their chemosensory abilities. After an acute exposure, crayfish could not localize a food odor source and these chemosensory deficits were long-term. Mac Loughlin *et al.*[53] showed that when juvenile red claw crayfish (*Cherax quadricarinatus*) were exposed to high concentrations of atrazine (2.5 mg L^{-1}) they exhibited decreased weight gain and decreased abdominal muscle protein. Further, they reported that the proportion of females increased progressively as the atrazine concentration increased, suggesting that atrazine is an endocrine disruptor in crayfish. Exposure to other triazines, such as prometryne (144, 1444 and 4320 μg L^{-1}), caused decreases to weight and growth rates, changes in the gills of marbled crayfish (*Procambarus fallax*), and a delay in development at the highest concentration.[54] Exposure of crayfish to prometryne, terbuthylazine, terbuthylazine-desethyl and metribuzin has been shown to cause histological damage, as well as changes in oxidative stress and antioxidant biomarkers (see Section 6.5).[54–57] Because triazines are among the most commonly used herbicides worldwide, it is important to study the effects of these substances on non-target organisms, like crayfish.

Other herbicides, such as chloroacetamide herbicides (*e.g.* metolachlor), phenoxy herbicides (*e.g.* 2,4-dichlorophenoxyacetic acid (2,4-D)), polychlorinated dioxins (*e.g.* 2,3,7,8-tetrachloro-dibenzo-*p*-dioxin (TCDD)), carbamates (*e.g.* Bolero) and glyphosate, have also been shown to cause changes in crayfish behavior and physiology after exposure to sublethal concentrations. Metolachlor is a herbicide that is heavily applied in agricultural areas in the United States.[58] Exposure of crayfish (*O. rusticus*) to 80 μg L^{-1} interfered with the ability of crayfish to receive or respond to chemical signals from conspecifics and thus affected certain agonistic behaviors.[59]

Wolf and Moore[60] also showed that when *O. rusticus* were exposed to metolachlor (25, 50 and 70 μg L^{-1}), they did not respond to alarm cues or food odors. A reduction in chemosensory abilities can be detrimental for animals that rely heavily on chemosensory information for survival. 2,4-D is another widely used agricultural herbicide that has been shown to affect crayfish. Benli *et al.*[61] showed that the 96 h LC_{50} value was calculated to be 32.6 mg L^{-1} in crayfish (*A. leptodactylus*) and also reported changes in behavior. Crayfish exposed to 30, 40 and 50 mg L^{-1} frequently stood in corners, had difficulty moving, began rocking, and walked in circles.[61] Browne and Moore[62] also noted changes in behavior of *O. rusticus* in a Y-maze following exposure to 7.65, 14.07 and 32.69 mg L^{-1} 2,4-D. They showed that 2,4-D exposure inhibits the ability of crayfish to locate food by causing crayfish to walk at more rapid speeds, spend less time in the correct arm of the maze, and take significantly longer to locate food. Consequently, they consumed less food.[62] The polychlorinated dioxin 2,3,7,8-tetrachloro-dibenzo-*p*-dioxin (TCDD) was toxic to crayfish (*P. leniusculus*), with an LD_{50} of 30–100 μg kg^{-1}; however, the toxicity of TCDD was delayed, with death occurring between 14 and 40 days after a single dose.[63] Changes in cytochrome P450 induction were also noted.[63] The 96 h LC_{50} for the carbamate Bolero (benthiocarb or thiobencarb) for crayfish *Orconectes nais* and *P. clarkii* was determined to be 2 mg L^{-1} and 6.5 mg L^{-1}, respectively. Bioconcentration occurred in the hepatopancreas and muscle tissue following exposure and elimination of Bolero from crayfish was slow.[64] Glyphosate acid, an active ingredient in Roundup®, is another heavily used aquatic and terrestrial herbicide that has been shown to affect crayfish following exposure. Avigliano *et al.*[65] found that when juvenile crayfish (*C. quadricarinatus*) were exposed for 60 days to 10 and 40 mg L^{-1} of glyphosate, that mortality was 33% at the highest concentration and that lipid and protein levels, as well as growth, were all reduced in the 40 mg L^{-1} treatment. Frontera *et al.*[66] also showed that exposing *C. quadricarinatus* for 50 days to glyphosate acid, either alone or in combination with polyoxyethylenamine, caused lower somatic growth and decreased muscle protein levels. Collectively, crayfish are highly sensitive to various herbicides and display a variety of behavioral and physiological responses to sublethal concentrations, and thus serve as an invaluable tool when monitoring environmental contamination.

6.3.2 Insecticides

Chlorinated pesticides, such as dichlorodiphenyl-tri-chloroethane (DDT), are persistent in the environment (high K_{ow}) and have been used to control insects. Additionally, dichlorodiphenyldichloroethylene (DDE), a breakdown product of DDT, is also found in the environment owing to its persistence.[47] Schilderman *et al.*[67] examined crayfish (*O. limosus*) from the river Meuse (The Netherlands) and showed that crayfish from the heavily polluted site in the river had an accumulation of DDT and DDE in their hepatopancreas and an elevated level of DNA adducts. They suggest that crayfish are exposed to persistent pollutants *via* the food chain or the sediment. Santerre *et al.*[68] also collected and analyzed crayfish (*P. clarkii*) tissue for the presence of organochlorines, organophosphates and pyrethroids. They found that 7% of the crayfish sampled from Louisiana and Texas (United States of America) contained detectable residues of DDT (average concentration detected 0.047 $mg\,L^{-1}$). Crayfish (*P. clarkii*) exposed to 1 $mg\,L^{-1}$ ^{14}C-ethion (an organophosphate insecticide) metabolized ethion into ethion mono-oxon, ethion dioxon, *O,O*-diethyl phosphorothioate, *O*-ethyl phosphorothioate and one unknown compound. These compounds accumulated in the gills and hepatopancreas.[69] Additionally, exposure to ethion concentrations of 0.36 $mg\,L^{-1}$ (1/4 96 h LC_{50}) caused extensive ultrastructural alterations to both hepatopancreas and gill epithelial cells.[69] Another organophosphate insecticide, fenitrothion, was shown to be toxic to the narrow-clawed crayfish (*A. leptodactylus*) at a low concentration (15.75 $\mu g\,L^{-1}$ 96 h LC_{50}).[70] Sarikaya *et al.*[70] showed that after a 24 h exposure, crayfish had decreased malondialdehyde levels at 5, 10 and 20 $\mu g\,L^{-1}$ due to oxidative stress, as well as decreased mobility, walking in circles and loss of equilibrium. Overall, they found that fenitrothion was highly toxic to crayfish and that lipid peroxidation markers, like malondialdehyde, levels can be used as biochemical biomarkers for environmental monitoring. Diazinon is another organophosphorus insecticide that is widely used on agricultural crops. Buřič *et al.*[71] tested the toxicity of diazinon on three size classes of crayfish (*O. limosus* and *P. leniusculus*). They found that young-of-the-year crayfish were found to be the most sensitive to diazinon (96 h $LC_{50} = 0.15$ $mg\,L^{-1}$), followed by juvenile crayfish (96 h $LC_{50} = 0.27$ $mg\,L^{-1}$) and adults (96 h $LC_{50} = 0.51$ $mg\,L^{-1}$). There was also a delayed effect of diazinon on adults was also detected (144 h $LC_{50} = 0.44$ $mg\,L^{-1}$). This suggests that sublethal concentrations may cause functional damage in the long-term.

Pyrethroids are the most commonly used insecticides to control rice water weevils (*Lissorhoptrus oryzophilus*) during rice farming; however, these pyrethroids (*e.g.* lambda-cyhalothrin and etofenprox) are very toxic to juvenile crayfish (*P. clarkii*), which are also farmed on the same land. Barbee and Stout[72] found that the 96 h LC_{50} for lambda-cyhalothrin was 0.16 $\mu g\,L^{-1}$ and for etofenprox was 0.29 $\mu g\,L^{-1}$. Crayfish are often exposed to these chemicals during crop rotation when *P. clarkii* are farmed on rice fields in Louisiana and pyrethroids are persistent in the soil after several applications to rice fields. Another pyrethroid, permethrin, caused 50–80% mortality in crayfish (*Procambarus* spp.) exposed in ponds to 1–3 $\mu g\,L^{-1}$.[73] Barbee and Stout[72] suggested that neonicotinoid insecticides should be used as an alternative to pyrethroids as they are two to three times less toxic (*e.g.* 96 h LC_{50} values: clothianidin, 59 $\mu g\,L^{-1}$; thiamethoxam, 967 $\mu g\,L^{-1}$; and dinotefuran, 2032 $\mu g\,L^{-1}$). Barbee *et al.*[74] also examined another potential replacement for pyrethroids for treating infestations of rice crops with rice water weevils, the insecticide chlorantraniliprole (anthranilic diamide insecticide). They showed that the 96 h LC_{50} was 951 $\mu g\,L^{-1}$ for *P. clarkii*, three times less toxic than the pyrethroids currently used, and suggested that they are more compatible with rice–crayfish crop rotations.[74] Overall, many insecticides used in agriculture are toxic to crayfish, and may bioaccumulate and cause physiological changes. Crayfish can be used as indicators for biomonitoring studies when insecticides are used so that alternatives may be investigated if necessary.

6.4 Metals

Heavy metals such as copper, cadmium, zinc, lead and mercury are among the most common anthropogenic chemical wastes that pollute aquatic systems. While trace levels of these metals are present in freshwater systems under normal conditions, discharges from agricultural, industrial and municipal sources can elevate heavy metal concentrations to levels that are lethal to crayfish. While crayfish are generally considered to be highly resistant to death caused by heavy metal contamination, sublethal concentrations have also been shown to have significant effects on crayfish through bioaccumulation in tissues and detrimental impacts on physiology and ecologically important behaviors. Generally, there are no significant differences between males and females in regards to affinity for heavy metal accumulation; however, differing accumulation trends in the tissues with the highest accumulation have been demonstrated between sexes in crayfish (*A. leptodactylus*) with males having more accumulation of metals in their tissues.[75,76]

6.4.1 Lethality

To investigate acute copper toxicity in the rusty crayfish (*O. rusticus*), Hubschman[77] used continuous flow exposure experiments for 24, 48, 72, and 96 h at concentrations up to 12 mg L^{-1}. Results from Hubschman's work showed that a copper concentration of 3 mg L^{-1} was sufficient enough to kill 50% of adult intermolt crayfish following a 96 h exposure. Additionally, Hubschman[77] reported that an exposure of 1 mg L^{-1} caused a 50% mortality rate in newly hatched crayfish within an exposure time that was 1/50th of that needed to cause the same rate in adults. Taylor *et al.*[78] used comparative toxicological responses of *C. robustus* from both a metal-contaminated site and unpolluted site to copper concentrations ranging from 0.61 to 24.06 mg L^{-1} in 24 h time intervals. This research revealed that crayfish that were reared in a polluted environment had higher tolerance to heavy metal exposures, reporting an LC_{50} value for crayfish from polluted sites of 4.07 mg L^{-1} and 3.48 mg L^{-1} for those from unpolluted sites for a 24 h exposure.[78]

Khan and Nugegoda[79] established heavy metal LC_{50} values for juvenile *C. destructor* following 96 h static-renewal exposures. The LC_{50} were reported as follows: nickel = 327 mg L^{-1}; iron = 50 mg L^{-1}; copper = 494 μg L^{-1}; and cadmium = 379 μg L^{-1}. Aguirre-Sierra *et al.*[80] established lethal concentrations of fluoride for the white-clawed crayfish *Autropotamobius pallipes* to be 93.0, 55.3, and 36.5 mg L^{-1} for 48, 72, and 96 h exposures, respectively. For the majority of crayfish species, heavy metal concentrations that occur in the environment are not high enough to directly cause death, which has led to studies focusing on bioaccumulation and behavioral response following sublethal exposures.[31]

6.4.2 Bioaccumulation

Like other crustaceans, crayfish are able to depurate and detoxify themselves of many heavy metals. The main organ that functions in metal detoxification is the hepatopancreas, which has the ability to concentrate metals from the hemolymph and digestive tract, storing them in intracellular vacuoles.[81] Consequently, the highest levels of bioaccumulation are seen in this organ.[82] However, once concentrations within the environment (either in sediment or water column) are elevated, crayfish may no longer be able to detoxify metals, leading to death. Alcorlo *et al.*[83] found that bioaccumulation of heavy metals in the tissues of *P. clarkii* was the result of the organism's interaction with contaminated substrates. A period of exposure of 6 to 12 days to a heavy metal-contaminated environment was sufficient enough to lead to significant bioaccumulation of metal in crayfish tissues.[83] For most heavy metals (copper, zinc, cadmium, lead, nickel, and mercury), bioaccumulation in various tissues, such as the hepatopancreas, abdominal muscles, green gland and digestive tract, exoskeleton, and/or gills, is time- and dose-dependent and may be indicative of metal concentrations in the surrounding environment.

Compared to other heavy metals, relatively high amounts of copper can be found in crayfish tissues given that copper is a crucial component of the respiratory metalloprotein (hemocyanin).[84] When copper bioavailability exceeds a high threshold, species-, concentration-, and exposure-period-specific accumulation has been documented in the hepatopancreas as well as other tissues, such as abdominal muscle, exoskeleton, and gills [Finerty *et al.*[85] (*P. clarkii* and *P. astacusacutus*); Madden *et al.*[86] (*P. clarkii*); Gherardi *et al.*[87] (*P. clarkii* and *Astacuspallipes*); Bruno *et al.*[82] (*C. destructor*); Hothem *et al.*[88] (*P. clarkii* and *P. leniusculus*)]. Maranhoão *et al.*[89] reported that following a 96 h exposure to copper concentrations of 0.125–0.500 mg L^{-1}, there was no significant accumulation of copper in tissues of *P. clarkii*. The same study reported that following an 8 week copper exposure of 5 mg L^{-1}, time-dependent copper accumulation was documented in the gills, exoskeleton, and abdominal muscles of *P. clarkii.* The levels of copper in *P. clarkii* gill, exoskeleton, and abdominal tissue were subsequently reduced when crayfish were placed in clean water following exposure.[90] *A. leptodactylus* demonstrated a similar pattern of accumulation and depuration following copper exposure.[91]

Like copper, zinc is an essential metal for crayfish metabolism and thus the content of zinc in crayfish bodies is naturally high regardless of environmental conditions. However, this metal has been shown to accumulate in the hepatopancreas and abdominal tissues of many species [Madden *et al.*[86] (*P. clarkii*); Gherardi *et al.*[87] (*P. clarkii* and *A. pallipes*); Bruno *et al.*[82] (*C. destructor*); Hothem *et al.*[88] (*P. clarkii* and *P. leniusculus*)]. Additionally, Bagatto and Alikhan[92] and Mackevičienė[93] found that zinc accumulated in the digestive tract tissues of *C. bartonii* and *A. astacus*, respectively. Zinc content in the hepatopancreas, gills, and abdominal muscle of *C. tenuimanus* was found to be highest in juveniles, owing largely in part to the higher permeability of the body surface, rendering them unable to regulate zinc as efficiently as older individuals.[94] Similar results were documented in *C. destructor* juveniles.[82]

Conversely to copper and zinc, cadmium and lead are not involved in crayfish metabolism and consequently tend to increase rapidly along with increasing environmental concentrations. Accumulation of cadmium has been shown to be positively correlated with the level of environmental cadmium and proximity to the source of cadmium pollution.[95,96] Crayfish are able to take up cadmium from both the surrounding environment and from food.[97,98] Chambers[99] and Bruno *et al.*[82] showed that cadmium accumulated largely in the hepatopancreas, followed by the gills, exoskeleton, and abdominal muscles in *C. tenuimanus* and *C. destructor*. Additional studies have reported that cadmium largely accumulates in the hepatopancreas and abdominal muscle tissue [Finerty *et al.*[85] (*P. clarkii* and *P. astacusacutus*); Madden *et al.*[86] (*P. clarkii*); Gherardi *et al.*[87] (*P. clarkii* and *A. pallipes*); Bruno *et al.*[82] (*C. destructor*); Hothem *et al.*[88] (*P. clarkii* and *P. leniusculus*)]. *C. destructor* adults were found to have higher levels of cadmium accumulation in the exoskeleton and muscle tissues as compared to juveniles, seemingly related to molting frequency.[82]

The accumulation of lead in crayfish tissue has been studied in both aquaculture and natural settings [Madden *et al.*[86] (*P. clarkii*); Gherardi *et al.*[87] (*P. clarkii* and *A. pallipes*); Bruno *et al.*[82] (*C. destructor*); Hothem *et al.*[88] (*P. clarkii* and *P. leniusculus*)]. The hepatopancreas has been documented as the main organ for lead accumulation, storing metal in vacuoles.[82,93] Accumulation has also been observed in the digestive tract, abdominal muscle, exoskeleton, and antennal (green) gland.[93,100] Following a 10 week exposure to 0.02 mg L^{-1} of lead, accumulation in *A. astacus* was primarily in the hepatopancreas, carapace, and gills.[101] Significant accumulation in the hepatopancreatic and gill tissues of *P. clarkii* was documented following 7 day exposure in a contaminated habitat.[95] Lead concentrations in tissues of *P. clarkii* were significantly decreased after a 3 week recovery period from a 7 week exposure to 0.15 mg L^{-1}.

Methylmercury represents approximately 90% of the total mercury accumulated in crayfish tissues.[88,102] In crayfish, mercury is largely accumulated in the abdominal muscle tissue [Finerty *et al.*[85] (*P. clarkii* and *P. astacusacutus*); Madden *et al.*[86] (*P. clarkii*); Simon *et al.*[103] (*A. astacus*); Loukola-Ruskeeniemi *et al.*[75] (*A. astacus*); Hothem *et al.*[88] (*P. clarkii* and *P. leniusculus*)]. Studies have demonstrated that mercury and methyl mercury also accumulated in the hepatopancreas, gills, and exoskeleton in *Orconectes propinquus* that were fed mercury and methyl mercury dosed pellets.[104] Crayfish have a marked tendency to accumulate methylmercury taken up from food and water, from which they also take up mercury.[105]

Like other heavy metals, nickel has been shown to accumulate in crayfish tissues based on the concentration in the surrounding environment. In both *A. astacus* and *C. bartonii*, nickel accumulation was documented to be the highest in the exoskeleton, suggesting that this type of tissue might be involved in the excretion of this metal.[92,93] Similarly, chromium was found to be at highest accumulation in the exoskeleton of *A. astacus* as well as at high levels in the hepatopancreas and abdominal muscle of *A. astacus* and *P. leniusculus*.[93,106] *P. clarkii* primarily accumulated chromium in the gills and hemolymph following a 7 day exposure in a contaminated environment.[95] Grosell *et al.*[107] reported that adult *C. diogenes diogenes* accumulated substantial amounts of silver in gill, hemolymph, and hepatopancreatic tissues following a 96 h silver exposure at a concentration of 8.41 $\mu g\,L^{-1}$. Fluoride was shown to accumulate primarily in the exoskeleton for *A. pallipes*.[80]

Research has shown that the amount of bioaccumulation that occurs in crayfish tissues is positively correlated to the concentration of heavy metal pollution in the animal's habitat.[83,91] Goretti *et al.*[108] suggested the use of *P. clarkii* as a potential bioindicator for heavy metal pollution in freshwater systems through establishing a toxic contamination index utilizing a ratio of bioaccumulation for cadmium, copper, lead, and zinc within the hepatopancreas and abdominal muscle. As the hepatopancreas has high detoxifying activity compared to the low activity of muscle, concentration values of heavy metals within those tissues can demonstrate the level of toxicity due to heavy metal pollution. This index was proposed to be used as an easy and useful tool that can assess the toxicity level of heavy metal-contaminated sites.[108] Thus, crayfish can be used to identify and monitor areas within aquatic systems that are under pollution stress.[83,88,109]

6.4.3 Physiological and Behavioral Impacts

In addition to accumulation in tissues, sublethal exposure to heavy metals can impact physiological processes of crayfish (for a complete review see Section 6.5). For example, cardiac arrhythmia followed by substantial levels of death was reported in *A. astacus* following exposure to mercury chloride at concentrations of 0.1–0.8 $mg\,L^{-1}$.[110,111] Inhibitory effects on the ovarian maturation in *P. clarkii* exposed to mercury were reported by Reddy *et al.*[112] Adult *C. diogenes diogenes* exhibited ionoregulatory disturbance and elevated production of metabolic ammonia following a 96 h exposure to 8.41 $\mu g\,L^{-1}$ of silver.[107] Rowe *et al.*[113] found that the standard metabolic rate of the crayfish *P. acutus* had approximately 30% higher standard metabolic rates following a chronic exposure to sediment contaminated with heavy metals that included arsenic, cadmium, chromium, copper, and selenium. This elevated metabolic rate supports that detoxifying and combating the deleterious effects of heavy metal contaminants to energetic costly to crayfish and could subsequently lead to reduced fitness.

Sublethal heavy metal exposures can also have severe consequences on ecologically important behaviors. Burba[114] investigated the effects of a 96 h exposure to a copper below the LC_{50} concentration on the exploratory and social behaviors of the noble crayfish (*A. astacus*). Crayfish exposed to copper showed decreased movement around a test arena and lack of thigmo- and chemo-tactic behaviors. Additionally, agonistic behavior was more frequent between individuals exposed to copper than those that were unexposed.[114] Escape responses (tail flips) in the white-clawed crayfish (*A. pallipes*) decreased significantly in animals exposed to 5.9–18.4 mg L^{-1} of fluoride.[80] Crayfish exposed to fluoride at 10.7 mg L^{-1} for 192 h also showed decreased food consumption.[80]

Sherba *et al.*[115] demonstrated that the freshwater crayfish *C. bartonii* exhibited increased foraging latency and was not able to localize food sources following a 120 h exposure to 0.02 or 0.2 mg L^{-1} of copper. Similarly, significant differences were found in the overall orientation ability of the rusty crayfish (*O. rusticus*) to locate an odor source when previously exposed to copper (4.5, 45 and 450 μg L^{-1}) for 120 h in combination with a source of copper pollution in the background of orientation trials. *O. rusticus* exposed to copper in any capacity showed altered orientation parameters as compared to unexposed crayfish, showing that copper impaired the crayfish's ability to detect, process, and/or respond appropriately to chemosensory information.[116] In another study, the antennular flicking rates of rusty crayfish exposed to 450 μg L^{-1} of copper for 120 h were significantly lower than unexposed animals, and were consequently less successful in locating a food odor source.[117] Once placed in clean water, crayfish demonstrated significant increases in both antennular flicking rates and successful localization of odors, indicating that *O. rusticus* can recover olfactory behaviors following clean-up of a copper polluted environment.

In addition to foraging and orientation behaviors, the overall population densities of crayfish have been shown to be impacted by elevated levels of heavy metal discharge. Allert *et al.*[118] reported that mean population densities of the crayfish *Orconectes hylas* were negatively correlated with sediment pore-water metal concentrations (lead, zinc, cadmium, nickel, and cobalt), with lower densities found at sites close to and downstream from lead–zinc mines. Similar results have been reported for *O. neglectus neglectus, O. luteus* and *O. virilis*.[119,120] During a 28 day *in situ* experiment, the survival and biomass of *O. hylas* were significantly lower in populations of crayfish located at sites in close proximity to historical lead–zinc mining activities.[121] Similar results following a 56 day *in situ* experiment were reported for *O. luteus*.[120]

6.5 Physiological Responses to Contaminants

When exposed to sublethal concentrations of contaminants in aquatic ecosystems, crayfish do not always display changes in behavior, reproductive effort, external morphology and/or death. They may have physiological changes that may be examined using molecular, histological and analytical methods. Typically, when organisms are exposed to contaminants, they experience reductions in physiological processes; however, in some cases they are increased.[32] Evaluating the 96 h LC_{50} or population size changes are techniques that are heavily used in toxicology research to test the lethality of a toxicant; however, concentrations that cause high mortality rates may be well above environmentally relevant concentrations and thus may be of little biological significance.[32] Sublethal concentrations of contaminants are more likely to be encountered in aquatic environments and lead to changes in the physiology of crayfish, rather than death. Moreover, crayfish can be used as a sentinel species as they display changes in tissue structure, cytochrome P450 levels and antioxidant activity, acetylcholinesterase activity, oxidative stressors, DNA structure, and cardiovascular, respiratory and metabolic function, as well as other physiological changes (*e.g.* enzymatic function) in response to changes in abiotic and biotic factors.

Crayfish absorb xenobiotics into their body *via* the gills and subsequently accumulate them in the hepatopancreas. This may lead to structural changes in the hepatopancreas (Figure 6.2). Desouky *et al.*[69] demonstrated that when crayfish (*P. clarkii*) were treated with the insecticide ethion (1 $mg\,L^{-1}$), it accumulated in both the gills and hepatopancreatic tissue. When the crayfish were then placed in clean water for 7 days, the concentrations of insecticide residues were decreased in both the hepatopancreas and gills. When *P. clarkii* where exposed to 1/4 the 96 h LC_{50} (0.36 $mg\,L^{-1}$), it caused ultrastuctural changes to both the gill and hepatopancreatic epithelial cells, which included vacuolation, degradation and distinct cell lysis in the hepatopancreas. Infiltration of hemocytes, cytoplasmic vacuolation and changes in the plasma membrane structure were visualized in gill tissue.[69] Exposure to environmentally relevant concentrations of the herbicide metribuzin for 30 days also caused histological changes in crayfish (*P. leniusculus*), which included disintegration of the tubular epithelium in the hepatopancreas.[56] Chupani *et al.*[122] also demonstrated that exposure to peracetic acid (2 $mg\,L^{-1}$), a powerful disinfectant used to eliminate zoospores (*Aphanomyces astaci*) that cause crayfish plague, produces slight damage to gill, hepatopancreatic and antennal gland tissue in crayfish (*P. leniusculus*). These changes were more pronounced when the environmentally relevant concentration (10 $mg\,L^{-1}$) was used. Gill tissues were infiltrated with hemocytes and had disorganized epithelial cells and malformed lamellar tips. Tissue changes were reduced to normal levels after a 7 day recovery period.[122] Exposure to xenobiotics does not always cause histopatological changes in the hepatopancreas. Stará *et al.*[57] found no changes in the hepatopancreas of *P. clarkii* after exposure to 0.51–1444 $\mu g\,L^{-1}$ of the herbicide prometryne; however, changes in antioxidant activity were observed for 11 and 25 day exposures. Overall, results suggest that histological changes in crayfish tissues vary depending on the type of xenobiotic they are exposed to. Changes in the histological organization of both the gills

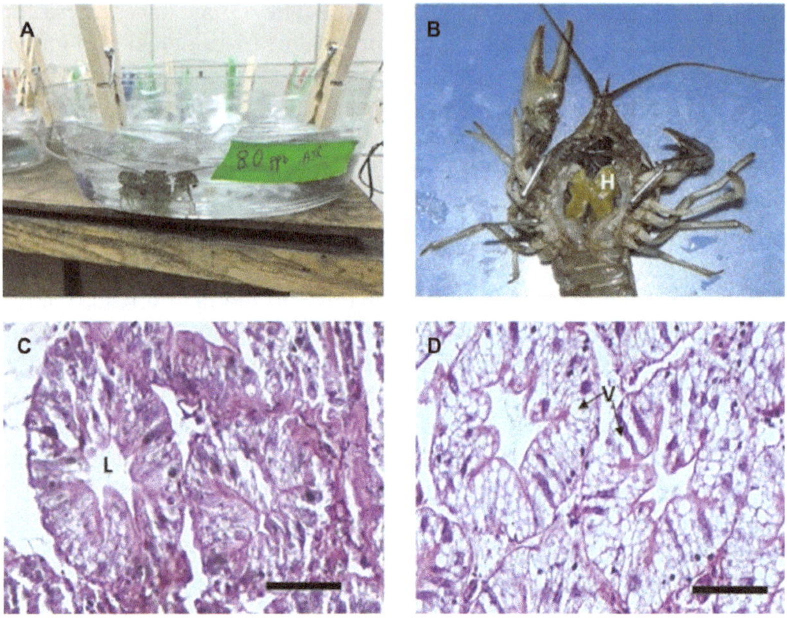

Figure 6.2 The hepatopancreas (or digestive gland) of crayfish can be used as a bioindicator for exposure to environmental contaminants as changes in histology are often seen post-exposure. (A) Crayfish can be exposed to contaminants in a controlled laboratory setting. (B) The hepatopancreas (H) can be removed, fixed and analyzed using standard hematoxylin and eosin staining procedures. Following exposure to atrazine, the hepatopancreas structure of control and atrazine-treated crayfish (*O. virilis*) can be compared. (C) The hepatopancreas from control crayfish contains a normal lumen (L) and a small number of vacuoles. (D) Increased vacuolation (V) in the hepatopancreas of atrazine-treated crayfish can be seen after a 15 day exposure to 1000 $\mu g\,L^{-1}$ atrazine. Scale bars are 100 μm.

and hepatopancreas of crayfish may be used a bioindicator of contamination, but should be examined in combination with other measures of toxicity (*e.g.* cytochrome P450 activation and antioxidant activity).

Cytochrome P450 oxidase are proteins that use a variety of small and large molecules as substrates in enzymatic reactions. Each isoform has a broad spectrum of catalytic activities and substrates.[123] These proteins have been shown to play an important role in the transformation of xenobiotics and endogenous chemicals in crayfish and other crustaceans, and are abundant in hepatopancreas microsomes.[124] The cytochrome P450 system has been characterized in the hepatopancreas and green gland of *P. clarkii*.[125] Ashley *et al.*[63] demonstrated that crayfish (*P. leniusculus*) exposed to an extremely low dose (3 $\mu g\,kg^{-1}$ body weight) of 2,3,7,8-tetrachloro-dibenzo-*p*-dioxin (TCDD) exhibited significant induction of cytochrome P450, measured spectrally, while TCDD-related histological changes were not observed. Fenitrothion is a powerful organophosphorus insecticide used to control rice stem bores. When crayfish (*P. clarkii*) were exposed to 200 μg L^{-1} for 1–2 weeks, induction ofcytochrome P450 activity was recorded, as well as the inhibition of acetylcholinesterase activity.[126,127] An increase in EROD activity (the catalytic form of cytochrome P4501A) was also increased after exposure of *P. clarkii* to fenitrothion.[127] This suggests that cytochrome P450 induction in crayfish can be used as a biomarker for exposure to environmental pollutants.

Antioxidant enzyme system activation (*e.g.* superoxide dismutase and glutathione reductase) can also be investigated to determine if these systems are activated during exposure to xenobiotics. Metribuzin is used in both agriculture and recreational areas, and runs off into local streams and rivers.[128] When crayfish (*P. leniusculus*) were exposed to 0.52 μg L^{-1} (environmentally relevant concentration) and 3.06 mg L^{-1} metribuzin for 10 and 30 days, changes in oxidative stress (thiobarbituric acid reactive substances) and antioxidant enzymes [total superoxide dismutase (SOD), catalase (CAT) and glutathione reductase (GR)] were observed in muscle, gill and hepatopancreatic tissues.[56] Crayfish (*P. clarkii*), exposed to prometryne (0.51 μg L^{-1}, 0.144 mg L^{-1}, and 1.144 mg L^{-1} for 11 and 25 days) also displayed changes in SOD, CAT and GR antioxidant enzyme activity in muscle and hepatopancreatic tissues.[57] *P. leniusculus* exposed to peracetic acid (2 mg L^{-1} and 10 mg L^{-1}) demonstrated decreases in SOD activity in both gill and hepatopancreatic tissues after a 3 day exposure to 10 mg L^{-1} when compared to controls and crayfish treated with 2 mg L^{-1}, while CAT activity remained unaffected in all treatment groups.[122] GR activity was significantly reduced in the gill tissue of crayfish treated with 10 mg L^{-1} peracetic acid and increased in crayfish exposed to 2 mg L^{-1}after a 7 day exposure. All enzyme levels returned to normal after a 7 day recovery period.[122] Malondialdehyde (MDA) activity can also be used as a marker of oxidative stress. Crayfish (*A. leptodactylus*) demonstrated decreased MDA levels when they were exposed to 5, 10, and 20 μg L^{-1} of fenitrothion.[70] Increased levels of oxidative stress can cause decreases in survival and reproductive rates and thus can be considered a marker for overall general health.[129] Taken together, these results suggest that antioxidant enzymes in crayfish can be evaluated and may serve as potential biomarkers for monitoring exposure to xenobiotics. Changes in antioxidant levels can be assessed in a sentinel species, like crayfish, and used to avoid toxic responses to aquatic animals before changes to overall population fitness occur.

For biomonitoring purposes, the presence of DNA adducts or DNA adduct formation can be used as a measure of the toxicity of substances in the aquatic environment. These adducts are formed when a segment of DNA bonds to a chemical, changing the structure of the DNA.[67] Crayfish have the ability to form DNA adducts when exposed to xenobiotics. For example, Schilderman *et al.*[67] examined crayfish (*O. limosus*) collected from the River Meuse (The Netherlands) to determine DNA adduct levels and correlated them with heavy metal residues, chlorinated pesticides and the seven indicator congeners of PCBs from four different locations. Overall, they found that DNA adduct levels can be used as a dosimeter for the internal dose of aromatic compounds, such as polycyclic aromatic hydrocarbons and PCBs, and that the highest hepatopancreatic DNA adducts were found to be significantly higher at the most polluted site.[67] Additionally, a Comet assay can be used to assess DNA damage or double strand breaks in crayfish exposed to changes in abiotic and biotic conditions. Malev *et al.*[130] showed that when *A. leptodactylus* were exposed environmental stressors (*i.e.* increased environmental temperatures), significantly more DNA damage was present in cells obtained from the hemolymph. These studies demonstrate that changes to crayfish DNA integrity can be used as a bioindicator of environmental health.

Exposure to xenobiotics has also been shown to cause cardiovascular, respiratory and metabolic changes in crayfish. Exposure of crayfish (*A. astacus*) to 0.1 $mg\,L^{-1}$ mercury ($HgCl_2$) produced cardiac arrhythmias, which were believed to be related to metabolic disturbances. These arrhythmias ultimately lead to death.[110,111] Copper exposure in *P. clarkii* lead to a reduction of both cardiac and ventilatory activity that was concentration- and time-dependent.[131] Lead (Pb) decreased the respiration rate in crayfish (*P. clarkii*). Oxygen uptake of the whole animal generally decreased with increasing Pb concentration, though not significantly. The histology of gill filaments of crayfish treated with 200 $mg\,L^{-1}$ Pb indicated a general disorganization.[132] Ahern and Morris[133] found that exposure of the crayfish *C. destructor* to 100 or 0.5 $mg\,L^{-1}$ Pb reduced oxygen consumption, along with reduced heart rate, although ventilation rate was unchanged. There was also a reduction in oxygen transfer factor across the gills after 21 days and a reduction in cellular metabolism.[134] Standard metabolic rates (SMR) in crayfish (*P. acutus*) were also altered following exposure to contaminants. Rowe *et al.*[113] found that crayfish that had chronic exposure in a site contaminated with trace elements had higher SMRs when compared to crayfish collected from a reference site (25.1 *vs.* 19.2 $J\,g^{-1}\,day^{-1}$). Growth of animals from contaminated sites was also lower than that for animals from control sites. The authors suggest that SMR comparisons can be used as a biomarker for contamination as many other species demonstrate elevations in SMRs when exposed to pollutants.[113] Generally, exposure to xenobiotics at environmentally relevant concentrations can slow growth and/or lead to death in crayfish when changes in the metabolic or cardiovascular system are involved.

In general, other physiological disturbances have been noted when crayfish are exposed to various contaminants from metals to pesticides. For example, digestive enzymes are inhibited by metal contaminants. Cadmium (Cd) exposure reduced amylase activity in gastric juice of the crayfish *P. clarkii.*[135] Changes in ion (Na^+) regulation, elevated metabolic ammonia production and substantial silver accumulation in the gills, hemolymph and hepatopancreas were also noted in crayfish (*C. diogenes diogenes*) when they were exposed to 8.41 ± 0.17 $\mu g\,L^{-1}$ of silver for 96 h.[107] A combination of metals and pesticides has also been shown to cause changes in esterase activity in the crayfish *P. clarkii*. Vioque-Fernández *et al.*[136,137] found that that examining acetylcholinesterase and carboxylesterase activity can be useful as a biomarker of pesticide exposure, including organophosphate and carbamate insecticides. Further, muscle pyruvate kinase activity was significantly lower in crayfish (*C. quadricarinatus*) exposed to 10 mg L^{-1} of the herbicide glyphosate [*N*-(phosphonomethyl)glycine] in freshwater. Further, no changes in lactate dehydrogenase were observed, indicating reduced metabolic intensity of cells.[65] Avigliano *et al.*[65] demonstrated that alanine and aspartate aminotransferase activities (ALAT and ASAT, respectively) can also be used as indicators of tissue damage following exposure to glyphosate. Crayfish (*C. quadricarinatus*) that were exposed for 60 days to 40 mg L^{-1} of pure glyphosate, an active ingredient in the herbicide formulation Roundup®, were found to have higher hemolymphatic ASAT : ALAT ratios then control crayfish. This suggests that there may be damage to several tissue types as crayfish may mobilize lipids and proteins under stressful situations. Overall, various physiological changes in crayfish can be used as biomarkers to determine if sublethal exposure to metals and/or pesticides is affecting the physiology of aquatic animals.

Physiological responses to environmental stressors are typically deleterious in nature. We see that exposure to environmentally relevant concentrations of xenobiotics can cause measureable physiological changes in the crayfish before death occurs. These endpoints can be easily examined using a variety of molecular, analytical and histological techniques. Any one or several of these physiological changes should be examined after exposing crayfish to potential environmental contaminants in order to evaluate their sublethal toxicity before they are released into the environment.

6.6 Conclusions

As more and more sources of freshwater are deemed unsuitable for consumption and/or agricultural use without some means of treatment, growing strain is placed on existing resources. Global focus has been aimed at finding efficient and effective methods of assessing, restoring, and/or maintaining healthy aquatic systems. This chapter has summarized crayfish ecology and how exposure to elevated levels of environmental contamination from anthropogenic sources can impact crayfish survival, physiology and behavior. The research included in this chapter establishes a solid case for the use of crayfish as a bioindicator for assessing and monitoring contaminated aquatic systems, as well as for using this invertebrate as a model organism in toxicological studies of aquatic pollutants. Foremost, crayfish are found in nearly every freshwater system worldwide, making toxicological studies using this organism broadly applicable. Abundant, widespread crayfish populations with narrow home ranges and low migration mean that crayfish tissue specimens are indicative of the locations from which animals were obtained. Crayfish are easily captured, can be easily maintained and cultured in laboratory settings, they produce large numbers of offspring from a single individual, and most species have body sizes that provide sufficient tissue for analyses.[79] These animals rely on chemical information from the surrounding environment for all of their behaviors, and possess olfactory receptors that are in direct contact with the fluid environment. Consequently, crayfish are highly sensitive to minute chemical fluctuations and modifications to their behaviors can serve as a more sensitive, less invasive means of environmental monitoring and assessment.

Research has also demonstrated that the uptake and toxic responses of many environmental pollutants are similar in crayfish and other aquatic species, such as teleost fish,[107,113] signifying that crayfish are a useful model for monitoring environmental conditions and determining remediation strategies. However; the LC_{50} values reported by Khan and Nugegoda[79] were higher than those documented for other aquatic species, providing evidence that juvenile *C. destructor* is less sensitive to heavy metals in the environment. This suggests that LC_{50} calculations should be coupled with physiological and behavioral assessments when determining toxicological endpoints and safe concentrations in the environment. Owing to the propensity for bioaccumulation in tissues, crayfish have the potential to transfer pollutants and toxins to organisms in higher trophic levels given their role in both aquatic and terrestrial food webs.[138] This propensity also provides means for establishing toxicity contamination indices as measures of environmental contamination. Further, crayfish are listed as the second most imperiled aquatic group in the United States, suggesting that research should focus on their conservation.[139] As this chapter has illustrated, crayfish are most often non-target organisms of the anthropogenic chemicals, providing urgency to studying the effects of xenobiotics on these animals. By continuing and increasing the use of crayfish as model organisms and bioindicators, knowledge can be gained for conservation and preservation of these crucially important organisms and their habitats. Lastly, crayfish may also serve as a sentinel organism for other aquatic organisms, allowing environmental managers to determine environmentally appropriate concentrations of xenobiotics.

References

1. D. M. Holdich, *Biology of Freshwater Crayfish*, Blackwell Science, Oxford, New York, NY, 2002.
2. J. Reynolds and C. Souty-Grosset, *Management of Freshwater Biodiversity: Crayfish as Bioindicators*, Cambridge University Press, 2011.
3. M. Longshaw and P. D. Stebbing, *Biology and Ecology of Crayfish*, CRC Press, Boca Raton, FL, 2016.
4. C. L. Owen, H. Bracken-Grissom, D. Stern and K. A. Crandall, A synthetic phylogeny of freshwater crayfish: insights for conservation, *Philos. Trans. R. Soc., B*, 2015, **370**, 1–10.
5. G. Vogt, Functional anatomy, in *Biology of Freshwater Crayfish*, ed. D. M. Holdich, 2002, pp. 53–151.

6. C. Souty-Gossert and J. W. Fetzner, Taxonomy and identification, in *Biology and Ecology of Crayfish*, ed. M. Longshaw and P. Stebbing, CRC Press, Boca Raton, 2016, pp. 1–30.
7. K. A. Crandall and J. E. Buhay, Global diversity of crayfish (Astacidae, Cambaridae, and Parastacidae––Decapoda) in freshwater, *Hydrobiologia*, 2008, **595**, 295–301.
8. K. A. Crandall, J. Reynolds and C. Souty-Gossert, Crayfish in the decapod lineage, their natural distribution, and their threatened status, in *Management of Freshwater Biodiversity: Crayfish as Bioindicators*, ed. J. Reynolds and C. Souty-Grosset, Cambridge University Press, New York, NY, 2011, pp. 14–44.
9. J. V. Kilian, R. J. Klauda, S. Widman, M. Kashiwagi, R. Bourquin, S. Weglein and J. Schuster, An assessment of a bait industry and angler behavior as a vector of invasive species, *Biol. Invasions*, 2012, **14**, 1469–1481.
10. S. Wutz and J. Geist, Sex-and size-specific migration patterns and habitat preferences of invasive signal crayfish (*Pacifastacus leniusculus* Dana), *Limnologica*, 2013, **43**, 59–66.
11. H. E. G. Ackefors, Freshwater crayfish farming technology in the 1990s: a European and global perspective, *Fish Fish.*, 2000, **1**, 337–359.
12. R.-Z. Guan, Burrowing behaviour of signal crayfish, *Pacifastacus leniusculus* (Dana) in the River Great Ouse, England, *Freshwater Forum*, 2010, **4**, 155–168.
13. A. N. Hendrix and W. F. Loftus, Distribution and relative abundance of the crayfishes *Procambarus alleni* (Faxon) and *P. fallax* (Hagen) in southern Florida, *Wetlands*, 2000, **20**, 194–199.
14. F. Jordan, C. J. DeLeon and A. C. McCreary, Predation, habitat complexity, and distribution of the crayfish *Procambarus alleni* within a wetland habitat mosaic, *Wetlands*, 1996, **16**, 452–457.
15. B. Hazlett, D. Rittschof and D. Rubenstein, Behavioral biology of the crayfish *Orconectes virilis* I. Home range, *Am. Midl. Nat.*, 1974, **92**, 301–319.
16. R. C. Guiasu, D. W. Barr and D. W. Dunham, Distribution and status of crayfishes of the genera Cambarus and Fallicambarus (Decapoda: Cambaridae) in Ontario, Canada, *J. Crustacean Biol.*, 1996, **16**, 373–383.
17. M. B. Griffith, L. T. Wolcott and S. A. Perry, Production of the crayfish *Cambarus bartonii* (Fabricius, 1798)(Decapoda, Cambaridae) in an acidic appalachian stream (USA), *Crustaceana*, 1996, **69**, 974–984.
18. C. Austin and B. Knott, Systematics of the freshwater crayfish genus *Cherax* Erichson (Decapoda: Parastacidae) in south-western Australia: electrophoretic, morphological and habitat variation, *Aust. J. Zool.*, 1996, **44**, 223–258.
19. C. L. McLay and A. M. van den Brink, Crayfish growth and reproduction, in *Biology and Ecology of Crayfish*, ed. M. Longshaw and P. Stebbing, CRC Press, Boca Raton, FL, 2016, pp. 62–116.

20. A. M. Hill and D. M. Lodge, Replacement of resident crayfishes by an exotic crayfish: the roles of competition and predation, *Ecol. Appl.*, 1999, **9**, 678–690.
21. W. T. Momot, Redefining the role of crayfish in aquatic ecosystems, *Rev. Fish. Sci.*, 1995, **3**, 33–63.
22. N. Usio, Endangered crayfish in northern Japan: distribution, abundance and microhabitat specificity in relation to stream and riparian environment, *Biol. Conserv.*, 2007, **134**, 517–526.
23. T. S. March and B. J. Robson, Association between burrow densities of two Australian freshwater crayfish (*Engaeus sericatus* and *Geocharax gracilis*: Parastacidae) and four riparian land uses, *Aquat. Conserv.*, 2006, **16**, 181–191.
24. D. Fernandes, J. Potrykus, C. Morsiani, D. Raldua, R. Lavado and C. Porte, The combined use of chemical and biochemical markers to assess water quality in two low-stream rivers (NE Spain), *Environ. Res.*, 2002, **90**, 169–178.
25. D. M. Holdich and I. D. Reeve, Distribution of freshwater crayfish in the British Isles, with particular reference to crayfish plague, alien introductions and water quality, *Aquat. Conserv.*, 1991, **1**, 139–158.
26. C. E. Boyd, *Water Quality*, Springer, New York, NY, 2000.
27. L. Füreder and J. Reynolds, Is *Austropotamobius pallipes* a good bioindicator?, *Bull. Fr. Pêche Piscic.*, 2003, **370–371**, 157–163.
28. D. J. H. Phillips and P. S. Rainbow, *Biomonitoring of Trace Aquatic Contaminants*, Elsevier Applied Science, 1993.
29. P. S. Rainbow, Biomonitoring of heavy metal availability in the marine environment, *Mar. Pollut. Bull.*, 1995, **31**, 183–192.
30. N. J. Dorn and J. M. Wojdak, The role of omnivorous crayfish in littoral communities, *Oecologia*, 2004, **140**, 150–159.
31. A. Kouba, M. Buřič and P. Kozák, Bioaccumulation and effects of heavy metals in crayfish: A review, *Water, Air, Soil Pollut.*, 2010, **211**, 5–16.
32. J. S. Weis, Some physiological responses of crustaceans to toxicants, in *Physiology*, ed. E. S. Chang and M. Thiel, Oxford University Press, New York, 2015, vol. 4, pp. 477–504.
33. T. Heberer, Occurrence, fate, and removal of pharmaceutical residues in the aquatic environment: a review of recent research data, *Toxicol. Lett.*, 2002, **131**, 5–17.
34. J. P. Bound and N. Voulvoulis, Pharmaceuticals in the aquatic environment—a comparison of risk assessment strategies, *Chemosphere*, 2004, **56**, 1143–1155.
35. J. A. Pedersen, M. Soliman and I. H. Suffet, Human pharmaceuticals, hormones, and personal care product ingredients in runoff from agricultural fields irrigated with treated wastewater, *J. Agric. Food Chem.*, 2005, **53**, 1625–1632.
36. C. G. Daughton and T. A. Ternes, Pharmaceuticals and personal care products in the environment: agents of subtle change? *Environ. Health Perspect.*, 1999, **107**, 907–938.

37. M. Crane, C. Watts and T. Boucard, Chronic aquatic environmental risks from exposure to human pharmaceuticals, *Sci. Total Environ.*, 2006, **367**, 23–41.
38. M. Cleuvers, Aquatic ecotoxicity of pharmaceuticals including the assessment of combination effects, *Toxicol. Lett.*, 2003, **142**, 185–194.
39. B. Ferrari, R. Mons, B. Vollat, B. Fraysse, N. Paxēaus, R. L. Giudice, A. Pollio and J. Garric, Environmental risk assessment of six human pharmaceuticals: are the current environmental risk assessment procedures sufficient for the protection of the aquatic environment? *Environ. Toxicol. Chem.*, 2004, **23**, 1344–1354.
40. G. K. Kulkarni, R. Nagabhushanam, G. Amaldoss, R. G. Jaiswal and M. Fingerman, In vivo stimulation of ovarian development in the red swamp crayfish, *Procambarus clarkii* (Girard), by 5-hydroxytryptamine, *Invertebr. Reprod. Dev.*, 1992, **21**, 231–239.
41. A. J. Tierney, K. N. Hanzlik, R. M. Hathaway, C. Powers and M. Roy, Effects of fluoxetine on growth and behavior in the crayfish *Orconectes rusticus*, *Mar. Freshwater Behav. Physiol.*, 2016, **49**, 133–145.
42. J. Goetz, C. Jones and M. Scott, Long term depression decreased in crayfish chronically exposed to Fluoxetine, *Pioneering Neurosci.*, 2006, 7, 31–33.
43. R. Sandodden and S. I. Johnsen, Eradication of introduced signal crayfish *Pasifastacus leniusculus* using the pharmaceutical BETAMAX VET, *Aquat. Invasions*, 2010, **5**, 75–81.
44. I. Kuklina, S. Sladkova, A. Kouba, S. Kholodkevich and P. Kozák, Investigation of chloramine-T impact on crayfish *Astacus leptodactylus* (Esch., 1823) cardiac activity, *Environ. Sci. Pollut. Res.*, 2014, **21**, 10262–10269.
45. USGS, The quality of our nation's waters: Pesticides in the nation's streams and ground water, 1992–2001, http://pubs.usgs.gov/circ/2005/1291/.
46. D. W. Connell and G. J. Miller, *Chemistry and Ecotoxicology of Pollution*, Wiley, New York, 1984.
47. J. C. Colombo, M. F. Khalil, M. Arnac, A. C. Horth and J. A. Catoggio, Distribution of chlorinated pesticides and individual polychlorinated biphenyls in biotic and abiotic compartments of the Rio de la Plata, Argentina, *Environ. Sci. Technol.*, 1990, **24**, 498–505.
48. P. K. Ghosh and L. Philip, Environmental significance of atrazine in aqueous systems and its removal by biological processes: an overview, *Global Nest J.*, 2006, **8**, 159–178.
49. M. Graymore, F. Stagnitti and G. Allinson, Impacts of atrazine in aquatic ecosystems, *Environ. Int.*, 2001, **26**, 483–495.
50. J. Velisek, A. Kouba and A. Stara, Acute toxicity of triazine pesticides to juvenile signal crayfish (*Pacifastacus leniusculus*), *Neuroendocrinol. Lett.*, 2013, **34**, 31–36.
51. R. M. Belanger, T. J. Peters, G. S. Sabhapathy, S. Khan, J. Katta and N. K. Abraham, Atrazine exposure affects the ability of crayfish (*Orconectes*

rusticus) to localize a food odor source, *Arch. Environ. Contam. Toxicol.*, 2015, **68**, 636–645.
52. R. M. Belanger, L. N. Mooney, H. M. Nguyen, N. K. Abraham, T. J. Peters, M. A. Kana and L. A. May, Acute atrazine exposure has lasting effects on chemosensory responses to food odors in crayfish (*Orconectes virilis*), *Arch. Environ. Contam. Toxicol.*, 2016, **70**, 289–300.
53. C. Mac Loughlin, I. S. Canosa, G. R. Silveyra, L. S. López Greco and E. M. Rodríguez, Effects of atrazine on growth and sex differentiation, in juveniles of the freshwater crayfish *Cherax quadricarinatus*, *Ecotoxicol. Environ. Saf.*, 2016, **131**, 96–103.
54. J. Velisek, A. Stará, D. Koutnik, E. Zuskova and A. Kouba, Effect of prometryne on early life stages of marbled crayfish (*Procambarus fallax* f. *virginalis*), *Neuroendocrinol. Lett.*, 2014, **35**(Suppl 2), 93–98.
55. A. Stará, E. Zuskova, A. Kouba and J. Velisek, Effects of terbuthylazine-desethyl, a terbuthylazine degradation product, on red swamp crayfish (*Procambarus clarkii*), *Sci. Total Environ.*, 2016, **566–567**, 733–740.
56. D. Koutnik, A. Stara, E. Zuskova, A. Kouba and J. Velisek, The effect of subchronic metribuzin exposure to signal crayfish (*Pacifastacus leniusculus* Dana 1852), *Neuroendocrinol. Lett.*, 2014, **35**(Suppl 2), 51–56.
57. A. Stará, A. Kouba and J. Velisek, Effect of chronic exposure to prometryne on oxidative stress and antioxidant response in red swamp crayfish (*Procambarus clarkii*), *BioMed Res. Int.*, 2014, **2014**, 1–6.
58. J. W. Frey, Occurrence, distribution, loads of selected pesticides in streams in the Lake Erie-St. Clair basin, 1996–1998, *USGS Water-Resource Investigations Report 00-4169*. National Water Quality Assessment Program, Branch of Information Services, US Geological Survey, Denver, Colorado, 2001, 69.
59. M. E. Cook and P. A. Moore, The effects of the herbicide metolachlor on agonistic behavior in the crayfish, *Orconectes rusticus*, *Arch. Environ. Contam. Toxicol.*, 2008, **55**, 94–102.
60. M. C. Wolf and P. A. Moore, Effects of the herbicide metolachlor on the perception of chemical stimuli by *Orconectes rusticus*, *J. N. Am. Benthol. Soc.*, 2002, **21**, 457–467.
61. A. Ç. K. Benli, R. Sarıkaya, A. Sepici-Dincel, M. Selvi, D. Şahin and F. Erkoç, Investigation of acute toxicity of (2,4-dichlorophenoxy) acetic acid (2,4-D) herbicide on crayfish (*Astacus leptodactylus* Esch. 1823), *Pestic. Biochem. Physiol.*, 2007, **88**, 296–299.
62. A. M. Browne and P. A. Moore, The effects of sublethal levels of 2,4-dichlorophenoxyacetic acid herbicide (2,4-D) on feeding behaviors of the crayfish *O. rusticus*, *Arch. Environ. Contam. Toxicol.*, 2014, **67**, 234–244.
63. C. M. Ashley, M. G. Simpson, D. M. Holdich and D. R. Bell, 2,3,7,8-tetrachloro-dibenzo-p-dioxin is a potent toxin and induces cytochrome P450 in the crayfish, *Pacifastacus leniusculus*, *Aquat. Toxicol.*, 1996, **35**, 157–169.

64. H. O. Sanders and J. B. Hunn, Toxicity, bioconcentration, and depuration of the herbicide bolero 8EC in freshwater invertebrates and fish, *Bull. Jpn. Soc. Sci. Fish.*, 1982, **48**, 1139–1143.
65. L. Avigliano, A. V. Fassiano, D. A. Medesani, M. C. Rios, de Molina and E. M. Rodriguez, Effects of glyphosate on growth rate, metabolic rate and energy reserves of early juvenile crayfish, *Cherax quadricarinatus* M, *Bull. Environ. Contam. Toxicol.*, 2014, **92**, 631–635.
66. J. L. Frontera, I. Vatnick, A. Chaulet and E. M. Rodriguez, Effects of glyphosate and polyoxyethylenamine on growth and energetic reserves in the freshwater crayfish *Cherax quadricarinatus* (Decapoda, Parastacidae), *Arch. Environ. Contam. Toxicol.*, 2011, **61**, 590–598.
67. P. A. E. L. Schilderman, E. J. C. Moonen, L. M. Maas, I. Welle and J. C. S. Kleinjans, Use of crayfish in biomonitoring studies of environmental pollution of the river Meuse, *Ecotoxicol. Environ. Saf.*, 1999, **44**, 241–252.
68. C. R. Santerre, R. Ingram, G. W. Lewis, J. T. Davis, L. G. Lane, R. M. Grodner, C. I. Wei, P. B. Bush, D. H. Xu, J. Shelton, E. G. Alley and J. M. Hinshaw, Organochlorines, organophosphates, and pyrethroids in channel catfish, rainbow trout, and red swamp crayfish from aquaculture facilities, *J. Food Sci.*, 2000, **65**, 231–235.
69. M. M. A. Desouky, H. Abdel-Gawad and B. Hegazi, Distribution, fate and histopathological effects of ethion insecticide on selected organs of the crayfish, *Procambarus clarkii*, *Food Chem. Toxicol.*, 2013, **52**, 42–52.
70. R. Sarikaya, A. Sepici-Dincel, A. C. K. Benli, M. Selvi and F. Erkoc, The acute toxicity of Fenitrothion on narrow-clawed crayfish (*Astacus leptodactylus* Eschscholtz, 1823) in association with biomarkers of lipid peroxidation, *J. Biochem. Mol. Toxicol.*, 2011, **25**, 169–174.
71. M. Buřič, A. Kouba, J. Machova, I. Mahovska and P. Kozak, Toxicity of the organophosphate pesticide diazinon to crayfish of differing age, *Int. J. Environ. Sci. Technol.*, 2013, **10**, 607–610.
72. G. C. Barbee and M. J. Stout, Comparative acute toxicity of neonicotinoid and pyrethroid insecticides to non-target crayfish (*Procambarus clarkii*) associated with rice-crayfish crop rotations, *Pest Manage. Sci.*, 2009, **65**, 1250–1256.
73. H. H. Jarboe and R. P. Romaire, Response of Procambarid crayfish populations to permethrin applications in earthen ponds, *Bull. Environ. Contam. Toxicol.*, 1995, **55**, 58–64.
74. G. C. Barbee, W. R. McClain, S. K. Lanka and M. J. Stout, Acute toxicity of chlorantraniliprole to non-target crayfish (*Procambarus clarkii*) associated with rice–crayfish cropping systems, *Pest Manage. Sci.*, 2010, **66**, 996–1001.
75. K. Loukola-Ruskeeniemi, M. Kantola, T. Halonen, K. Seppänen, P. Henttonen, E. Kallio, P. Kurki and H. Savolainen, Mercury-bearing black shales and human Hg intake in eastern Finland: impact and mechanisms, *Environ. Geol.*, 2003, **43**, 283–297.

76. E. Tunca, E. Ucuncu, A. D. Ozkan, Z. E. Ulger, A. E. Cansızoğlu and T. Tekinay, Differences in the accumulation and distribution profile of heavy metals and metalloid between male and female crayfish (*Astacus leptodactylus*), *Bull. Environ. Contam. Toxicol.*, 2013, **90**, 570–577.
77. J. H. Hubschman, Effects of copper on the crayfish *Orconectes rusticus* (Girard), *Crustaceana*, 1967, **12**, 141–150.
78. R. M. Taylor, G. D. Watson and M. A. Alikhan, Comparative sub-lethal and lethal acute toxicity of copper to the freshwater crayfish, *Cambarus robustus* (Cambaridae, Decapoda, Crustacea) from an acidic metal-contaminated lake and a circumneutral uncontaminated stream, *Water Res.*, 1995, **29**, 401–408.
79. S. Khan and D. Nugegoda, Sensitivity of juvenile freshwater crayfish *Cherax destructor* (Decapoda: Parastacidae) to trace metals, *Ecotoxicol. Environ. Saf.*, 2007, **68**, 463–469.
80. A. Aguirre-Sierra, A. Alonso and J. A. Camargo, Fluoride bioaccumulation and toxic effects on the survival and behavior of the endangered white-clawed crayfish *Austropotamobius pallipes* (Lereboullet), *Arch. Environ. Contam. Toxicol.*, 2013, **65**, 244–250.
81. J. D. Icely and J. A. Nott, Digestion and absorption: digestive system and associated organs, in *Microscopic Anatomy of Invertebrates, vol. 10. Decapod Crustacea*, ed. F. W. Harrison and A. G. Humes, Wiley-Liss, Inc., New York, 1992, pp. 147–201.
82. G. Bruno, M. G. Volpe, G. De Luise and M. Paolucci, Detection of heavy metals in farmed *Cherax destructor*, *Bull. Fr. Pêche Piscic.*, 2006, **380–381**, 1341–1349.
83. P. Alcorlo, M. Otero, M. Crehuet, A. Baltanas and C. Montes, The use of the red swamp crayfish (*Procambarus clarkii*, Girard) as indicator of the bioavailability of heavy metals in environmental monitoring in the River Guadiamar (SW, Spain), *Sci. Total Environ.*, 2006, **366**, 380–390.
84. P. S. Rainbow, Trace metal concentrations in aquatic invertebrates: why and so what? *Environ. Pollut.*, 2002, **120**, 497–507.
85. M. W. Finerty, J. D. Madden, S. E. Feagley and R. M. Grodner, Effect of environs and seasonality on metal residues in tissues of wild and pond-raised crayfish in southern Louisiana, *Arch. Environ. Contam. Toxicol.*, 1990, **19**, 94–100.
86. J. D. Madden, R. M. Grodner, S. E. Feagley, M. W. Finerty and L. S. Andrews, Minerals and xenobiotic residues in the edible tissues of wild and pond-raised Louisiana crayfish, *J. Food Saf.*, 1991, **12**, 1–15.
87. F. Gherardi, S. Barbaresi, O. Vaselli and A. Bencini, A comparison of trace metal accumulation in indigenous and alien freshwater macro-decapods, *Mar. Freshwater Behav. Physiol.*, 2002, **35**, 179–188.
88. R. L. Hothem, D. R. Bergen, M. L. Bauer, J. J. Crayon and A. M. Meckstroth, Mercury and trace elements in crayfish from northern California, *Bull. Environ. Contam. Toxicol.*, 2007, **79**, 628–632.

89. P. Maranhoão, J. C. Marques and V. Madeira, Copper concentrations in soft tissues of the red swamp crayfish *Procambarus clarkii* (Girard, 1852), after exposure to a range of dissolved copper concentrations, *Freshwater Crayfish*, 1995, **10**, 282–286.
90. M. S. Naqvi, I. Devalraju and H. N. Naqvi, Copper bioaccumulation and depuration by red swamp crayfish, *Procambarus clarkii*, *Bull. Environ. Contam. Toxicol.*, 1998, **61**, 65–71.
91. U. Guner, Freshwater crayfish *Astacus leptodactylus* (Eschscholtz, 1823) accumulates and depurates copper, *Environ. Monit. Assess.*, 2007, **133**, 365–369.
92. G. Bagatto and M. A. Alikhan, Copper, cadmium, and nickel accumulation in crayfish populations near copper-nickel smelters at Sudbury, Ontario, Canada, *Bull. Environ. Contam. Toxicol.*, 1987, **38**, 540–545.
93. G. Mackevičienė, Bioaccumulation of heavy metals in noble crayfish (*Astacus astacus* L.) tissues under aquaculture conditions, *Ekologija*, 2002, **2**, 79–82.
94. M. G. Bennet-Chambers and B. Knott, Does the freshwater crayfish *Cherax tenuimanus* (Smith)[Decapoda: Parastacidae] regulate tissues zinc concentrations? *Freshwater Crayfish*, 2002, **13**, 405–423.
95. M. B. Bitner Anderson, P. Reddy, J. E. Preslan, M. Fingerman, J. Bollinger, L. Jolibois, G. Maheshwarudu and W. J. George, Metal accumulation in crayfish, *Procambarus clarkii*, exposed to a petroleum-contaminated Bayou in Louisiana, *Ecotoxicol. Environ. Saf.*, 1997, **37**, 267–372.
96. C. J. Schmitt, W. G. Brumbaugh, G. L. Linder and J. E. Hinck, A screening-level assessment of lead, cadmium, and zinc in fish and crayfish from Northeastern Oklahoma, USA, *Environ. Geochem. Health*, 2006, **28**, 445–471.
97. J. P. Giesy, J. W. Bowling and H. J. Kania, Cadmium and zinc accumulation and elimination by freshwater crayfish, *Arch. Environ. Contam. Toxicol.*, 1980, **9**, 683–697.
98. M. Devi, D. A. Thomas, J. T. Barber and M. Fingerman, Accumulation and physiological and biochemical effects of cadmium in a simple aquatic food chain, *Ecotoxicol. Environ. Saf.*, 1996, **33**, 38–43.
99. M. G. Chambers, The effect of acute cadmium toxicity on marron, *Cherax tenuimanus* (Smith, 1912)(Family Parastacidae), *Freshwater Crayfish*, 1995, **10**, 209–220.
100. B. M. Roldan and R. R. Shivers, The uptake and storage of iron and lead in cells of the crayfish (*Orconectes propinquus*) hepatopancreas and antennal gland, *Comp. Biochem. Physiol., C: Comp. Pharmacol.*, 1987, **86**, 201–214.
101. W. Meyer, M. Kretschmer, A. Hoffmann and G. Harisch, Biochemical and histochemical observations on effects of low-level heavy metal load (lead, cadmium) in different organ systems of the freshwater crayfish, *Astacus astacus* L. (crustacea: Decapoda), *Ecotoxicol. Environ. Saf.*, 1991, **21**, 137–156.

102. C. M. Pennuto, O. P. Lane, D. C. Evers, R. J. Taylor and J. Loukmas, Mercury in the northern crayfish, *Orconectes virilis* (Hagen), in new England, USA, *Ecotoxicology*, 2005, **14**, 149–162.
103. O. Simon, F. Ribeyre and A. Boudou, Comparative experimental study of cadmium and methylmercury trophic transfers between the asiatic clam *Corbicula fluminea* and the crayfish *Astacus astacus*, *Arch. Environ. Contam. Toxicol.*, 2000, **38**, 317–326.
104. D. A. Wright, P. M. Welbourn and A. V. M. Martin, Inorganic and organic mercury uptake and loss by the crayfish *Orconectes propinquus*, *Water, Air, Soil Pollut.*, 1991, **56**, 697–707.
105. O. Simon and A. Boudou, Simultaneous experimental study of direct and direct plus trophic contamination of the crayfish *Astacus astacus* by inorganic mercury and methylmercury, *Environ. Toxicol. Chem.*, 2001, **20**, 1206–1215.
106. L. Jorhem, J. Engman, B. Sundström and A. M. Thim, Trace elements in crayfish: regional differences and changes induced by cooking, *Arch. Environ. Contam. Toxicol.*, 1994, **26**, 137–142.
107. M. Grosell, C. J. Brauner, S. P. Kelly, J. C. McGeer, A. Bianchini and C. M. Wood, Physiological responses to acute silver exposure in the freshwater crayfish (*Cambarus diogenes diogenes*)–a model invertebrate? *Environ. Toxicol. Chem.*, 2002, **21**, 369–374.
108. E. Goretti, M. Pallottini, M. Ricciarini, R. Selvaggi and D. Cappelletti, Heavy metals bioaccumulation in selected tissues of red swamp crayfish: An easy tool for monitoring environmental contamination levels, *Sci. Total Environ.*, 2016, **559**, 339–346.
109. J. P. Hagen and J. Sneddon, Determination of copper, iron, and zinc in crayfish (*Procambrus clarkii*) by inductively coupled plasma-optical emission spectrometry, *Spectrosc. Lett.*, 2009, **42**, 58–61.
110. B. Styrishave, A. D. Rasmussen and M. H. Depledge, The influence of bulk and trace metals on the circadian rhythm of heart rates in freshwater crayfish, *Astacus astacus*, *Mar. Pollut. Bull.*, 1995, **31**, 87–92.
111. B. Styrishave and M. H. Depledge, Evaluation of mercury-induced changes in circadian heart rate rhythms in the freshwater crab, Potamon potamios and the crayfish, *Astacus astacus* as an early predictor of mortality, *Comp. Biochem. Physiol., Part A: Mol. Integr. Physiol.*, 1996, **115**, 349–356.
112. P. S. Reddy, S. R. Tuberty and M. Fingerman, Effects of cadmium and mercury on ovarian maturation in the red swamp crayfish, *Procambarus clarkii*, *Ecotoxicol. Environ. Saf.*, 1997, **37**, 62–65.
113. C. L. Rowe, W. A. Hopkins, C. Zehnder and J. D. Congdon, Metabolic costs incurred by crayfish (*Procambarus acutus*) in a trace element-polluted habitat: further evidence of similar responses among diverse taxonomic groups, *Comp. Biochem. Physiol., Part C: Toxicol. Pharmacol.*, 2001, **129**, 275–283.

114. A. Burba, The effect of copper on behavioural reactions of noble crayfish *Astacus astacus* L, *Acta Zool. Litu.*, 1999, **9**, 30–36.
115. M. Sherba, D. Dunham and H. Harvey, Sublethal copper toxicity and food response in the freshwater crayfish *Cambarus bartonii* (Cambaridae, Decapoda, Crustacea), *Ecotoxicol. Environ. Saf.*, 2000, **46**, 329–333.
116. S. E. Lahman, K. R. Trent and P. A. Moore, Sublethal copper toxicity impairs chemical orientation in the crayfish, *Orconectes rusticus*, *Ecotoxicol. Environ. Saf.*, 2015, **113**, 369–377.
117. S. E. Lahman and P. A. Moore, Olfactory sampling recovery following sublethal copper exposure in the rusty crayfish, *Orconectes rusticus*, *Bull. Environ. Contam. Toxicol.*, 2015, **95**, 441–446.
118. A. L. Allert, J. F. Fairchild, R. J. Distefano, C. J. Schmitt, J. M. Besser, W. G. Brumbaugh and B. C. Poulton, Effects of lead-zinc mining on crayfish (*Orconectes hylas*) in the Black River watershed, Missouri, USA, *Freshwater Crayfish*, 2008, **16**, 97–111.
119. A. L. Allert, R. J. DiStefano, C. J. Schmitt, J. F. Fairchild and W. G. Brumbaugh, Effects of mining-derived Metals on riffle-dwelling crayfish in southwestern Missouri and southeastern Kansas, USA, *Arch. Environ. Contam. Toxicol.*, 2012, **63**, 563–573.
120. A. L. Allert, R. J. DiStefano, J. F. Fairchild, C. J. Schmitt, M. J. McKee, J. A. Girondo, W. G. Brumbaugh and T. W. May, Effects of historical lead-zinc mining on riffle-dwelling benthic fish and crayfish in the Big River of southeastern Missouri, USA, *Ecotoxicology*, 2013, **22**, 506–521.
121. A. L. Allert, J. F. Fairchild, R. J. Distefano, C. J. Schmitt, W. G. Brumbaugh and J. M. Besser, Ecological effects of lead mining on Ozark streams: In-situ toxicity to woodland crayfish (*Orconectes hylas*), *Ecotoxicol. Environ. Saf.*, 2009, **72**, 1207–1219.
122. L. Chupani, E. Zuskova, A. Stará, J. Velisek and A. Kouba, Histological changes and antioxidant enzyme activity in signal crayfish (*Pacifastacus leniusculus*) associated with sub-acute peracetic acid exposure, *Fish Shellfish Immunol.*, 2016, **48**, 190–195.
123. L. M. Tompkins and A. D. Wallace, Mechanisms of cytochrome P450 induction, *J. Biochem. Mol. Toxicol.*, 2007, **21**, 176–181.
124. M. O. James and S. M. Boyle, Cytochromes P450 in crustacea, *Comp. Biochem. Physiol., Part C: Pharmacol., Toxicol. Endocrinol.*, 1998, **121**, 157–172.
125. C. S. E. Jewell and G. W. Winston, Characterization of the microsomal mixed-function oxygenase system of the hepatopancreas and green gland of the red swamp crayfish, *Procambarus clarkii*, *Comp. Biochem. Physiol.*, 1989, **92**, 329–339.
126. C. Porte and E. Escartín, Cytochrome P450 system in the hepatopancreas of the red swamp crayfish *Procambarus clarkii*: a field study, *Comp. Biochem. Physiol., Part C: Pharmacol., Toxicol. Endocrinol.*, 1998, **121**, 333–338.

127. E. Escartín and C. Porte, Bioaccumulation, metabolism, and biochemical effects of the organophosphorous pesticide, Fenitrothion, in *Procambarus clarkii*, *Environ. Toxicol. Chem.*, 1996, **15**, 915–920.
128. F. J. Fairchild and C. L. Sappington, Fate and effects of the triazinone herbicide metribuzin in experimental pond mesocosms, *Arch. Environ. Contam. Toxicol.*, 2002, **43**, 198–202.
129. M. Beaulieu and D. Costantini, Biomarkers of oxidative status: missing tools in conservation physiology, *Conserv. Physiol.*, 2014, **2**, 1–16.
130. O. Malev, M. Srut, I. Maguire, A. Stambuk, E. A. Ferrero, S. Lorenzon and G. I. V. Klobucar, Genotoxic, physiological and immunological effects caused by temperature increase, air exposure or food deprivation in freshwater crayfish *Astacus leptodactylus*, *Comp. Biochem. Physiol., Part C: Toxicol. Pharmacol.*, 2010, **152**, 433–443.
131. G. Bini and G. Chelazzi, Acclimatable cardiac and ventilatory responses to copper in the freshwater crayfish *Procambarus clarkii*, *Comp. Biochem. Physiol., Part C: Toxicol. Pharmacol.*, 2006, **144**, 235–241.
132. A. Torreblanca, J. Diaz-Mayans, J. D. Ramo and A. Núñez, Oxygen uptake and gill morphological alterations in *Procambarus clarkii* (Girard) after sublethal exposure to lead, *Comp. Biochem. Physiol., Part C: Pharmacol., Toxicol., Endocrinol.*, 1987, **86**, 219–224.
133. M. D. Ahern and S. Morris, Respiratory, acid–base and metabolic responses of the freshwater crayfish *Cherax destructor* to lead contamination, *Comp. Biochem. Physiol., Part A: Mol. Integr. Physiol.*, 1999, **124**, 105–111.
134. S. Morris, W. J. van Aardt and M. D. Ahern, The effect of lead on the metabolic and energetic status of the Yabby, *Cherax destructor*, during environmental hypoxia, *Aquat. Toxicol.*, 2005, **75**, 16–31.
135. P. S. Reddy and M. Fingerman, Effect of cadmium chloride on amylase activity in the red swamp crayfish, *Procambarus clarkii*, *Comp. Biochem. Physiol., Part C: Pharmacol., Toxicol. Endocrinol.*, 1994, **109**, 309–314.
136. A. Vioque-Fernández, E. A. de Almeida, J. Ballesteros, T. García-Barrera, J. L. Gómez-Ariza and J. López-Barea, Doñana National Park survey using crayfish (*Procambarus clarkii*) as bioindicator: Esterase inhibition and pollutant levels, *Toxicol. Lett.*, 2007, **168**, 260–268.
137. A. Vioque-Fernández, E. A. de Almeida and J. López-Barea, Esterases as pesticide biomarkers in crayfish (*Procambarus clarkii*, Crustacea): Tissue distribution, sensitivity to model compounds and recovery from inactivation, *Comp. Biochem. Physiol., Part C: Toxicol. Pharmacol.*, 2007, **145**, 404–412.
138. A. J. Wigginton and W. J. Birge, Toxicity of cadmium to six species in two genera of crayfish and the effect of cadmium on molting success, *Environ. Toxicol. Chem.*, 2007, **26**, 548–554.
139. D. S. Wilcove and L. L. Master, How many endangered species are there in the United States? *Front. Ecol. Environ.*, 2005, **3**, 414–420.

CHAPTER 7

The Crayfish Cambarellus montezumae *as a Possible Freshwater Non-conventional Biomonitor*

EDUARDO MADRIGAL-BUJAIDAR,*[a]
ISELA ÁLVAREZ-GONZÁLEZ,[a] EUGENIA LÓPEZ-LÓPEZ,[b]
JACINTO ELÍAS SEDEÑO-DÍAZ[c] AND
RICARDO ARTURO RUIZ-PICOS[b]

[a] Laboratorio de Genética, Escuela Nacional de Ciencias Biológicas, Instituto Politécnico Nacional. Av. Wilfrido Massieu s/n. Col. Zacatenco, México City 07738, México; [b] Laboratorio de Evaluación de la Salud de los Ecosistemas Acuáticos. Departamento de Zoología, Escuela Nacional de Ciencias Biológicas, Instituto Politécnico Nacional, Prol. de Carpio y Plan de Ayala, Col. Santo Tomás, México City 11340, México; [c] Coordinación Politécnica para la Sustentabilidad, Instituto Politécnico Nacional, Av. Instituto Politécnico Nacional s/n esq Av. Wilfrido Massieu. Col. Zacatenco, México City 07738, México
*Email: eduardo.madrigal@lycos.com

7.1 Introduction

A global estimation of freshwater resources is that they comprise only 1% of the world's water bodies and occupy approximately 0.8% of the Earth's surface. Nevertheless, this tiny fraction of water harbors more than 100 000

Issues in Toxicology No. 33
Ecotoxicology and Genotoxicology: Non-traditional Aquatic Models
Edited by Marcelo L. Larramendy

Published by the Royal Society of Chemistry, www.rsc.org

species, which represents almost 6% of the total described species.[1] Unfortunately, these water bodies can be strongly impacted by a variety of anthropic modifications leading to highly endangered ecosystems.[2] During the last four decades, researchers have been focusing their interest on the assessment of ecosystem health, and thus on finding suitable ecological indicators (in the aquatic biota) so as to properly identify the status of the involved ecosystem in order to take the necessary measures.[3] Among the aquatic biota, the crayfish fauna constitute a particular macroinvertebrate group that has been considered for such a purpose. They are among the largest invertebrates in many freshwater systems, occupy an important position in the food webs, can attain a high biomass, modify their habitat through foraging activities, and direct the flow of energy in freshwater ecosystems owing to their polyphagous and opportunistic feeding habits; moreover, they may eventually reduce the macrophytic biomass, and may even determine the habitat's suitability for other organisms because of the induction of changes in sediment transport, and in the resulting increase of spatio-temporal habitat diversity.[4]

Owing to their remarkable diversity, crayfish are able to occupy several types of freshwater habitats (from seasonal flooded wetlands to permanent water bodies), even in caves, subterranean streams and pools associated with ground water.[1,4] A wide range of habitat-related variables have been associated with the distribution of crayfish; some include substratum type, shading by the riparian canopy, presence of woody debris, aquatic macrophytes and water velocity.

Biomonitoring refers to techniques that can be used in living organisms in order to provide information about the effects of the abiotic and biotic components of an environment. This strategy may provide valuable information with respect to acute or chronic exposures, whose effects as a consequence may disturb the integrity or health of the involved ecosystem, that is, its capability to support and maintain a balanced, integrated and adaptive community with species diversity, composition and functional organization, comparable to that of the natural habitats of the region.[5,6] Figure 7.1 depicts *Podarcis sicula*, a Mediterranean lizard that has expanded to different regions of the world, including the south of the USA and Mexico, living in grassy areas, road and wood edges, sandy areas near the sea or domestic gardens.[7]

Figure 7.1 Mediterranean lizard *Podarcis sicula*.

The main principle of biomonitoring centers on the concept that living organisms reflect the conditions in which they live, and therefore any change in the studied biota implies a change in their particular environment. Thus, these types of studies may compensate the drawbacks of water chemical examination related to the expression of conditions only at the moment of sampling, the difficulty in quantifying the hundreds of substances usually present in a water body, and the fact that toxic agents may be present in minute amounts.[6]

A number of successful aquatic biomonitors sustain the advantages of this method. There have been studies on animals or plants belonging to different phyla that have demonstrated physiological and genetic alterations as a result of exposure to a variety of stressors. Such studies, however, have indicated some essential requirements for a proposed biomonitor. Initially, clear knowledge is required regarding the organism's taxonomy and biology, followed by specific structural, behavioral and population characteristics that would support its usefulness. There are several conventional biomonitors, such as *Daphnia magna* or *Hyalella azteca*, that have a tested protocol for bioassays in the laboratory.[8] They are inter-calibrated, allowing comparisons between the effects of water from different water bodies, to the study of sites along time or specific contaminants at different concentrations. However, in field studies there are few sentinel organisms that have been used to assess the effects of contaminants; in these cases, the sentinel species should be native since they are facing adverse environmental conditions. In this sense, there are several organisms that are not subject to specific and tested protocols, but despite this they are able to reflect the environmental conditions and therefore can be used as non-conventional biomonitors. According to the information expressed earlier, the aim of the present chapter is to analyze the biology and particular characteristics of the freshwater crayfish group, and especially those of *Cambarellus montezumae*, so that they can sustain their usefulness for monitoring purposes in their habitat, which corresponds to a variety of water bodies, such as ponds, rivers, lakes, channels, and reservoirs of central Mexico, places that can be moderately or severely impacted by natural effects or anthropogenic activities.

7.2 Freshwater Biomonitors: Characteristics and Relevance for the Ecosystem

Aquatic ecosystems are exposed to a variety of stressors, including chemicals that are discharged to water bodies and form complex mixtures, which are difficult to assess. Marine environments in particular have a great capacity for dilution, an advantage that is not usually found in freshwater ecosystems, and is therefore one of the problems that contribute to their ecological risk. Moreover, their ecological risk is increased because of their exposure to human activities, such as agriculture, cattle raising, urbanization, municipal and industrial wastewater discharges, and deforestation.[1,9] The hazards or improper use of freshwater bodies can be grouped into five categories: overexploitation, water pollution, flow modification, degradation of habitat, and invasion of exotic species. While the first category mainly refers to fish, reptiles, and amphibians, the other four categories may have consequences for all freshwater biodiversity. Pollution problems in particular are widespread and increasing, in spite of efforts to reduce the distinct types of discharges that may damage ecosystems, efforts that are especially made in the more industrialized countries.[1] At present, one of the most endangered ecosystems is inland waters, which are extraordinarily abundant and have around 100 000 species of animals and plants.[1,10]

Traditionally, water resources have been assessed using physical and chemical parameters,[11–13] an approach that cannot consider the induction of chemical reactions or the antagonistic, synergistic or potentiation effects that may have an impact on aquatic organisms. More recently, therefore, attention has been focused on the effects of pollutants on aquatic biota since their exposure to environmental contaminants can affect the survival of the studied organism(s), and therefore this may potentially affect the structure and function of aquatic ecosystems.[14]

Biomonitoring is a scientific technique for assessing aquatic ecosystems exposed to natural and synthetic chemicals through the use of an organism or a group of organisms in which one could examine biological responses useful for identifying and monitoring changes in the aquatic environment.[12] In accordance with John,[15] Zhou *et al.*,[12] Hamza-Chaffai,[16] and López-López and Sedeño-Díaz,[17] an ideal biomonitor must have the following characteristics:

(a) Capacity to accumulate high levels of pollutants without death, or be too sensible.
(b) Sedentary lifestyle or very low mobility.
(c) Sufficient abundance and wide distribution for repetitious sampling and comparison.
(d) Sufficiently long life for the comparison between various ages.
(e) Suitable target cell/tissue provided for further research at micro-cosmic level.
(f) Easy sampling and culturing in the laboratory.
(g) Ability to show a dose–effect relationship.
(h) Usefulness for both long- and short-term monitoring.
(i) Ability to play a significant ecological role in the food web.

Biomonitoring can be practiced from any level of biological organization (suborganismal, organismal, population, community, and ecosystem)[18] and it has a broad relationship with exposure time. In the suborganismal and organismal levels, the use of biomarkers is the best approach to identify early warning responses, before the damage by pollutants becomes evident in superior organization levels. The kind of damage identified and measured by biomarkers has the advantage that it can be evident in a short time of exposure (from seconds to days). A biomarkers is defined as a xenobiotically induced variation in cellular or biochemical components or processes, structures, or functions that can be measured in a biological system or samples.[19] There are diverse types of biomarkers and many of them are sensitive to a particular class of contaminants. Biomarkers must also meet certain requirements to afford reliable information about the aquatic ecosystem health. Van der Oost *et al.*[20] proposed some of these requirements:

(a) They must be reliable, relatively cheap and easy to perform.
(b) Their effect or response to pollutants should be sensitive in order to serve as an early warning parameter.
(c) The baseline data they present should be well defined in order to distinguish between natural variability and contaminant-induced stress.
(d) The impacts of confounding factors should be well established.
(e) The underlying mechanism of the relationship between biomarker response and pollutant exposure (dosage and time of exposure) should be established.
(f) Their toxicological significance should be established.

In addition, Fossi and Marsili[21] proposed that biomarkers should preferentially be non-invasive or non-destructive, to allow or facilitate environmental monitoring of pollution effects on protected or endangered species.

Because biomarker results are not easy to extrapolate from suborganismal and organismal levels to ecosystem effects, regulatory agencies have not completely accepted this approach. However, they have generally recognized that the use of biomarkers is a highly reliable method to examine exposure to environmental stressors, or to determine adverse health effects on biota from pollutant exposures.[14]

Early warning behavioral responses are another way of assessing aquatic ecosystem health at the level of organisms, and they consist of measuring aspects such as locomotion, gill ventilation, and feeding depression or avoidance. Behavioral responses are extremely sensitive to chemical stress and have the advantage of not being invasive or destructive tools to assess the environment.[12,22]

Macroinvertebrates are another interesting group for biomonitoring purposes since they meet almost all criteria required for an adequate biomonitor. Diverse phyla of this large group can be used for such a purpose, including Platyhelminthes, Annelida, Mollusca, and Arthropoda, with Insecta and Crustacea as the more representative groups of the last mentioned phyla. Biological indices based on macroinvertebrates to assess freshwater ecosystems have been employed. Among these, one of the most commonly used is the Biological Monitoring Working Party, which was designed in the United Kingdom, and has been modified for use in several countries. Likewise, considering the concept of biotic integrity proposed by Karr,[23] studies have used diverse indices taking into account biological and ecological attributes from macroinvertebrates.

7.3 Distribution and Taxonomy of Freshwater Crayfishes

Freshwater crayfishes are distributed around the world, with the exception of the Indian and the Antarctic continents. Two centers of diversity have been described for the organisms: the southeastern Appalachian Mountains in the Northern Hemisphere and southeast Australia in the Southern Hemisphere. There are currently over 640 described species of these organisms, which are taxonomically organized in two superfamilies, the Northern Hemisphere Astacoidea and the Southern Hemisphere Parastacoidea. The first superfamily contains two families: the Cambaridae, characterized by a high species diversity, which includes over 420 species contained within 12 genera, and the Astacidae family with six genera and 39 species. On the other hand, the Parastacoidea superfamily comprises a single family (Parastacidae) consisting of 15 genera and over 170 species.[24,25]

The freshwater crayfish has been considered as a monophyletic group and a sister taxon to the clawed lobsters from the superfamily Nephropoidea. However, studies on the Cambaridae family have not necessarily supported this type of phylogeny, particularly because of the position of the genus Cambaroides, which could be associated with Astacidae; because of this, the Cambaridae family is suggested to require a relatively complete taxonomic sampling in order to reach reasonable conclusions.[25,26]

The wide species diversity of the Cambaridae family is distributed in North America east of the Rocky Mountains, northward into southern Canada, southward through Mexico, and also in Asia (Figure 7.2). In Mexico there is an estimated presence of around 11.9% of the world's diversity of Decapodas. Of that amount, 89.9% corresponds to marine species, and the

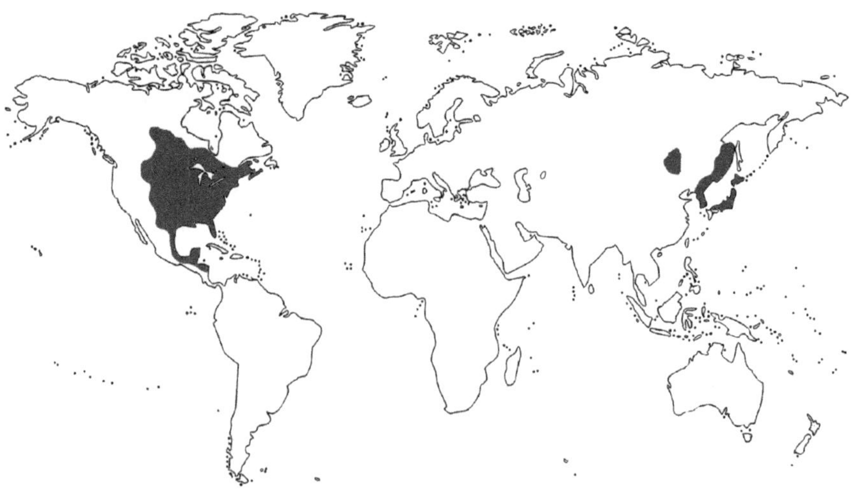

Figure 7.2 Distribution of the Cambaridae family around the world. Modified from Crandall *et al.*[26]

remainder corresponds to freshwater species. In these last groups, a high degree of endemism has been described; for example, this characteristic is found in at least 4 of 12 species and 21 of 83 species of the families Atyidae and Palemonidae, respectively; besides, from organisms of the Cambaridae family, 55 are endemic in México and represent about 13% of the described world species.[27] The Cambaridae family of crayfishes is distributed from the south of Canada through the eastern part of the USA, and in Mexico in the slopes of the Gulf of Mexico and through the Transversal Neovolcanic Axis; however, some species can be found in Guatemala, Belize, Honduras and Cuba.[28,29]

The Cambaridae family includes three genera: *Orconectes, Procambarus*, and *Cambarellus*, whose species are widely distributed from the north of Florida to the south of Texas. In Mexico we found endemism in the northern states of Chihuahua, Coahuila, and Nuevo León, as well as in the states of Sinaloa, Nayarit, Colima, Jalisco, and Michoacán. In the central region, endemism is also observed in parts of Querétaro, Guanajuato, Tlaxcala, Morelos, State of México, Mexico City, Hidalgo and Puebla.[30,31]

7.4 General Characteristics of *Cambarellus montezumae*

According to the Integrated Taxonomic Information System,[32] the taxonomic position of the studied crayfish is as follows:

Kingdom: Animalia
Phylum: Arthropoda
Class: Malacostraca
Order: Decapoda
Superfamily: Astacoidea
Family: Cambaridae
Genus: *Cambarellus*
Species: *montezumae*

Figure 7.3 shows the distribution of Cambaridae and *Cambarellus montezumae* in Mexico. *C. montezumae* is an endemic crayfish species in Mexico, mostly inhabiting seven states in the central part of the country: Querétaro, Tlaxcala, México, Puebla, Jalisco, Michoacán, and Mexico City. In these places, the decapod is found abundantly in rivers, lakes, reservoirs, ponds and channels, usually living at a pH from 7.6 to 9.0, oxygen concentrations from 5.0 to 7.5 $mg\,L^{-1}$, and temperatures from 10 to 25 °C; however, the growing of the organism is also adequate in temperatures somewhat lower than 20 °C in a reduced environment, where a large number of sediments and microorganisms may develop.[33,34]

The crayfish lives along the first 50 cm of water depth associated with the riparian vegetation (for example, plants of the families Cyperaceae,

Figure 7.3 Distribution of the Cambaridae family and *Cambarellus montezumae* in México.

Gentianaceae, Umbelliferae, Onagraceae, Polygonaceae, and Lemnaceae), which is important in maintaining the sediments, and in providing organic material, as well as in acting as protection against predators and environmental factors.[31] Besides, crayfish usually construct refuges that they may abandon for variable periods (Figure 7.4).

Although mainly known as detritivores and herbivores, they are in fact voracious polythropic omnivores that may feed on the larvae of other macroinvertebrates, eggs and immature forms of fishes and amphibians, besides detritus, algae and vascular macrophytes. This behavior can play a significant role in regulating thropic chains, and crayfishes are therefore considered one of the largest invertebrates to act as a keystone species in many ecosystems, influencing energy flow and community structure.[35] On the other hand, they are consumed by different predators, including amphibians, reptiles, birds and fishes, for example species of the genus *Cyprinus, Oreochromis* as well as *Chirostoma estor and Amiurus mexicanus.*[31,34,36] In this sense, the prey–predator equilibrium is relevant to avoid disturbance that may affect the ecosystem.[37] Moreover, *C. montezumae* usually interact and compete for space and food with other organisms, mainly annelids, insects, and mollusks.[33]

In regard to the size of the studied crayfish (Figure 7.5), its length in adulthood has been reported to usually vary from 30 to 50 mm, and its body is organized in three segments: the head, followed by the thorax, which together are known as the cephalothorax, and finally the abdomen. The cephalotorax includes a pair of pedunculated eyes, the buccal apparatus,

Figure 7.4 Representative habitats of *Cambarellus montezumae* in the Basin of México: (a) Xochimilco Lake in Mexico City, (b) Tecocomulco Lake, Hidalgo State.

Figure 7.5 Picture of *Cambarellus montezumae* collected in Xochimilco Lake, 2016.

branchial chambers, and the main parts of the nervous, circulatory, digestive and reproductive systems. The thorax specifically includes five pairs of appendices called pereiopods, three of them modified for attack and defense and two others that are thinner and are used for locomotion, feeding and cleanliness. In the last segment, the abdomen, appendices called pleopods are present, as well as the anus and telson.[38,39]

The crayfish has clear sexual dimorphism, shown by the larger size of the females, which also have a larger and wider abdomen than males, a feature related to the storage and transportation of the eggs. In the male the state of sexual maturity is observed in the first pair of pleopods, which are also modified for the copula, while the function of the second pair of pleopods is related to the transportation of filamentous spermatophorous coming from the three testicles situated under the heart, which at the appropriate time are deposited around the female annulus ventralis.[39,40]

With respect to its systemic organization, the animal is known to possess a nervous system formed by a pair of cerebroid ganglia, which are followed by an esophagic collar and a ventral chain of ganglia that give rise to the innervation of the whole body. The animal possesses a small heart situated in the dorsal part of the thorax, where the hemolymph is expulsed through six ramified arteries, to finally reach the different organs. Respiration in the crayfish is carried out by means of feathery gills placed in the branchial chambers at each side of the body; the digestive apparatus starts in the ventral part of the head, and continues with a short esophagus ending in a grinding stomach, which in turn continues in a long intestinal tube where the digestive juices act, and finishes at the base of the telson. As regards excretion, this process is carried out by two glands situated at the ventral zone of the head, called green glands, which collect the wastes carried by the hemolymph and discharge them through the bladder (Figure 7.6).[39,40]

Reproduction of *C. montezumae* occurs throughout the whole year; however, the coupling is low and difficult. After various trials, the male can

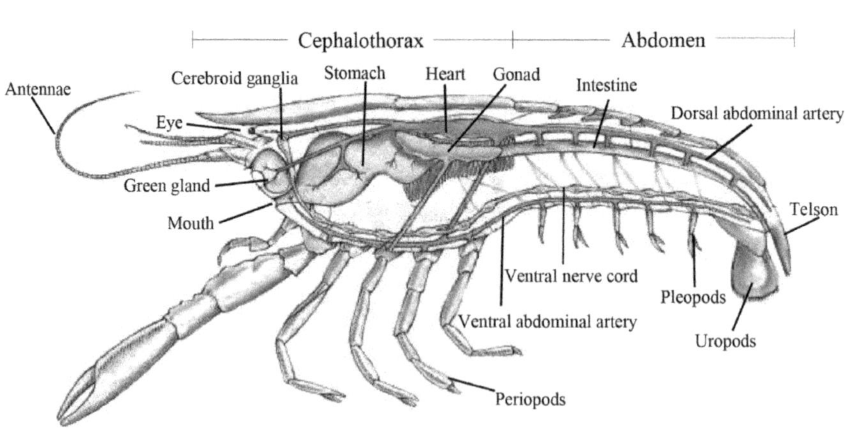

Figure 7.6 Main anatomical features of *Cambarellus montezumae*.

deposit the spermatophorous sac, which may remain in the female for up to six months before fecundation takes place. Eggs are incubated for between 35 to 40 days, and their liberation is gradually made over 2 to 5 days. At the time of hatching, a small crayfish of 2–3 mm long with a soft shell is observed. During this initial period, the organism is mostly adhered to the mother for about 10 days, a time when it is fed with its own reserves from its cephalotorax and can pass through two molting changes; after that, the crayfish adopts a free life and grows through various molting steps.[31,41]

The crayfish grows from 2.5 to 3 mm per month. Males go through a juvenile phase and then to a mature sexual state called form I that can be reached from 4 to 9 months. This is followed by a non-mature sexual state named form II. However, these forms can be alternated in relation to the molting. Mortality was studied in the channels of Xochimilco, where a mean half-life of 15 months was determined for the crayfishes; however, in laboratory conditions animals may live up to 2 years.[31]

7.5 Ecotoxicology as a Tool to Assess Aquatic Ecosystem Health

The biological health of ecosystems is not always an obvious feature of the environment, and owing to their complexity it is not easy to find suitable ecological indicators. A summary of the conceptual definition of ecosystem health includes homeostasis, absence of disease, diversity or complexity, stability or resilience, vigor or scope for growth, and balance between the components of the system.[42,43] This complexity clearly shows the need for multi-view descriptors to capture all the features in order to give a fully informative assessment of the condition of an ecosystem; however, although the use of a particular species or component is not the optimum way to reflect the complexity of an ecosystem, this procedure is very useful in communicating such complex information to resource managers, as well as to public and private institutions related to the evaluation and improvement of the environment.[43,44] Then, the use of aquatic organisms for biomonitoring purposes is highly relevant to examine the integrity of a freshwater ecosystem, more so when it is accompanied by chemical and physicochemical water analysis, the study of riparian vegetation, and other determinations.

Ecotoxicity tests in the monitor organism usually determine acute toxicity or more chronic effects, such as those on growth and reproduction. For these purposes, the use of specific biomarkers is relevant, of which two general classes can be utilized. One class, the exposure biomarkers, can be exemplified by measuring the induction of CYP450 isoenzymes, while the other class, biomarkers of effect, can be exemplified by determining DNA damage indices, acetylcholinesterase inhibition values, or oxidative stress levels, among other parameters; however, biomarkers that integrate exposure and effects have also been described.[14]

In crayfish, studies that provide information about the immunologic, biochemical and physiological effects induced by xenobiotics have been performed. With these aims in mind, authors have measured modifications in the levels of glucose, lactate, and protein content, as well as in the amount of hemocytes, thereby gaining more specific information about the state of the circulatory and excretory systems, and about the level of immunity and disease resistance of the organism. Besides, a number of oxidative stress parameters have been quantified, mainly those related to oxidation in proteins and lipids, along with alterations in antioxidant enzymes.[45] Moreover, histological determinations in gills or other tissues have also been made and they have shown concordance with the damaging effect of the involved stressor.[45]

Ecogenotoxicity is a branch of ecotoxicology that deals with effects on DNA structure and function induced by different agents of the studied ecosystem. Research in this area has been rapidly gaining acceptance because of its close relationship with the health of the involved organism (including a possibly carcinogenic process), and because of the possibility of germline alterations. Another concomitant points toward explaining that such impulse refers to the development of more sensitive methods, which may be used as early biomarkers, as well as to the highly documented knowledge that contaminants may affect the genetic material either directly or indirectly, through active metabolites, or by affecting cellular functions such as replication or transcription.[46,47] Moreover, the consequences of environmental contamination may be reflected in the size and structure of the population, including genetic drift and genetic adaptation.[48]

Studies in ecogenotoxicology were initially developed in fishes, where genotoxins were known to induce different types of DNA damage that could be expressed in malignancies, reduced growth, abnormal development and reduced survival of embryos, larvae and adults.[49] Fishes were therefore considered sensitive indicators of water quality that could highlight the danger of toxicants in the aquatic environment.[50] The determination of excessive DNA breakage was implemented in fishes and turtles to successfully monitor genotoxicity in places receiving a variety of industrial wastes.[51–53] Investigation in freshwater crayfishes has followed this line of research, although more slowly; however, a number of studies have demonstrated the usefulness of these organisms as monitors for contamination disturbances, and for the analysis of physiological and ecological stressors.

The use of the single cell gel electrophoresis (Comet) assay, and the evaluation of micronuclei are two methods that have been successfully implemented in crayfishes; they have also demonstrated their applicability and relevance to determine DNA damage in organisms exposed to different stressors. Therefore, these methods are expected to be more commonly applied in the near future.

The Comet assay is a sensitive and relatively rapid method for detecting DNA damage at the level of individual cells. The damage is mainly manifested as single (or double) strand breaks, alkali-labile sites, and incomplete excision repair sites. The test has the advantage of requiring a small number of cells from practically any tissue, and it is based on the ability of negatively charged loops/fragments to be drawn through an agarose gel in response to an electric field. The extent of DNA migration depends directly on the DNA damage present in the cells.[54] (Figure 7.7). Besides, the damage identification can be carried out in different ways, which range from the arbitrary classification of non-damaged nucleoids to the presence of large Comets, the measurement of Comet length, the length to width Comet index, or the determination of momentum, among other methods.[55]

Micronuclei correspond to small chromatin corpuscles that represent chromosome fragments or complete chromosomes that fail to incorporate into the daughter nuclei during the anaphase step of cell division and remain in the cytoplasm throughout the life cycle of the cell.[56] Micronuclei can

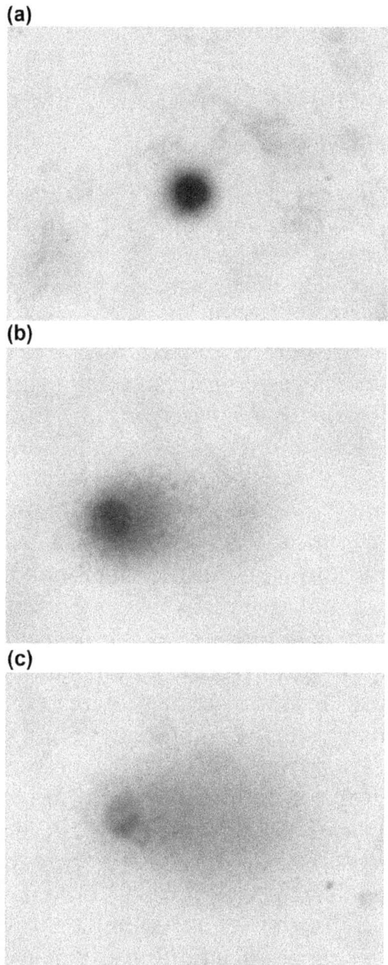

Figure 7.7 Cells treated with the single cell gel electrophoresis (Comet) assay: (a) control (DNA without damage), (b) Comet with moderate DNA damage, and (c) Comet with high DNA damage.

be determined in the hemocytes and stained with appropriate indicators, such as Giemsa or orcein, or by applying the Feulgen technique. On account of the genotoxic information provided by the test, the relative ease with which it is performed, and its reproducibility, the assay has partially replaced the more time-consuming examination of specific chromosome aberrations. However, more sophisticated assays have also been applied in crayfishes. For example, Jiang *et al.*[57] considered that nitrite toxicity involves reactive oxygen species and would demand an increased level of energy that leads to a change in the amount of metabolic enzymes. Therefore, they analyzed the levels of mRNA of ten key metabolic and antioxidant enzymes, and found that nitrite did indeed induce oxidative and metabolic stress in the gills of *Cherax quadricarinatus*, implying that gene expression of the evaluated enzymes was strongly modified.

7.6 Use of Crayfishes in Aquatic Ecosystem Biomonitoring

Freshwater crayfishes have been successfully used to determine the effect of a number of physiological and contaminant stressors. In regard to this field, a number of investigations will be briefly described to support our statement. An interesting example corresponds to the use of *Astacus leptodactylus* in Armenian water bodies.[58,59] This organism gave rise to relevant information that we summarize here. Caged reference organisms were placed for a week in three selected water bodies, and then a genotoxic examination of their hemocytes was carried out by applying the Comet assay and the micronucleus test. The results showed that crayfishes are able to have different intensities of DNA damage that are clearly correlated with the level of contamination in the studied location. Sediment contaminants were identified as polycyclic aromatic hydrocarbons, mineral oil hydrocarbons, heavy metals (Cu, Zn, Ni, Pb, Cd, Co and Hg) and the metalloid As. Therefore, the conclusion of the study was that the crayfish has high sensitivity for monitoring polluted areas. Moreover, the effect of pollution also showed a certain increase in the hemocyte count and in the protein content, a finding related to alterations in metabolic routes. Another interesting point, related to the behavior of the organism, was the adaptation of non-damaged crayfishes to mildly polluted sites. Moreover, non-polluted animals were collected and transported to laboratory conditions in order to evaluate the effect of temperature increase and air exposure for up to 7 days, and food deprivation for up to 14 days. During this time the authors observed the presence of changes in the levels of glucose, lactate, and protein content, and in the amount of hemocytes; however, after a recovery period of 7 days the evaluated animals tended to return to normal values, a finding that suggests the adaptability of the organisms. Furthermore, the strongest effect was produced with the temperature increase, which was also concordant with an increase in DNA damage.

The ability of crayfishes to tolerate low doses of chemicals, besides their ability to adapt to normality after a recovery period, was observed in a report on the effect of peracetic acid, a therapeutic agent used to cure a crayfish plague caused by the oomycete *Aphanomyces astaci.*[45] In the study using *Pacifastacus leniusculus*, the authors established the safe use of the chemical at the recommended therapeutic level after a number of histologic observations in the gill, hepatopancreas, and antenna gland, as well as the determination of catalase and superoxide dismutase.

Accumulation of pollutants by crayfishes is a feature that has also been confirmed in *P. leniusculus* living in lakes and streams of southern Sweden.[60] The organism, which mainly feeds on other invertebrates, primary producers and detritus, was found to accumulate persistent organic pollutants, such as polychlorinated biphenyls and DDT. Moreover, in the River Meuse in The Netherlands, which receives agricultural and urban pollutants, researchers detected aromatic DNA adducts, heavy metal residues, polychlorinated biphenyls, and organochlorine pesticides in the hepatopancreas of crayfishes, findings that indicate that these animals can be used as biological indicators of exposure to both organic and inorganic pollution in aquatic systems.[61]

The ability of crayfishes to respond to pollutants has also been demonstrated in *Procambarus clarkii*. In this case, de la Sienra *et al.*[62] exposed experimental animals to three concentrations of potassium dichromate for 7 days (from 50 to 400 $\mu g L^{-1}$) and found a significant induction of micronuclei in their hemocytes at the high dose; furthermore, such damage was correlated with the amount of the chemical in the water and with the amount of chromium found in the gills. Moreover, the accumulation of metals in the same species was demonstrated in animals exposed to petroleum contamination.[63]

7.7 *Cambarellus montezumae* as a Biomonitor in Freshwater Ecosystems

The selection of a biomonitor represents a significant challenge. Initially, thorough knowledge is required regarding its feeding habits, life cycle, reproductive behavior and population structure, among other aspects.[64,65] As demonstrated in the previous section, this knowledge is available with respect to *Cambarellus montezumae.* Other important characteristics required for an appropriate monitor are also present in *C. montezumae*. These include contaminant accumulation, dose–response relationship, and sedentary life style with low migration in the area. What is more, a central property is that they can act as bottom dwellers, keeping much of their bodies in contact with surrounding objects, and that they are abundant organisms with simple feeding habits, relatively tolerant to environmental changes, and easy to transport and adapt to laboratory facilities.[39,66]

Studies on *C. montezumae* were started by Villalobos[38] with a description of the taxonomy, morphology, and distribution of the Cambaridae in

Mexico. In Mexico, the crayfish has been consumed since the period of the Aztec civilization to the present time. With respect to this consideration, various authors have demonstrated the significant nutritive value of the animal; for example, Santos *et al.*[67] reported the following values: humidity 77.57%, protein 44.04%, carbohydrates 20.17%, lipids 3.56%, fiber 8.93%, and minerals 23.3%. Moreover, the organism is rich in the polyunsaturated lipids omega 3 and omega 6.[68] Owing to this nutritional advantage, a number of studies on the crayfish have been focused on evaluating parameters that may confirm its cultivation for both nutritional and commercial purposes, and yet other studies have analyzed a variety of *in situ* or *in vitro* scientific parameters. Unfortunately, most of them have been reported in academic theses or in local journals, a situation that suggests the need for more international diffusion, and an effort coherent with the present chapter. In this section in particular, our attempt is to summarize findings on *C. montezumae* that may support its quality for biomonitoring purposes.

A large amount of varied research related to the growing and reproduction of *C. montezumae* in different conditions of temperature and diet has been performed. Temperature is a primal factor for the behavior, tolerance, and resistance of species; in the presently studied animal, in spite of the more common and optimal range for its development (from 12 to 26 °C), an ample range of temperature tolerance (from −2 to 30 °C) has been documented, a feature that suggests high ecological plasticity reflected in a capacity for adaptation to diverse environments.[39,69] A further well-studied aspect of the crayfish is the effect of different types of food consumption on its growth, survival, and efficiency of energy transfer.[28,37,39,70] The extent of these assays has varied from a few weeks to several months, and the results have demonstrated the generally good acceptance by the crayfish of different feeding, as reflected in normal growing and reproduction. Some foods used in these studies include: microbiologically enriched detritus of *Egeria densa*, macerated *Poecillia serrata* combined with carrot,[71] mixture of amaranth, beer yeast and chicken feathers,[72] rabbit food mixed with meat flour,[72] *Daphnia pulex*,[73] detritus of *Eichhornia crassipes* and *Typha latifolia*,[74] and squid flour plus fish oil.[39]

Studies in *in situ* conditions have reported that juvenile organisms are mainly hunters, consuming animal protein, while in adulthood this activity decreases and is partially substituted by the ingestion of vegetal detritus. Besides, two peaks of feeding have been described, probably related with light intensity, an intense peak between 18:00 and midnight, and another peak at dawn.[70,75,76]

Population dynamic studies have also been made in various water bodies, and have shown reproductive activity throughout the year, although preferentially in the warmest months. An almost equal amount of males and females with significant numbers of juvenile animals have been found, and a maximum size of 48.7 mm, according to von Bertalanffy's growing model.[77]

Physicochemical parameters besides temperature have shown that crayfishes usually live in a pH between 7 and 9, and energetic balance determinations have shown a calorie consumption of about 1440 per individual per day in warm months and of 876 in cold months, a 2% loss of energy by evacuation, 46–51% by respiration and 7% by nitrogen excretion. Thus, a calculation of 45% energy used for growing in warm months and 40% in cold months was made.[39]

As to the presence of xenobiotics in *C. montezumae*, a number of heavy metals have been determined in organisms of some water bodies, places that are known to be affected by anthropogenic activities, and by the introduction of residual treated water.[31,78–80] The largest number of detected metals included Fe, Cu, Mn, Zn, Pb, Cr, Co, Ni, and Cd. In one of the reports, researchers observed the differential capacity of the crayfish to accumulate metals depending on the three different selected sizes of the organisms, and that the macroinvertebrates were in the accepted range for human ingestion. However, two reports concluded that according to the Mexican and international norms, the crayfish were not appropriate for human consumption: While this discrepancy could be related to the place and month of collection, method of quantification, and other variables, the noteworthy capacity of the organism to accumulate and apparently live normally with such wasteful charge is interesting; however, the use of early warning markers could surely reveal the molecular or cellular damage induced by the metal exposure.

The described characteristics and responses of *C. montezumae* suggest its capability to be used for freshwater biomonitoring purposes, a possibility also supported by a report that evaluated its toxic and genotoxic response toward the effect of contaminants.[81] The study was made in laboratory facilities using reconstituted water as well as water from the reservoir where crayfishes were captured. The LC_{50} of the pesticides dieldrin and chlorpyrifos was initially determined, followed by the determination of their capacity to induce DNA damage (using the Comet assay) in the brain and hepatopancreas of the crayfish. The authors additionally determined the lipid peroxidative effect of the pesticides in hepatopancreas tissue. Two concentrations of each chemical were applied in the water and observations were made at 24, 48 and 72 h. Dieldrin produced DNA damage with the two tested concentrations (0.05 and 0.5 $\mu g\,L^{-1}$) at the three evaluated times in the brain and the hepatopancreas, while chlorpyrifos was similarly genotoxic in the brain with the high tested concentration at 72 h of exposure, and in the hepatopancreas with the two concentrations at the three evaluated times. A significant lipid peroxidation increase was also determined with the two concentrations, suggesting a certain correlation with the observed DNA damage. Moreover, comet length was somewhat larger in animals exposed to pesticides in water from the reservoir. The obtained results were interesting given that the low concentration was approximately a hundred times lower than the obtained LC_{50}. Such results are also encouraging to further evaluate the effect of anthropogenic activities near the involved water bodies, or for testing the involved (or other) chemical under laboratory conditions. The investigations described earlier, therefore, support our suggestion that

C. montezumae is a good candidate for freshwater biomononitoring purposes.

References

1. D. Dudgeon, A. H. Arthington, M. O. Gessner, Z. Kawabata, D. J. Knowler, C. Lévêque, R. J. Naiman, A. H. Prieur-Richard, D. Soto, M. L. Stiassny and C. A. Sullivan, Freshwater biodiversity: importance, threats, status and conservation challenges, *Biol. Rev.*, 2006, **81**, 163–182.
2. J. L. Nel, D. J. Roux, R. Abell, P. J. Ashton, R. M. Cowling, J. V. Higgins, M. Thieme and J. H. Viers, Progress and challenges in freshwater conservation planning, *Aquat. Conserv. Mar. Freshwater Ecosyst.*, 2009, **19**, 474–485.
3. F. L. Xu, S. E. Jørgensen and S. Tao, Ecological indicators for assessing freshwater ecosystem health, *Ecol. Modell.*, 1999, **116**, 77–106.
4. K. Johnston and B. J. Robson, Habitat use by five sympatric Australian freshwater crayfish species (Parastacidae), *Freshwater Biol.*, 2009, **54**, 1629–1641.
5. J. R. Karr and D. R. Dudley, Ecological perspective on water quality goals, *Environ. Manage.*, 1981, **5**, 55–68.
6. J. Day, Biomonitoring: appropriate technology for the 21st century, *1 st WARFSA/WaterNet Symposium: sustainable use of water resources*, Maputo, South Africa, 2000, p. 8.
7. C. M. Donihue, M. Lambert and G. Watkins-Colwell, Italian wall lizard (*Podarcis sicula*): The first population found in New England, *Herpetol. Rev.*, 2014, **45**, 661–662.
8. J. R. Elphick, K. D. Bergh and H. C. Bailey, Chronic toxicity of chloride to freshwater species: effects of hardness and implications for water quality guidelines, *Environ. Toxicol. Chem.*, 2011, **30**, 239–246.
9. J. E. Sedeño-Díaz and E. López-López, Freshwater fish as sentinel organisms: from the molecular to the population level, a review, in: *New advances and contributions to fish biology*, ed. H. Turker, Intech, Rijeka, 2013, vol. 4, pp. 151–173.
10. D. J. Hawksworth and M. T. Kalin-Arrollo, Magnitude and distribution of biodiversity. in: *Global biodiversity assessment*, ed. V. H. Heywood, Cambridge University Press, Cambridge U.K., 1995, pp. 107–191.
11. J. E. Sedeño-Díaz and E. López-López, Water quality in the Río Lerma, Mexico: an overview of the last quarter of the twentieth century, *Water. Resour. Manag Ser.*, 2007, **21**, 1797–1812.
12. Q. Zhou, J. Zhang, J. Fu, J. Shi and G. Jiang, Biomonitoring: an appealing tool for assessment of metal pollution in the aquatic ecosystem, *Anal. Chim. Acta*, 2008, **606**, 135–150.
13. T. Espinal-Carreón, J. E. Sedeño-Díaz and E. López-López, Evaluación de la calidad del agua en la Laguna de Yuriria, Guanajuato, México, mediante técnicas multivariadas: un análisis de valoración para dos épocas 2005, 2009-2010, *Rev. Int. Contam. Ambient.*, 2013, **29**, 147–163.

14. S. E. Hook, E. P. Gallagher and G. E. Batley, The role of biomarkers in the assessment of aquatic ecosystem health, *Integr. Environ. Assess. Manage.*, 2014, **10**, 327–341.
15. J. John, Bioassessment of health of aquatic systems by the use of diatoms, in: *Modern trends in applied aquatic ecology*, ed. R. S. Ambasht and N. K. Ambasht, Kluwer Academic/Plenum Publishers, New York, U.S.A., 2003, vol. 1, pp. 1–20.
16. A. Hamza-Chaffai, Usefulness of bioindicators and biomarkers in pollution biomonitoring, *Int. J. Biotechnol. Wellness Ind*, 2014, **3**, 19–26.
17. E. López-López and J. E. Sedeño-Díaz, Biological indicators of water quality: the role of fish and macroinvertebrates as indicators of water quality, in: *Environmental Indicators*, ed. R. H. Armon and O. Hänninen, Springer, Netherlands, 2015, p. 1068.
18. L. Li, B. Zheng and L. Liu, Biomonitoring and bioindicators used for river ecosystems: definitions, approaches and trends, *Procedia Environ. Sci.*, 2010, **2**, 1510–1524.
19. L. R. Shugart, J. F. McCarthy and R. S. Halbrook, Biological markers of environmental and ecological contamination: an overview, *Risk Anal.*, 1992, **12**, 353–360.
20. R. van der Oost, J. Beyer and N. P. Vermeulen, Fish bioaccumulation and biomarkers in environmental risk assessment: a review, *Environ. Toxicol. Pharmacol.*, 2003, **13**, 57–149.
21. M. C. Fossi and L. Marsili, The use of non-destructive biomarkers in the study of marine mammals, *Biomarkers*, 1997, **2**, 205–216.
22. A. Gerhardt, Behavioural early warning responses to polluted water. Performance of *Gammarus pulex* L. (Crustacea) and *Hydropsyche angustipennis* (Curtis) (Insecta) to a complex industrial effluent, *Environ. Sci. Pollut. Res.*, 1996, **3**, 63–70.
23. J. R. Karr, Biological integrity: a long-neglected aspect of water resource management, *Ecol. Appl.*, 1991, **1**, 66–84.
24. H. H. Hobbs, An illustrated checklist of the American crayfish (Decapoda: Astacidae, Cambaridae, and Parastacidae), *Smithson. Contr. Zool.*, 1989, **480**, 8–10.
25. K. A. Crandall and J. E. Buhay, Global diversity of crayfish (Astacidae, Cambaridae and Parastacidae- Decapoda) in freshwater, in: *Freshwater animal diversity assessment*, ed. E. V. Balian, C. Leveque, H. Segers and K. Martens, Springer, Netherlands, 2008, vol. 595, pp. 295–301.
26. K. A. Crandall, D. J. Harris and J. W. Fetzner Jr., The monophyletic origin of freshwater crayfish estimated from nuclear and mitocondrial DNA sequences, *Proc. Biol. Sci.*, 2000, **267**, 1679–1686.
27. F. Álvarez, J. L. Villalobos, M. E. Hendrickx, E. Escobar-Briones, G. Rodríguez-Almaraz and E. Campos, Biodiversidad de crustáceos decápodos (Crustacea: Decapoda) en México, *Rev. Mex. Biodiv.*, 2014, **85**, S208–S219.
28. J. R. Latournerié, Y. Nacif, R. J. Cárdenas and J. Romero, Crecimiento, producción y eficiencias de energía de crías de acocil *Cambarellus*

montezumae (Saussure) alimentadas con detritus de *Egeria densa*, *Revista electrónica de veterinaria,* 2006, 7, 12.

29. F. Álvarez, J. L. Villalobos, G. Armendáriz and C. Hernández, Relación biogeográfica entre cangrejos dulceacuícolas y acociles a lo largo de la zona mexicana de transición: revaluación de la hipótesis de Rodríguez (1986), *Rev. Mex. Biodiv.*, 2012, **83**, 1073–1083.
30. Y. Rojas, F. Álvarez and J. L. Villalobos, Morphological variation in the crayfish *Cambarellus (Cambarellus) montezumae* (Decapoda: Cambaridae), in: *Modern approaches to the study of Crustacea*, ed. E. Escobar-Briones and F. Álvarez, Kluwer Academic/Plenum, New York, 2002, pp. 311–317.
31. F. Álvarez and R. Rangel, Estudio poblacional del acocil *Cambarellus montezumae* (Crustacea: Decapoda: Cambaridae) en Xochimilco, México, *Rev. Mex. Biodiv.*, 2007, **78**, 431–437.
32. www.itis.gov (last accessed May 2016).
33. F. Arana-Magallón, R. Pérez-Rodríguez and A. Malpica-Sánchez, Cambáridos de tres embalses del estado de Tlaxcala, México (Crustacea: Decapoda), *Rev. Soc. Mex. Hist. Nat.*, 1998, **48**, 23–35.
34. A. N. Cerón-Ortiz, O. Moctezuma-Resendiz, M. A. Ángeles-Monroy, E. Montufar-Serrano and J. A. León-Escamilla, Efecto interactivo del alimento y la calidad del agua en el crecimiento y sobrevivencia de postlarvas de acocil de río *Cambarellus montezumae*, *Rev. Mex. Biodiv.*, 2015, **86**, 131–142.
35. N. J. Wilson and C. R. Williams, A critical review of freshwater crayfish as amphibian predators: capable consumers of toxic prey? *Toxicon*, 2014, **82**, 9–17.
36. M. Rodríguez-Serna, Aspectos sobre el crecimiento y reproducción del acocil *Cambarellus (Cambarellus) montezumae* (Saussure, 1857). Un análisis ecofisiológico, PhD thesis, Universidad Autónoma Metropolitana, México, 1999.
37. J. L. Arredondo-Figueroa, A. Vásquez-González, L. G. Núñez-García, I. A. Barriga-Sosa and J. T. Ponce-Palafox, Aspectos reproductivos del acocil *Cambarellus* (*Cambarellus*) *montezumae* (Crustacea: Decápoda: Cambaridae) en condiciones controladas, *Rev. Mex. Biodiv.*, 2011, **82**, 169–178.
38. F. A. Villalobos, Cambarinos de la fauna Mexicana (Crustacea:Decapoda). PhD thesis, Facultad de Ciencias, Universidad Nacional Autónoma de México, México, 1955, p. 290.
39. G. M. García Padilla, Aspectos sobre el crecimiento y reproducción del acocil *Cambarellus (Cambarellus) montezumae* (Saussure, 1857). Un análisis ecofisiológico, MSc thesis, Facultad de Ciencias, Universidad Nacional Autónoma de México, México, 2014. p. 199.
40. J. H. Thorp and A. P. Covich, *Ecology and Classification of North American Freshwater Invertebrates*, Academic Press, USA, 2009, p. 1036.
41. J. W. Avault and J. V. Huner, Crawfish culture in the United States. in: *Crustacean and Mollusk Aquaculture in theUnited States*, ed. J. V. Huner and E. E. Brows, AVI. Publ. Co., Westpent, Connecticut, 1985, pp. 1–61.

42. R. Costanza, Toward an operational definition of ecosystem health, in: *Ecosystem Health: New Goals for Enviromental Management*, ed. R. Costanza, B. G. Norton and B. D. Haskell, Island Press, Washinton DC, U.S.A., 1992, pp. 239–256.
43. F. L. Xu, S. E. Jørgensen and S. Tao, Ecological indicators for assessing freshwater ecosystem health, *Eco. Mod.*, 1999, **116**, 77–106.
44. B. D. Haskell, B. G. Norton and R. Costanza, What is ecosystem health and why should we worry about it? ed. R. Costanza, B. G. Norton and B. D. Haskell, Island Press, Washinton DC, U.S.A., 1992, pp. 3–20.
45. L. Chapani, E. Zuskova, A. Stora, J. Velisek and A. Konba, Histological changes and antioxidant enzyme activity in signal crayfish (*Pacifastacus leniusculus*) associated with sub-acute peracetic acid exposure, *Fish Shellfish Immunol.*, 2016, **48**, 190–195.
46. N. M. Belfiore and S. L. Anderson, Genetic patterns as a tool for monitoring and assessment of environmental impacts: the example of genetic ecotoxicology, *Environ. Toxicol. Pharmacol.*, 1998, **51**, 465–479.
47. G. Mohanty, J. Mohanty, S. D. Jena and S. K. Dutta, Genotoxicity testing in pesticide safety evaluation, in: *Pesticides in the modern world - Pests control and pesticides exposure and toxicity assessment*, ed. M. Stoytcheva, Intech, Rijeka, 2011, pp. 403–423.
48. B. U. Akpoilih, Fish ecogenotoxicology: an emerging science, an emerging tool for environmental monitoring and risk assessment, *J. Environ. Anal. Toxicol.*, 2013, **3**, 1–8.
49. P. J. Singh, S. Pandey and S. Sharma, Micronucleus assay for evaluation of in vivo genotoxicity in fishes: Training in genotoxic assays in fishes. TOGAIF-2005, 2005, pp. 1–107.
50. S. H. Mitchell and S. Kennedy, Tissue concentrations of organochlorine compounds in common seals from the coast of Northern Ireland, *Sci. Total Environ.*, 1992, **115**, 163–177.
51. S. De Flora, L. Viganò, F. D'Agostini, A. Camoirano, M. Bagnasco, C. Bennecelli, F. Melodia and A. Arillo, Multiple genotoxicity biomarkers in fish exposed in situ to polluted river water, *Mutat. Res.*, 1993, **319**, 167–177.
52. M. O. Obiakor, C. D. Ezeonyejiaku, C. O. Ezenwelu and G. C. Ugochukwu, Aquatic genetic biomarkers of exposure and effect in catfish (*Clarias gariepinus*, Burchell, 1822), *Am-Euras. J. Toxicol. Sci.*, 2010, **2**, 196–202.
53. M. O. Obiakor, J. C. Okonkwo, P. C. Nnabude and C. D. Ezeonyejiaku, Ecogenotoxicology: micronucleus assay in fish erythrocytes as in situ aquatic pollution biomarker: a review, *J. Anim. Sci. Adv.*, 2012, **2**, 123–133.
54. www.cometassay.com (last accessed May 2016).
55. E. Rojas, M. C. Lopez and M. Valverde, Single cell gel electrophoresis assay: methodology and applications, *J. Chromatogr. B. Biomed. Sci. Appl.*, 1999, **722**, 225–254.
56. I. Udroiu, The micronucleus test in piscine erythrocytes, *Aquat. Toxicol.*, 2006, **79**, 201–204.

57. Q. Jiang, W. Zhang, H. Tan, D. Pan, Y. Yang, Q. Ren and J. Yang, Analysis of gene expression changes, caused by exposure to nitrite, in metabolic and antioxidant enzymes in the red claw crayfish, *Cherax quadricarinatus*, *Ecotoxicol. Environ. Saf.*, 2014, **104**, 423–428.
58. G. I. Klobučar, O. Malev, M. Šrut, A. Štambuk, S. Lorenzo, Ž. Cvetković, E. A. Ferrero and I. Maguire, Genotoxicity monitoring of freshwater environments using caged crayfish (*Astacus leptodactylus*), *Chemosphere*, 2012, **87**, 62–67.
59. O. Malev, M. Šrut, I. Maguire, A. Štambuk, E. A. Ferrero, S. Lorenzo and G. I. Klobučar, Genotoxic, physiological and immunological effects caused by temperature increase, air exposure or food deprivation in freshwater crayfish *Astacus leptodactylus*, *Comp. Biochem. Physiol. C Toxicol. Pharmacol.*, 2010, **152**, 433–443.
60. N. Holmqvist, P. Stenroth, O. Berglund, P. Nyström, W. Graneli and P. Larsson, Persistent organic pollutants (POP) in a benthic omnivore - A comparison between lake and stream crayfish populations, *Chemosphere*, 2007, **66**, 1070–1078.
61. P. A. Schilderman, E. J. Moonen, L. M. Maas, I. Welle and J. C. Kleinjans, Use of crayfish in biomonitoring studies of environmental pollution of the river Meuse, *Ecotoxicol. Environ. Saf*, 1999, **44**, 241–252.
62. E. De la Sienra, M. A. Armienta and M. E. Gonsebatt, Potassium dichromate increases the micronucleus frequency in the crayfish *Procambarus clarkii*, *Environ. Pollut.*, 2003, **126**, 367–370.
63. M. B. Anderson, P. Reddy, J. E. Preslan, M. Fingerman, J. Bolliger, L. Jolibois, G. Maheshwarudu and W. J. George, Metal accumulation in crayfish, *Procambarus clarkii,* exposed to a petroleum-contaminated bayou in Louisiana, *Ecotoxicol. Environ. Saf.*, 1997, **37**, 267–272.
64. S. Augusto, C. Máguas and C. Branquinho, Guidelines for biomonitoring persistent organic pollutants (POPs), using lichens and aquatic mosses - A review, *Environ. Pollut.*, 2013, **180**, 330–338.
65. D. Kulkarni, A. Gergs, U. Hommen, H. T. Ratte and T. G. Preuss, A plea for the use of copepods in freshwater ecotoxicology, *Environ. Sci. Pollut. Res. Int.*, 2013, **20**, 75–85.
66. F. Páez-Osuna and C. Osuna-Martínez, Biomonitors of coastal pollution with reference to the situation in the Mexican coasts: a review on the utilization of organisms, *Hidrobiológica*, 2011, **21**, 229–238.
67. S. M. Santos, A. Celis and M. García Rodríguez, Consumo de acocil *Cambarellus montezumae* como fuente importante de macronutrientes, *Rev. Latinoam. Quim.*, 2011, **38**, 75–76.
68. A. M. Monroy, J. Carabias, C. Mapes and C. Toledo, Análisis de parámetros físico-químicos que afectan la población en medio natural del acocil (*Cambarellus montezumae*) en Tezontepec de Aldama, Hildago, I Reunión de Innovación Acuícola y Pesquera Campeche, Nutrición Acuícola, 2010, p.43.
69. E. Aguilar-Román, Efecto de la temperatura en el metabolismo energético y crecimiento de crías de acocil *Cambarellus montezumae*

(Saussure), Theses BsC, Facultad de Ciencias, Universidad Nacional Autónoma de México, México, 2011.

70. R. Serna, Influencia de la dieta en la eficiencia de asimilación. Pérdidas de energía por respiración y productos nitrogenados en el balance energético del acocil *Cambarellus montezumae* (Saussure) (Crustaceae: Cambaridae), Tesis BSc Universidad Autónoma Metropolitana, México, 1991, p. 43.
71. B. F. Morones, Aspectos reproductivos bajo condiciones de laboratorio de *Cambarellus montezumae* (Saussure) proveniente de la zona lacustre de Xochimilco, D.F. Informe servicio social. Universidad Autónoma Metropolitana-Xochimilco, México, 1991, p. 94.
72. B. L. Barcena, Evaluación de dietas regionales en el crecimiento (longitud y peso) del acocil *Cambarellus montezumae*, bajo condiciones de cultivo. BSc thesis, Universidad Autónoma de Tlaxcala, México, 2000, p. 96.
73. O. A. Cerón, M. M. Ángeles and E. J. León, Evaluación de la influencia del alimento y calidad del agua en el crecimiento de post-larvas del acocil *Cambarellus montezumae*. I Reunión Nacional de Innovación Acuícola y Pesquera de Campeche. Nutrición Acuícola, México, 2010, p. 41.
74. G. E. Escalante, Crecimiento de crías de acocil *Cambarellus montezumae* (Suassure) con diferentes dietas vegetales, BSc, Facultad de Ciencias. Universidad Nacional Autónoma de México, México, 2013, p. 99.
75. R. Serna, Biología y sistemática de los cambáridos del sudeste de México y su potencial aprovechamiento en la acuacultura, PhD thesis, Universidad Nacional Autónoma de México, México, 1999, p. 118.
76. F. Gherardi, Behaviour, in: *Biology of Freshwater Crayfish*, ed. D. M. Holdich, Blackwell Science UK, 2002, pp. 258–290.
77. R. Rangel, Ecología poblacional de *Cambarellus montezumae* en Xochimilco, México, D.F., MSc thesis, Facultad de Ciencias. Universidad Nacional Autónoma de México, México, 2009, p. 72.
78. V. E. González, Presencia de metales pesados en *Cambarellus montezumae, Ambistoma mexicanum* y *Chirostona jordani*, especies endémicas comestibles del área lacustre de Xochimilco, BSc thesis, Universidad Autónoma Metropolitana, 1997, p. 65.
79. M. P. Sánchez, Aprovechamiento de los ambientes reducidos en los canales de Xochimilco para el cultivo del acocil *Cambarellus montezumae*, para consumo humano, MSc thesis, Universidad Autónoma Metropolitana, México, 2007, 44.
80. F. Galloso, Evaluación de metales pesados en acocil y carpa del lago de Xochimilco, BSc thesis, Facultad de Estudios Superiores Zaragoza, Universidad Nacional Autónoma de México, México, 2009.
81. S. Díaz-Barriga, L. Martínez-Tabche, I. Álvarez-González, E. López-López and E. Madrigal-Bujaidar, Toxicity induced by dieldrin and chlorpyrifos in the freshwater crayfish *Cambarellus montezumae* (Cambaridae), *Rev. Biol. Trop.*, 2015, **63**, 83–96.

CHAPTER 8

Freshwater Crabs Potamonautes *spp. (Malacostraca, Potamonautidae) as a Model in Nanotoxicity Studies*

C. R. WALTERS,*[a] V. S. SOMERSET[a] AND E. POOL[b]

[a] Natural Resources and the Environment (NRE), Council for Scientific and Industrial Research (CSIR), Stellenbosch, South Africa; [b] Department of Medical Biosciences, University of the Western Cape (UWC), Bellville, South Africa

*Email: cwalters@csir.co.za

8.1 Introduction

Freshwater crabs (Potamonautidae) have been used as prominent model invertebrates in a variety of biological disciplines. *Potamonautes* spp. has been extensively used to study biological processes, including biomonitoring,[1–4] genomics,[5] cell biology,[6] and toxicology,[3,7] because of its well established biology and a high sensitivity to environmental stressors.[8] Furthermore, *Potamonautes* spp. studies have already fostered a better understanding of the underlying mechanisms of toxicity.

In light of the use of *Potamonautes* spp. in conventional toxicology and biomonitoring studies, there has been increased interest in the use of *Potamonautes* in nanomaterial-mediated ecotoxicology. The escalation of nanotechnologies and nano-enabled products in several fields is projected to

Issues in Toxicology No. 33
Ecotoxicology and Genotoxicology: Non-traditional Aquatic Models
Edited by Marcelo L. Larramendy

Published by the Royal Society of Chemistry, www.rsc.org

result in a \$1.5 trillion industry by 2015.[9] Nanoparticles have a wide variety of functions and therefore have been used in several fields of discipline, including agriculture, electronics, biomedicine, manufacturing, pharmaceuticals and cosmetics.[10] As a result of the widespread escalation of nanomaterials (NMs) applications, there has been an increase in the exposure of both humans and the environment to nanoparticles.

Nanomaterials, materials having at least one dimension in the size range of 1–100 nm,[11,12] differ from their bulk counterparts in several ways, including high surface/volume ratio. They exhibit novel physical and chemical properties, such as high conductivity, strong optical scattering properties, strong absorbance and ease of functionalization, which govern their toxicity. The incorporation of NMs into consumer products,[13] in imaging[14] and drug delivery systems[15,16] are cause for concern as NMs are often associated with toxicity.

The hazardous effects of NMs have been reported to include the generation of reactive oxygen species (ROS),[7,17–19] lipid peroxidation,[20–22] mitochondrial dysfunction,[23–26] genotoxicity,[27] apoptosis,[28,29] and changes in cell morphology.[30,31] With the rapid increase in nanotechonology and nanoscience over recent years, the research pertaining to toxicology has also advanced. Unfortunately, most of the nanotoxicity studies have been conducted on a limited number of organisms, such as *Daphnia magna*,[26,27,32] *Chironomus tentans*[33,34] and rainbow trout *Oncorhynchus mykiss.*[35–39] Recently, attention has been focused on *Potamonautes* spp. as a model organism for toxicity evaluation of nanoparticles (NPs).[7]

In the field of ecotoxicology, biomarkers are used to assess the health of ecosystems, using sentinel organisms to determine risk associated with environmental chemicals. Crustacean crabs show a high sensitivity to environmental stressors[8] and have therefore been used in several biomonitoring studies. A recent editorial has thus encouraged the use of *Potamonautes* spp. (specifically *P. perlatus*) in nanotoxicity research in view of the possible novel scientific knowledge *Potamonautes* can bring about.[7] The purpose of this chapter is to illustrate the potential of using *Potamonautes* spp. as important freshwater model organisms for nanotoxicology studies. Additionally, the gaps in nano-toxicological research and recommendations for future research initiatives are addressed.

8.2 Life History of *Potamonautes* spp.

Freshwater crabs of the genus *Potamonautes* (Milne-Edwards, 1837) belong to the class Malacostraca, which is of high ecological significance. Malacostraca is the largest of the six classes of crustaceans, containing about 40 000 living species, divided among 16 orders. *Potamonautes* is one of 18 genera of family Potamonautidae, a family of freshwater crabs endemic to tropical parts of Africa that includes four extensively studied model species, *P. perlatus, P. loveni, P. neumanni, and P. warreni.* The colour of *Potamonautes* spp. varies from deep orange to dark brown to moulted green (Figure 8.1).[40]

Figure 8.1 *Potamonautes* spp. (A: *P. perlatus*; B: *P. loveni*; C: *P. neumanni*; D: *P.* spp.; E: *P.* spp., F: *P. warreni*).
Figures B–E reproduced with permission from M. Dobson, Freshwater crabs in Africa, *Freshwater Forum*, 2004, **21**, 3–26. © Freshwater Biological Association 2004.

They typically have nine pairs of gills, which lie in branchial chambers enclosed in a carapace.[41] The foregut (mouth, oesophagus and stomach), midgut (anterior and posterior caecum) and hindgut (hepatopancreas) form the digestive system. Paired gonads open onto the ventral surface of the trunk form the reproductive system.[41]

Potamonautes spp. have a wide geographical distribution and are widespread in sub-Saharan Africa. Southern African (Angola, Botswana, Lesotho, Mozambique, Namibia, South Africa, Swaziland, Zambia and Zimbabwe) is home to 19 species of freshwater crabs of the genus *Potamonautes*, commonly found under boulders in the middle and lower reaches of rivers and other freshwater water courses. For example, *P. lovenii* have been reported in western Kenya and *P. niloticus*is are largely present in the Nile catchment and Upper Miocene of Lake Albert, Uganda.[42] *Potamonautes warreni* and *P. perlatus* are prevalent in intertidal regions of rivers of the south-western region of the Western Cape of South Africa,[3] where they play a pivotal role in the processing of organic material.[43] *Potamonautes* spp. occupy an important niche in freshwaters in Africa, serving as the main component of the diet of several predators, such as otters,[44,45] water mongoose,[45] kingfishers[46] and eels.[47] The moulting cycle of *Potamonautes* spp. is typically divided into four stages, including: (1) pre-moult, (2) moult, (3) post-moult, and (4) inter-moult.

8.3 Significance of Crustacean Crabs *Potamonautes* spp. as Test Species

The use of *Potamonautes* spp. in ecotoxicology and, more specifically, nanotoxicology studies have been fairly limited, with most nanotoxicological studies being focussed largely on the use of other aquatic invertebrates, such as *Daphnia magna* and *Chironomus tentans*, as test species. Freshwater crabs (Potamonautidae) have been a prominent model invertebrate in a variety of biological disciplines. For example, *Potamonautes* spp. have been used as an *in vivo* model organism for trace metal monitoring.[1,2,48–50] Recently, the freshwater crab genus (specifically *P. perlatus*) was also developed as an *in vitro* model organism for toxicology and nanotoxicology studies.[7] As such, there has been enhanced interest in the use of *Potamonautes* spp. for the understanding of nanomaterial-mediated toxicity.

Potamonautes shows favourable characteristics for use as a model organism for environmental studies. Freshwater crabs are the largest invertebrate species and constitute the largest biomass in several of South Africa's rivers.[43] In addition they play an important role in the freshwater food web, both as food for fish and predators themselves (Figure 8.2). *Potamonautes* are benthic aquatic organisms and therefore are likely to interact with aggregated NPs deposited in sediment. These true freshwater crabs show a high sensitivity to environmental stressors.[50] In crustacean crabs, exposure routes are largely *via* ingestion or direct passage across the gill and other external surface epithelia.

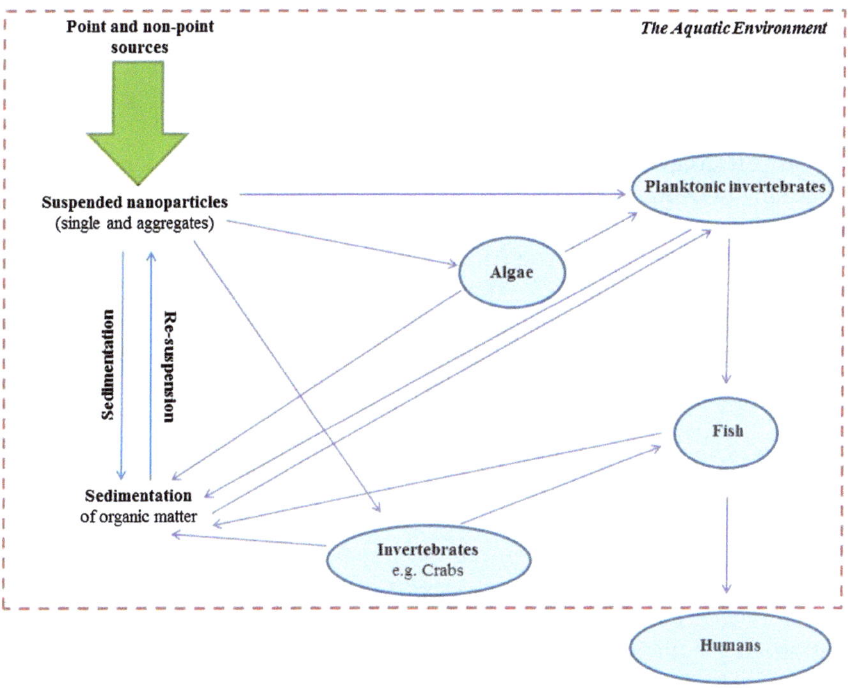

Figure 8.2 Significance of *Potamonautes* spp. in the aquatic food chain and possible routes of exposure to nanoparticles from anthropogenic sources.

8.4 Toxicity of NMs

The toxicity of NPs to a variety of organisms has been demonstrated in a number of recent studies. The physical and chemical properties (size, shape, high conductivity *etc.*) of NMs (categorized into carbon-based, silicon, metal and metal oxide) are significant drivers of their toxicity. For instance the toxicity of AgNPs, one of the most common NMs used in consumer products, was found to be size-dependent.[51] Another study reported differential toxicity of TiO_2NPs owing to different crystalline structures.[52,53] Navarro *et al.*[54] reported acute toxicity of AgNPs in *Chlamydomonas reinhardtii*, while Fabrega *et al.*[55] reported similar results in the bacteria *Pseudomonas fluorescens.*

Mechanisms for the toxicity of NPs to aquatic organisms have been identified. NPs are known to induce the production of ROS[17–19] and free radicals, causing oxidative stress in the form of activation/inhibition of the antioxidant defence system, lipid peroxidation and DNA damage. For example, studies have reported on the production of ROS leading to mortality and reproductive and growth effects.[28,54–56] It is believed that the toxicity due to AgNPs is largely related to the release of ionic Ag during oxidation and dissolution. This was supported by several reports.[57–61] Another metal NP, *i.e.* AuNP, was reported to induce generation of oxidative stress and inhibit antioxidant enzymes.[21] Carbon-based NPs are also known to cause toxicity in aquatic organisms. Commercial nano-sized carbon black (10 $\mu g\,mL^{-1}$) induced significant decreases in mitochondrial mass/number and membrane potential in the blue mussel *Mytilus galloprovincialis.*[62] Studies have also been reported on the toxicity of SiO_2NPs[63–65] and ROS formation by CuONPs has also been reported.[66–70]

Lipid peroxidation, a process whereby the double bonds on fatty acid tails of membrane phospholipids are oxidized, is regarded as a major consequence of ROS production and is known to disturb vital cellular functions.[71] Lipid peroxidation increases membrane permeability, makes cells susceptible to stress and limits nutrient uptake.[72] These effects have been reported by several authors.[73–75] Nanotoxicity studies reporting on genotoxicity such as chromosomal abnormalities,[76,77] DNA strand breakages,[22,78,79] mutations,[80–82] oxidative DNA adducts and alterations in gene expression profiles[83] have increased.

8.5 Fate and Behaviour of NPs in the Aquatic Environment

Concerns regarding the environmental fate and effects of these NMs have motivated studies to forecast concentrations in various environmental matrices (water, soil, sediment, biota) and also to establish threshold concentrations for their toxicological effects on aquatic organisms. In the environment (both terrestrial and aquatic), NPs undergo several transformation processes that are dependent on both the properties of the NP itself and those of the local chemistry of the receiving medium. These transformations generally include chemical and physical processes, as well as biodegradation of surface coatings used to stabilize the NM, and photochemical and biological reactions within organisms.[84] Nanoparticles are inevitably released into the soil and water courses. Once in the aquatic environment, NMs may aggregate and settle to sediments, or be transported within the water flow. The key processes governing NP behaviour and toxicity in the aquatic environment are aggregation and dissolution. The former results in augmentation of NPs, often to sizes in excess of the NP size definition, *i.e.* >100 nm, restricting their transport and bioavailability, ultimately leading to sedimentation. Toxicity is thus affected by reducing the rate of dissolution, uptake by aquatic organisms, and stability of the NPs against aggregation. Nanoparticle aggregation is dependent on several factors, including surface charge, pH, particle size, ionic strength, and the presence or absence of a capping layer.[85,86] Walters *et al.*[87] showed temperature driven aggregation/disaggregation with smaller aggregates at higher temperatures. Similar results are reported for NP aggregation at high pH.[88] However, this augmentation can be averted by the presence of surfactants and other surface coatings, as well as the presence of natural humic substances. Studies have shown that NP aggregation may affect its toxicity.[7,87]

Dissolution of NPs is an important property that influences their mode of action, uptake pathway and toxicity mechanisms, and will contribute to the concentration of silver ions in the environment. A study by Elzey *et al.*[89] showed that size dependent dissolution at low pH. The authors reported complete AgNP dissolution at a pH of 0.5. Similar results were also supported by Liu and Millero.[90] Walters *et al.*[7] reported dissolution of AgNPs at elevated temperatures, while Li *et al.*[91] reported a reduction of dissolution by photoreduction.

Sedimentation is regarded as an important process governing NP toxicity. In the aquatic environment, sedimentation largely depends on the type of NM and its properties, as well as the condition of the aqueous environment. For example, Wiench *et al.*[92] visually investigated the sedimentation of TiO_2 NPs and reported differential rates depending on the experimental media used. In another study, Baalousha[93] reported differential sedimentation at higher NP concentrations. Nanoparticle aggregation typically leads to sedimentation,[94,95] which removes NMs from the water column and reduces their mobility, thereby limiting the potential uptake of NPs. However, this process allows for settled NMs to be taken up by filter feeders and benthic organisms.[96] Sediments are therefore regarded as a major sink for NPs. Supporting this suggestion is a recent study by Walters *et al.*,[87] which showed that AgNPs that had undergone sedimentation had contributed to a higher availability and toxicity of AgNPs (and Ag^+) to *P. perlatus.*

8.6 *Potamonautes* spp.: Uptake, Tissue Localization and Toxicity

The physical and chemical properties of NMs, as well as the characteristics of the environmental media, govern the localization of NMs in the target

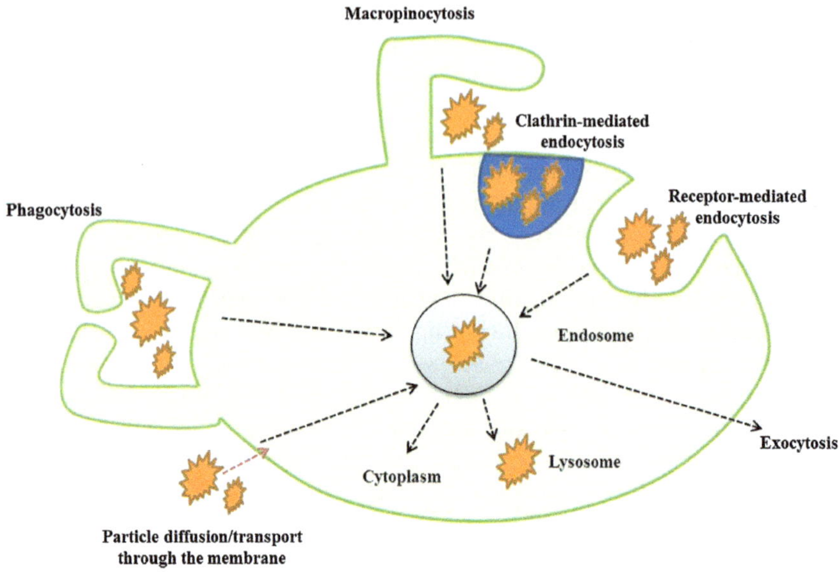

Figure 8.3 Mechanisms of cellular uptake of NPs.

cells.[97,98] Endocytosis is a cellular uptake mechanism involving the uptake of particulate matter into eukaryotic cells *via* enclosure by the cell membrane. Forms of endocytotic uptake mechanisms include phagocytosis, macropinocytosis, and clathrin- and caveolin-mediated endocytosis (Figure 8.3). Most of the currently available ecotoxicological data regarding NPs are limited to species. As such, only a few studies have to date have reported on tissue localization, internal distribution and NP uptake. In crabs, the hepatopancreas is regarded as the main site for cellular internalization, toxicant metabolism and biotransformation of ROS.[99] In a freshwater crab species, the Cape river crab *P. perlatus*, exposed to various concentrations of AgNPs, Walters *et al.*[7] reported higher levels of enzymatic activities (superoxide dismutase, catalase, glutathione *S*-transferase) in the hepatopancreas when compared to the gills. This was also confirmed in a similar study by Pinho *et al.*[100] when investigating the effects of AgNP in the estuarine crab *Chasmagnathus granulatus*. In another study by Walters,[101] the authors reported the haemolymph as being more susceptible to oxidative stress originating by AgNPs, whereas the gills constitutes the main storage organ for Ag.

8.7 Sensitivity of *Potamonautes* spp. to Common Environmental Stressors

Aquatic ecosystems are increasingly coming under pressure of various anthropogenic pollutants, and aquatic organisms are therefore often subjected to these stressors. Heavy metal pollution of the aquatic ecosystem is a global problem, and their effects are well-documented. Schuwerack *et al.*[2]

investigated the potential of *P. warreni* as a biondicator species of heavy metal pollution. The authors reported that Cu was largely sequestered in the gills and to a lesser extent in the digestive gland. Similarly, Steenkamp *et al.*[102] reported the accumulation of Fe to be largely in the gills. In another study, Steenkamp *et al.*[103] reported Cu concentrations in various tissues of *P. warreni* generally ranking: carapace $<$ muscle $<$ gonads $<$ midgut gland $<$ gills. Sanders *et al.*[1] determined that Fe and Pb bioaccumulation was a function of sex, with higher levels measured in females. The authors also reported that the bioaccumulation of Cu, Mn, Pb, Ni, and Zn were influenced by the size; the latter was also reported by Steenkamp *et al.*[102] for Fe. Somerset *et al.*[104] reported uptake of heavy metals associated with mining activities by *P. warreni*. These studies concluded that *P. warreni* were useful bioaccumulative indicators of metal pollution in the aquatic environment.

Abiotic factors are also significant inducers of stress in aquatic organisms. For instance, temperature is regarded as an important abiotic factor effecting the survival and functioning of aquatic organisms. Owing to the intertidal habitat of *Potamonautes* spp., they are prone to frequent temperature fluctuations. Van Aardt,[6] investigated the effects of temperature on haemocyanin function and oxygen uptake in *P. warreni*, and reported a significant decrease in VO_2 ($110\ ml\ O_2\ kg^{-1}\ hr^{-1}$ at 35 °C) and further decreasing as temperature reached 40 °C). Another important environmental factor is salinity. This was investigated by Morris and Van Aardt,[105] who reported that *P. warreni* was able to regulate salt loss in water containing up to 40% sea water, but not in 80% sea water.

8.8 Conclusion

In recent years, the potential of *Potamonautes* spp. as an indicator organism has been investigated in a variety of environmental studies. As such, as a model organism for conventional environmental toxicity studies, there is an increasing body of evidence to support the significant role that *Potamonautes* could play in assessing the impacts of NM-mediated toxicity.

Acknowledgements

This research was supported by the Council for Scientific and Industrial Research (CSIR).

References

1. M. J. Sanders, Masters Thesis, University of Johannesburg, 1997.
2. P.-M. M. Schuwerack, J. W. Lewis and P. W. Jones, Pathological and physiological changes in the South African freshwater crab *Potamonautes warreni* Calman induced by microbial gill infestations, *J. Invertebr. Pathol.*, 2001, 77, 269–279.

3. R. G. Snyman, A. J. Reinecke and J. A. J. Nel, Uptake and distribution of copper in the freshwater crab, *Potamonautes perlatus* (Crustacea) in the Eerste River, South Africa, *Afr. Zool.*, 2002, **37**(1), 81–89.
4. A. J. Reinecke, R. G. Snyman and J. A. J. Nel, Uptake and Distribution of Lead (Pb) and Cadmium (Cd) in the Freshwater Crab *Potamonautes perlatus* in the Eerste River, South Africa, *Water, Air, Soil Pollut.*, 2003, **145**(1), 395–408.
5. E. E. Phiri and S. R. Daniels, Disentangling the divergence and cladogenesis in the freshwater crab species (Potamonautidae: *Potamonautes perlatus sensulato*) in the Cape Fold Mountains, South Africa, with the description of two novel cryptic lineages, *J. Linn. Soc. London, Zool.*, 2014, **170**, 310–322.
6. W. J. van Aardt, Oxygen uptake and haemocy- anin oxygen affinity of *Potamonautes warreni* Caiman after exercise, *S. Afr. J. Zool.*, 1990, **25**, 11–17.
7. C. R. Walters, P. Cheng, E. Pool and V. S. Somerset, Effect of temperature on oxidative stress parameters and enzyme activity in tissues of Cape River crab (*Potamanautes perlatus*) following exposure to silver nanoparticles (AgNP), *J. Toxicol. Environ. Health A*, 2016, **79**(2), 61–70.
8. Q. Qin, S. Qin, L. Wang and W. Lei, Immune repsonses and ultrastructural changes of hemocytes in freshwater crab *Sinopotamonhenanense* exposed to elevated cadmium, *Aquat. Toxicol.*, 2012, **106–107**, 140–146.
9. A. Nel, T. Xia, L. Madler and T. Li, Toxic potential of materials at the nanolevel, *Science*, 2006, **11**, 622–627.
10. J. Fabrega, S. N. Luoma, C. R. Tyler, T. S. Galloway and J. R. Lead, Silver nanoparticles: behaviour and effects in the aquatic environment, *Environ. Int.*, 2011, **37**, 517–531.
11. M. C. Powell and M. S. Kanarek, Nanomaterial health effects–Part 2: Uncertainties and recommendations for the future, *Wis. Med. J.*, 2006, **105**, 18–23.
12. SCENIR (Scientific Committee on Emerging and Newly Identified Health Risks), European Comission, European Union, 2010.
13. F. R. Khan, S. K. Misra, J. García-Alonso, B. D. Smith, S. Strekopytov, P. S. Rainbow, S. N. Luoma and E. Valsami-Jones, *Environ. Sci. Technol.*, 2012, **46**, 7621–7628.
14. S. Yamamoto and H. Watarai, Surface-enhanced Raman spectroscopy of dodecanethiol-bound silver nanoparticles at the liquid/liquid interface, *Langmuir*, 2006, **22**(15), 6562–6569.
15. N. L. Rosi, D. A. Giljohann, C. S. Thaxton, A. K. R. Lytton-Jean, M. Han and C. A. Mirkin, Oligonucleotide-modified gold nanoparticles for intracellular gene regulation, *Science*, 2006, **312**, 1027–1030.
16. M. Rangasamy and K. G. Parthiban, Recent advances in novel drug delivery systems, *Int. J. Res. Ayurveda Pharm.*, 2012, **1**(2), 316–326.
17. M. Ahamed, R. Posgai, T. J. Gorey, M. Nielsen, S. M. Hussain and J. J. Rowe, Silver nanoparticles induced heat shock protein 70, oxidative

stress and apoptosis in *Drosophila melanogaster*, *Appl. Pharmacol.*, 2010, **242**, 263–269.

18. C. Levard, E. M. Hotze, G. V. Lowry and G. E. Brown, Environmental Transformations of silver nanoparticles: impact on stability and toxicity, *Environ. Sci. Technol.*, 2012, **46**, 6900–6914.
19. M. J. Piao, K. A. Kang, I. K. Lee, H. S. Kim, S. Kim, J. Y. Choi, J. Choi and J. W. Hyun, Silver nanoparticles induce oxidative cell damage in human liver cells through inhibition of reduced glutathione and induction of mitochondria-involved apoptosis, *Toxicol. Lett.*, 2011, **201**, 92–100.
20. E. Oberdörster, Manufactured nanomaterials (fullerenes, C60) induce oxidative stress in the brain of juvenile largemouth bass, *Environ. Health Perspect.*, 2004, **112**, 1058–1062.
21. Y. H. Siddique, A. Fatima, S. Jyoti, F. Naz, W. Rahul, B. R. Khan, Singh and A. H. Naqvi, Evaluation of the toxic potential of graphene copper nanocomposite (GCNC) in the third instar larvae of transgenic *Drosophila melanogaster* (hsp70-lacZ)Bg(9.), *PLoS One*, 2013, **8**(12), e80944.
22. L. A. Maranho, R. M. Baena-Nogueras, P. A. Lara-Martín, T. A. DelValls and M. L. Martín-Díaz, Bioavailability, oxidative stress, neurotoxicity and genotoxicity of pharmaceuticals bound to marine sediments. The use of the polychaete *Hedistediversicolor* as bioindicator species, *Environ. Res.*, 2014, **134**, 353–365.
23. M. Knee, Sensitivity of ATPases to silver ions suggests that silver acts outside the plasma membrane to block ethylene action, *Phytochemicals*, 1992, **31**, 1093–1096.
24. R. Weissleder, Molecular imaging in cancer, *Science*, 2006, **312**, 1168–1171.
25. N. Lewinski, V. Colvin and R. Drezek, Cytotoxicity of nanoparticles, *Small*, 2008, **4**(1), 26–49.
26. M. Heinlaan, A. Kahru, K. Kasemets, B. Arbeille, G. Prensier and H.-C. Dubourguier, Changes in the *Daphnia magna* midgut upon ingestion of copper oxide nanoparticles: a transmission electron microscopy study, *Water Res.*, 2011, **45**, 179–190.
27. S.-Y. Park and J. Choi, Geno- and ecotoxicity evaluation of silver nanoparticles infreshwater crustacean *Daphnia magna*, *Environ. Eng. Res.*, 2010, **15**(1), 23–27.
28. P. V. Asharani, Y. L. Wu, Z. Gong and S. Valiyaveettil, Toxicity of silver nanoparticles in zebrafish models, *Nanotechnology*, 2008, **19**, 1–8.
29. J. Garcia-Alonso, F. R. Khan, S. K. Misra, M. Turmaine, B. D. Smith, P. S. Rainbow, S. N. Luoma and E. Valsami-Jones, Cellular internalization of silver nanoparticles in gut epithelia of the estuarine polychaete Nereisdiversicolor, *Environ. Sci. Technol.*, 2011, **45**, 4630–4636.
30. A. Shvedova, V. Castranova, E. Kisin, D. Schwegler-Berry, A. Murrav, V. Gandelsman, A. Maynard and P. Baron, Exposure to carbon nanotube material: assessment of nanotube cytotoxicity using human keratinocyte cells, *J. Toxicol. Environ. Health*, 2003, **66**, 1909–1926.

31. S. Arora, J. Jain, J. M. Rajwade and K. M. Paknikar, Interactions of silver nanoparticles with primary mouse fibroblasts and liver cells, *Toxicol. Appl. Pharm.*, 2009, **236**(3), 310–318.
32. B. K. Gaiser, A. Biswas, P. Rosenkranz, M. A. Jepson, J. R. Lead, V. Stone, C. R. Tyler and T. F. Fernandes, Effects of silver and cerium dioxide micro- and nano-sized particles on *Daphnia magna*, *J. Environ. Monit.*, 2011, **13**, 1227–1235.
33. P. J. Oberholster, N. Musee, A.-M. Noth, P. K. Chelule, W. W. Focke and P. J. Ashton, Assessment of the effect of nanomaterials on sediment-dwelling invertebrate *Chironmustentans* larvae, *Ecotoxicol. Environ. Saf.*, 2011, **74**, 416–423.
34. S. Y. Au and S. J. Klaine, Behavioral changes and oxidative stress resulting fromexposure to nano-TiO2 in *Chironomustentans*. *Presented at the annual meeting of the Carolina's Society of Environmental Toxicology and Chemistry*, March 2013, Raleigh, NC.
35. C. J. Smith, B. J. Shaw and R. D. Handy, Toxicity of single walled carbon nanotubes to rainbow trout, (*Oncorhynchus mykiss*): respiratory toxicity, organ pathologies, and other physiological effects, *Aquat. Toxicol.*, 2007, **82**, 94–109.
36. A. Baun, S. N. Sorensen, R. F. Rasmussen, N. B. Hartmann and C. B. Koch, Toxicity and bioaccumulation of xenobiotic organic compounds in the presence of aqueous suspensions of aggregates of nano-C(60), *Aquat. Toxicol.*, 2008, **86**(13), 379–387.
37. R. D. Handy, T. B. Henry, T. M. Scown, B. D. Johnston and C. R. Tyler, Manufactured nanoparticles: their uptake and effects on fish-a mechanistic analysis, *Ecotoxicol.*, 2008, **17**, 396–409.
38. A. G. Cattaneo, R. Gornati, M. Chiriva-Internati and G. Bernardini, Ecotoxicology of nanomaterials: the role of invertebrate testing, *Invert. Surviv. J.*, 2009, **6**, 78–97.
39. J. Farkas, P. Christian, J. A. G. Urread, N. Roose, M. Hassellöv, K. E. Tollefsen and K. V. Thomas, Effects of silver and gold nanoparticles on rainbow trout (*Oncorhynchus mykiss*) hepatocytes, *Aquat. Toxicol.*, 2010, **96**, 44–52.
40. M. Dobson, Freshwater crabs in Africa, *Freshwater Forum*, 2004, **21**, 3–26.
41. N. Cumberlidge, The freshwater crabs of West Africa. Family Potamonautidae, *IRD*, 1999, p. 382.
42. R.-P. Carriol and S. Secretan, Présencedans le Bassin du Lac Albert (Ouganda), dès le Miocènesupérieur, de *Potamonautes (Acanthothelphusa) niloticus* (Crustacea, Brachyura), *ComptesRendus de l'Academie des Sci.*, 1992, **314**, 411–417.
43. M. P. Hill and J. H. O'Keeffe, Some aspects of the ecology of the freshwater crab (*Potamonautes perlatus* Milne Edwards) in the upper reaches of the Buffalo River, Eastern Cape Province, South Africa, *S. Afr. J. Aquat. Sci.*, 1992, **18**, 42–50.

44. J. R. A. Butler and J. T. du Toit, Diet and conservation status of Cape clawless otters in eastern Zimbabwe, *S. Afri. J. Wildlife Res.*, 1994, **24**, 41–47.
45. M. G. Purves, H. Kruuk and J. A. J. Nel, Crabs Potamonautes perlatus in the diet of otter *Aonyxcapensis* and water mongoose *Atilaxpaludinosus* in a freshwater habitat in South Africa, *Z. Saugetierkunde*, 1994, **59**, 332–341.
46. J. B. F. Arkell, Aspects of the feeding and breeding biology of the Giant Kingfisher, *Ostrich*, 1979, **50**, 176–181.
47. J. R. A. Butler and B. E. Marshall, Resource use within the crab eating guild of the upper Kairezi River, Zimbabwe, *J. Trop. Ecol.*, 1996, **12**(4), 475–490.
48. A. Vosloo, W. J. van Aardt and L. J. Mienie, Sublethal effects of copper on the freshwater crab *Potamonautes warreni*, *Comp. Biochem. Physiol., Part A: Mol. Integr. Physiol.*, 2002, **133**, 695–702.
49. S. Thawley, S. Morris and A. Vosloo, Zn and Cd accumulation in *Potamonautes warreni* from sites in the North-West Province of South Africa, *Int. Congr. Ser.*, 2004, **1275**, 180–188.
50. P. M. Schuwerack, J. W. Lewis and P. W. Jones, Interactive cellular and physiological responses of the freshwater crab *Potamonautes warreni* to cadmium and microbial gill infestations, *Ecotoxicol. Environ. Saf.*, 2007, **67**(2), 302–310.
51. D. P. K. Lankvel, A. G. Oomen, P. Krystek, A. Neigh, A. Troost-deJong, J. C. H. Noorlander, R. E. Geertsma and W. H. De Jong, The kinetics of the tissue distribution of silver nanoparticles of different sizes, *Biomaterials*, 2010, **31**, 8350–8361.
52. C. M. Sayes, R. Wahi, P. A. Kurian, Y. Liu, J. L. West, K. D. Ausman, D. Warheit and V. L. Colvin, Correlating nanoscale titania structure with toxicity: a cytotoxicity and inflammatory response study with human dermal fibroblasts and human lung epithelial cells, *Toxicol. Sci.*, 2006, **92**, 174–185.
53. J. Petkovic, B. Zegura, M. Stevanovic, N. Drnovsek, D. Uskokovic, S. Novak and M. Filipic, DNA damage and alterations in expression of DNA damage responsive genes induced by TiO2 nanoparticles in human hepatoma HepG2 cells, *Nanotoxicology*, 2011, **5**, 341–353.
54. E. Navarro, F. Piccapietra, B. Wagner, F. Marconi, R. Kaegi and N. Odzak, Toxicity of silver nanoparticles to *Chlamydomonas reinhardtii*, *Environ. Sci. Technol.*, 2008, **42**, 8959–8964.
55. J. Fabrega, S. R. Fawcett, J. C. Renshaw and J. R. Lead, Silver nanoparticle impact on bacterial growth: effect of pH, concentration, and organic matter, *Environ. Sci. Technol.*, 2009, **43**, 7285–7290.
56. O. Choi and Z. Hu, Size dependent and reactive oxygen species related nanosilvertoxicity to nitrifying bacteria, *Environ. Sci. Technol.*, 2008, **42**, 4583–4588.

57. E. Navarro, F. Piccapietra, B. Wagner, F. Marconi, R. Kaegi, N. Odzak, L. Sigg and R. Behra, Toxicity of Silver Nanoparticles to *Chlamydomonas reinhardtii*, *Environ. Sci. Technol.*, 2008, **42**, 8959–8964.
58. I. Sondi and B. Salopek-Sondi, Silver nanoparticles as antimicrobial agent: a case study on *E. coli* as a model for Gram-negative bacteria, *J. Colloid Interface Sci.*, 2004, **275**, 177–182.
59. J. R. Monrones, J. L. Elechiguerra, A. Camacho, K. Holt, J. B. Kouri, J. T. Ramirez and M. J. Tacaman, The bactericidal effect of silver nanoparticles, *Nanotechnology*, 2005, **16**(10), 2346–2353.
60. M. Farre, K. Gajda-Schrantz, L. Kantiani and D. Barcelo, Ecotoxicology and analysis of nanomaterials in the aquatic environment, *Anal. Biochem.*, 2009, **393**(1), 81–85.
61. S. Elzey and V. H. Grassian, Agglomeration, isolation and dissolution of commercially manufactured silver nanoparticles in aqueous environments, *J. Nanopart. Res.*, 2010, **12**, 1945–1958.
62. L. Canesi, C. Ciacci, M. Betti, R. Fabbri, B. Canonico, A. Fantinati, A. Marcomini and G. Pojana, Immunotoxicity of carbon black nanoparticles to blue mussel hemocytes, *Environ. Int.*, 2008, **34**, 1114–1119.
63. M. Chen and A. von Mikecz, Formation of nucleoplasmic protein aggregates impairs nuclear function in response to SiO_2 nanoparticles, *Exp. Cell. Res.*, 2005, **305**(1), 51–62.
64. L. K. Adams, D. Y. Lyon and P. J. J. Alvarez, Comparative eco-toxicity of nanoscale TiO_2, SiO_2, and ZnO water suspensions, *Water Res.*, 2006, **40**(1), 3527–3532.
65. W. Lin, Y. W. Huang, X. D. Zhou and Y. Ma, *In vitro* toxicity of silica nanoparticles in human lung cancer cells, *Toxicol. Appl. Pharmacol.*, 2006, **217**(3), 252–259.
66. M. Heinlaan, A. Ivask, I. Blinova, H.-C. Dubourguier and A. Kahru, Toxicity of nanosized and bulk ZnO, CuO and TiO_2 to bacteria *Vibrio fischeri* and crustaceans *Daphnia magna* and *Thamnocephalus platyurus*, *Chemosphere*, 2008, **71**(7), 1308–1316.
67. H. K. Karlsson, P. Cronholm, J. Gustafsson and L. Moller, Copper oxide nanoparticles are highly toxic: a comparison between metal oxide nanoparticles and carbon nanotubes, *Chem. Res. Toxicol.*, 2008, **21**(9), 1726–1732.
68. K. Kasemets, A. Ivask, H.-C. Dubourguier and A. Kahru, Toxicity of nanoparticles of ZnO, CuO and TiO_2 to yeast *Saccharomyces cerevisiae*, *Toxicol. In Vitro*, 2009, **23**(6), 1116–1122.
69. I. Blinova, A. Ivask, M. Heinlaan, M. Mortimer and A. Kahru, Ecotoxicity of nanoparticles of CuO and ZnO in natural water, *Environ. Poll.*, 2010, **158**(1), 41–47.
70. M. Heinlaan, A. Kahru, K. Kasemets, B. Arbeille, G. Prensier and H.-C. Dubourguier, Changes in the *Daphnia magna* midgut upon ingestion of copper oxide nanoparticles: a transmission electron microscopy study, *Water Res.*, 2011, **45**, 179–190.

71. G. Oberdörster, E. Oberdörster and J. Oberdörster, Nanotoxicology: an emerging discipline evolving from studies of ultrafine particles, *Envrion. Health Perspect.*, 2005, **113**, 823–839.
72. E. Cabiscol, J. Tamarit and J. Ros, Oxidative stress in bacteria and protein damage by reactive oxygen species, *Int. Microb.*, 2000, **3**, 3–8.
73. E. Oberdorster, S. Zhu, T. M. Blickley, P. McClellan-Green and M. L. Haasch, Ecotoxicology of carbon-based engineered nanoparticles: Effects of fullerene (C-60) on aquatic organisms, *Carbon*, 2006, **44**, 1112–1120.
74. T. M. Scown, E. M. Santaos, B. D. Johnston, B. Gaiser, M. Baalousha, S. Mitov, J. R. Lead, V. Stone, T. F. Fernandes, M. Jepson, R. van Aerle and C. R. Tyler, Effects of aqueous exposure to silver nanoparticles of different sizes in rainbow trout, *Toxicol. Sci.*, 2010, **115**(2), 521–534.
75. J. E. Choi, S. Kim, J. H. Ahn, P. Youn, J. S. Kang, K. Park, J. Yi and D.-Y. Ryu, Induction of oxidative stress and apoptosis by silver nanoparticles in the liver of adult zebrafish, *Aquat. Toxicol.*, 2010, **11**, 151–159.
76. J. Maluszynska and J. Juchimiuk, Plant genotoxicity: a molecular cytogenetic approach in plant bioassays, *Arh. Hig. Rada. Toksikol.*, 2005, **56**, 177–184.
77. M. Yıldız, I. HakkıCigerci, M. Konuk, A. F. Fidan and H. Terzi, Determination of genotoxic effects of copper sulphate and cobalt chloride in *Allium cepa* root cells by chromosome aberration and comet assays, *Chemosphere*, 2009, **75**, 934–938.
78. N. P. Singh, Microgels for estimation of DNA strand breaks, DNA protein crosslinks and apoptosis, *Mutat. Res.*, 2000, **455**, 111–127.
79. S. K. Bopp, H. K. Abicht and K. Knauer, Copper-induced oxidative stress in rainbow trout gill cells, *Aquat. Toxicol.*, 2008, **86**, 197–204.
80. G. Golling, A. Amsterdam, Z. Sun, M. Antonelli, E. Maldonado and W. Chen, Insertional mutagenesis in zebrafish rapidly identifies genes essential for early vertebrate development, *Nat. Genet.*, 2002, **31**, 135–140.
81. J. F. Yu, S. Fukamachi, H. Mitani, H. Hori and A. Kanamori, Reduced expression of vps11 causes less pigmentation in medaka, *Oryziaslatipes*, *Pigm. Cell Res.*, 2006, **19**, 628–634.
82. M. L. López-Moreno, G. de la Rosa, J. A. Hernández-Viezcas, H. Castillo-Michel, C. E. Botez, J. R. Peralta-Videa and J. L. Gardea-Torresdey, Evidence of the differential biotransformation and genotoxicity of ZnO and CeO_2 nanoparticles on soybean (*Glycine max*) plants, *Environ. Sci. Technol.*, 2010, **44**, 7315–7320.
83. N. Singh, B. Manshian, G. J. S. Jenkins, S. M. Griffiths, P. M. Williams, T. G. G. Maffeis, C. J. Wright and S. H. Doak, NanoGenotoxicology: the DNA damaging potential of engineered nanomaterials, *Biomaterials*, 2009, **30**, 3891–3914.
84. Y. Ju-Nam and J. R. Lead, Manufactured nanoparticles: an overview of their chemistry, interactions and potential environmental implications, *Sci. Tot. Environ.*, 2008, **400**(1–3), 396–414.

85. A. Baun, N. B. Hartmann, K. Grieger and K. O. Kusk, Ecotoxicity of engineered nanoparticles to aquatic invertebrates: a brief review and recommendations for future toxicity testing, *Ecotoxicol.*, 2008, **17**, 387–395.
86. G. E. Batley, J. K. Kirby and M. J. McLaughlin, Fate and risks of nanomaterials in aquatic and terrestrial environments, *Acc. Chem. Res*, 2011, **46**(3), 854–862.
87. C. R. Walters, E. J. Pool and V. S. Somerset, Ecotoxicity of silver nanomaterials in the aquatic environment: A review of literature and gaps in nanotoxicological research, *J. Environ. Sci. Health*, 2014, **49**(13), 1588–1601.
88. B. Gilbert, G. Lu and C. S. Kim, Stable cluster formation in aqueous suspensions of iron oxyhydroxide nanoparticles, *J. Coll. Int. Sci.*, 2007, **313**(1), 152–159.
89. S. R. Elzey, Phd thesis, University of Iowa, 2010.
90. X. Liu and F. J. Millero, The solubility of iron hydroxide in sodium chloride solutions, *Geochim. Cosmochim. Acta*, 1999, **63**, 3487–3497.
91. X. Li and J. J. Lenhart, Aggregation and dissolution of silver nanoparticles in natural surface water, *Environ. Sci. Technol.*, 2012, **46**, 5378–5386.
92. K. Wiench, W. Wohlleben, V. Hisgen, K. Radke, E. Salinas, S. Zok and R. Landsiedel, Acute and chronic effects of nano- and non-nano-scale TiO_2 and ZnO particles on mobility and reproduction of the freshwater invertebrate *Daphnia magna*, *Chemosphere*, 2009, **76**(10), 1356–1365.
93. M. Baalousha, Aggregation and disaggregation of iron oxide nanoparticles: Influence of particle concentration, pH and natural organic matter, *Sci. Tot. Environ.*, 2009, **407**, 2093–2101.
94. H. Hyung, J. D. Fortner, J. B. Hughes and J. H. Kim, Natural organic matter stabilizes carbon nanotubes in the aqueous phase, *Environ. Sci. Technol.*, 2007, **41**, 179–184.
95. A. A. Koelmans, B. Nowack and M. R. Wiesner, Comparison of manufactured and black carbon nanoparticle concentrations in aquatic sediments, *Environ. Pollut.*, 2009, **157**, 1110–1116.
96. B. Nowack and T. D. Bucheli, Occurrence, behaviour and effects of nanoparticles in the environment, *Environ. Pollut.*, 2007, **150**, 5–22.
97. C. Brandenberger, C. Mühlfeld, Z. Ali, A.-G. Lenz, O. Schmid, W. J. Parak, P. Gehr and B. Rothen-Rutishauser, Quantitative evaluation of cellular uptake and trafficking of plain and polyethylene glycol-coated gold nanoparticles, *Small*, 2010, **6**, 1669–1678.
98. C. Schweiger, R. Hartmann, F. Zhang, W. Parak, T. H. Kissel and P. Rivera Gil, Quantification of the internalization patterns of superparamagnetic iron oxide nanoparticles with opposite charge, *J. Nanobiotechnology*, 2012, **10**, 28–38.
99. D. R. Livingstone, The fate of organic xenobiotics in aquatic ecosystems: quantitative and qualitative differences in biotransformation by invertebrates and fish, *Comp. Biochem. Physiol., Part A: Mol. Integr. Physiol.*, 1998, **120**, 43–49.

100. G. L. L. Pinho, C. M. Da Rousa, F. E. Maciel, A. Bianchini, J. S. Yunes, L. A. O. Proenca and J. M. Moserrat, Antioxidant responses and oxidative stress after microcystin exposure in the hepatopancreas of an estuarine crab species, *Ecotoxicol. Environ. Saf.*, 2005, **61**, 353–360.
101. C. R. Walters, PhD thesis, University of the Western Cape, 2016.
102. V. E. Steenkamp, H. H. Du Preez, H. J. Schoonbee, A. J. B. Wiid and M. M. Bester, Bioaccumulation of iron in three freshwater crab (*Potamonautes warreni*) populations and notes on the physical-chemical parameters of the water, *Water S.A.*, 1993, **19**(4), 281–290.
103. V. E. Steenkamp, H. H. Du Preez and H. J. Schoonbee, Bioaccumulation of copper in the tissues of *Polamonautes warreni* (Caiman) (Crustacea, Decapoda. Branchiura), from industrial, mine and sewage-polluted freshwater ecosystems, *S. Afri. J. Zool.*, 1994, **29**, 152–161.
104. V. Somerset, C. Van der Horst, B. Silwana, C. Walters and E. Iwuoha, Biomonitoring and evaluation of metal concentrations in sediment and crab samples from the North-West Province of South Africa, *Water, Air, Soil Pollut.*, 2015, **226**, 1–25.
105. S. Morris and W. J. van Aardt, Salt and water relations, and nitrogen excretion in the amphibious freshwater crab *Potamonautes warreni* in water and in air, *J. Exp. Biol.*, 1998, **201**, 883–898.

CHAPTER 9

Freshwater Prawns Macrobrachium borellii *and* Palaemonetes argentinus *(Crustacea: Palaemonidae) as Valid Indicators of Organophosphate Pesticide Exposure. A Biochemical Assessment*

C. F. GARCÍA*[a] AND R. R. SOTELO-MUNDO[b]

[a] Instituto de Investigaciones Bioquímicas de La Plata. Prof. Dr. R. R. Brenner (INIBIOLP), (CONICET-UNLP)- Fac de Cs Naturales y Museo, UNLP. Calles 60 y 120- 1900-, La Plata, Argentina; [b] Laboratorio de Estructura Biomolecular, Centro de Investigación en Alimentación y Desarrollo, A.C. Carretera a Ejido La Victoria km 0.6. Hermosillo, Sonora 83304, Mexico
*Email: cfgarcia1123@yahoo.com.ar

Issues in Toxicology No. 33
Ecotoxicology and Genotoxicology: Non-traditional Aquatic Models
Edited by Marcelo L. Larramendy

Published by the Royal Society of Chemistry, www.rsc.org

9.1 Introduction

9.1.1 Crustaceans

Crustaceans are one of the arthropod groups with greater diversity in the animal kingdom. They are characterized by their legs being formed from articulated segments and a thick chitin exoskeleton from which their name is derived.

Much of the current interest and recent research is owing to the need to understand nutrition, reproduction and immunity in intensive culture, namely aquaculture. Among these, nutrition, immune system, and reproduction are the most studied areas related to crustacean rearing owing to their growth rate.

Decapods form part of a large group of crustaceans found from the tidal zone to the abyssal depths, whereas fresh water species are frequently found in creeks and rivers. These are the most popular crustaceans and constitute the so-called crabs, lobsters, shrimp, prawns and spider crabs.

Within the benthonic littoral community of South America, fresh water prawns *Macrobrachium borellii* and *Palaemonetes argentinus* are considered the most frequent and important ones.[1,2] These animals are characterized by being omnivorous,[3] with plants as the most important part of their diet.[4]

Macrobrachium borellii (Nobili, 1896) (Decapoda: Palaemonidae) (Figure 9.1, bottom) is a small species with a smooth shell. It is characterized by a branchiostegal line originated in the anterior part toward the hepatic region.

According to the distribution data, this species in found in Argentina, Paraguay, and Uruguay, It inhabits exclusively fresh water environments (20°S–40°S). In the Republic of Argentina it is found in all the northern provinces and the central region of the country until Mendoza in closed basins, small rivers, lakes, reservoirs, and so on. It has been introduced to many environments and has easily thrived.

The length measured from the end of the face to the end of the telson is 4–5.5 cm in females and 5.7–6.5 cm in males, with exceptional specimens

Figure 9.1 Two female adult specimens of river prawn. *Macrobrachium borellii* (top) and *Palaemontes argentinus* (bottom).

with up to 8 cm. This species has a brown color that ranges from light to dark, depending on environmental conditions, age and season.

Palaemonetes argentinus (Nobili, 1901) (Decapoda: Palaemonidae) (Figure 9.1, top) is distributed in the north and central areas of Argentina, Uruguay and south of Brazil[5,6] (15°S–45°S). It is found in lakes, rivers and creeks, but as opposed to *M. borellii* this crustacean inhabits bodies of water of variable salinity.[7] Its size in relation to *M. borellii* is considerably smaller. Its colouring, though variable, is more transparent than the previous one, commonly named as ghost shrimps or glass shrimps.

9.1.2 Pollution/Toxicology

Aquatic animals, mainly those living near farms, factories or cities, are subject to pollution caused by pesticides and/or chemicals. Several authors have reported that pesticides, even at low concentrations, affect the metabolism of aquatic organisms.[8–11] Since many pesticides are liposoluble and of widespread use, they accumulate in adipose tissue and provoke alterations in lipid metabolism.[12,13] Although studies related to this issue have been performed mainly in terrestrial mammals, some research has been done in aquatic animals, mainly in fish,[14] and it has been demonstrated that both chlorinated and phosphorated insecticides alter the lipid metabolism.[15–17] With respect to terrestrial invertebrates, most studies on the toxicity and biochemical action of diverse pesticides were performed in insects.

Xenobiotic actions occur once they are within the interior of the cell, and therefore cell membranes have a poorly recognized role in the understanding of xenobiotic metabolism. Considerable efforts have been made to elucidate the molecular mechanisms for the toxic effects of insecticides such as DDT, lindane, and parathion. Many organophosphorus pesticides are hydrophobic and accumulate mainly in membranes, modifying their mobility as well as the physicochemical and physiological properties of their constituting lipids. Numerous studies have been reported with relation to the variations of the physicochemical properties of membranes upon interaction with insecticides.[18–23] In previous years Antunes-Madeira and collaborators showed how insecticides affect the properties of different models of natural and artificial membranes.[24–26]

Within phosphorated insecticides, fenitrothion (acid, *O,O*-dimethyl *O*-(3-methyl-4-nitrophenyl) phosphorothioate ester) (FS) (Figure 9.2) is one of the most used in recent years. It is hydrophobic and affects many organisms that are not the target of action.

Fenitrothion is as an example of an organophosphorus pesticide that is degraded by photolysis[27] and hydrolysis.[28] In 1980, Greenhalgh *et al.* studied the fenitrothion hydrolysis in model and natural aquatic systems. From this study, they concluded that degradation occurred above pH 8 thus forming 3-methyl-4-nitrophenol. Below pH 7, a second dealkylation also took place to produce desmethylfenitrothion. The average time for fenitrothion degradation at 23 °C and pH 7.5 in natural lakes was 49.5 days. The authors

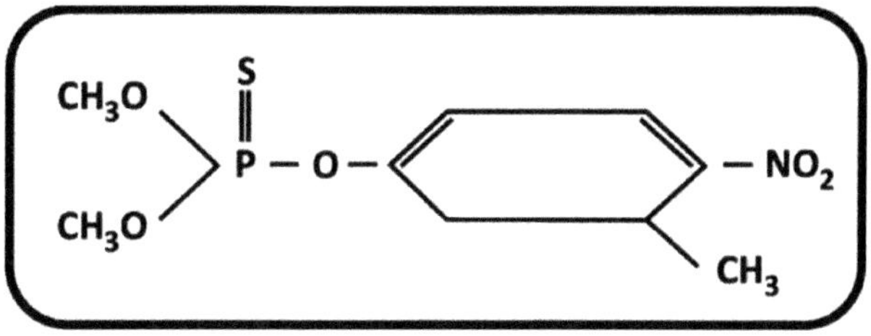

Figure 9.2 Chemical structure of the organophosphorus fenitrothion (acid, *O,O*-dimethyl *O*-(3-methyl-4-nitrophenyl) phosphorothioate ester).

suggested that the photolysis and microbiota activity are mainly responsible for the higher degradation velocity in the natural aquatic environments.[29]

Unfortunately, routine use in agriculture of this and other organophosphorus pesticides affect many non-target aquatic organisms. They can accumulate in organisms (bioaccumulation) or enter into different trophic chains (biomagnification). Basal links such as crustaceans are a great centre of interest in this sense as they are particularly sensitive. The potential damages to the organisms are varied, from different intoxication degrees to alterations of the immune system and endocrine system, among others.

9.2 Results

9.2.1 Pesticides and Membranes

Some studies have been performed with reference to insecticide toxicity on crustaceans.[15,30,31] Johnston and Corbett[16] performed assays on the lobster *Callinectes sapidus* and determined that fenitrothion is more quickly absorbed at higher temperatures and at higher salinity. The authors suggested that the increase in the absorption may increment its toxicity. Kobayashi's group[30] also investigated how fenitrothion was metabolized *in vivo* by the prawn *Penaus japonicus*, thus observing that the metabolites found were similar to those encountered in *Callinectes*.

The inhibiting action caused by organophosphorus pesticides on acetylcholinesterase, neuroreceptors and membrane proteins suggested that their toxic activity could be influenced by the compositions and properties of the membranes. This triggered research into the alterations produced when liposoluble pesticides enter the phospholipidic bilayer of a biological membrane. Related studies were carried out with natural membranes of vertebrates[32,33] and invertebrates,[34] and artificial membranes,[35,36] in which it was demonstrated that pesticides change the physicochemical properties of biomembranes.

The effect of fenitrothion on the physical condition of microsomal membranes of the hepatopancreas of *M. borellii* was previously determined using fluorescence polarization techniques. In this case, it was shown not only in microsomal membranes of animals exposed to insecticides but also in liposomal membranes.[34,37]

On the other hand, the effect of fenitrothion on artificial bilayers of dipalmitoylphosphatidylcholine (DPPC) and 1-palmitoyl-2-oleoylphosphatidylcholine (POPC) was also studied at different temperatures, also using fluorescence spectroscopy. In this case not only was the partition of fenitrothion in this type of models shown but it was also observed that this toxin influences the gel–transparent liquid phase transition of these membranes and builds microdomains with different physicochemical properties in the membrane in its lipid–water interphase.[36]

In 2001 it was observed that fenitrothion could enter into the circulating lipoproteins of *M. borellii*, decrease their fluidity, increase the order of the lipid phase, and produce defects in the packing, affecting their structure and function. This fact has great importance for two main reasons related to physiology and toxicology. On the one hand, it demonstrates that the presence of pesticides may alter the normal exchange of lipids between lipoproteins and tissues. On the other hand, it shows that invertebrate lipoproteins may act as transport vehicles and distribute toxins to the different target tissues and organs.[38]

9.2.2 Pesticides and Cholinesterases

As previously mentioned, it is well-known that organophosphorus pesticides induce their toxicity by cholinesterase inhibition and as a consequence generate a disruption in the transmission of the nerve impulse. The structure of organophosphorus compounds is similar to the endogenous neurotransmitter acetylcholine. Cholinesterases are typically subdivided into two classes, acetylcholinesterases (AChE; EC. 3.1.1.7) and butyrylcholinesterases (BChE; EC. 3.1.1.8).

These enzymes are described in different organs, for example in fish mostly AChEs are found in muscle tissue and brain, while plasma and liver contain mostly BchEs.[39,40]

Many studies have been performed using the inhibition of cholinesterase as a tool to determine organophosphorus poisoning in vertebrates (for example[41,42]) and in invertebrates.[43] In the latter, the reported works are considerably scarcer.[44]

AChE activity has been measured in a wide variety of tissues with varied effects, for instance brain, ventral ganglion, hepatopancreas (principal metabolic organ of crustaceans), nervous system, muscle, and whole body of adults and larvae.[45–50] In 1998 Lignot observed that fenitrothion affected AchE activity of the shrimp *P. stylirostris* but not *P. lannamei*.[51] For the case of *P. clarkii* treated with 0.1–15 mg l^{-1} organophosphorus trichlorfon, a change in AChE was observed.[52] Later using the same organophosphorus it was observed in *M. rosenbergii* that concentrations of 0.2 and 0.4 mg l^{-1} affected AChE activities.[53]

Table 9.1 shows an example of the value of AChE activity (expressed as enzyme units per milligram of protein) in the muscular and hemolymphatic tissues of *M. borellii* and *P. argentinus*. It also demonstrates how fenitrothion

Table 9.1 Effect of different concentrations of FS on the enzymatic activity of AchE of the muscle and hemolymph of *Macrobrachium borrellii* and *Palaemonetes argentinus*. Each value represents the mean ± SD of 12 determinations. Significantly different from the corresponding control.

		Effect of FS on the enzymatic activity of AChE (U mg protein^{-1})							
Shrimp	Time	0 $\mu g\,l^{-1}$ of FS	0.2 $\mu g\,l^{-1}$ of FS	0.3 $\mu g\,l^{-1}$ of FS	0.6 $\mu g\,l^{-1}$ of FS	0.8 $\mu g\,l^{-1}$ of FS	0.9 $\mu g\,l^{-1}$ of FS	1.4 $\mu g\,l^{-1}$ of FS	2.0 $\mu g\,l^{-1}$ of FS
M. borellii hemolymph	24 h	29.5 ± 3.9	30.1 ± 4.8				27.5 ± 8.7	28.2 ± 3.8	
M. borellii hemolymph	48 h	35 ± 12.6	51.8 ± 10				33.4 ± 7.3	19.8 ± 3.5	
M. borellii hemolymph	72 h	41.2 ± 21.9	41.6 ± 1.4				22.6 ± 6.7	20 ± 7.9	
M. borellii hemolymph	96 h	48 ± 7.8	27.5 ± 7.8[a]				19.4 ± 11.8[a]	20.3 ± 5.6[a]	
M. borellii muscle	96 h	19.4 ± 7.5			16.5 ± 5.5	17.1 ± 4.8			17.5 ± 2.6
P. argentinus hemolymph	96 h	53.3 ± 4.8		40.4 ± 2.3[a]	38.8 ± 0.4[a]	27.6 ± 6.1[a]			
P. argentines muscle	96 h	4.8 ± 1.6		4.9 ± 0.8	4.3 ± 1.1	7.9 ± 2.4			

[a] $p < 0.05$.

affects their activity in sub lethal concentrations,[54,55] [unpublished results Lavarias and Garcia].

The methodology was used according to the modified version of the colorimetric technique described by Ellman *et al.* (1961).[55] AChE activity was expressed as enzyme units/mg protein. One AChE unit was the amount of enzyme that hydrolyzed 1 nmol of acetylcholine/min.

When the muscle and hemolymph isoforms of the two freshwater palemonids presented in Table 9.1 were analyzed comparatively, it was observed that the hemolymph isoform was the only one affected, which may be related to the values obtained from the initial activity. The hemolymph isoform has twice the activity of the muscular isoform. This analysis is hard to generalize since the data related to this type of study are scarce and dispersed. In 2010 Yaqin and Hansen observed that the sensitivity of ChE activity in mussels *Mytilus edulis* could be discriminated into three clusters, which are high (gill), moderate (foot and mantle) and low sensitivities (hemolymph, digestive gland).[56] For the case of the muscle *Litopenaeus stylirostris*, FS affected AChE activity.[51]

Finally, the use of hemolymph as a biomarker of pollution through the analysis of AChE activity may have an advantage over other organs/tissues (gonads, hepatopancreas, and even whole body). This method is non-invasive for the animal specimen and the analysis can be performed with only a few microlitres of hemolymph sample.

9.2.3 Pesticide and Antioxidant Enzymes

As previously mentioned, organophosphorus compounds generate free radicals, which are harmful/toxic for the organisms. To avoid these free radicals, organisms have developed mechanisms that remedy this problematic biochemical situation. One method is by the use of antioxidant enzymes, such as superoxide dismutase (SOD; EC 1.15.1.1), an enzyme that is present in almost all cells. SOD converts O_2 into H_2O_2; the velocity of the non-enzymatic peroxide degradation is of 2×10^5 $M^{-1}\,s^{-1}$, whereas the reaction catalyzed by the SOD-Cu/Zn increases by several orders of magnitude (2×10^9 $M^{-1}\,s^{-1}$). SOD is considered a first line defence against reactive oxygen species. Hydrogen peroxide produced by this reaction is then degraded by other detoxifying enzymes, such as catalase (see below).

SODs are classified into three groups according to the metal content. The so-called FeSOD are described in prokaryotes and in plants; mitochondrial and bacterial are MnSODs; and cytosolic SODs have Cu and Zn as cofactors, CuZnSOD. For the case of *P. argentinus* we identified four SODs in the transcriptome (manuscript in preparation). After a BLAST search against non-redundant protein sequences (nr) using Blastp, we found two isoform sequences that corresponded to intracellular CuZnSOD and two isoforms of MnSODs (manuscript in preparation).

Table 9.2 shows the effect of sublethal FS doses on the activity of SOD in *M. borellii* and *P. argentinus*.[54,55] The enzyme activity was determined according to the methodology described by Misra and Fridovich. This

Table 9.2 Effect of different concentrations of FS on the enzymatic activity of superoxide dismutase of the hepatopancreas of *Macrobrachium borrellii* and *Palaemonetes argentinus*. Each value represents the mean ± SD of 12 determinations. Significantly different from the corresponding control.

		Effect of FS on the enzymatic activity of superoxide dismutase (U mg protein^{-1})					
Shrimp	Time	0 μg l^{-1} of FS	0.2 μg l^{-1} of FS	0.3 μg l^{-1} of FS	0.6 μg l^{-1} of FS	0.9 μg l^{-1} of FS	1.4 μg l^{-1} of FS
M. borellii hepatopancreas	24 h	1.48 ± 0.08	1.61 ± 0.12			1.66 ± 0.2	1.58 ± 0.19
M. borellii hepatopancreas	48 h	0.99 ± 0.2	1.15 ± 0.05			1.1 ± 0.31	1.55 ± 0.14
M. borellii hepatopancreas	72 h	1.13 ± 0.13	1.1 ± 0.11			1.35 ± 0.09	1.41 ± 0.08
M. borellii hepatopancreas	96 h	1.33 ± 0.12	1.18 ± 0.13			1.44 ± 0.16	1.14 ± 0.18
P. argentinus hepatopancreas	96 h	0.85 ± 0.21		1.37 ± 0.37^{a}	1.06 ± 0.09	1.09 ± 0.31	

$^{a}p < 0.05$.

method is based on the inhibition of the auto-oxidation of epinephrine. Results were expressed as units of SOD per mg of protein. One SOD unit was considered as the amount of enzyme necessary to inhibit the rate of auto-catalytic adrenochrome formation by 50%.[54] From these results we can conclude that in *M. borellii* high concentrations of organophosphorus significantly change the enzymatic activity; at intermediate times this may occur since at these times the enzymatic activity is capable of removing all the reactive groups of O_2.[55] On the other hand, with the use of RT-PCR it could be determined that the expression levels of this enzyme in the shrimp *M. borellii* increased when exposed to fenitrothion; this was a dose-dependent increase.[54] With relation to the effect exerted by the insecticide in *P. argentinus*, it was observed that there was a variation only at lower concentrations. For this model it can be concluded that the enzyme may only take part at these concentrations because it becomes inactive at higher organophosphorus concentrations.[55] This enzyme was also affected when the effect of another organophosphate (trichlorfon) was analyzed using the crustacean *M. rosenbergii* as a model.[53]

The enzyme catalase (CAT; EC 1.11.1.6) is widely studied and has numerous cellular types. As aforementioned, it functions after SOD and catalyzes the conversion of hydrogen peroxide to water at a velocity of 200 000 transformations s^{-1} subunit^{-1}. CAT is classified into three separate families: Mn-catalases, catalase-peroxidases and mono-functional catalases. The monofunctional CAT are the best characterized. Through the RNA-Seq analysis of the transcriptome of the prawn *P. argentinus*, sequence of a CAT could be determined the, which in the future will be of great importance to complement the activity analysis with the expression analysis (manuscript in preparation).

Table 9.3 shows the effect of FS in sublethal concentrations on CAT action in *M. borellii* and *P. argentinus.*[54,55] The type of test methodology used involves spectrophotometric absorbance measurements performed at 240 nm.[57] One CAT unit was the amount of enzyme required to catalyze the degradation of 1 pmol of H_2O_2 per minute. In the case of *M. borellii* and *P. argentinus*, all concentrations tested generated an increase in the catalase enzyme activity. These results demonstrate that this enzyme is a valuable tool to help monitor fenitrothion exposure of these two shrimps, but its activity is increased early after exposure to fenitrothion.

Glutathione *S*-transferases (GSTs, EC 2.5.1.18) belong to a large family of enzymes present in eukaryotes and prokaryotes and are known as key components of phase II xenobiotic metabolism. They are classified with relation to their cellular localization: cytosolic, mitochondrial, and microsomal. These enzymes catalyze nucleophilic attack of a physiological substrate (reduced glutathione or GSH) on the electrophilic centre of a great number of toxic structures (organophosphates among them), thus they remove a great quantity of xenobiotics from different organisms. The cytosolic family (enzyme mainly linked to detoxification) is classified into seven

Table 9.3 Effect of different concentrations of FS on the enzymatic activity of catalase of *Macrobrachium borrellii* and *Palaemonetes argentinus*. Each value represents the mean ± SD of 12 determinations. Significantly different from the corresponding control.

		Effect of FS on the enzymatic activity of catalase (U mg protein^{-1})					
Shrimp	Time	0 μg l^{-1} of FS	0.2 μg l^{-1} of FS	0.3 μg l^{-1} of FS	0.6 μg l^{-1} of FS	0.9 μg l^{-1} of FS	1.4 μg l^{-1} of FS
M. borellii hepatopancreas	24 h	8.3 ± 0.8	9.2 ± 2			9.2 ± 0.9	10.7 ± 0.7[a]
M. borellii hepatopancreas	48 h	8.5 ± 1.1	12.9 ± 0.9[a]			13.7 ± 1.6[a]	13.3 ± 1.3[a]
M. borellii hepatopancreas	72 h	7.7 ± 0.5	8.9 ± 1.6			8.6 ± 1.4	13.1 ± 1[a]
M. borellii hepatopancreas	96 h	10.9 ± 0.9	9.6 ± 1.2			16.3 ± 2.6[a]	11.5 ± 1.4
P. argentinus hepatopancreas	96 h	8.8 ± 0.6		12.8 ± 3[a]	13.8 ± 1.6[a]	19.5 ± 0.5[a]	

[a] $p < 0.05$.

Table 9.4 Effect of different concentrations of FS on the enzymatic activity of GST of *Macrobrachium borrellii* and *Palaemonetes argentinus*. Each value represents the mean ± SD of 12 determinations. Significantly different from the corresponding control.

		Effect of FS on the enzymatic activity of glutathione *S*-transferase (U mg protein^{-1})					
Shrimp	Time	0 μg l^{-1} of FS	0.2 μg l^{-1} of FS	0.3 μg l^{-1} of FS	0.6 μg l^{-1} of FS	0.9 μg l^{-1} of FS	1.4 μg l^{-1} of FS
M. borellii hepatopancreas	24 h	19.9 ± 1.3	22 ± 3.2			19.8 ± 0.8	16.7 ± 2.5
M. borellii hepatopancreas	48 h	16.3 ± 1.4	16.5 ± 2.4			21.1 ± 1.9[a]	20.7 ± 0.9[a]
M. borellii hepatopancreas	72 h	10.7 ± 1.4	12.6 ± 2.3			13.8 ± 0.8[a]	13 ± 1.2[a]
M. borellii hepatopancreas	96 h	16.8 ± 1.5	15.2 ± 1.6			22 ± 2.1[a]	19.3 ± 1.5[a]
P. argentinus hepatopancreas	96 h	5.6 ± 1.9		6.1 ± 1.7	11.5 ± 1.9[a]	15.2 ± 3[a]	

[a] $p < 0.05$.

families (alpha, kappa, mu, pi, sigma, theta, and zeta), which are differentiated in their sequence, immune properties and physiological role.

Using RNA-Seq techniques in *P. argentinus*, only two subclasses were found, the so-called mu-GST cytosolic and theta-GST cytosolic, which are considered the most ancient isoforms (manuscript in preparation).

Table 9.4 shows the effect of FS in sublethal concentrations on the GST action in *M. borellii* and *P. argentinus*.[54,55] For this type of assay, the methodology described by Habig *et al.* was used.[54] 1-chloro-2,4-dinitrobenzene (CDNB) is used as substrate and is measured spectrophotometrically at 340 nm. One GST unit represented the amount of enzyme required to conjugate GSH with 1 μmol of 1-chloro-2,4-dinitrobenzene per minute.

In the case of *M. borellii* an increase in FS activity was observed, mainly when exposed for times longer than 48 h and at higher concentrations. For the case of *P. argentinus* the same as *M. borellii* was noted since the last two concentrations had an effect. These results suggest that GST is active in the process of toxic metabolization at late stages and that only concentrations of FS higher than 0.3 $\mu g\,l^{-1}$ cause an effect. This effect was also observed for the case of the crustacean called American red crayfish *Procambarus clarkii* when exposed to FS[58] and in *Chasmagnathus granulatus* when it was exposed to methyl parathion.[59]

9.3 Conclusion

In conclusion, we can say that there is great controversy with relation to water molecules and to bodies of water. On the one hand, historically water was considered as a vehicle for the removal of contaminants and/or toxic residues. On the other hand, water is considered the most vital resource for the biota. This is a great conflict between nature and technology (so as not to specifically consider the anthropogenic actions). According to Krüger,[60] the paradigm to establish a homeostatic state for natural systems must include a monitoring process to be able to apply corrective actions in time. A biomarker may be the solution to generate these corrective actions in the future. One of the possible definitions of a biomarker is the use of body fluids, cells or a set of them (tissue) that indicate the presence of pollutants/xenobiotics[61] in cellular or biochemical terms. This definition can also be widened considering behavioural parameters, such as reproductive and social aspects.

The measurements at the biochemical level can quickly and specifically detect the presence of several toxic compounds, which enables the early identification of a change, before the deleterious effects reach higher levels of organizations, such as morphological/histological changes.

One of the objectives of the present chapter is to report the scarce data with reference to the basal activities of biomarker enzymes, which can be used as tools for ecotoxicological studies and/or monitoring. Another objective is to give a clear example of the biomarker functions of different proteins (enzymes) in to two crustaceans of freshwater Argentine environment when exposed to an organophosphate. Cholinesterase activity is an example of a bioindicator of organophosphate exposure.[62] SOD, CAT, and GST are examples of bioindicators because different pollutants can directly or indirectly modify the balance of the concentration of pro- and anti-oxidants.[63] At this point it is important to mention that some issues in this chapter still need to be discussed, such as DNA damage, protein oxidation or lipid peroxidation, which are strongly linked to the detoxifying responses of different organisms in general and of crustaceans in particular.

Acknowledgements

We apologize to authors whose work has not been cited or mentioned. Authors are thankful for the support from CONICET (National Council of Scientific and Technical Research, Argentina) grant COOP-CTES-D680 and CONACYT (Mexico's National Council for Science and Technology) for Mexico-Argentina grant C0005-2012-189333 to C.F. García and R. Sotelo-Mundo. We are grateful for the grants from Consejo Nacional de Investigaciones Científicas y Técnicas (CONICET-PIP 2014-2016) and Agencia Nacional de Promoción Científica y Tecnológica (PICT 2014-2580), Argentina. The authors are grateful to Rosana del Cid for the revision of the English, and Mario Ramos for the figure design.

References

1. P. Collins and J. C. Paggi, Feeding ecology of *Macrobrachium borellii* (Nobili) (Decapoda: Palaemonidae) in flood valley of river Paraná Argentina, *Hydrobiologia*, 1998, **362**, 21–30.
2. P. Collins, Feeding of *Palaemonetes argentinus* (Nobili) (Decapoda: Palaemonidae) in flood valley of river Paraná Argentina, *J. Crustacean Biol.*, 1999, **19**, 485–492.
3. P. Collins, Ritmo diario de alimentación del camarón *Macrobrachium borellii* (Decapoda, Palaemonidae), *Iheringia, Sér. Zool.*, 1997, **82**, 19–24.
4. F. Giri, V. Williner and P. Collins, Tiempo de evacuación del camarón dulceacuícola *Palaemonetes argentinus* (Crustacea, Decapoda) alimentado con larvas de mosquito *Culex pipiens*, FABICIB, Santa Fe, 2002.
5. E. D. Spivak, K. Anger, T. A. Luppi, C. C. Bas and D. Ismael, Distribution and habitat preferences of two grapsid crab species in Mar Chiquita Lagoon (Province of Buenos Aires, Argentina), *Helgoländer Meeresunters*, 1994, **48**, 59–78.
6. P. Collins, A coexistence mechanism for two freshwater prawns in the Paraná river floodplain, *J. Crustacean Biol.*, 2005, **25**, 219–225.
7. E. D. Spivak, Life history of a brackish-water population of *Palaemonetes argentinus* (Decapoda: Caridea) in Argentina, *Ann. Limnol.*, 1997, **33**, 179–190.
8. O. T. Sanders, R. Zepp and L. Kirkpatrik, Effect of PCB ingestion on sleeping times, organ weights, food consumption, serum corticosterone

and survival of albino mice, *Bull. Environ. Contam. Toxicol.*, 1974, **12**, 394–399.

9. R. Ashauer, A. Boxall and C. Brown, Predicting effects on aquatic organisms from fluctuating or pulsed exposure to pesticid, *Environ. Toxicol. Chem.*, 2006, **7**, 1899–1912.
10. J. M. Giddings, W. M. Williams, K. R. Solomon and J. P. Giesy, Risks to aquatic organisms from use of chlorpyrifos in the United States, *Rev. Environ. Contam. Toxicol.*, 2014, **231**, 119–162.
11. R. Annett, H. R. Habibi and A. Hontela, Impact of glyphosate and glyphosate-based herbicides on the freshwater environment, *J. Appl. Toxicol.*, 2014, **34**, 458–479.
12. M. Eto, *Organophosphorous Pesticides; Organic and Biological Chemistry*, CRC Press, Cleveland, 1974.
13. T. Purshottam and R. K. Srivastava, Parathion toxicity in relation to liver microsomal oxidases, lipid composition and fluidity, *Pharmacology*, 1987, **35**, 227–233.
14. B. Lal and T. P. Singh, Impact of pesticides on lipid metabolism in the freshwater catfish, *Clarias batrachus* during the vitellogenic phase of its annual reproductive cycle, *Ecotoxicol. Environ. Saf.*, 1987, **13**, 13–23.
15. R. Sarojini, M. Nagabhushanam and S. Avelin, Effect of fenitrothion on reproduction of the freshwater prawn *Macrobrachium lamerri*, *Ecotoxicol. Environ. Saf.*, 1986, **11**, 243–250.
16. J. Johnston and M. Corbett, The uptake and in vivo metabolism of the organophosphate insecticide fenitrothion by the blue crab, *Callinectes sapidus*, *Toxicol. Appl. Pharmacol.*, 1986, **85**, 181–188.
17. K. Sarma, A. K. Pal and K. Grinson-George, Effect of sub-lethal concentration of endosulfan on lipid and fatty acid metabolism of spotted murrel, *Channa punctatus*, Baruah, *Environ. Biol.*, 2015, **36**(2), 451–454.
18. G. M. Omann and J. R. Lakowicz, Interactions of chlorinated hydrocarbon insecticides with membranes, *Biochim. Biophys. Acta*, 1982, **684**, 83–85.
19. K. J. Stelzer and M. A. Gordon, Interactions of pyrethroids with phosphatidylcholine liposomal membranes, *Biochim. Biophys. Acta*, 1985, **812**, 361–368.
20. M. A. Pérez-Albarsanz, P. López-Aparicio, S. Senar and M. N. Recio, Effects of lindane on fluidity and lipid composition in rat renal cortex membranes, *Biochim. Biophys. Acta*, 1991, **1066**, 124–130.
21. P. López-Aparicio, M. N. Recio, J. C. Prieto, M. J. Carmena and M. A. Pérez-Albarsanz, Effects of lindane on fluidity and lipid composition in rat renal cortex membranes, *Life Sci.*, 1991, **49**(1141), 124–130.
22. J. Blasiak, Changes in the fluidity of model lipid membranes evoked by the organophosphorus insecticide methylbromfenvinfos, *Acta Biochim. Pol.*, 1993, **40**, 39–41.
23. M. R. Moya-Quiles, E. Munoz-Delgado and C. J. Vidal, Interactions of the pyrethroid insecticide allethrin with liposomes, *Arch. Biochem. Biophys.*, 1994, **312**, 95–100.

24. M. C. Antunes-Madeira and V. M. Madeira, Effects of DDE on the fluidity of model and native membranes: implications for the mechanisms of toxicity, *Biochim. Biophys. Acta*, 1993, **1149**(1), 86–92.
25. M. C. Antunes-Madeira, L. M. Almeida and V. M. Madeira, Depth-dependent effects of DDT and lindane on the fluidity of native membranes and extracted lipids. Implications for mechanisms of toxicity, *Bull. Environ. Contam. Toxicol.*, 1993, **51**(6), 787–794.
26. R. A. Videira, M. C. Antunes-Madeira, J. B. Custodio and V. M. Madeira, Partition of DDE in synthetic and native membranes determined by ultraviolet derivative spectroscopy, *Biochim. Biophys. Acta*, 1995, **1238**(1), 22–28.
27. H. Ohkawa, N. Mikami and J. Miyamoto, Photodecomposition of sumithion ((o,o-dimethyl methyl-4-notrophenyl) phosphorothioate), *Agric. Biol. Chem.*, 1974, **38**, 2247–2255.
28. Y. Nishizawa, K. Fujii, T. Kadota, J. Miyamoto and H. Sakamoto, Studies on the organophosphorus insecticides. VII. Chemical and biological properties of new low toxic organophosphorus insecticide. O,O-dimethyl-O-(3-methyl-4-nitrophenyl) phosphorothioate, *Agric. Biol. Chem.*, 1961, **25**(8), 605–610.
29. R. Greenhalgh, K. L. Dhawan and P. Weinberger, Hydrolysis of fenitrothion in model and natural aquatic systems, *J. Agric. Food Chem.*, 1980, **28**(1), 102–105.
30. K. Kobayashi, Y. Nakamura and N. Imada, Metabolism of an organophosphorus insecticide, fenitrothion, in tiger shrimp, *Pennies japonicus*, *Bull. Jpn. Soc. Sci. Fish.*, 1985, **51**, 599–603.
31. M. C. Montagna and P. A. Collins, Survival and growth of *Palaemonetes argentinus* (Decapoda; Caridea) exposed to insecticides with chlorpyrifos and endosulfan as active element, *Arch. Environ. Contam. Toxicol.*, 2007, **53**(3), 371–378.
32. T. Purshottan and R. K. Srivastava, Parathion toxicity in relation to liver microsomal oxidases, lipid composition and fluidity, *Pharmacology*, 1987, **35**(4), 227–233.
33. M. C. Antunes-Madeira, R. A. Videira and V. M. Madeira, Effects of parathion on membrane organization and its implications for the mechanisms of toxicity, *Biochim. Biophys. Acta*, 1994, **1190**, 149–154.
34. M. R. González-Baró, H. Garda and R. J. Pollero, Effect of fenitrothion on hepatopancreas microsomal membrane fluidity in *Macrobrachium borellii*, *Pest. Biochem. Physiol.*, 1997, **58**(2), 133–143.
35. J. Blasiak, Changes in the fluidity of model lipid membranes evoked by the organophosphorus insecticide methylbromfenvinfos, *Acta Biochim. Pol.*, 1993, **40**(1), 39–41.
36. M. R. González-Baró, H. Garda and R. J. Pollero, Effect of fenitrothion on dipalmitoyl and 1-palmitoyl-2-oleoylphosphatidylcholine bilayers, *Biochim. Biophys. Acta*, 2000, **14688**(1–2), 304–310.
37. C. F. García, M. Gonzalez-Baró and R. Pollero, Effect of fenitrothion on the acylglyceride exchange in crustacean lipoproteins, *Pestic. Biochem. Physiol.*, 2005, **82**(3), 177–184.

38. C. F. García, M. Gonzalez-Baró, M. Cunningham, H. Garda and R. Pollero, Effect of fenitrotion on the physical properties of crustacean lipoproteins, *Lipids*, 2002, **37**, 673–679.
39. C. Habig and R. D. DiGiulio, *Biochemical Characteristics of Cholinesterases in Aquatic Organisms*, ed. P. Mineau, Elsevier, New York, USA, 1991, pp. 19–33.
40. J. V. Rao, G. Begum, R. Pallela, P. K. Usman and R. N. Rao, Changes in behavior and brain acetylcholinesterase activity in mosquito fish, *Gambusia affinis* in response to the sub-lethal exposure to chlorpyrifos, *Int. J. Environ. Res. Public Health*, 2005, **2**(3–4), 478–483.
41. S. Bretaud, J. P. Toutant and P. Saglio, Effects of carbofuran, diuron and nicosulfuron on acetylcholinesterase activity in goldfish (*Carassius auratus*), *Ecotoxicol. Environ. Saf.*, 2000, **47**(2), 117–124.
42. D. W. Beyers and P. J. Sikoski, Acetylcholinesterase inhibition in federally endangered Colorado squawfish exposed to carbaryl and malathion, *Environ. Toxicol. Chem.*, 1994, **13**, 935–939.
43. K. E. Day and I. M. Scott, Use of acetylcholinesterase activity to detect sublethal toxicity in stream invertebrates exposed to low concentrations of organophosphate insecticides, *Aquat. Toxicol.*, 1990, **18**, 101–113.
44. E. Escartin and C. Porte, Acetylcholinesterase inhibition in the crayfish *Procambarus clarkii*, *Ecotoxicol. Environ. Saf.*, 1995, **34**, 160–164.
45. C. Habig, R. T. DiGiulio, A. A. Nomeir and M. B. Abou-Donia, Comparative toxicity, chlinergic effects and tissue levels of s,s,s-tri-butyl phosphorotrithioate (DEF) to channel catfish (*Ictalurus punctatus*) and blue crabs (*Callinectes sapidus*), *Aquat. Toxicol.*, 1986, **9**, 193–206.
46. J. Chambers, J. Heitz, F. McCorkle and J. Yarbrough, Enzyme activities following chronic exposure to crude oil in a simulated ecosystem, *Environ. Res.*, 1979, **20**, 133–139.
47. M. S. Reddy and K. R. Rao, *In vivo* recovery of acetylcholinesterase activity from phosphamidon and methylparathion induced inhibition in the nervous tissue of penaeid prawn (*Metapenaeus monoceros*), *Bull. Environ. Contam. Toxicol.*, 1988, **40**(5), 752–758.
48. F. Galgani and G. Bocquene, *In vitro* inhibition of acetylcholinesterase from four marine species by organophosphates and carbamates, *Bull. Environ. Contam. Toxicol.*, 1990, **45**(2), 243–249.
49. A. Abdulah, A. Kumar and J. Chapman, Inhibition of acetylcholinesterase in the Australian freshwater shrimp (*Paratya australiensis*) by profenofos, *Environ. Toxicol. Chem.*, 1994, **13**, 1861–1866.
50. P. B. Key and M. H. Fulton, Lethal and sublethal effects of chlorpyrifos exposure on adult and larval stages of the grass shrimp *Palaemonetes pugio*, *J. Environ. Sci. Health B*, 1993, **28**, 621.
51. J. H. Lignot, G. Charmantier and J. C. Cochard, J. Effect of an organophosphorus insecticide, fenitrothion, on survival, osmoregulation, and acetylcholinesterase activity in different life stages of two penaeid

shrimps: *Penaeus stylirostris* and *Penaeus vannamei* (Crustacea, Decapoda), *Shell Fish Res.*, 1998, **17**, 1251–1258.
52. G. Repetto, P. Sanz and M. Repetto, *In vivo* and *in vitro* effect of trichlorfon on esterases of the red crayfish *Procambarus clarkii*, *Environ. Contam. Toxicol.*, 1988, **41**, 597–603.
53. C. Chang, P. Lee, C. Liu and W. Cheng, Trichlorfon, an organophosphorus insecticide, depresses the immune responses and resistance to *Lactococcus garvieae* of the giant freshwater prawn *Macrobrachium rosenbergii*, *Fish Shellfish Immunol.*, 2006, **20**, 574–585.
54. S. Lavarias, C. García, R. Crespo, N. Pedrini and H. Heras, Study of biochemical biomarkers in freshwater prawn *Macrobrachium borellii* (*Crustacea: Palaemonidae*) exposed to organophosphate fenitrothion, *Ecotoxicol. Environ. Saf.*, 2013, **96**, 10–16.
55. S. Lavarias and C. F. García, Acute toxicity of organophosphate fenitrothion on biomarkers in prawn *Palaemonetes argentinus* (Crustacea: Palaemonidae), *Environ. Monit. Assess.*, 2015, **65**, 187–197.
56. K. Yaqin and P. D. Hansen, The use of cholinergic biomarker, cholinesterase activity of blue mussel *Mytilus edulis* to detect the effects of organophosphorous pesticides, *Afr. J. Biochem. Res.*, 2010, **4**, 265–272.
57. H. Aebi, Catalase *in vitro*, *Methods Enzymol.*, 1984, **105**, 121–126.
58. A. Blat, M. M. Almar and F. J. Romero, The effect of two sulphur-containing pesticides, fenitrothion and endosulfan, on glutathione (GSH) content and on GSH S-transferase and gamma-glutamyl transpeptidase activities in midgut gland of the American red crayfish *Procambarus clarkii*, *Drug Metab. Drug Interact.*, 1988, **6**(3–4), 383–394.
59. A. Bianchini and J. M. Monserrat, Effects of methyl parathion on *Chasmagnathus granulatus* hepatopancreas: Protective role of sesamol, *Ecotoxicol. Environ. Saf.*, 2007, **67**(1), 100–108.
60. E. L. Krüger, Desenvolvimento e Meio Ambiente, UFPR, Paraná, 2001.
61. D. R. Livingstone, Biotechnology and pollution monitoring: use of molecular biomarkers in the aquatic environment, *J. Chem. Technol. Biotechnol.*, 1993, **57**, 195–211.
62. R. V. Hyne and W. A. Maher, Invertebrate biomarkers: links to toxicosis that predict population decline, *Ecotoxicol. Environ. Saf.*, 2003, **54**, 366–374.
63. L. A. Geracitano, J. M. Monserrat and A. Bianchini, Oxidative stress in *Laeonereis acuta* (Polychaeta, Nereididae): environmental and seasonal effects, *Mar. Environ.*, 2004, **58**(2–5), 625–630.

CHAPTER 10

The Crab Ucides cordatus *(Malacostraca, Decapoda, Brachyura) and Other Related Taxa as Environmental Sentinels for Assessment and Monitoring of Tropical Mangroves from South America*

M. A. A. PINHEIRO,*[a] C. A. SOUZA,[a] F. P. ZANOTTO,[b] R. A. TORRES[c] AND C. D. S. PEREIRA[d]

[a] UNESP – Universidade Estadual Paulista, Instituto de Biociências (IB), Campus do Litoral Paulista (CLP), Praça Infante Dom Henrique, s/n° – Parque Bitaru, 11330-900, São Vicente, SP, Brazil; [b] USP – Universidade de São Paulo, Departamento de Fisiologia, Cidade Universitária, 05508-090, São Paulo, SP, Brazil; [c] UFPE – Universidade Federal de Pernambuco, Centro de Ciências Biológicas (CCB), Departamento de Zoologia, Av. Professor Moraes Rego, Cidade Universitária, 50670-420, Recife, PE, Brazil; [d] UNIFESP – Universidade Federal de São Paulo, Departamento de Ciências do Mar, Campus Baixada Santista. Avenida Almirante Saldanha da Gama, 89 – Ponta da Praia, 11030-400, Santos, SP, Brazil
*Email: pinheiro@clp.unesp.br

Issues in Toxicology No. 33
Ecotoxicology and Genotoxicology: Non-traditional Aquatic Models
Edited by Marcelo L. Larramendy

Published by the Royal Society of Chemistry, www.rsc.org

10.1 Anthropic History: Actions *vs.* Reaction

Since early times, man has been settling near to rivers and other freshwater sources, which are used as water supplies for their own maintenance, as well as for animal husbandry and agronomic purposes. Therefore, human densification has occurred worldwide on riverbanks, as well as in coastal regions where water flows, promoting relevant areas for many activities. Among them, there are the establishment of seaport, food stock/transportation, and others, that are responsible for city infrastructures. Actually, more than 50% of the human population lives less than 60 km from the sea,[1] in a narrow territorial strip. With the advent of the industrial revolution, this uncontrolled occupation caused harmful effects to the environment, firstly through the diversification of pollutants generated and second by the damages promoted by contact or inadequate disposal of slag resulting from anthropic activities. Among these, there are seaport facilities and the traffic of ships, generally associated with large industrial complexes, where products are stocked, many of them with strong pollutant power (*e.g.*, gas, liquids, oils, greases and other petrochemicals).

Several regions of the world stand out in terms of industrial development, often with no regard to the neutralization of the pollutants generated, which inevitably reach the adjacent environments through the action of the water cycle, and are finally dispersed, reaching the oceans. Among developing countries, the greatest concern as to the target levels of pollution is in China, where the diversity of the products generated and the myriad of xenobiotics from different origins have a large contamination potential.[2,3]

Industrialization led many European countries through alarming advents of pollution. An example occurred in the 1950s in London (UK), where remarkable industrial atmospheric pollution (smog = smoke + fog) provoked the death of at least 4000 people and a range of diseases.[4] More recently, a serious atmospheric accident occurred in the 1960s in Cubatão municipality (Southeast Brazil); a large industrial complex caused contamination of air, rivers, estuary and forests, increasing the incidence of pulmonary diseases, congenic malformations,[5] and high mortality of aquatic organisms, making them unfit for human consumption.[6]

10.2 Coastal Environments: Biodiversity and Conservation

Coastal environments are highly affected by pollutants, carried from terrestrial areas to rivers and estuarine systems, in densely occupied areas where anthropic activities occur with no monitoring. Many accidents have occurred and here we highlight some significant ones: oil spills, *e.g.*, 'Deepwater Horizon' drilling platform accident,[7] at a depth of 1500 m, occurred in April in the at Gulf of Mexico, where approximately 780 000 m^3 of crude oil was released to the sea over the course of three months; mine slag contamination, *e.g.*, the Fundão dam collapse by the mining company

Samarco (a subsidiary of Vale & BHP)[8] in the central region of Brazil in November 2015, leading to pollution of the Doce river and the death of millions of fishes in the coastal ecosystems, including the marine life of the Abrolhos archipelago; and chemical accidents, *e.g.*, a leak of 400 thousand liters of firefighting liquid foam generator, used to eradicate fire accidents, occurred with six fuel tanks at Ultracargo[9] in April 2015, and this was dispersed to the estuarine area of Santos-São Vicente, Southeast Brazil. Therefore, coastal areas are subjected to elevated environmental risk of anthropic origin, despite the existence of security plans that unfortunately are insufficient, inefficient and dysfunctional.

The release of these products promotes the death of organisms responsible for primary productivity (microalgae in aquatic ecosystems, and aquatic vegetation from wetlands *e.g.*, mangroves), to other trophic levels, such as filter feeders (*e.g.*, oysters and clams) up to benthonic organisms with different food habits, among them herbivores, omnivores, detritivores (*e.g.*, turtles, fishes, shrimps and crabs), and carnivores (*e.g.*, aquatic birds, porpoises, raccoons).

10.3 Mangrove Ecosystem: Importance and Threats

Mangroves are unique coastal ecosystems, found in shallow areas in intertidal zones present in estuarine systems, in an ecotone that involves terrestrial and aquatic environments, the last one being made up of marine and freshwater areas.[10,11] Mangroves are registered in tropical and subtropical regions of the world, and total 137 760 km^2, mainly near the equatorial zone.[12] In these areas, deposition of fine sediments occurs, mainly silt (0.05–0.002 mm) and clay (<0.002 mm), but very fine sand (0.1–0.05 mm) can also occur, resulting in banks of fluid mud.[13] The fine granulometric characteristics of mangrove sediments make this environment susceptible to contamination by oil residues, toxic metals, and organic/inorganic compounds.[14] Therefore, these areas act as efficient biogeochemical barriers, holding back these xenobiotics (mainly toxic metals), blocking their free circulation and making them unavailable to plants and animals of these areas.[15,16] This sedimentary matrix involves an association of organic matter and nutrients, varying with flooding area and tides, and promoting different scenarios of redox potential, availability and action of chemical elements over the course of the day.[16] This specific natural oscillation of physical and chemical parameters (*e.g.*, salinity, pH and oxygen content) acts on the diversity and density of micro/macro organisms that live in this environment, which show specific structures and physiological conditions to survive in these areas. Owing to the high productivity of mangroves, this ecosystem has an important role during nutrient cycling, affecting adjacent environments.[10,11] Therefore, mangrove areas attract animal species that use this environment for food, protection, and reproduction purposes.[17] The muddy sediment of mangroves is flooded by tides daily, and the interstitial water of the sediment is characterized by a significant change of the salinity[7] and low oxygen because of the high organic matter content.[18] These stressful environmental conditions allow the colonization of facultative halophyte plants called mangue,[10,11] adapted to live there, with suitable morphology for fixation in the sediment (rhizophores), oxygen extraction by aerial roots (pneumatophores) and physiological strategies to control salt load.[16] These characteristics attenuate the impacts of flood tides and are fundamental for the retention of high concentrations of organic and inorganic substances,[18] many of which are not essential to metabolic processes and are considered pollutants.

In this nutrition-rich environment, organic matter (particulate and dissolved) is available to microorganisms, where a diversity of zooplankton and juveniles of many invertebrates, such as molluscs, crustaceans and fishes, can survive.[11,17] The mangrove is composed of resident species (*e.g.* decapod crustaceans), semi-residents (occupying these areas in specific moments of life, *e.g.* during reproduction), regular visitors and opportunistic visitors.[17] In Brazil, a few species live permanently in mangroves, among them decapod crustaceans (Figure 10.1), found in diverse strata or in adjacent environments, as follows: (A) *uçá* crab *Ucides cordatus* (Linnaeus, 1763), a herbivorous crab that uses mangrove litter and actively digs mangrove sediments; (B) grapsid crab *Goniopsis cruentata* (Latreille, 1803), which lives among roots and holes in trees, and is omnivorous; (C) arboreal crab *Aratus pisonii* (H. Milne Edwards, 1837), which lives in canopies and branches of trees and is uniquely herbivorous, feeding on green leaves; (D) species of the genus *Uca* (Leach, 1814), here represented by *Uca maracoani* (Latreille, 1802), which are deposit feeders, using organic matter in the sediment for their survival, and digging galleries in the sediment; (E) *guaiamú* crab *Cardisoma guanhumi* (Latreille, 1828), generally occurring in sandy estuaries, associated with freshwater in transition with tropical forests, where the topography is more elevated and flooding tides are not frequent; and (F) ghost crab *Ocypode quadrata* (Fabricius, 1787), with omnivorous habit and that use macroinvertebrates present in the intertidal zone of sandy beaches, dig their galleries in supralittoral areas, and are generally associated with sandy dunes. These representatives of the infraorder Brachyura (brachyuran crabs) are important to many biological processes, where scavengers promote bioturbation of the sediment, increasing aeration, stratum mixture, and the content of organic matter and other nutrients content. Thus, they have been called ecosystem engineers.[19,20]

Despite the economic and ecological importance of mangroves, they are subjected to high and systematic anthropic pressure, with significant decline (approximately 1–2% per year), leading to the disappearance of this ecosystem in about 100 years.[21] Environmental contamination, as a result from disorderly and rapid urbanization and industrialization, has led to concerns from researchers and environmental analysts because the residues produced (*e.g.*, by estuarine dredging, dumping of liquid effluents, indiscriminate use of fertilizers and pesticides in cultivable fields), are cumulative, toxic and persistent, as seen with toxic metals.[22] Information about this situation reveals the urgent necessity for more basal studies and

Figure 10.1 Brachyuran species (crabs) and their occurrence on the Brazilian coast, with potential use for evaluation and monitoring of mangroves [A: *Ucides cordatus* (Linnaeus, 1763), photo by Delson Gomes; B: *Goniopsis cruentata* (Latreille, 1803), photo by Marcelo Pinheiro; C: *Aratus pisonii* (H. Milne-Edwards, 1837), photo by Marcelo Pinheiro; D: genus *Uca*, represented by *Uca maracoani* (Leach, 1814), photo by Luis Ernesto Bezerra], *restingas* [E: *Cardisoma guanhumi* (Latreille, 1828), photo by Carlos Cantareli], and sandy beaches [F: *Ocypode quadrata* (Fabricius, 1787), photo by Alexandre Almeida]. Graphic design by Gustavo Pinheiro.

monitoring programs to increase knowledge about mangrove contamination levels, and an integration with coastal and fishery management.[23] In this sense, application of models involving many areas, among them, morphology, physiology, geological processes, as well as those resulting from climatic changes of oceans water levels is highly desirable.[24]

10.4 Contaminants: Main Types, Origins and Effects on Biota

The liberation and dispersion of chemical compounds in aquatic ecosystems can occur through incorrect discharge. This can include disposal of untreated domestic sewage, industrial effluents, indiscriminate (or unnecessary) use of pesticides, and others that result from accidents with petroleum (or their organic derivatives), and atmospheric pollution from fossil fuel burning.[25] These compounds can lead to metabolic alterations, and strongly compromise vital physiological processes of the biota, with effects in respiration, reproduction and growth.[26]

In Brazil, the southeast coastal region of São Paulo state, comprises nine municipalities forming the Metropolitan Region of Baixada Santista (MRBS), and the most central portion (São Vicente, Cubatão and Santos) hosts one of the most industrialized regions of the Brazilian coast and the largest seaport of Latin America, with a human density of approximately 450 million people (see review in ref. 27). According to these authors, eight toxic metals, four organochlorine pesticides, 12 PAHs and one PCB contaminated this central coast of São Paulo state. These numbers are much higher than the southern coast, where the diversity of contaminants is 50% lower, with reduced presence of toxic metals (37.5%), similar amounts of pesticides and the absence of the other two pollutants. As a result of the higher human density on the central coast, resulting from the historic use of this area since the discovery of Brazil, pollutant amounts are higher than those cited in guidelines used by environmental agencies.[28] Among the identified substances present in water, sediment and local biota, there are dangerous toxic metals (*e.g.*, As, Cd, Pb, Cu, Cr, Hg, Ni, Zn, and others), PCBs widely used in plastic products and the paint and fluid industries, as well as PAHs, synthesized during incomplete burning of organic matter.[27,29,30]

Release of xenobiotic compounds in water can increase waterborne diseases, chemical contamination and accumulation/magnification effects of toxic metals.[31] As a consequence, decreased fishery stocks and reduced resource quality are obvious consequences of pollutant effects. Many of these substances present a toxicity period, persistence, mobility and bioaccumulation, leading to risks to higher trophic levels, especially to human health. As an example, accumulation of organochlorine, found in agricultural pesticides such as DDT and others organochlorine insecticides, favours bioaccumulation and biomagnification along the trophic chain.[32]

Bioaccumulation processes due to absorption and discharge of chemical substances through water and food ingestion have been a particular concern in estuarine regions, where high contamination indexes are frequently found (*e.g.*, Baixada Santista in Brazil). The bioaccumulation leads to biomagnification, a process where a substance is absorbed by aquatic organisms through the trophic chain, accumulating at higher trophic levels.[33] Bioaccumulation of xenobiotics depends on factors such as biological processes (feeding, physical/chemical composition and lifestyle) and environmental chemical compounds (pH, salinity, sediment composition, *etc.*).[34] Sediments reflect this variability owing to changes occurring in abiotic factors and promoting water quality alteration due to pollutant dispersion to the water column, followed by the distribution of potential toxic trace elements to biota and human populations.[29] It is important to highlight those studies using biomarkers and environmental impacts should take into account differences promoted by a range of abiotic parameters as a function of climatic seasons. Among them, local variations of abiotic parameters (*e.g.*, pH, salinity, and others) can cause distinct contamination levels in biota, influencing the health of these organisms.

Invertebrates play an important role influencing many biological functions in wetland ecosystems,[35] especially as food for birds and mammals. Generally, the diversity and richness of species decreases with local stressors (*e.g.*, wetland hydrology, vegetation complexity, and water quality). Considering aquatic biota, a more intensive effect occurs in freshwater as a function of the strong effect of pollutants from terrestrial and reclaimed lands generated by agronomical contamination sources (*e.g.*, organochlorine and organophosphate pesticides, petroleum products and other xenobiotics), which accumulate in water bodies, lagoons, rivers, estuaries and the ocean. It is clear that a higher concentration of contaminants in smaller water bodies causes a stronger impact compared to in the marine environment. In this sense, an evaluation of the conservation status of 255 species of Brazilian decapod crustaceans, conducted from 2010 to 2014 by the Chico Mendes Institute for Biodiversity Conservation (ICMBio), from the Brazilian Environmental Ministry (MMA), revealed that of 11% of threatened species ($N = 28$), 93% originated from freshwater, 4% from estuaries, and only 3% from marine origin.[36]

As previously mentioned, abiotic and biotic factors contribute to changes in the concentrations of some pollutants, and their complexation to other toxic molecules. An example is the presence of environmental mercury and its transformation to methylmercury, a known neurotoxin that is produced by anaerobic bacteria, with significant health risk to humans.[37,38] This is of relevance because most bioassays are specific to only one type of contaminant and, because of the chemical complexation that occurs with a set of contaminants, the results obtained with biomarkers are not clear enough to explain the contamination registered in a particular area.

The dynamics of bioaccumulation differ among contaminants, with a response variation according to the organisms and organs of the same individual.[39] Little is known about the physiological and biochemical modifications that affect organisms[40] or the changes in animals' homeostasis in chronically polluted areas. Some populations acquire tolerance mechanisms under moderate pollutant levels, developing mechanisms of excretion and other processes to deal with chemical compounds.[41] These mechanisms grant the animals in affected areas the ability to minimize and/or repair the effects of environmental pollution, especially through a detoxification ability, compartmentalization of metals in certain organs, higher excretion rates and even escape behavior.[39]

Crabs, in particular, are able to store toxic metals in intracellular organelles when these metals are above physiologically tolerable limits through detoxification processes that render the metals inactive.[42] High concentrations of metals can induce the cellular production of intracellular proteins, such as metallothioneins.[43] Toxic metals may also be captured by lysosomes,[42] or even compartmentalized into granules in detoxification organs, such as the hepatopancreas.[44] Gills, on the other hand, as the first organ affected by pollutants, represent a selective organ that is in contact with both the internal and external environments, acting as a temporary store for accumulated metals.[45]

The damage involves a cascade of events promoted by stressors[46] where bioindicator responses act at several levels of biological organization to show sublethal stress effects, serving as early warning signals in a causal relationship between stressors and effects, later manifested at higher levels of biological organization. Therefore, stress responses measured at a biochemical or physiological level can represent possible (and irreparable) future damage to the population and community levels (Figure 10.2). Therefore, biomarkers reflecting the health status at lower organizational levels show an immediate response to stress and have a high toxicological relevance, while biomarkers that reflect health conditions at higher organizational levels respond slowly to stress and have decreased toxicological relevance, but are more relevant ecologically.

10.5 Environmental Monitoring Based on Biomarkers

Faced with a growing panorama of industrial and reckless occupation of important ecosystems on our planet, there is an acceleration and intensification of environmental degradation, with risks and consequences requiring continued vigilance. Biota exposed to pollutants react to the exposure, and the response produced can be identified and measured using biomarkers.[34] The use of these biological parameters reflects the behaviour and the interactions between toxic agents and the biological systems, and can be used as powerful tools for environmental monitoring. Their responses can be used as a signal of toxic effects to the organism, involving perturbations of biochemical and molecular nature inside the cells, later leading to effects at higher levels of organization.[46]

Biomarkers can be used in a variety of studies and types of chemical exposition. The general idea of their application is the possibility of precocious detection of environmental perturbations that can contribute to the decline of populations or whole communities, through sublethal effects, not those responsible for an immediate animal death, but rather for alterations of biological processes, as in endocrine and reproductive systems, for example.[34] In general biomarkers can be classified into three types:

(1) Exposure: covering the detection and measurement of an exogenous substance or its metabolite or the product of an interaction between a xenobiotic agent and some target molecule or cell that is measured in a compartment within an organism.

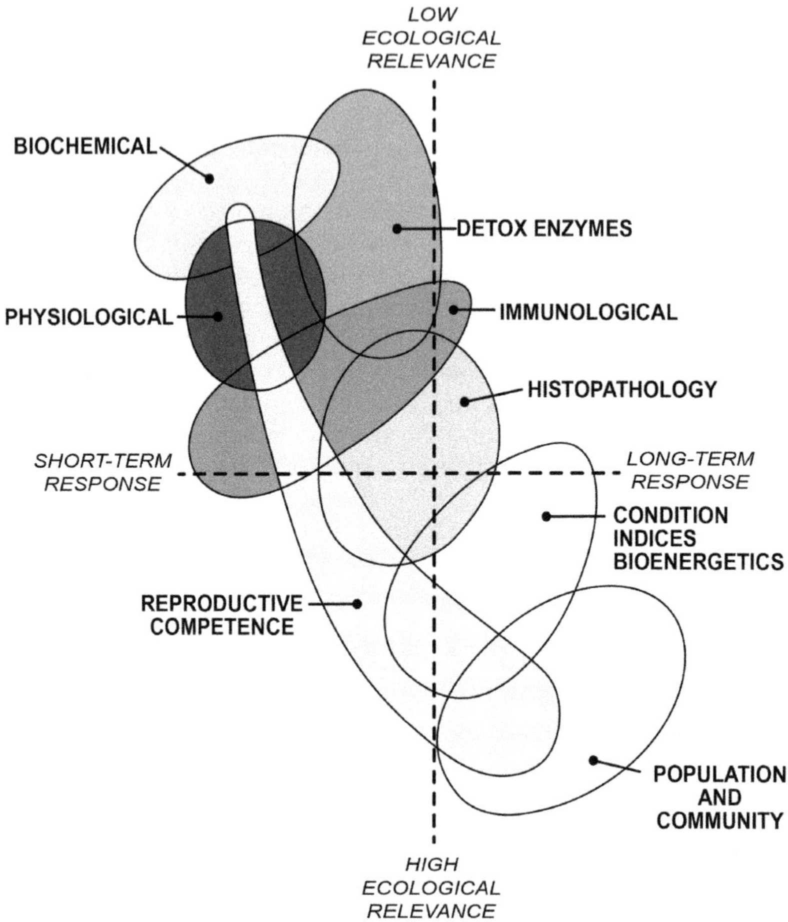

Figure 10.2 Levels of biological responses to pollutant stress in aquatic organisms, representing the continuum of these responses along time gradients and according to toxicological and ecological relevance.[46]
Reproduced from *Mar. Environ. Res.*, **28**, S. M. Adams, K. L. Shepard, M. S. Greeley Jr, B. D. Jimenez, M. G. Ryon, L. R. Shugart, J. F. McCarthy. The use of bioindicators for assessing the effects of pollutant stress on fish, 459–464, Copyright 1990, with permission from Elsevier. Redrawn by Gustavo Pinheiro.

(2) Effect: including measurable biochemical, physiological or other alterations within tissues or body fluids of an organism that can be recognized as being associated with an established or possible health impairment or disease.
(3) Susceptibility: indicating the inherent or acquired ability of an organism to respond to the challenge of exposure to a specific xenobiotic substance, including genetic factors and changes in receptors which alter the susceptibility of an organism to that exposure.[34,39]

Since the perturbations occurred by exposure to contaminants are evidenced by changes in different biochemical processes (biotransformation and excretion), cellular damages can occur, with special attention to protocols involving DNA and organelles damages. The action of xenobiotics on DNA is called 'genotoxic disease syndrome', where DNA damage can promote a strong effect at the individual and population levels, and genotoxic effects in marine organisms exposed to organic compounds, such as aromatic hydrocarbons.[47–49]

Environmental monitoring *in situ* is widely used for some crustacean species, such as Micronucleus Assay (MN‰), Comet Assay (CO) and the Neutral Red Retention Time (NRRT), all indicating the cellular, genetic and physiological integrity of the organism.

Micronuclei are structures formed from snippets of chromosomes that are not incorporated into the nucleus of the daughter cells during cellular division, with subsequent encapsulation of these DNA snippets, promoting a depletion of genic expression, endangering the production of proteins by cells, and causing DNA damage.[50] Thus, the generation of micronucleus cells in high frequency when compared with basal values in the mollusc *Mytillus galloprovincialis* (>4 MN‰),[51] and persistence in the cytoplasm are indicative of extended interaction with pollutants that are compromising DNA integrity. The micronucleus assay has been used for a long time in environmental monitoring to evaluate genotoxicity, as well as in toxicity tests (acute and chronic), owing to use equipment requirements and low cost. A single hemolymph drop (in the case of arthropods, such as crustaceans) can produce thousands of valid cells for analysis without any special treatment, except for fixation (Carnoy solution), coloration (Giemsa solution), and micronucleated cell counting under a microscope, in relation to 1000 cells analysed. The test does not need a trained cytogeneticist, and hence the protocol is considered an excellent tool to be applied in management projects of conservation units (CUs).

The Comet assay (single cell gel electrophoresis) is a technique that is able to detect DNA damage in single cells,[52] and has wide use in biomonitoring of genotoxic agents.[53] The advantages of this assay include its simplicity, quick performance and high sensibility to many types of DNA damage,[54] confirmed by alteration of the damaged DNA during its migration in electrophoresis gel. Cells in eukaryote organisms have DNA of a few centimetres in length, which needs to be strongly condensed inside the cell nuclei. Damaged DNA is less condensed and, occasionally, structural breaks appear,[55] with differential migration in the gel slide as a function of the size of the snippets. Cells with non-fragmented DNA will have a preserved nucleus during electrophoretic migration, maintaining its circular form, whereas DNA with minor damage tends to migrate more quickly. When the DNA is very damaged, many snippets with distinct sizes are formed with DNA migration at different velocities, generating a typical figure of a Comet tail.[56] This is a more sensitive method to detect genomic microlesions, allowing individual identification of damage in most eukaryote cells,[57] and is used as an ecological indicator for diagnosis and monitoring of coastal areas.[58,59] Success has been obtained with this assay during environmental monitoring using vertebrates such as fishes, amphibians and mammals, as well as invertebrates such as decapod crustaceans.[60–62]

The integrity of the lysosome membrane (neutral red retention time, NRRT) is an *in vitro* test that has been used as a chemical stress biomarker for the detection of xenobiotics' harmful effects in organelles. This test is low cost, efficient and easily replicated, based on the observation of the retention time of the neutral red dye by the lysosome membrane, evaluating its selective permeability.[63] The first animals that NRRT was used in were in bivalves, where their hemocytes were applied for detection of alterations in the lysosome membrane as a function of pollutant exposure, causing autophagy and cellular degeneration processes.[63] Evidence suggests that NRRT is a biomarker for a large range of chemical stressors owing to its high sensitivity to a set of contaminants and it is thus recommended in environmental monitoring.[64] Animals subject to constant stress in their environment show damage in their cellular membrane through the action of free radicals, affecting lysosome functions, such as recycling of other cellular organelles, cellular constituents and particles from the external environment. Therefore, a reduction of the lysosomal membrane stability is considered a general indicator of the physiological condition of the individuals,[65] affecting cellular nutrition, immunological defence processes, and reproductive efficiency of marine invertebrates,[66] and can predict future damages to higher levels of biological organization.[64,67]

Finally, it is important to highlight that some biomarkers are specific for a set of pollutants. As an example, we can mention the protein metallothionein (MTs), a biomarker specially synthetized by aquatic organisms (*e.g.*, crustaceans) and used exclusively to bind some essential and non-essential toxic metals (*e.g.*, arsenic, cadmium, copper, mercury, selenium and zinc),[68] but not all metals. Other physiological (*e.g.*, NRRT) and genetic biomarkers (*e.g.*, CO and MN‰) are more representative for environmental monitoring because they respond to a range of environmental pollutants. This indicates the greater importance of some biomarkers when compared to others; some are more relevant for monitoring environmental quality.

10.6 Sampling Design: Spatial Distribution, Replicates, and Other Parameters

Studies conducted using biomarkers for environmental monitoring need a careful sampling design to obtain reliable data to be used in coastal management within a study area. When the objective is to determine the response of biota to specific pollutants, it is necessary to carry out an initial screening test to avoid expenditure of financial resources because some methods are very expensive to apply (*e.g.*, concentration of toxic metals, PAH levels, and others). A bibliographic consultation can provide details of common contaminants in a selected area to be studied, as well as knowledge about industrial facilities present in the area, all of them providing information that is important to reveal the set of xenobiotics that can influence the local biota, specially sessile and/or resident species.

During the screening phase, other aspects to be considered are information about the total size of the area to be studied, as well as the spatial distribution of the species, which are important measures to obtain an optimum response. A comparison between estuarine systems also involves concerns with the selection of subareas where the organisms occur, which should have similar characteristics. Considering different ecosystems pertaining to estuarine systems (*e.g.*, mangrove forests, estuarine channel), selected subareas should have similar abiotic (*e.g.*, granulometry of the sediment and flooding level) or biotic factors (*e.g.*, arboreal composition and/or dominance of mangrove forest). However, some factors are considered determinant in some cases owing to the higher variation in estuarine systems (*e.g.*, salinity and water pH), where different physiological processes can affect the biota subjected to a specific pollutant (*e.g.*, toxic metals), as mentioned earlier. Therefore, a good sampling design leads to a good spatial area representation, with a selection of similar subareas and, whenever possible, separating haloclines with freshwater, brackish water and seawater characteristics (Figure 10.3). According to the goal of the study (*e.g.*, a specific river, a region inside an estuary or an entire estuarine system) it is necessary to establish a reduced number of subareas as a function of size and objectives (B, C and D) or a wider number of subareas to be sampled that represent an entire mangrove peninsula region (A) or estuarine system (E).

Another important fact is that the subareas should represent true replicates, with the capture of a minimum (and adequate) number of specimens in each one. As an example, in studies conducted in São Paulo State, we used a minimum of five individuals per subarea to perform CO and MN‰ assays, generally with three subareas representing the whole studied area ($N = 15$). For the NRRT assay, a greater number of individuals is required (10 individuals per subarea; $N = 30$ per area). However, the number of subareas depends on the total size of the area to be studied. Following this procedure, the reliability of the data obtained is increased, avoiding pseudoreplication, which is not statistically independent.[69,70]

10.7 Case Study of Mangrove Crab *Ucides cordatus* and Other Semi-terrestrial Brachyuran Crabs

The mangrove crab *Ucides cordatus* (Linnaeus, 1763), called the *uçá* crab, is a semi-terrestrial brachyuran species endemic in mangroves, with economic relevance in coastal areas of the occidental Atlantic coast of the south American continent, generating employment and income to riverside communities.[71] This species has a geographic distribution from Florida state

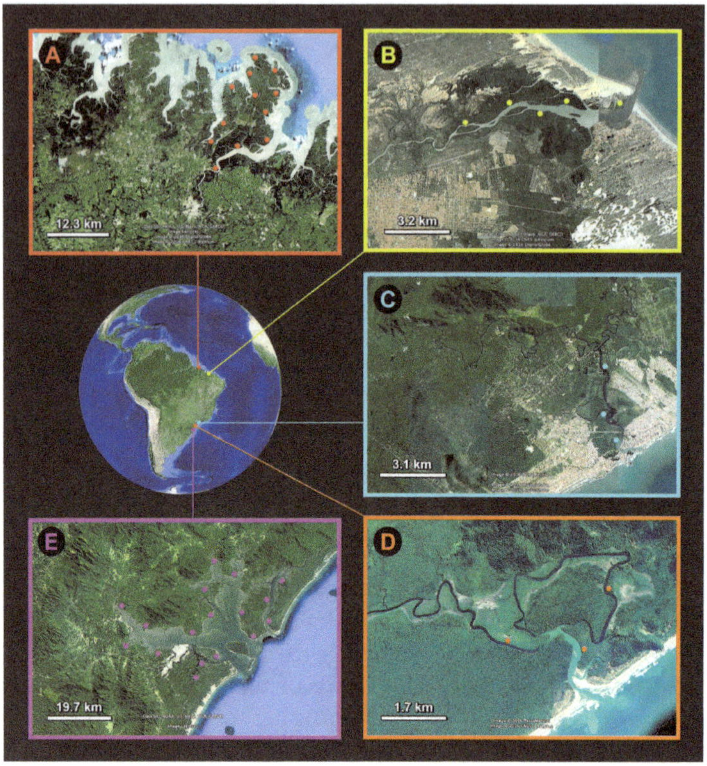

Figure 10.3 Examples of some Brazilian estuarine systems with indication of subarea sampling in relation to size, more important in greater than in smaller areas. A: Bragantina peninsula in Pará state (near Bragança city), very large and jagged, with expressive size and an integral mangrove vegetation cover ($N=12$). B: Delta of the Parnaíba river (near Parnaíba city), in Piauí state, with a medium size ($N=5$). C: Itanhaém river (in Itanhaém city), in São Paulo state, with reduced mangrove fragments and reduced size, with two small tributary rivers (left and right in the upper position), where the salinity wedge does not penetrate ($N=3$). D: Estuarine system of Una river at Juréia-Itatins Ecologic Station (22.5 km from Peruíbe city), with a pristine mangrove but with a reduced size ($N=3$). E: Paranaguá Estuarine system, with a large and expressive size and a complex structure, formed by Paranaguá Bay, two small estuarine cities (Antonina at W, and Guaraqueçaba at N) and conservation units at Superagui National Park, requiring more sampling ($N=17$). In the last example, the sampling design followed the complexity of the environment with different anthropic pressures. Photos from Google Earth and (A) from Data SIO, NOAA, U.S. Navy, NGA, GEBCO, Image Landsat. Image copyright 2016 DigitalGlobe, image copyright 2016 CNES/Astrium. Photo (B) from Data SIO, NOAA, U.S. Navy, NGA, GEBCO, image copyright 2016 CNES/Astrium, image copyright 2016 DigitalGlobe. Photo (C) image copyright 2016 TerraMetrics, image copyright 2016 DigitalGlobe. Photo (D) image copyright 2016 TerraMetrics, image copyright 2016 CNES/Astrium. Photo (E) from Data SIO, NOAA, U.S. Navy, NGA, GEBCO, Image Landsat. Graphic design by Gustavo Pinheiro.

(USA) to the south of Brazil (Laguna city).[72] The crab reaches a large size in the adult phase,[71] building burrows in the muddy sediment, mainly in the intertidal mangrove zone, feeding on senescent leaves and propagules of the arboreal species in this ecosystem,[73] representing an ecological relevance in recycling nutrients in this environment.[71,74] Due to their economical importance, there is an intense capture above the growth capacity, which affects population renewal, mainly in estuaries of the North Brazilian region.[75] The capture is artisanal and occurs through the introduction of the arm of crab catchers or using fishing tackle, some of them traditional (*e.g.*, catch hook) and others prohibited by Brazilian law (*e.g.*, a trap called a *redinha*).[71,75] *U. cordatus* has a slow growth rate and it reaches its maximum size by 10 years, with maturity age at 3 years old and commercial size at 4 years old for males (CW, carapace width $\geq$60 mm).[76] Capture of females is prohibited by law during the reproductive season (December to May), while males cannot be captured during the *andada* phenomenon, when individuals of this species actively walk on mangrove sediment during reproduction.[75,77] Owing to these biological limitations, this species was included in Annex II of the Normative Instruction n°5/2003,[78] where it is categorized as underexplored or threatened by overexploitation, with other species of invertebrates and vertebrates used for fishing. Since 2011, the National Plan of Management indicated the need for monitoring populations of three species of the infraorder Brachyura with commercial importance (*U. cordatus, Cardisoma guanhumi* and *Callinectes sapidus*).[79] The National Plan Management reinforced the importance of maintaining the quality of the estuarine environment, with the establishment of a Monitoring Program to evaluate water and sediment quality, as well as the meat of these crustaceans, aiming to elaborate recovery plans.[79]

Populations of *uçá* crab have been subjected to a range of anthropic environmental pressures. Therefore, this globally distributed ecosystem of mangroves is continually affected by mining activities, effluent discharge, deforestation, grounding, and improper occupation by aquaculture tanks, among others.[11] In this reality, all mangrove species have been influenced by the synergic action of a range of pressures; among them are those caused by pollutants, which are damaging at many biological levels.[80] Metal contamination and the bioaccumulation that occur in local fauna is based on contact with different environmental matrices (water and sediment) and by trophic pathway.[42] In this sense, *U. cordatus* is highlighted as a bioindicator species due to many characteristics that are considered relevant in monitoring studies *in situ*, based on its capacity of dealing with metals and xenobiotics by different pathways,[30] such as:

(1) Trophic position: *uçá* crab use as food the leaf litter composed of senescent leaves and propagules of arboreal species in mangroves (*e.g., Rhizophora mangle, Laguncularia racemosa* and *Avicennia* spp. in Brazil).[73] The transference of toxic metals and other pollutants from the sediment to vegetation is a result of metabolic processes of each

resident vegetal species, having a special role in water and sediment depuration of high quantities of these pollutants.[81] The decomposition process is very common in mangroves as an effect of decomposers (fungi and bacteria) on senescent leaves and other organisms, and contributes to reduce dissolved oxygen in the sediment, directly affecting the mobility of toxic metals. This organic matter (particulate or dissolved) is transported by tides and accumulates toxic metals, promoting contamination of the trophic chain in adjacent coastal ecosystems.[82] This accumulated biomass transfers metals to the sediment, and also to higher trophic levels by nutrient recycling promoted by mangrove crabs, increasing bioavailability, for example, through filtering and through decomposing organisms. Studies about the relationship between consumption of *R. mangle* leaves and assimilation of metals by *U. cordatus* have been conducted in Brazilian regions (north and southeast), proving that the *uçá* crab accumulate metals in dangerous quantities for human consumption.[30,83]

(2) Bioturbation of sediment: occurs when this species digs the muddy sediment and incorporates organic matter and nutrients.[73,84] Due to the mineralogical characteristics of this sediment, being anoxic with fine granulometry, it is a typical reducer, acting as a true 'sink' of heavy metals, petroleum, and other residues (organics and inorganics).[14,85] However, in natural conditions, without the influence of pollutant sources, these sediment characteristics ensure the reduction of potential deleterious effects caused by metals to biota, impeding the remobilization and availability of these pollutants.[14] When inside burrows, *U. cordatus* maintains close contact with the sediment, the water in the burrow, and contaminants present in ingested leaf litter;[73,74] as a consequence this species is subjected to three sources of contamination.

(3) Biological limitations: due to slow growth rate and long life cycle, expressive abundance, wide distribution and a low mobility during their life.[76] The minimum capture size of 60 mm carapace width by law, and a minimum age of 3 years old, is a relevant standardization to allow comparisons. Another important characteristic of decapod crustaceans is the selection of intermoult exemplars to use in biomarker evaluations because in the pre- and post-moult stages there are many metabolic demands, compromising the dynamics of absorption of pollutants and their defence system.[58] The larval development of *U. cordatus* comprises six instars of a zoea stage and one stage of megalopa, that last one moulting to juvenile. During the first juvenile stage the animals have reduced size, being attracted by conspecific odours that are left by adults in the sediment, aiming to attract individuals for recruitment in mangroves.[86–88] Therefore, polluted mangrove areas can affect the attractiveness of the *uçá* crab and recruitment, leading to a reduction of population density.[27] Otherwise, this species can be used as a testimonial of mangrove

health, which has been proved in ecotoxicological studies that indicate the reliability of the contamination status of mangrove ecosystem using the *uçá* crab as a bioindicator. Thus, in Brazil *U. cordatus* has been used as an important bioindicator of environmental quality and is strongly responsive to many pollutants (*e.g.*, oils, polycyclic aromatic hydrocarbons, toxic metals, and others).[30,89] Many studies have identified the accumulation of some pollutants in greater or lesser amounts in environmental matrices (*e.g.*, sediment, water and food), as well as accumulation in different tissues of the crab. In the case of decapod crustaceans, such as *U. cordatus*, some contaminants such as toxic metals can accumulate in the muscles (meat), gills, midgut gland (hepatopancreas),[30,89–91] including the carapace (*e.g.*, Pb), and even in eggs carried in the female abdomen.[92]

On the São Paulo state coast, Brazil (see Figure 10.4), there are highly contaminated mangrove areas, contrasting with pristine areas inside conservation areas managed by governmental institutions (*e.g.*, Juréia-Itatins Ecologic Station, supervised by the Florestal Foundation, an environmetal agency). In general, the north coast of São Paulo has a few mangrove fragments, with greater abundance occurring on the central and south coasts of this state. On the central coast of São Paulo state the first Brazilian colony (São Vicente) was established in 1532, and historically it is the most anthropic region, where the Santos port and Cubatão industrial complex facilities were established, both with a high population, contrasting with the south coast of this state, which have preserved mangroves with no human influence (317 600 and 2606 habitants km^{-2}, respectively).[93] Studies using two biomarkers (micronucleus and neutral red, see Figures 10.5 and 10.6, respectively) were conducted in *U. cordatus*, indicating that mangroves of the south coast of São Paulo state have a better conservation status, considered for categorization purposes[27] as areas with a Probable No Impact (PNI), where these biomarkers were <3 MN‰ and >120 minutes NRRT. Juréia and Cananéia are examples of these mangrove areas, both localized in the large Environmental Protected Area (EPA) of Cananéia, Iguape and Peruíbe. Considering the same categorization purpose, Cubatão and São Vicente, both pertaining to the Estuarine Complex of Santos-São Vicente, were considered as Probable High Impact (PHI) to the health of biota in that area, based on the *U. cordatus* response, with values >5 MN‰ and <60 minutes for NRRT.[27] Iguape was categorized as a Probable Low Impact (PLI) area, where intermediate values were found for each biomarker. Another study captured a specimen of *U. cordatus* with cheliped malformation in São Vicente (in the east of SAV1 subarea, see Figure 10.4), presenting a large quantity of micronuclei (11.5 ± 2 MN‰), explained by a set of industrial contaminants from Cubatão and close to two public dumps (called *Alemoa* and *Sambaiatuba*).[94] The last influence was caused by a high concentration of leachate produced by these dump sites that carries great concentrations of many xenobiotics, mainly toxic metals.[95]

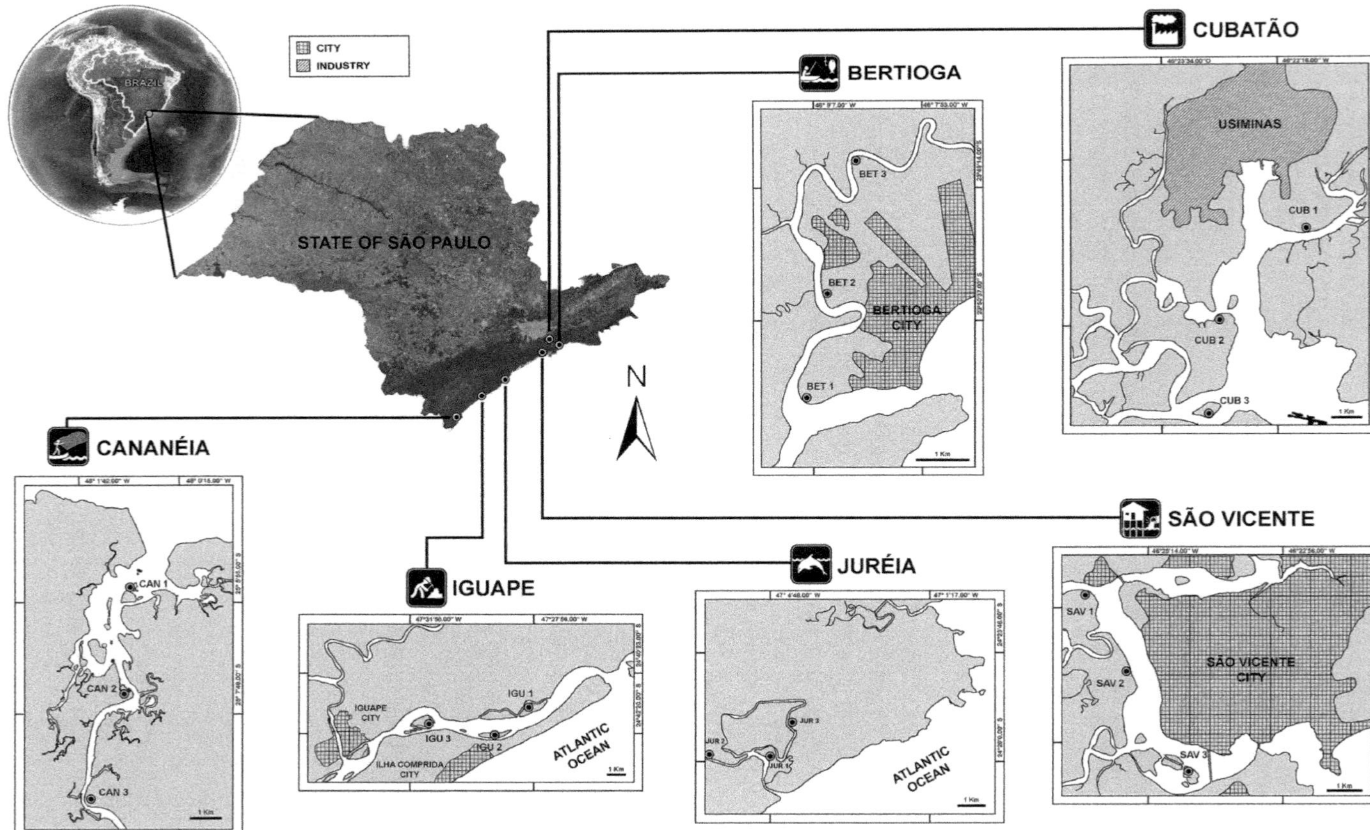

Figure 10.4 Location of six mangrove areas sampled in the state of São Paulo (Brazil), represented by 18 subareas located in the central and southern coasts.
Figure created using data from L. F. A. Duarte, C. A. Souza, C. R. Nobre, C. D. S. Pereira, M. A. A. Pinheiro. Multi-level biological responses in *Ucides cordatus* (Linnaeus, 1763) (Brachyura, Ucididae) as indicators of conservation status in mangrove areas from the western Atlantic, *Ecotoxicol. Environ. Saf.*, 2016, **133**, 176–187.[27] Graphic design by Gustavo Pinheiro.

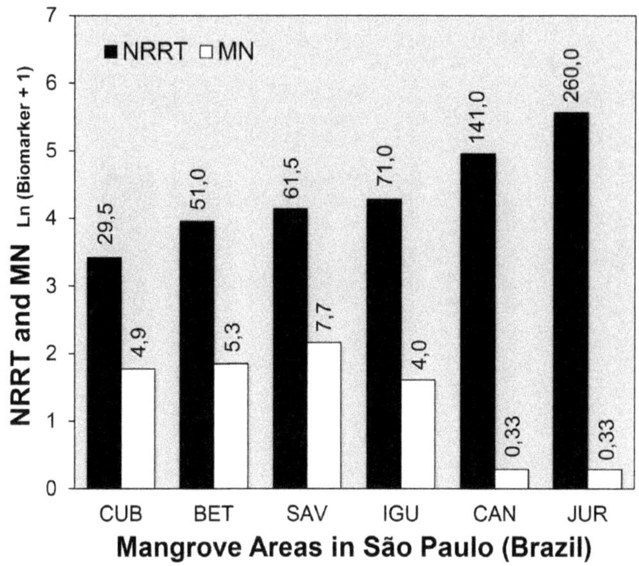

Figure 10.5 Mean values of Neutral Red Retention Time (NRRT, in minutes) and Micronucleus assay (MN‰, micronucleus cells in 1000 analysed) obtained from the crab *Ucides cordatus* (Linnaeus, 1763) from six mangrove areas sampled in the state of São Paulo (Brazil), represented in a logarithmic scale. Values were from three subareas, each one with replicates for NRRT ($n = 10$ per subarea and 30 per area) and MN‰ ($n = 5$ per subarea and 15 per area). Data were obtained during the conduction of the Project Uçá (Phase III – FAPESP 2009/14725-1).[27,115] Graphic design by Gustavo Pinheiro. Where: BET, Bertioga; CAN, Cananéia; CUB, Cubatão; IGU, Iguape; JUR, Ecologic Station of Juréia-Itatins; and SAV, São Vicente.

Ucides cordatus has been used for physiological studies related to toxic metal transport using both gills and hepatopancreas cells as tissue targets.[96–98] In general, both Cu and Cd are transported through cell membrane transporters that involve carriers associated with calcium transport.[97,98] Posterior gills, which are responsible for ion transport, compared to the respiratory role of anterior gills, transport and accumulate more toxic metals compared to anterior gills.[98] These should be target organs for study of pollution effects, together with the hepatopancreas, which is known in these crabs as a detoxifying organ, similar to the pancreas and liver in vertebrates. Studies with discarded drugs containing iron (Fe) showed that this metal is transported in *U. cordatus* hepatopancreatic cells, rendering these crabs as good models to study the deleterious effects of discarded metal-rich wastes.[99]

Ucides cordatus responds well to biomarkers such as the neutral red retention test and the micronucleus assay.[27,89,100] Additional stress biomarkers, such as metallothionein (MT),[96] a protein found in the cell cytoplasm of a range of animals, respond specifically to pollutants known as

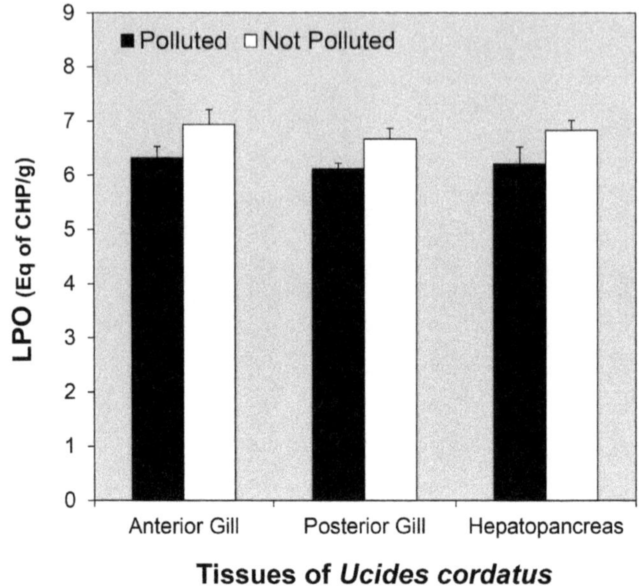

Figure 10.6 Mean values of lipid peroxidation (LPO, in Eq CHP$\times$g^{-1}) using three tissues (anterior gill, posterior gill and hepatopancreas) of the crab *Ucides cordatus* (Linnaeus, 1763) in a polluted mangrove (Itanhaém river) and a non-polluted one (Ecologic Station of Juréia-Itatins), measured from four individuals in triplicate.
Graphic design by Gustavo Pinheiro. Data obtained from M. G. Sá, F. P. Zanotto. Characterization of copper transport in gill cells of a mangrove crab *Ucides cordatus, Aquat. Toxicol.*, 2013, **144–145**, 275–283.[96]

toxic metals, and for specific ones like Cu, Cd and Zn.[101,102] Another stress biomarker of environmental pollution is lipid peroxidation (LPO), which is also associated with the presence of metals[103–105] and oxidative processes in general, caused by a myriad of other pollutants. Several studies with invertebrates have shown that with rising pollution levels there is an increase in the antioxidant systems, including lipid peroxidation.[104] The crab *U. cordatus* from polluted and unpolluted regions in Brazil was responsive to LPO and there was a good correlation between LPO and pollution levels (Figure 10.6), but not when metallothionein levels were evaluated, which were not directly related to metal pollution for *U. cordatus*.[96] These results suggest that *U. cordatus* from chronically polluted regions do not increase metallothionein production as a defence mechanism. Therefore, in crabs from chronically polluted regions, it is possible that the main mechanism of detoxification occurs through the formation of vacuoles and/or accumulation of metals in cellular organelles, leaving the metals in a non-toxic state.[96] Such mechanism has already been seen in *U. cordatus*, with respect to Zn accumulation in the hepatopancreas.[45] Moreover, there are tissue-related variations in MT levels and the relative influence of contamination

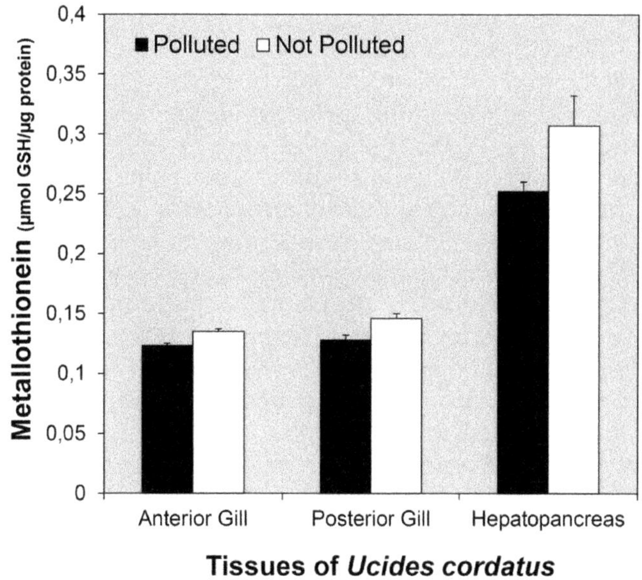

Figure 10.7 Mean values of metallothioneins (MT, µmol GSH µg of protein^{-1}) using three tissues (anterior gill, posterior gill and hepatopancreas) of the crab *Ucides cordatus* (Linnaeus, 1763) in a polluted mangrove (Itanhaém river) and a non-polluted one (Ecologic Station of Juréia-Itatins), measured from four individuals in triplicates.
Graphic design by Gustavo Pinheiro. Data obtained from M. G. Sá, F. P. Zanotto. Characterization of copper transport in gill cells of a mangrove crab *Ucides cordatus, Aquat. Toxicol.*, 2013, **144–145**, 275–283.[96]

factors limits the possibility of using MT levels as a reliable biomarker of metal exposure (Figure 10.7).[96]

Although crabs in general are able to metabolize and get rid of some of the toxic metals loaded into cells through vacuole disposal,[42] and possibly through the shedding of the exoskeleton, they are, as stated above, good and abundant models to indicate environmental contamination, mainly for polluted mangrove areas where they live, in addition to the established filter-feeding models, such as mussels.

Often a unique bioassay is not sufficient to reveal the effect of pollutants in a studied area. As an example, *U. cordatus* were subjected to two biomarkers assays (MN‰, micronucleus in hemocytes; and PM, pyrene metabolites in urine) related to oil-derived PAHs in mangrove sediments; no significant results were found for the MN‰ assay but excellent results were obtained for the PM assay, showing that *U. cordatus* is an excellent bioindicator of mangrove quality related to the concentration of individual or total PAHs.[61,106] In such cases, physiological analyses combined with biomarkers evaluation (*e.g.*, NRRT) can be very effective for quantification of environmental contamination and to establish their contamination category.[27]

Another aspect to be considered is the number of environmental matrices (*e.g.*, water, sediment, and contaminated food items) that each crab contacts in their habitat. Some crabs are exclusively aquatic (*e.g.*, Portunidae species, frequently called swimming crabs), while other species inhabit terrestrial environments (Figure 10.1), and have more matrices to interact with, as well as more pollutants sources to get contaminated by. These semi-terrestrial species have similar characteristics when compared with *U. cordatus*, and can provide good testimonials of environmental quality, but sometimes they interact to greater or lesser degrees with environmental matrices. Examples include the Occidental Atlantic mangroves of South America, where the species *A. pisonii* (Figure 10.1C) interacts with water and green leaves, while the red mangrove crab *G. cruentata* (Figure 10.1B) explores and interacts with more mangrove matrices (*e.g.*, water, sediment and plants/animals used as food items).[107] However, studies involving these species are still incipient, but reveal a promising application using the micronucleus assay and enzymatic activity. Therefore, the use of these species allows a comparison between polluted and pristine areas, and can provide valuable information about the status of diverse environmental matrices, including different trophic levels.[58,107] In addition, *U. cordatus* (Figure 10.1A) has an ecological similarity when compared with *C. guanhumi* (Figure 10.1E), considering the large contact with many environmental matrices in mangrove and *restinga* areas, respectively. However, *guaiamú* crabs (*Cardisoma guanhumi*) build their galleries several hundred meters offshore, particularly along estuaries and river banks, composed of sand, associated with adjacent coastal forests (*e.g.*, Atlantic forest), 5 km from rivers.[108,109] Owing to its occupation of coastal habitats, *C. guanhumi* has been studied from a genotoxic point of view in the northeast region of Brazil, where it is abundant and used as food by man, and the results obtained revealed its importance as a sentinel species as well as the future use in diagnosis and environmental monitoring.[110]

The ghost crab *Ocypode quadrata* (Figure 10.1F) has also been considered an important indicator of anthropic impacts, with population levels varying according to different recreational uses in sandy beaches, mainly related to density.[111–113] Therefore, this species could be used as a model to represent changes in these environments, with potential applications in studies involving biomarkers, although its longevity is only around 3 years,[114] comprising 30% of the entire lifespan of *U. cordatus*.[76]

Among all crab species studied, *U. cordatus* can be considered as a good bioindicator of mangrove quality. Results obtained from genetic and physiological biomarkers are comparable with local contamination, especially NRRT results, which indicate a reliable response to pollutant effects based on a set of contaminants in mangrove areas. In conclusion, these crab species are very useful bioindicators of environmental contamination owing to their abundance and easy capture, and because some of them show a relatively long life, an uncommon feature in macroinvertebrates.

Acknowledgements

Thanks to graphic designer Gustavo H. S. Pinheiro for the figures and graphics added to the chapter. Many thanks to 'Escrever Ciência' (www.assessoriaciencia.com), for valuable English technical revision.

References

1. J. A. Dias, J. A. Carmo and M. Polette, As zonas costeiras no contexto dos recursos marinhos, *J. Integr. Coast. Zone Manage.*, 2009, **9**(1), 3–5.
2. Y. Guo, H. Zeng, R. Zheng, S. Li, A. G. Barnett, S. Zhang, X. Zou, R. Huxley, W. Chen and G. Williams, The association between lung cancer incidence and ambient air pollution in China: a spatiotemporal analysis, *Environ. Res.*, 2016, **144**, 60–65.
3. Z. Li, Z. Ma, T. J. van der Kuijp, Z. Yuan and L. Huang, A review of soil metal pollution from mines in China: pollution and health risk assessment, *Sci. Total Environ.*, 2014, **468–469**, 843–853.
4. E. T. Wilkins, Air pollution aspects of the London fog of December 1952, *Q. J. R. Meteorol. Soc.*, 1954, **80**(344), 267–271.
5. L. C. Ferreira, Os fantasmas do Vale: conflitos em torno do desastre ambiental de Cubatão, SP, *Revista Política e Trabalho*, 2006, **25**, 165–188.
6. J. Gutberlet, *Cubatão: Desenvolvimento, Exclusão social e degradação ambiental*, Editora Edusp, FAPESP, 2006, p. 248.
7. H. C. Trannum and T. Bakke, Environmental effects of the deepwater horizon oil spill – foccus on effects on fish and effects for dipersants, *NIVA Report SNO* 6283-2012.
8. M. Marta-Almeida, R. Mendes, F. N. Amorim, M. Cirano and J. M. Dias, Fundão Dam colapse: Oceanic dipersion of River Doce after the greatest Brazilian environmental accident, *Mar. Pollut. Bull.*, 2016, **112**(1–2), 359–364.
9. C. F. Gillam, Effects of Social and Environmental Inequalities on the Wellbeing of a Slum Community: The case of Vila dos Pescadores in Southeast Brazil. Thesis Submitted to Saint Mary's University, Halifax, Nova Scotia, 2016.
10. Y. Schaeffer-Novelli, *Manguezal: ecossistema entre a terra e o mar*, Caribbean Ecological Research, 1995, p. 64.
11. M. Spalding, M. Kainuma and L. Collins, *World Atlas of Mangroves*, Earthscan, Washington, DC, 2010, vol. 1, pp. 23–43.
12. C. Giri, E. Ochieng, L. L. Tieszen, Z. Zhu, A. Singh, T. Loveland, J. Masek and N. Duke, Status and distribution of mangrove forests of the world using Earth observation satellite data, *Global Ecol. Biogeogr.*, 2011, **20**, 154–159.
13. V. S. Souza-Junior, P. Vidal-Torrado, M. G. Tessler, L. C. R. Pessenda, T. O. Ferreira, X. L. Otero and F. Macías, Evolução quaternária,

distribuição de partículas nos solos e ambientes de sedimentação em manguezais do Estado de São Paulo, *Rev. Bras. Cienc. Solo*, 2007, **31**, 753–769.
14. S. M. Saifullah, S. H. Khan and S. Ismail, Distribution of nickel in a polluted mangrove habitat of the Indus delta, *Mar. Pollut. Bull.*, 2002, **44**(6), 551–576.
15. L. D. Lacerda, R. C. Campos and R. E. Santelli, Metals in water, sediments, and biota of an offshore oil exploration area in Potiguar Basin, Northeastern Brazil, *Environ. Monit. Assess.*, 2013, **185**(5), 4427–4447.
16. E. Bernini, M. A. B. Silva, T. M. S. Carmo and G. R. F. Cuzzuol, Composição química do sedimento e de folhas das espécies do manguezal do estuário do Rio São Mateus, Espírito Santo, Brasil, *Rev. Bras. Bot.*, 2006, **29**(4), 689–699.
17. M. A. A. Pinheiro, T. M. Costa, O. B. F. Gadig and F. S. C. Buchman, in *Panorama Ambiental da Baixada Santista*, ed. A. J. F. C. Oliveira, M. A. A. Pinheiro and R. F. C. Fontes, Universidade Estadual Paulista – Campus Experimental do Litoral Paulista, São Vicente, 1ª Edição, 2008, vol. 2, pp. 7–26.
18. H. R. Harvey, R. D. Fallon and J. S. Patton, The effect of organic matter and oxygen on the degradation of bacterial membrane lipids in marine sediments, *Geochim. Cosmochim. Acta*, 1986, **50**, 795–804.
19. E. Kristensen, Mangrove crabs as ecosystem engineers; with emphasis on sediment processes, *J. Sea Res.*, 2008, **59**(1–2), 30–43.
20. G. Penha-Lopes, F. Bartolini, S. Limbu, S. Cannici, E. Kristensen and J. Paula, Are fiddler crabs potentially useful ecosystem engineers in mangrove wastewater wetlands?, *Mar. Pollut. Bull.*, 2009, **58**(11), 1694–1703.
21. N. C. Duke, J. Meynecke, A. M. Dittmann, A. M. Ellison, K. Anger, U. Berger, S. Cannici, K. Diele, K. C. Ewel, C. D. Field, N. Koedam, S. Y. Lee, C. Marchand, I. Nordhaus and F. Daudouh-Guebas, A world without mangroves?, *Science*, 2007, **317**, 41–43.
22. B. E. Udechukwu, A. Ismail, S. Z. Zulkifli and H. Omar, Distribution, mobility, and pollution assessment of Cd, Cu, Ni, Pb, Zn, and Fe in intertidal surface sediments of Sg. Puloh mangrove estuary, Malaysia, *Environ. Sci. Pollut. Res. Int.*, 2015, **22**(6), 4242–4255.
23. L. C. M. Santos and M. D. Bitencourt, Remote sensing in the study of Brazilian mangroves: review, gaps in the knowledge, new perspectives and contributions for management, *J. Integr. Coast. Zone Manage.*, 2016, **16**(3), 245–261.
24. Y. Schaeffer-Novelli, E. J. Soriano-Sierra, C. C. Vale, E. Bernini, A. S. Rovai, M. A. A. Pinheiro, A. J. Schmidt, R. Almeida, C. Coelho Júnior, R. P. Menghini, D. I. Martinez, G. M. O. Abuchahla, M. Cunha-Lignon, S. Charlier-Sarubo, J. Shirazawa-Freitas and G. Cintrón-Molero, Climate changes in mangrove forests and salt marshes, *Braz. J. Oceanogr.*, 2016, **64**(sp2), 37–52.

25. C. H. Walker, S. P. Hopkin, R. M. Sibly and D. B. Peakall, *Principles of Ecotoxicology*, Taylor & Francis, Londres, 1996, p. 312.
26. J. J. Stegeman, M. Brouwer, R. Di Giulio, L. Forlin, B. A. Fowler, B. M. Sanders and P. A. Van Veld, *Biomarkers Biochemical. Physiological Markers of Anthropogenic Stress*, Lewis Publishers, 2005, pp. 235–334.
27. L. F. A. Duarte, C. A. Souza, C. R. Nobre, C. D. S. Pereira and M. A. A. Pinheiro, Multi-level biological responses in *Ucides cordatus* (Linnaeus, 1763) (Brachyura, Ucididae) as indicators of conservation status in mangrove areas from the western Atlantic, *Ecotoxicol. Environ. Saf.*, 2016, **133**, 176–187.
28. Cetesb, Companhia de Tecnologia de Saneamento Ambiental, Relatório do Programa de Controle de Poluição, 2001, São Paulo, p. 137.
29. W. Luis-Silva and W. Machado, Diluição geoquímica entre contaminantes sedimentares do estuário do rio Morrão, sistema estuarino de Santos-Cubatão, Brasil, *Geochim. Bras.*, 2012, **26**(1), 39–48.
30. M. A. A. Pinheiro, P. P. G. Silva, L. F. A. Duarte, A. A. Almeida and F. P. Zanotto, Accumulation of six metals in the mangrove crab *Ucides cordatus* (Crustacea: Ucididae) and its food source, the red mangrove *Rhizophora mangle* (Angiosperma: Rhiszophoraceae), *Ecotoxicol. Environ. Saf.*, 2012, **81**, 114–121.
31. P. S. Pompeu and C. B. M. Alves, The effect of urbanization on biodiversity and water quality in the Rio das Velhas Basin, Brazil, *Am. Fish. Soc. Symp.*, 2005, **47**, 11–22.
32. C. D'Amato, J. P. M. Torres and O. Malm, DDT (Dicloro Difenil Tricloroetano): toxicidade e contaminação ambiental – uma revisão, *Quim. Nova*, 2002, **25**(6), 995–1002.
33. K. Borga and A. V. Souza, Biomagnification of organochlorines along a Barents Sea food chain, *Environ. Pollut.*, 2001, **113**(2), 187–198.
34. F. A. Azevedo and A. A. M. Chasin, *As bases toxicológicas da ecotoxicologia*, Rima, São Carlos, 2003, p. 340.
35. M. D. Meyer, C. A. Davis and D. Dvorett, Response of wetland invertebrate communities to local and landscape factors in North Central Oklahoma, *Wetlands*, 2005, **35**(3), 533–546.
36. Portaria 445/2014, http://www.icmbio.gov.br/cepsul/legislacao/portaria/427-2014.html (accessed December 2016).
37. H. Hun, H. Lin, W. Zheng, S. J. Tomanicek, A. Johs, D. A. Elias, L. Liang and B. Gu, Oxidation and methylation of dissolved elemental mercury by anaerobic bacteria, *Nat. Geosci.*, 2013, **6**, 751–754.
38. J. M. Parks, A. Johs, M. Podar, R. Bridou, R. A. Hurt-Jr., S. D. Smith, S. J. Tomanicek, Y. Qian, S. D. Brown, C. C. Brandt, A. V. Palumbo, J. C. Smith, J. D. Wall, D. A. Elias and L. Liang, The genetic basis for bacterial mercury methylation, *Science*, 2013, **339**(6125), 1332–1335.
39. S. N. Luoma and P. S. Rainbow, *Metal Contamination in Aquatic Environments: Science and Lateral Management*, Cambridge University Press, 2008, p. 573.

40. A. A. Otitoloju and K. N. Don-Pedro, Integrated laboratory and field assessments of heavy metals accumulation in edible periwinkle, *Tympanotonus fuscatus* var *radula* (L.), *Ecotoxicol. Environ. Saf.*, 2004, **57**(3), 354–362.
41. P. L. Klerks and R. C. Swartz, in *Ecotoxicology*, ed. S. A. Levin, Springer-Verlag, Berlin, 1987, vol. 3, pp. 41–67.
42. G. A. Ahearn, P. K. Mandal and A. Mandal, Mechanisms of heavy-metal sequestration and detoxification in crustaceans: a review, *J. Comp. Physiol., B*, 2004, **174**(6), 439–452.
43. C. L. Bayne, K. R. Clarke and J. S. Gray, Background and rationale to a practical workshop on biological effects of pollutants, *Mar. Ecol.: Prog. Ser.*, 1988, **46**, 1–5.
44. J. D. Corrêa-Júnior, S. Allodi, G. M. Amado-Filho and M. Farina, Zinc accumulation in phosphate granules of *Ucides cordatus* hepatopancreas, *Braz. J. Med. Biol. Res.*, 2000, **33**, 217–221.
45. A. Soegianto, M. Charmantier-Daunes, J. P. Trilles and G. Charmantier, Impact of copper on the structure of gills and epipodites of the shrimp *Penaeus japonicus*, *J. Crustacean Biol.*, 1999, **19**(2), 209–223.
46. S. M. Adams, K. L. Shepard, M. S. Greeley Jr, B. D. Jimenez, M. G. Ryon, L. R. Shugart and J. F. McCarthy, The use of bioindicators for assessing the effects of pollutant stress on fish, *Mar. Environ. Res.*, 1989, **28**, 459–464.
47. V. V. Cheung, R. J. Wedderburn and M. H. Depledge, Molluscan lysosomal responses as a diagnostic tool for the detection of a pollution gradient in Tolo Harbor, Hong Kong, *Mar. Environ. Res.*, 1997, **46**, 237–241.
48. R. Van der Oost, J. Beyer and N. P. E. Vermeulen, Fish bioaccumulation and biomarkers in environmental risk assessment: a review, *Environ. Toxicol. Pharmacol.*, 2003, **13**(2), 57–149.
49. B. Kurelec, The genotoxic disease syndrome, *Mar. Environ. Res.*, 1993, **35**, 341–348.
50. P. I. Countryman and J. A. Heddle, The production of micronuclei from chromosome aberrations in irradiated cultures of human lymphocytes, *Mutat. Res.*, 1976, **41**, 321–332.
51. R. Scarpato, L. Migliore, G. Alfinito-Cognetti and R. Barale, Induction of micronucleus in gill tissue of *Mytillus galloprovincialis* exposed to polluted marine waters, *Mar. Pollut. Bull.*, 1990, **21**(2), 74–80.
52. G. Speit and A. Hartmann, *Methods in Molecular Biology: DNA Repair Protocols: Mammalian Systems*, Humana Press Inc, Totowa, NJ, 2006.
53. P. A. White and J. B. Rasmussen, The genotoxic hazards of domestic wastes in surface waters, *Mutat. Res.*, 1998, **410**(3), 223–236.
54. J. Silva, B. Erdtamnn and J. A. P. Henriques, *Genética Toxicológica*, Editora Alcance, Porto Alegre, 2003, p. 424.
55. E. Rojas, M. C. Lopez and M. Valverde, Single cell gel electrophoresis assay: methodology and applications, *J. Chromatogr. B: Biomed. Sci. Appl.*, 1999, **722**(1–2), 225–254.

56. P. L. Olive, J. P. Banath and R. E. Durand, Heterogeneity in radiation-induced DNA damage and repair in tumor and normal cells measured using the "comet" assay, *Radiat. Res.*, 1990, **122**(1), 86–94.
57. N. P. Singh, M. T. McCoy, R. R. Tice and E. L. Schneider, A simple technique for the quantification of low levels of DNA damage in individual cells, *Exp. Cell Res.*, 1988, **175**(1), 184–191.
58. M. B. Davanso, L. B. Moreira, M. F. Pimentel, L. V. Costa-Lotufo and D. M. S. Abessa, Biomarkers in mangrove root crab *Goniopsis cruentata* for evaluating quality of tropical estuaries, *Mar. Environ. Res.*, 2013, **21**, 80–88.
59. A. J. S. Rocha, M. T. Botelho, F. M. Hasue, M. J. A. C. R. Passos, C. P. Vignardi and V. Gomes, Genotoxicity of shallow waters near the Brazilian Antartic station "Comandante Ferraz" (EACF), Admiralty Bay, King George Island, Antarctica, *Braz. J. Oceanogr.*, 2005, **61**(1), 63–70.
60. G. Frenzilli, M. Nigro and B. P. Lyons, The comet assay for the evaluation of genotoxic impact in aquatic environments, *Mutat. Res.*, 2009, **681**(1), 80–92.
61. A. H. Nudi, A. L. R. Wagener, E. Francioni, C. B. Sette, A. V. Sartori and A. L. Scofield, Biomarkers of PAH exposure in crabs *Ucides cordatus*: laboratory assay and filed study, *Environ. Res.*, 2010, **110**, 137–145.
62. C. R. Arcaute, J. M. Pérez-Iglesias, N. Nikoloff, G. S. Natale, S. Soloneski and M. L. Larramendy, Genotoxicity evaluation of the insecticide imidacloprid on circulating blood cells of Montevideo tree frog *Hypsiboas pulchellus* tadpoles (Anura, Hylidae) by comet and micronucleus bioassays, *Ecol. Indic.*, 2014, **45**, 632–639.
63. D. M. Lowe, V. U. Fossato and M. H. Depledge, Contaminant-induced lysosomal membrane damage in blood cells of mussels *Mytilus galloprovincialis* from the Venice Lagoon: an in vitro study, *Mar. Ecol.: Prog. Ser.*, 1995, **129**, 189–196.
64. M. N. Moore, J. I. Allen and A. McVeigh, Environmental prognostics: an integrated model supporting lysosomal stress responses as predictive biomarkers of animal health status, *Mar. Environ. Res.*, 2006, **61**(3), 278–304.
65. OSPAR Comission: Background Document on Biological Effects Monitoring Techniques. Assessment and Monitoring Series. Available at: http://www.ospar.org/documents/dbase/publications/p00333_Background%20Document%20of%20biologiocal%20effects.pdf, 2007.
66. A. H. Ringwood, J. Hoguet, C. Keppler and M. Gielazyn, Linkages between cellular biomarker responses and reproductive success in oysters – *Crassostrea virginica*, *Mar. Environ. Res.*, 2004, **58**(2–5), 151–155.
67. C. D. S. Pereira, D. M. S. Abessa, R. B. Choueri, V. Almagro-Pastor, A. Cesar, L. A. Maranho, M. L. Martín-Diaz, R. J. Torres, P. K. Gusso-Choueri, J. E. Almeida, F. S. Cortez, A. A. Mozeto, H. L. N. Silbiger, E. C. P. M. Souza, T. A. Del Valls and A. C. D. Bainy, Ecological relevance

of sentinels biomarkers responses: a multi-level approach, *Mar. Environ. Res.*, 2014, **96**, 118–126.
68. A. Siegel, H. Siegel and R. K. O. Siegel, *Metallothioneins and Related Chelators*, Royal Society of Chemistry, Cambridge, 1st edn, 2009, p. 536.
69. S. H. Hurlbert, Pseudoreplication and the design of ecological field experiments, *Ecol. Monogr.*, 1984, **54**(2), 187–211.
70. A. J. Underwood, *Experiments in Ecology: Their Local Design and Interpretation using Analysis of Variance*, Cambridge University Press, Cambridge, 1997, p. 524.
71. A. G. Fiscarelli and M. A. A. Pinheiro, Perfil sócio-econômico e conhecimento etnobiológico do catador de caranguejo-uçá, *Ucides cordatus* (Linnaeus, 1763), nos manguezais de Iguape (24°41'S), SP, Brasil, *Actual. Biol.*, 2002, **24**(77), 129–142.
72. G. A. S. Melo, *Manual de Identificação dos Brachyura (Caranguejos e Siris) do Litoral Brasileiro*, Editora Plêiade, São Paulo, 1996, p. 604.
73. R. A. Christofoletti, G. Y. Hattori and M. A. A. Pinheiro, Food selection by a mangrove crab: temporal changes in fasted animals, *Hydrobiologia*, 2013, **702**, 63–72.
74. I. Nordhaus, K. Diele and M. Wolff, Activity patterns, feeding and burrowing behavior of the crab *Ucides cordatus* (Ucididae) in a high intertidal mangrove forest in North Brazil, *J. Exp. Mar. Biol. Ecol.*, 2009, **379**, 104–112.
75. K. Diele, V. Koch and U. Saint-Paul, Population structure, catch composition and CPUE of the artisanally harvested mangrove crab *Ucides cordatus* (Ocypodidae) in the Caeté estuary, North Brazil: Indications for overfishing?, *Aquat. Living Resour.*, 2005, **18**, 169–178.
76. M. A. A. Pinheiro, A. G. Fiscarelli and G. Y. Hattori, Growth of the mangrove crab *Ucides cordatus* (Brachyura, Ocypodidae), *J. Crustacean Biol.*, 2005, **25**(2), 293–301.
77. A. C. Wunderlich, M. A. A. Pinheiro and A. M. T. Rodrigues, Biologia do caranguejo-uçá, *Ucides cordatus* (Crustacea, Decapoda: Brachyura), na Baía de Babitonga, Santa Catarina, Brasil, *Rev. Bras. Zool.*, 2008, **25**(2), 188–198.
78. MMA. Ministério do Meio Ambiente. Instrução Normativa n° 5, de 21 de maio de 2004. Diário Oficial da União - Seção 1. Brasília, 2004, http://www.mma.gov.br/estruturas/179/_arquivos/179_05122008033927.pdf (accessed December 2016).
79. IBAMA – Instituto Brasileiro do Meio Ambiente e dos Recursos Renováveis, in *Proposta de Plano Nacional de Gestão para o uso sustentável do caranguejo-uçá, do guaiamum e do siri-azul*, J. Dias-Neto, IBAMA, Brasília, 2011, p. 156.
80. V. E. Forbes, A. Palmqvist and L. Bach, The use and misuse of biomarkers in ecotoxicology, *Environ. Toxicol. Chem.*, 2006, **25**, 272–280.
81. W. J. Zheng, X. Y. Chen and P. Lin, Accumulation and biological cycling of heavy-metal elements in *Rhizophora stylosa* mangroves in Yingluo Bay, China, *Mar. Ecol.: Prog. Ser.*, 1997, **159**, 293–301.

82. C. A. Ramos e Silva, A. P. Silva and S. R. Oliveira, Concentration, stock and transport rate of heavy metals in a tropical red mangrove, Natal, Brazil, *Mar. Chem.*, 2006, **99**, 2–11.
83. M. S. P. Vilhena, M. L. Costa and J. F. Berredo, Accumulation and transfer of Hg, As, Se, and other metals in the sediment-vegetation-crab-human food chain in the coastal zone of the northern Brazilian state of Pará (Amazonia), *Environ. Geochem. Health*, 2013, **35**(4), 477–494.
84. I. I. Nordhaus, M. Wolff and K. Diele, Litter processing and population food intake of the mangrove crab *Ucides cordatus* in a high intertidal forest in northern Brazil, *Estuar. Coast. Shelf Sci.*, 2006, **67**, 239–250.
85. D. L. Semensatto-Júnior, G. C. L. Araújo, R. H. F. Funo, J. Santa-Cruz and D. Dias-Brito, Metais e não-metais em sedimentos de um manguezal não-poluído, Ilha do Cardoso, Cananéia (SP), *Rev. Pesq. Geociênc.*, 2007, **34**(2), 25–31.
86. K. Diele and D. J. B. Smith, Effects of substrata and conspecific odour on the metamorphosis of mangrove crab megalopae, *Ucides cordatus* (Ocypodidae), *J. Exp. Mar. Biol. Ecol.*, 2007, **348**, 174–182.
87. D. J. B. Smith, F. A. Abrunhosa and K. Diele, Chemical induction in mangrove crab megalopae *Ucides cordatus* (Ucididae): Do young recruits emit metamorphosis triggering odours as do conspecific adults?, *Estuar. Coast. Shelf Sci.*, 2013, **131**, 264–270.
88. D. J. B. Smith and K. Diele, Metamorphosis of mangrove crab megalopae *Ucides cordatus* (Ocypodidae): Effects of interspecific versus intraspecific settlement cues, *J. Exp. Mar. Biol. Ecol.*, 2008, **362**(2), 101–107.
89. M. A. A. Pinheiro, L. F. A. Duarte, T. R. Toledo, M. L. Adam and R. A. Torres, Habitat monitoring and genotoxicity in *Ucides cordatus* (Crustacea: Ucididae), as tools to manage a mangrove reserve in southeastern Brazil, *Environ. Monit. Assess.*, 2013, **185**, 8273–8285.
90. R. R. Harris and M. C. F. Santos, Heavy metal contamination and physiological variability in the Brazilian mangrove crabs *Ucides cordatus* and *Callinectes danae* (Crustacea: Decapoda), *Mar. Biol.*, 2000, **137**, 691–703.
91. J. M. C. Araújo Júnior, T. O. Ferreira, M. Suarez-Abelenda, G. N. Nóbrega, A. G. B. M. Albuquerque, A. C. Bezerra and X. L. Otero, The role of bioturbation by *Ucides cordatus* crab in the fractionation and bioavailability of trace metals in tropical semiarid mangroves, *Mar. Pollut. Bull.*, 2016, **111**(1–2), 194–202.
92. E. V. Almeida, V. T. Kutter, E. D. Marques and E. V. Silva-Filho, First assessment of trace metal concentration in mangrove crab eggs and other tissues, SE Brazil, *Environ. Monit. Assess.*, 2016, **188**, 421.
93. IBGE Brasil, Instituto Brasileiro de Geografia e Estatística, *Censo demográfico 2010*. http://www.ibge.gov.br (accessed December 2016).
94. M. A. A. Pinheiro and T. R. Toledo, Malformation in the crab *Ucides cordatus*, (Linnaeus, 1763) (Crustacea, Brachyura, Ocypodidae), in São

Vicente, State of São Paulo, Brazil, *Rev. CEPSUL – Biodivers. Conserv. Mar.*, 2010, **1**(1), 61–65.

95. P. Kjeldsen, M. A. Barlaz, A. P. Rooker, A. Baun, A. Ledin and T. H. Christensen, Present and long-term composition of MSW landfill leachate: a review, *Crit. Rev. Environ. Sci. Technol.*, 2002, **32**(4), 297–336.
96. P. Ortega, H. A. Vitorino, R. G. Moreira, M. A. A. Pinheiro, A. A. Almeida, M. R. Custódio and F. P. Zanotto, Physiological diferences in the crab *Ucides cordatus* from two populations inhabiting mangroves with different levels of cadmium contamination, *Environ. Toxicol. Chem.*, 2017, **36**(2), 361–371.
97. M. G. Sá and F. P. Zanotto, Characterization of copper transport in gill cells of a mangrove crab *Ucides cordatus*, *Aquat. Toxicol.*, 2013, **144–145**, 275–283.
98. P. Ortega, M. R. Custódio and F. P. Zanotto, Characterization of cadmium plasma membrana transport in gills of a mangrove crab *Ucides cordatus*, *Aquat. Toxicol.*, 2014, **157**, 21–29.
99. H. A. Vitorino, P. Ortega, R. Y. Pastrana, F. P. Zanotto and B. P. Esposito, Iron loading in hepatopancreatic cells of the mangrove crab *Ucides cordatus* through magnetite nanoparticles, ferrocene derivatives and iron supplements (unpublished work).
100. C. A. Souza and M. A. Pinheiro, Mangrove conservation monitoring by genocytotoxic biomarkers of the 'uçá'-crab (*Ucides cordatus*): The seasonal effect on micronucleus (MN) and neutral red (NR) assays (unpublished work).
101. F. Silvestre, C. Duchene, G. Trausch and P. Devos, Tissue-specific cadmium accumulation and metallothionein-like protein levels during acclimation process in the Chinese crab *Eriocheir sinensis*, *Comp. Biochem. Physiol., Part C: Toxicol. Pharmacol.*, 2005, **140**(1), 39–45.
102. J. C. Amiard, C. Amiard-Triquet, S. Barka, J. Pellerin and P. S. Rainbow, Metallothioneins in aquatic invertebrates: their role in metal detoxification and their use as biomarkers, *Aquat. Toxicol.*, 2006, **76**(2), 160–202.
103. M. Hermes-Lima, W. G. Willmore and K. B. Storey, Quantification of lipid peroxidation in tissue extracts based on Fe (III) xylenol orange complex formation, *Free Radical Biol. Med.*, 1995, **19**(3), 271–280.
104. J. M. Monserrat, L. A. Geracitano, G. L. L. Pinho, T. M. Vinagre, M. Faleiros, J. C. Alciati and A. Bianchini, Determination of lipid peroxides in invertebrates tissues using the Fe(III) xylenol orange complex formation, *Arch. Environ. Contam. Toxicol.*, 2003, **45**, 177–183.
105. H. A. Vitorino, L. Mantovanelli, F. P. Zanotto and B. P. Espósito, Iron metallodrugs: stability, redox activity and toxicity against *Artemia salina*, *PLoS One*, 2015, **10**(4), e0121997.
106. A. H. Nudi, A. L. R. Wagener, E. Francioni, A. L. Scofield, C. B. Sette and A. Veiga, Validation of *Ucides cordatus* as a bioindicator of oil contamination and bioavailability in mangroves by evaluating sediment and crab PAH records, *Environ. Int.*, 2007, **33**, 315–327.

107. N. Kriegler, C. A. Souza and M. A. A. Pinheiro, presented in part at *Abstracts of 5° Congresso Brasileiro de Biologia Marinha*, Porto de Galinhas (PE), May, 2015.
108. C. A. Gifford, Some observations on the general biology of the land crab, *Cardisoma guanhumi* (Latreille), in South Florida, *Biol. Bull.*, 1962, **123**(1), 207–223.
109. M. E. Hostetler, F. J. Mazzotti and A. K. Taylor, Blue Land Crab (*Cardisoma guanhumi*). University of Florida, IFAS Extension, 2003, http://edistt.ifas.ufl.edu/pdffiles/UW/UW01300.pdf (accessed December 2016).
110. C. B. R. Falcão, M. A. A. Pinheiro, R. A. Torres and M. L. Adam, Spatio-temporal genotoxicity in tropical estuarine systems of the Ocidental Atlantic: Potential use of the blue crab (*Cardisoma guanhumi*) as biological indicator of climate oscillations (unpublished work).
111. A. J. Steiner and S. P. Leatherman, Recreational impacts on the distribution of ghost crabs *Ocypode quadrata* fab., *Biol. Conserv.*, 1981, **20**(2), 111–122.
112. A. Blankensteyn, O uso do caranguejo maria-farinha *Ocypode quadrata* (Fabricius) (Crustacea, Ocypodidae) como indicador de impactos antropogênicos em praias arenosas da Ilha da Santa Catarina, Santa Catarina, Brasil, *Rev. Bras. Zool.*, 2006, **23**(3), 807–876.
113. F. M. Neves and C. E. Bemvenuti, The ghost crab *Ocypode quadrata* (Fabricius, 1787) as a potential indicator of anthropic impact along the Rio Grande do Sul coast, Brazil, *Biol. Conserv.*, 2006, **133**(4), 431–435.
114. R. M. F. Alberto and N. F. Fontoura, Distribuição e estrutura etária de *Ocypode quadrata* (Fabricius, 1787) (Crustacea, Decapoda, Ocypodidae) em praia arenosa do litoral sul do Brasil, *Rev. Bras. Zool.*, 1999, **59**(1), 95–108.
115. M. A. A. Pinheiro. Genotoxic impact in populations of 'uçá'-crab, *Ucides cordatus* (Linnaeus, 1763) (Crustacea, Brachyura, Ucididae): Evaluation and correlation with heavy metals in five mangroves of São Paulo state, http://www.bv.fapesp.br/en/auxilios/26954/project-uca-iii-genotoxic-impact-on-population-of-uca-crab-ucides-cordatus-linnaeus-1763-crus/ (accessed December 2016).

Section II: Aquatic Vertebrates as Experimental Models

CHAPTER 11

The Use of Fish as Model Aquatic Organisms in Genotoxicity Studies

TOLGA ÇAVAŞ

Uludag University, Faculty of Arts & Sciences Department of Biology, 16059, Bursa, Turkey
Email: tcavas@uludag.edu.tr

11.1 Introduction

As a result of the increasing input of genotoxic agents into environment, growing attention has recently been paid to the assessment of the effect of genotoxic different agents on organisms and ecosystems. Agents that induce alterations in genetic material are called as genotoxic agents or genotoxins. Genetic toxicology studies have mainly two aims: (1) to detect and analyze the potential harmful effects of physical and chemical agents capable of damaging DNA; and (2) to investigate the mechanisms of action of those agents. Exposure to genotoxic agents might result in changes in DNA sequences and, if not repaired quickly and correctly, these changes can lead to heritable changes *i.e.* mutations, which may lead to cancer. Mutations may affect the organism itself through changes in body cells, or they may be passed on to other generations through alteration of the germ cells (Figure 11.1).

The aquatic environment is the ultimate recipient of man-made contaminants, *i.e.* industrial, agricultural and domestic wastes. At this point, fish serve as the best model for detecting genotoxic effects in the aquatic

Issues in Toxicology No. 33
Ecotoxicology and Genotoxicology: Non-traditional Aquatic Models
Edited by Marcelo L. Larramendy

Published by the Royal Society of Chemistry, www.rsc.org

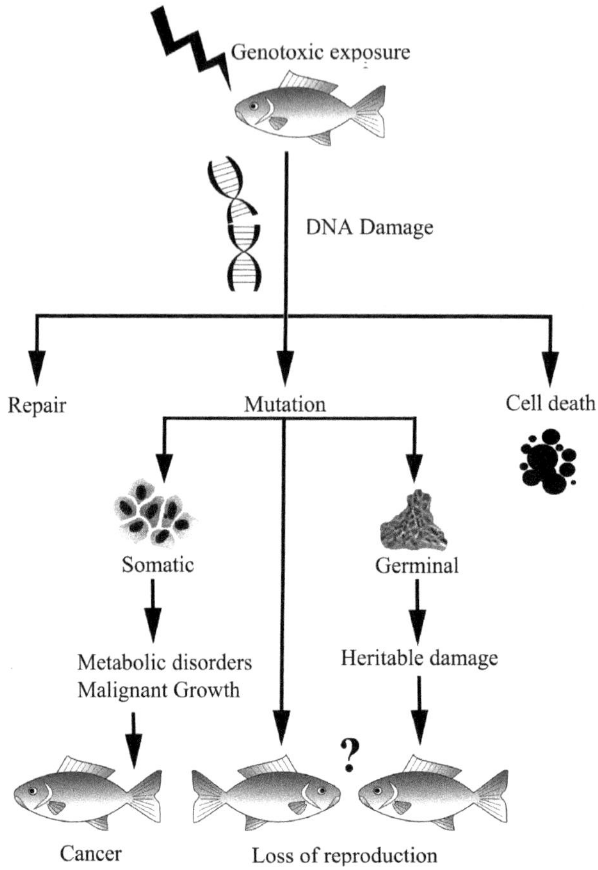

Figure 11.1 Schematic presentation of anticipated consequences of genotoxic exposure in fish.

environment. The assessment of the ecological risks associated with environmental genotoxic exposure is usually based on individual responses. Genotoxic damage induced in somatic cells can lead to cell death or malignancies, which might be manifested as different types of disorders in the organism itself. On the other hand, some genotoxins are prone to affect offspring generations directly or indirectly through the germ cells and reproductive process, which may eventually cause loss of reproduction and changes in population dynamics.[1,2]

11.2 Fish as Model Organisms in Genetic Toxicology

Fish may act as sentinel organisms in evaluation of genotoxicity as they often respond to toxicants in a manner similar to higher vertebrates.[3,4] They are capable of inhabiting different environmental zones in aquatic habitats. Teleost fish, especially small aquarium species, can be easily held in the laboratory and exposed to chemicals under controlled conditions. Furthermore, they therefore offer a large number of models of adaptation and response to a wide variety of natural and anthropogenic environmental conditions. Another advantage of using fish as a model organism is that they are directly and effectively exposed to waterborne contaminants through their gills and gastro-intestinal system. Fish can be used to screen for chemicals that have the potential to induce genotoxic damage or carcinogenic effects in humans and other organisms or to determine the distribution and effects of pollutants in the aquatic environment. Thus, especially within the last two decades, the use of fish as suitable models for genotoxicity monitoring in aquatic environments has become popular. Literature surveys indicate that more than 40 marine and freshwater fish species have been used to assess genotoxicity, including widespread, endemic, tropical or even aquarium fish, mainly depending on the location and availability of specimens. Some of the most commonly used model fish species are listed in Table 11.1.

In aquatic genotoxicity studies, the main cell types used in fish are: peripheral erythrocytes, gill epithelial cells, liver hepatocytes, kidney cells and fin epithelial cells (Figure 11.2). Among these cells, the use of erythrocytes has gained importance in biomonitoring of aquatic genotoxicity in fish thanks to their nucleated structure as well as the practicality of obtaining and processing blood samples. The use of fin and blood cells has further advantages over other cell types as it is not necessary to kill fish to obtain these cell samples.

11.3 Genotoxicity Techniques

The genotoxicity of different factors in fish has been assessed by using different test systems. These assays differ in their end points and specificities to explore the genotoxic effects induced by toxicants. In this chapter, some of the most commonly used *in vivo* genotoxicity techniques are described, namely chromosome aberrations, micronucleus, nuclear abnormalities and the Comet assay.

11.3.1 Cytogenetic Techniques

Genotoxicity studies in fish started with the use cytogenetic techniques originally developed for mammals.[15–18] However, the use of metaphase techniques, such as sister chromatid exchange (SCE) and chromosomal aberrations (CA), is not practical for most fish species and only a limited number of fish species have suitable chromosomes for metaphase analyses. This is mainly because the karyotype of most fish consists of a large number of small and irregular chromosomes. Furthermore, karyological data is available for about only 10% of approximately 25 000 taxonomically known species and individual specimens with different karyotype compositions can be seen in the same species or population.[19–21] Therefore, a fish species to be used should have a well-established karyotype with a small number of large

Table 11.1 The most commonly used model fish species in aquatic genotoxicity studies. Images used from references.[5–14]

Fish species	Common Name	Photo	Ref.
Danio rerio	Zebrafish		5
Oreochromis niloticus	Nile Tilapia		6
Carassius auratus	Goldfish		7
Cyprinus carpio	Common carp		8
Oncorhynchus mykiss	Rainbow trout		9
Oryzias latipes	Medaka		10
Platichthys flesus	Flounder		11
Channa punctatus	Spotted snakehead		12
Mugil cephalus	Mullet		13
Dicentrarchus labrax	Sea bass		14

chromosomes. Some fish species with a suitable karyotype that can be used in chromosomal aberration analyses are given in Table 11.2. On the other hand, the micronucleus test provides an alternative to the metaphase

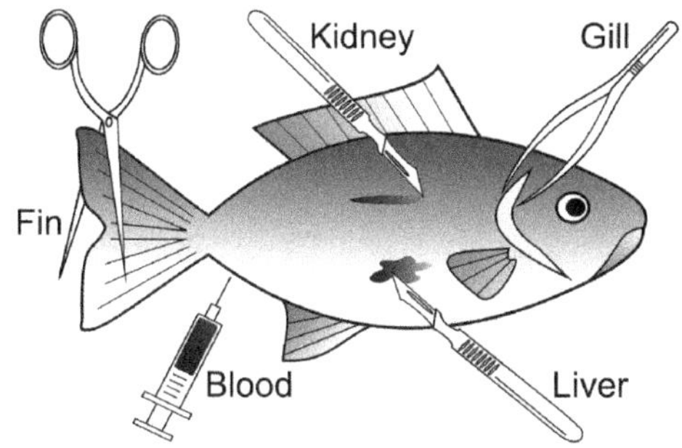

Figure 11.2 The main fish tissues used in genotoxicity assessments (drawing by T. Cavas).

Table 11.2 Some fish species with suitable karyotype for chromosomal aberration tests.

Species	Family	Chromosome number
Ameca splendens (Miller and Fitzsimons, 1971)	Goodeidae	$2n=26$
Aphyosemion christyi (Boulenger, 1915)	Nothobranchiidae	$2n=18$
Apteronotus albifrons (Linnaeus, 1766)	Apteronotidae	$2n=24$
Galaxias maculatus (Jenyns, 1842)	Galaxiidae	$2n=22$
Nothobranchius rachowi (Ahl, 1926)	Nothobranchiidae	$2n=16$
Umbra limi (Kirtland, 1840)	Umbridae	$2n=22$
Umbra pygmaea (DeKay, 1842)	Umbridae	$2n=22$

chromosomal aberration test as it examines cells at interphase and thus provides faster and less subjective results.[22]

11.3.2 The Micronucleus Test

The micronucleus (MN) test was first introduced by Evans *et al.*[23] in radiation-exposed *Vicia faba* root tips and subsequently developed and improved by several authors.[24–26] Today, MN test has become a well-defined genotoxicity endpoint used in the assessment of structural and numerical chromosomal damages in different types of cells. The MN test is based on the detection of small membrane-bound DNA fragments (micronuclei) in interphase cells to measure structural and numerical chromosomal aberrations (Figure 11.3).

A MN is composed of small chromatin fragments that arise as a result of chromosome breaks after clastogenic action or whole chromosomes that do not migrate during the anaphase as a result of aneugenic effects.[27] These lagged chromosome fragments or chromosomes are enclosed by a nuclear membrane at the end of cell division and can be visualized by conventional staining in interphase cells (Figure 11.4). As a MN is originated during cell division, at least one cycle of cell division is necessary for the expression of MN in interphase cells.

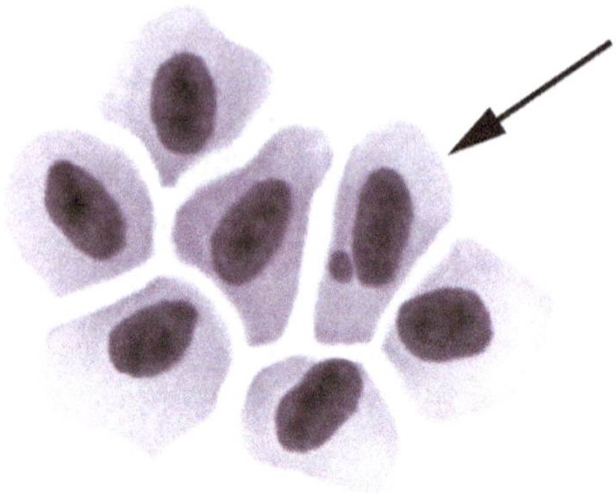

Figure 11.3 Photomicrograph of a micronucleated erythrocyte of *O. niloticus* (1000×; photo by T. Cavas).

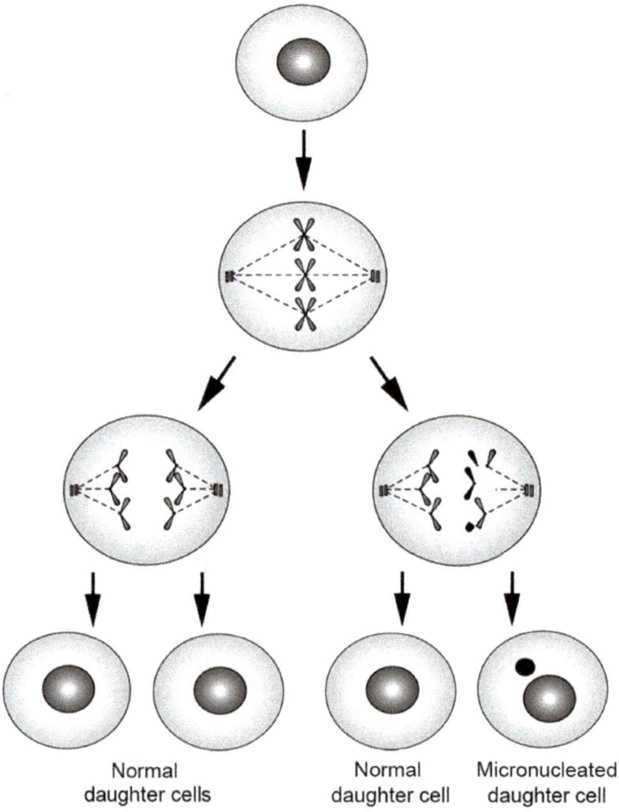

Figure 11.4 A schematic illustration of the mechanism of micronucleus formation in mononucleated cells following genotoxic exposure.
Reprinted from *Mutat. Res.*, **343**, K. Al-Sabti and C. D. Metcalfe, Fish micronuclei for assessing genotoxicity in water, 121–135, Copyright (1995), with permission from Elsevier.

The criteria used for the identification of cells containing MN are as follows: (a) the diameter of the MN should be smaller than one-third of the main nucleus; (b) the MN should be clearly separated from the main nucleus, unless there is a clear identification of nuclear boundaries; and (c) the MN should not be refractive and should have a similar color and staining pattern as the main nucleus.

MN induction in fish was first reported by Hooftman and de Raat[28] in peripheral erythrocytes of eastern mudminnow (*Umbra pygmaea*) treated with ethyl methanesulfonate. Furthermore, Al-Sabti[29] demonstrated induction of micronuclei in three cyprinid fish following exposure to known mutagen substances. On the other hand, Manna and Sadhukhan[30] presented a methodology for the micronucleus test in gill and kidney cells of tilapia (*Oreochromis mossambicus*) treated with X-rays and cadmium chloride. Furthermore, Williams and Metcalfe[31] developed an *in vivo* hepatic micronucleus assay protocol with rainbow trout (*Oncorhynchus mykiss*). The possibility of the use of fin cells, which are in direct contact with water, was first demonstrated by Arkhipchuk and Garanko[32] on common carp (*Cyprinus carpio*). Furthermore, Faßbender and Braunbeck[5] successfully applied the micronucleus test on gonad cells of zebrafish (*Danio rerio*) exposed to methyl methanesulfonate.

Giemsa is the most commonly used staining technique in MN tests worldwide. However, in this staining method used in bright-field microscopy, some non-nucleic artifacts can also be stained and these artifacts could be mistakenly recorded as micronuclei. AO staining is better adapted to piscine erythrocyte MN assays than classical Giemsa staining. On the other hand, DNA-specific fluorescent staining reduces false-positive MN scoring due to artefacts. It was suggested that this specificity of OA provides more reliable and error-free scoring of MN. The acridine orange (AO) fluorescent staining method has been widely used in mammalian erythrocyte MN tests to discriminate immature polychromatic (PCE) and mature normochromatic (NCE) erythrocytes.[33] AO is a nucleic acid-specific fluorochrome that stains DNA bright yellowish-green and RNA orange/reddish. Under a fluorescent microscope, micronuclei and main nuclei exhibit strong yellow-green fluorescence and among all cells immature erythrocytes can be easily identified by their RNA-containing and red-fluorescing cytoplasm (Figure 11.5). The AO staining technique was first modified for fish species by Ueada *et al.*[34] than successfully used by Hayashi *et al.*[35] on fish erythrocyte and gill cells, and followed by other authors.[36–39]

One other advantage of using AO staining is that it allows calculation of PCE/NCE ratios. A decrease in the proportion of PCE to NCE is considered an indicator of cytotoxicity and PCE/NCE calculation is routinely included in the micronucleus test with mammalian test models.[25,40,41] Decreases in PCE ratios in fish *Anguilla anguilla* following exposure to benzo[*a*]pyrene, dehydroabietic acid and bleached kraft paper mill effluent was demonstrated by Pacheco and Santos.[42] Similar results were reported in *Oreochromis niloticus* treated with metronidazole,[36] in *Carassius auratus auratus* treated with

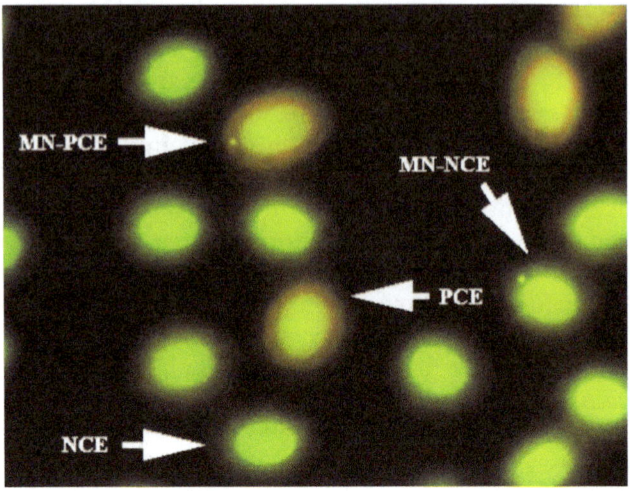

Figure 11.5 Acridine orange stained peripheral blood erythrocytes of *C. auratus*. Arrows indicate: NCE: normochromatic (mature) erythrocyte; MN-NCE: micronucleated normochromatic erythrocyte; PCE: polychromatic (young) erythrocyte with RNA-containing cytoplasm; MN-PCE: micronucleated polychromatic erythrocyte.
Reprinted from *Food. Chem. Toxicol.*, **46**, T. Cavas, *In vivo* genotoxicity of mercury chloride and lead acetate: Micronucleus test on acridine orange stained fish cells, 352–358, Copyright (2008), with permission from Elsevier.

mercury chloride and lead acetate,[43] and in *Gambusia affinis* treated with lambda-chyhalothrin.[44]

11.3.3 Nuclear Abnormalities in Fish Erythrocytes

In fish erythrocytes, the appearance of some morphological nuclear abnormalities (NAs), apart from the micronucleus formations was first described by Carrasco *et al.* in white croaker (*Genyonemus lineatus*) collected from polluted and reference areas along the California coast.[45] Although they did not observe a consistent correlation between erythrocytic nuclear abnormalities and the levels of chemical contaminants, these abnormalities were later further evaluated, validated and included by many authors in assessment of genotoxicity as a complementary test to the micronucleus assay.[46–53]

Although the literature survey indicates the lack of uniform scoring criteria for nuclear abnormalities in fish erythrocytes, the main types of nuclear abnormalities other than micronuclei can be can be classified into six groups (Figure 11.6). Briefly, nuclear blebs (BL) present a relatively small evagination of the nuclear membrane, which contains euchromatin whereas evaginations larger than the blebbed nuclei, which could have several lobes, are classified as lobed nuclei (LB). It is suggested micronuclei and nuclear

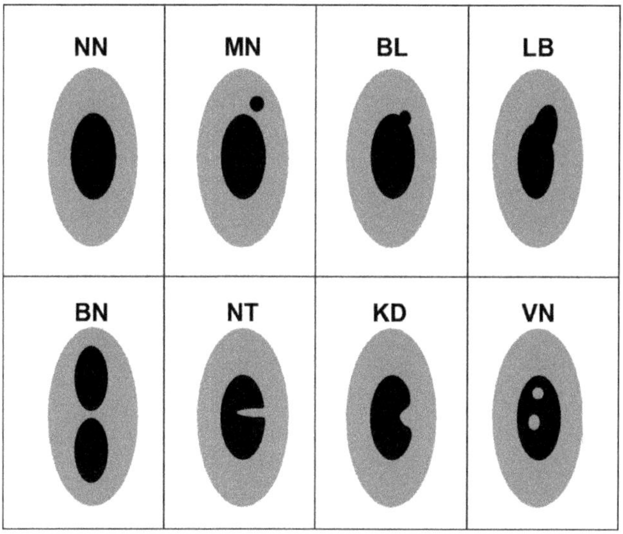

Figure 11.6 Schematic illustration of nuclear abnormalities (NAs) in fish erythrocytes. NN: normal nucleus; MN: micronucleus; BL: blebbed nucleus; BN: binucleus; NT: notched nucleus; KD: kidney-shaped nucleus; VN: vacuolated nucleus (drawing by T. Cavas).

buds are the result of related genotoxic events and that these types of NAs may result from problems segregating tangled and attached chromosomes.[54,55] It was also suggested that gene amplification *via* the breakage–fusion–bridge cycle could cause LBs or BLs during the elimination of amplified DNA from the nucleus.[56,57] In fact, statistical evaluation of nuclear BL formations further revealed that their frequencies were positively and significantly correlated with those of micronuclei (Figure 11.7), suggesting the importance of this abnormality as a genotoxicity parameter in the piscine micronucleus test.[37,58]

Cells with two nuclei are considered as binuclei or binucleate cells (BN). These types of cells usually arise as a result of an incomplete process of cell division. It is suggested that the formation of binucleate cells during the cleavage divisions is due to erroneous mitosis, where karyokinesis is not synchronized with cytokinesis.[59,60] Furthermore, nuclei with slight invaginations are generally called kidney-shaped nuclei (KD) whereas nuclei with sharp and appreciable depth into a nucleus that does not contain nuclear material are classified as notched nuclei (NT). A nucleus containing vacuoles is termed a vacuolated nucleus (VN). These types of nuclei having different amount of invaginations are considered as indicators of cytotoxicity. Although the mechanisms underlying the formation of some nuclear abnormalities have not been fully explained, significant increases in the frequencies of total or individual NAs have been demonstrated in fish erythrocytes following exposure to genotoxic stress under both field and laboratory conditions (Table 11.3).

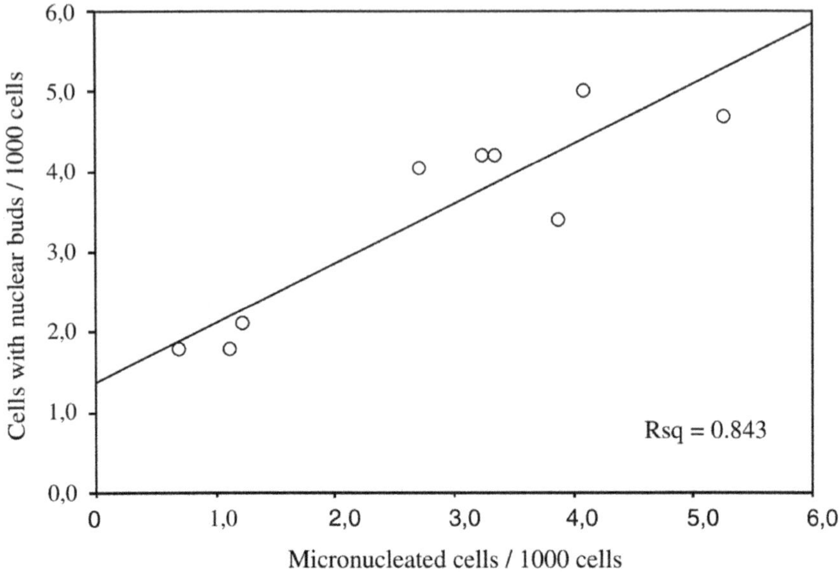

Figure 11.7 Correlation between micronucleated erythrocytes per 1000 cells and cells with nuclear buds per 1000 cells of fish captured from Goksu Delta, Turkey ($R^2 = 0.843$, $P = 0.001$).
Reproduced from *Ecotoxicology*, Monitoring of nuclear abnormalities in peripheral erythrocytes of three fish species from the Goksu Delta (Turkey): genotoxic damage in relation to water pollution, **16**, 2007, 385–391, S. Ergene, T. Cavas, A. Celik, N. Koleli, F. Kaya and A. Karahan, (© Springer Science + Business Media, LLC 2007) With permission of Springer.

11.3.4 Micronucleus Test Protocols with Different Fish Tissues

11.3.4.1 MN and NA Test in Giemsa-stained Erythrocytes

- Collect peripheral blood of fish using a selected procedure.
- Smear blood on pre-cleaned glass slides (at least two slides per fish)
- Dry slides at room temperature
- Fix cells with methanol for 10 min.
- Stain slides with 5% Giemsa prepared in Sorenson's buffer at pH 6.8 for 5 min.
- Wash slides with distilled water to remove excessive stain
- Mount slides and analyze under a light microscope.
- Score at least 1000 cells per slide for MN and NA.

11.3.4.2 MN Test in Acridine Orange-stained Erythrocytes

- Collect peripheral blood of fish using a selected procedure.
- Dilute blood with approximately the same amount of fetal bovine serum (FBS).

Table 11.3 Summary of studies reporting induction of erythrocyte nuclear abnormalities (NAs) in fish exposed to genotoxic stress and under field and laboratory conditions. BL: blebbed nucleus; LB: Lobed nucleus; NT: notched nucleus; BN: binucleus; VN: vacuoled nucleus; KD: kidney-shaped nucleus), +: significant increase; Total NA: total nuclear abnormality.

Fish species	Test substance/location	Nuclear abnormalities	Results	Ref.
Aequidens metae *Astyanax bimaculatus*	Ocoa River water, Colombia	LB, BL, NT, BN	+ for all NAs in comparison to reference area	53
Oreochromis niloticus	Textile effluent	BN, VN, BL, NT, BN	+ for all NAs in comparison to control group	52
Cnesterodon decemmaculatus	2,4-D-based commercial herbicide	LB, BL, NT, BN	+ for BL and NT in comparison to control group	51
Oreochromis niloticus	Textile effluent	LB, BL, NT, BN	+ for total NA in comparison to control group	61
Oreochromis niloticus	Petroleum refinery and chromium plant effluents	LB, BL, NT, BN	+ for all NAs in comparison to control group	46
Mugil cephalus	Mediterranean sea coast, Mersin, Turkey	LB, BL, NT, BN	+ for all NAs in contaminated areas in comparison to reference area	47
Carassius auratus	Herbicide gesaprim	LB, BL, NT, BN	+ for total NA in comparison to control group	7
Oreochromis niloticus	Berdan river, Turkey	LB, BL, NT, BN	+ for all NAs in polluted downstream samples comparison to clean upstream	62
Bathygobius soporator	Salvador city coast, Brazil	LB, BL, NT, BN	+ for all NAs in contaminated areas in comparison to reference area	63
Phoxinus phoxinus	Colchicine, mitomycin-C, cyclophosphamide	LB, BL, NT	+ for total NA in treatment groups in comparison to control group	64

Table 11.3 (*Continued*)

Fish species	Test substance/location	Nuclear abnormalities	Results	Ref.
Poecilia latipinna	Cadmium, mercury	LB, BL, NT	+ for total NA in treatment groups in comparison to control group	64
Dicentrarchus labrax	Adriatic sea coast	BL, LB, NT,	+ for total NA in polluted areas in comparison to cleaner areas	49
Mugil cephalus	Ave, Douro and Mondengo estuaries, Portugal	LB, KD, BL, NT, BN, VC	+ for all NAs in more contaminated areas in comparison to less contaminated area	50
Scophthalmus maximus *Gadus morua*	Crude oil, nonylphenol	BN, LB	+ for all NAs in crude oil treatment group in comparison to control group	65

- Smear blood FBS mixture on pre-cleaned glass slides (at least two slides per fish)
- Dry slides overnight at room temperature
- Fix cells with methanol for 10 min.
- Immediately before microscopy stain cells with 0.003% acridine orange in Sorenson's buffer at pH 6.8 for 2–3 min.
- Mount slides and analyze under an epifluorescent microscope using an appropriate filter.
- Score at least 1000 PCE (red fluorescent cytoplasm) per slide for micronucleus test.
- Calculate the PCE/NCE ratio using the following equation, by scoring 1500 erythrocytes per fish:

$$\text{PCE frequency}\,\% = \frac{\text{No. PCEs}}{\text{No. PCEs} + \text{NCEs}} \times 100 \tag{11.1}$$

11.3.4.3 MN Test in Gill Cells

- Remove both gill arches and place them in petri dishes containing Ca^{++} and Mg^{++} free phosphate buffered saline (CMF-PBS) and wash to remove the artefacts.
- Gently crush gill arches and scrape off cells using a pair of tweezers in a small amount of fresh CMF-PBS.
- Pipette the cell suspension into a 1.5 ml microtube and discard remaining tissue.
- Centrifuge cell suspension at 800 rpm for 5 min.
- Remove supernatant and re-suspend pellet in 0.07 M KCL solution for 1–2 min for hypotonic treatment (this step can be omitted for some fish species).
- Centrifuge again at 800 rpm for 5 min, remove the supernatant and re-suspend cells in Carnoy's fixative (3 : 1 methanol acetic acid) for 10 min. Repeat this step twice.
- Re-suspend cells in Carnoy's fixative, spread onto pre-cleaned slides and air dry.
- Stain with 5% Giemsa for 5 min.
- Score at least 1000 cells per slide to obtain frequencies of micronucleated cells.

11.3.4.4 MN Test in Liver Cells

- Before the experiments, fish can be injected with a non-genotoxic hepatic necrogen allyl formate to induce tissue damage/regeneration in the liver (this step can also be omitted).
- Remove liver tissue and place into vials containing Carnoy's fixative.

- Before the preparation transfer small pieces of livers into vials containing 45% acetic acid for 30 min.
- After chemical maceration with acetic acid, gently mince tissue pieces and filter to obtain a cell suspension.
- Smear obtained cells on clean slides (at least two slides per fish) and air dry
- Fix cells in Carnoy's fixative for 20 min.
- Stain with 5% Giemsa solution for 20 min.
- Score at least 1000 cells per slide to determine the MN frequencies.

11.3.4.5 MN Test in Fin Cells

- At the beginning of experiment, cut the edge of the caudal fin at a depth of 2–3 mm (depending on the body size of fish) to induce regeneration in fin tissue.
- Place fish into aquaria containing test substance to allow the formation of new cells under the impact of test material. (At this point best regeneration/exposure duration could be pre-determined for different fish species.)
- Collect regenerated piece of fin edge and proceed to slide preparation.
- Fix regenerated tissue pieces in 1 : 1 : 1 acetic acid–glyceric acid–distilled water fixative for at least 1 day.
- After fixation smear cells on the clean slides and air dry.
- Stain cells with 5% Giemsa solution for 15 min.
- Score at least 1000 cells per fish to determine the MN frequencies

11.3.5 Comet Assay

The Comet or single cell gel electrophoresis (SCGE) assay is a rapid, simple, visual and sensitive technique used for analysis of DNA strand breaks.[66–68] This technique allows the analysis of genotoxic damage in individual cells embedded in a thin agarose gel on a microscope slide. In brief, cells are placed on microscope slides covered with layers of agarose and lysed to remove cellular proteins. Following unwinding in alkaline conditions, the DNA undergoes electrophoresis in order to allowing the broken DNA fragments or damaged DNA to migrate away from the nucleus. The broken ends of the negatively charged fragments freely migrate towards the anode in the electric field of the electrophoresis solution. After neutralization and staining with a DNA-specific fluorescent dye, such as ethidium bromide, the resulting image looks like a comet with a distinctive head and tail (Figure 11.8). The extent of DNA liberated from the head of the comet is directly proportional to the amount of DNA damage.

The Comet assay has several advantages over other genotoxicity test systems: (1) It can be virtually applied to any eukaryotic cell population regardless of the extent of mitotic activity. (2) Only a small amount of tissue is required (*i.e.* 25 μl of blood or 10^4–10^5 cells). (3) In addition to manual

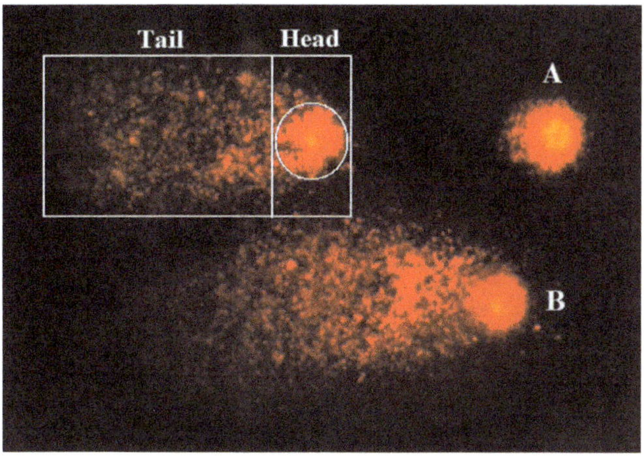

Figure 11.8 Comet images from erythrocytes of *O. niloticus* treated with ethyl methanesulfonate. (A): Non-damaged intact cell nucleus. (B): Damaged erythrocyte nucleus (1.000×; photo by T. Cavas).

scoring[69] automated scoring of comets is also possible. (4) The results can be obtained in a single day.

The Comet assay was first introduced by Ostling and Johanson[66] as a microelectrophoretic technique for direct visualization of DNA damage in individual cells based on neutral conditions, which allows the detection of only double stranded DNA breaks. Later alkaline version of this assay which could assess both double and single strand DNA breaks was adapted by Singh *et al.*[67] In the last two decades, the Comet assay has become one of the most commonly used genotoxicity endpoints in assessment of DNA damage in many types of cells and organisms.[70]

The Comet assay procedure was first adapted to fish by Pandrangi *et al.*[71] to assess the genotoxic effects of marine pollution in erythrocytes of bullheads (*Ameiurus nebulosus*) and carp (*Cyprinus carpio*). In the following years, this technique was successfully applied on liver and gill cells of *Brachydanio rerio*[72] as well as hepatocytes of *Salmo trutta*[73] and *Peluronectes americanus*.[74]

11.3.5.1 Comet Assay in Fish Erythrocyte, Gill and Liver Tissues

- For blood processing, collect blood samples using a preferred method (*i.e.* from a caudal vein using a heparinized syringe and dilute with an equal amount of CMF-PBS).
- For gill and liver tissues remove them and gently wash in cold CMF-PBS solution.
- Transfer tissues into homogenization buffer (1× Hank's balanced salt solution, 20 mM EDTA, 10% dimethyl sulfoxide, pH 7–7.5).
- Cut tissues into small pieces using scissors and homogenize to obtain a cell suspension.
- Centrifuge the cell suspension at 3000 rpm at 4 °C for 5 min.

- Discard supernatant and re-suspend the pellet in 1 ml of CMF-PBS.
- Pipette 60 μl of the cell suspension and mix with 200 μl of 0.65% low melting point agarose.
  - Pipette 75 μl of blood–agarose mixture and layer on glass slides pre-coated with normal melting point agarose.
- Cover with a coverslip and keep in the refrigerator for 10 min to solidify.
- Gently remove coverslips and immerse slides in cold lysing solution (2.5 M NaCl, 100 mM Na_2 EDTA, 10 mM Tris, pH 10, with 10% DMSO and 1% Triton X-100) and keep in refrigerator (+4 °C) for 2 hours.
- After lysis, place slides into a horizontal electrophoresis tank side by side.
- Fill the tank with freshly prepared cold electrophoresis solution (1 mM Na_2 EDTA, 300 mM NaOH, and pH 13.5) to a level about 0.25 cm above the slides.
- Left slides in electrophoresis solution for 30 min to allow unwinding.
- Perform electrophoresis using the same solution at 25 V, 300 mA for 25 min.
- Neutralize slides gently with 0.4 M Tris buffer at pH 7.5.
- Stain slides with a DNA-specific fluorescent stain *i.e.* 75 μl ethidium bromide (20 $\mu g\, ml^{-1}$).
- Prepare at least two slides per fish and analyze at least 50 cells per slide.

In the absence of a fluorescent microscope facility, the staining of the comet slides can be performed using silver nitrate staining method. With this method, silver stained comets can be analyzed by light microscope.[75] To perform silver staining;

- Put air-dried slides into fixing solution (15% trichloroacetic acid, 5% zinc sulfate heptahydrate, 5% glycerol) for 10 min.
- Wash twice with deionized water and dry overnight at room temperature.
- Rehydrate for 5 min in deionized water.
- Prepare staining solutions A (0.1% ammonium nitrate, 0.1% silver nitrate, 0.25% tungstosilicic acid, 0.15% formaldehyde, v/v) and B (5% sodium carbonate).
- Prepare 100 ml of fresh final staining solution by mixing 34 ml of staining solution A and 66 ml of staining solution B in a Coplin jar.
- Stain slides in final staining solution in dark conditions with a shaker for 35 min.
- Wash slides with deionized water twice
- Immerse slides into stop solution (acetic acid 1%) for 5 min.
- Air dry and analyze under a light microscope.

11.3.5.2 Comet Scoring

In visual comet analysis, the amount of DNA damage is calculated and expressed as arbitrary units based on the classification of comets into five (0–4)

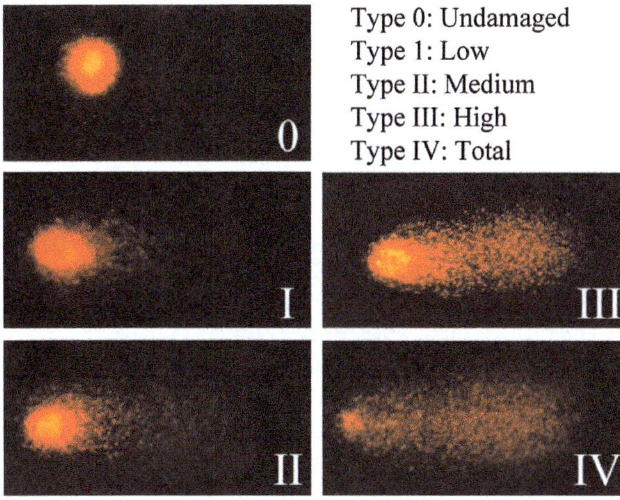

Figure 11.9 Classification of the comet formations in fish (*O. niloticus*) erythrocytes treated with EMS. 0: Class 0 (undamaged); I: Class 1 (low damage); II: Class 2 (medium damage); III: Class 3 (large damage); IV: Class 4 (complete damage). (1.000×; photo by T. Cavas).

categories according to tail length (Figure 11.9). Undamaged: Type 0, low-level damage: Type I, medium-level damage: Type II, high-level damage: Type III and complete damage: Type IV.[76] The percentage of damaged (% damage) cells is calculated as the sum of types II, III and IV.[77] Furthermore, the genetic damage index (GDI) value is calculated from the arbitrary values assigned to different categories (type 0 = 0 to type IV = 4).[78]

In automated Comet assay analysis, the software can calculate more than 20 comet parameters using live or previously taken images.[79] However, as a general trend, three comet parameters, the tail length, tail% DNA and Olive tail moment, are used to express DNA damage.[74] Tail percent DNA means the percent of total DNA in the comet tail and is calculated as 100 − head% DNA. Tail length is the distance of DNA migration from the body of the nuclear core. Olive tail moment is defined as the product of the tail length and the fraction of total DNA in the tail and is calculated by using the following formula:[68]

$$\text{Olive Tail Moment} = (\text{tail mean} - \text{head mean}) \times \text{tail\% DNA}/100. \quad (11.2)$$

All these parameters can be automatically calculated by using commercially available or open access software, such as OpenComet[80] and NIH image.[81]

11.4 Experimental Designs

Fish have been used as models in genotoxicity experiments under both laboratory and field conditions. In field conditions, the induced genotoxic effects often exist as a result of synergistic and antagonistic interactions among different types of physical and chemical factors that cannot be that reproduced under laboratory conditions Therefore, genotoxicity evaluations on or near the impact sites have greater relevancy for risk assessment in comparison to the data derived from laboratory analyses only.[82,83] On the other hand, data obtained from well-designed and combined laboratory and field-scale experiments can improve the value of environmental genotoxicity risk assessments.[84]

Under laboratory conditions, fish are exposed to test chemicals in controlled conditions *via* different routes, such as intracoelomic injection, oral feeding or by directly dissolving in aquarium water, depending on the properties of the studied test substance (Figure 11.10).

In general, three exposure doses or concentrations are selected for the experiments.[47] It is suggested that at least five fish must be used for each test group to be able to generate sufficient genotoxicity data for statistical analyses. The result of previously performed studies indicated that an exposure duration of 24 h is enough to induce micronucleus formation in fish tissues.[46,48] Furthermore, study design must include a positive control group containing fish treated with a known genotoxic agent (*i.e.* benzene, cyclophosphamide, methyl or ethyl methanesulfonate) to prove that the test system is working. If the used test substance is non-soluble in water, a separate solvent control test group should also be included in addition to the negative and positive control groups. Thus, at least 30 fish are required for a genotoxicity experiment to be conducted with a single test substance for one exposure duration (Figure 11.11). The number of necessary fish for multiple

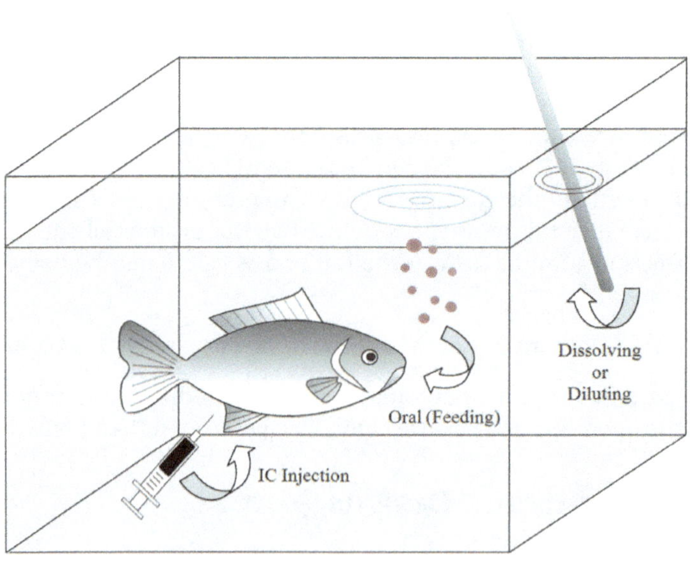

Figure 11.10 Main routes of exposure to test materials in genotoxicity experiments with fish under laboratory conditions (drawing by T. Cavas).

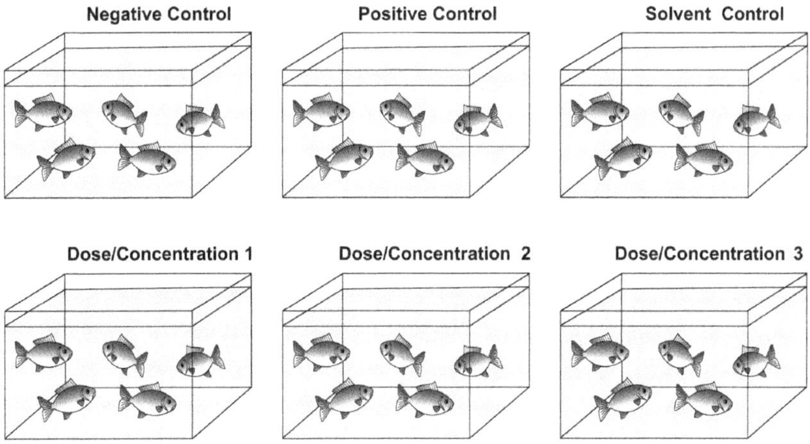

Figure 11.11 Experimental groups for a fish genotoxicity assay with a single substance and exposure duration (drawing by T. Cavas).

exposure durations can be calculated by multiplying 30 by the number of exposure periods.

In situ genotoxicity analyses in-field are generally performed in marine, river and lake environments under the influence of domestic, industrial and agricultural activities, considered as the main sources of aquatic contamination. Fish are sampled from the selected sites of interest by using nets, electroshocking or other suitable methods.

There are three main approaches used in field studies to evaluate potential genotoxic impacts of environmental pollution on fish:

i. Evaluation of genotoxicity in fish following exposure to environmental water samples under controlled laboratory conditions.
ii. Evaluation of genotoxicity in fish *in situ* collected from native populations.
iii. Evaluation of genotoxicity in fish *in situ* exposed to environmental water in cages.

In river systems, basic sampling is carried out at downstream points from the affected area (Figure 11.12A) as they have a tendency to become more affected as a result of the input of different types of contaminants along the river's flow-path. On the other hand, upstream samples serve as a control group. (Figure 11.12B). If the studied river system is more complex and has multiple branches, an increased number of sampling points can be selected accordingly. For instance, Ergene *et al.* studied the genotoxic potential of Berdan and Kusun rivers (Mersin, Turkey) under the influence of numerous different industrial activities.[62] The authors reported significantly increased genotoxic damage in gill and erythrocytes of *O. niloticus* exposed to polluted downstream water samples in comparison with those of exposed to upstream samples.

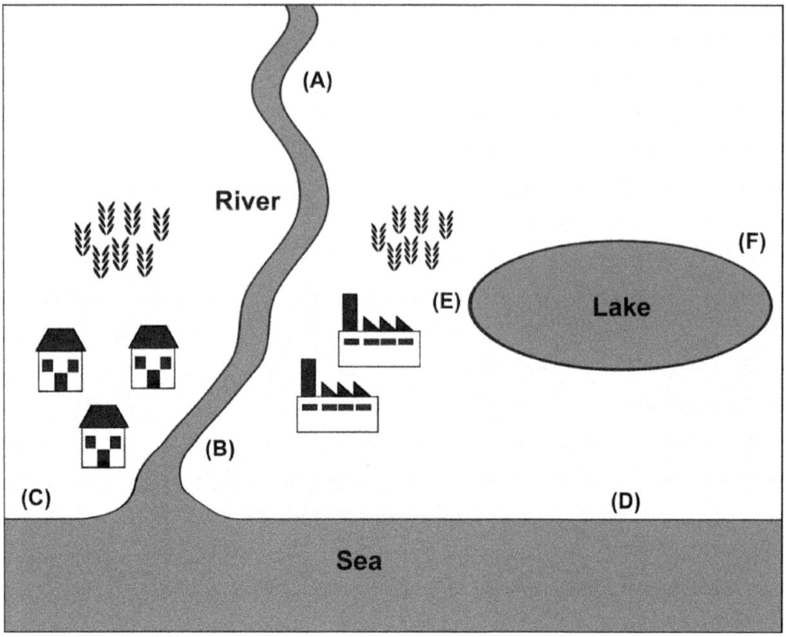

Figure 11.12 Map demonstrating example study areas receiving pollutants from different sources and the potential fish sampling points for assessment of aquatic genotoxicity in river (A: upstream; B: downstream), sea (C: contaminated; D: reference) and lake (E: contaminated; F: reference) waters (drawing by T. Cavas).

In sea and lake environments, the sampling stations are selected on the basis of pollution levels, such as at or near discharge points of domestic and industrial effluents or close to agricultural areas (Figure 11.12C, E), whereas control stations must be considerably far from these points and lack such discharges or activities (Figure 11.12D, F). In a study performed by Cavas and Ergene-Gozukara, analysis of micronucleus frequencies in erythrocytes of *M. cephalus* sampled from Mediterranean coast of Turkey revealed higher genotoxic damage in fish captured from Mersin and Karaduvar harbors receiving discharges from chromium, manure and soda factories in comparison to those captured from Erdemli area about 45 km west of the Mersin and Karaduvar harbors, which receives no industrial point source discharges and is a relatively unpolluted area.[47] Recently, Hyllanda *et al.* performed a large scale analysis in marine environments using dab (*Limanda limanda*) as a sentinel species and the Comet assay as a genotoxicity endpoint.[85] Fish samples were collected at five stations in the North Sea that were influenced by inputs from industrial and maritime activities as well as oil and gas platforms. On the other hand, two other stations in Iceland were selected as reference areas and fish collected from these areas served as controls. The authors reported that the genotoxic response was higher in fish collected from polluted locations in comparison to those collected in reference areas.

Furthermore, Simonyan *et al.* performed an *in situ* genotoxicity analysis in the basin of Lake Sevan (Armenia) using the Comet assay on erythrocytes of *Carassius auratus gibelio.*[86] Fish were sampled from three polluted and a relatively unpolluted location approximately 10 km from the affected area. The obtained data revealed higher DNA damage in fish collected from areas polluted with agricultural and urban wastes.

An inherent weakness of an *in situ* study using wild fish is the lack of a true negative control. In fact, finding a suitable negative control might not always be possible when assessing the effects of genotoxic pollution on native fish populations under field conditions. This is mainly due to factors such as seasonality, migration, unavailability of certain species or lack of sufficient/homogeneous specimens.[14,47,50,61] To overcome this problem, two approaches can be used. First, *in situ* exposure of caged fish. Second, transfer and recovery of field-collected native fish in clean water under laboratory conditions.

The use of the *in situ* caging approach has advantages as it provides the opportunity to conduct more controlled experiments under field conditions.[14,87–89] Fish from the same species/sex/age groups can be used as well as exposures can be performed at the desired location and duration. Furthermore, the cages can be placed at the desired depth according to the study design (*i.e.* surface water or sediment exposure) or to the biology of the selected species (Figure 11.13).

In a study performed by Srut *et al., Dicentrarchus labrax* was used as a sentinel organism to evaluate pollution-related genotoxicity in the eastern Adriatic Sea.[14] The fish were exposed to Necujam Bay (clean reference area) and Kastela Bay (polluted area) waters in polyethylene cages (1.5 m×1 m

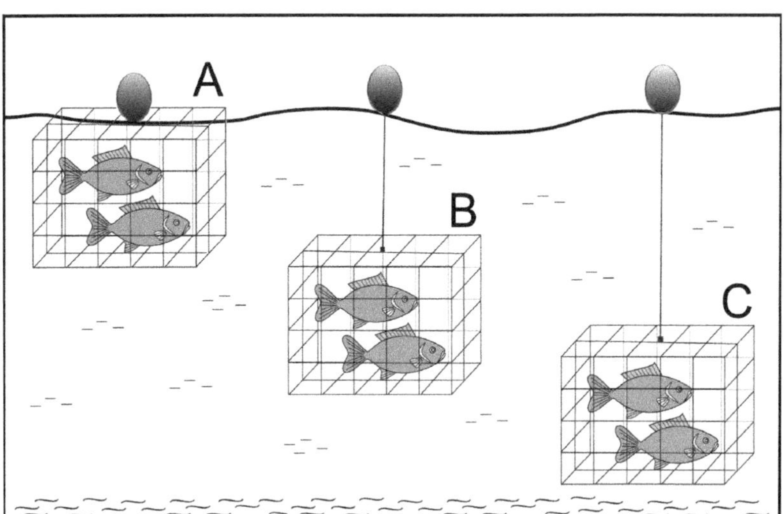

Figure 11.13 *In situ* exposure of caged fish to environmental waters. (A): pelagic exposure; (B): benthopelagic exposure; (C): benthic exposure (drawing by T. Cavas).

with 12 mm mesh size) at a population density of 50 fish per cage for 4 weeks. The authors reported significantly higher genotoxic damage in fish reared in the polluted areas, as determined by the micronucleus test and Comet assay. Similar findings were reported by Klobucar *et al.*, who assessed genotoxicity of Drava river (Croatia) water *in situ* using caged *Cyprinus carpio.*[90] Higher genotoxic damage was found in erythrocytes of fish exposed to downstream water in comparison to those reared in the clean reference area. The authors also mentioned that an exposure period of 3 weeks was sufficient to observe genotoxic damage in caged fish.

Another alternative approach for producing negative control data in *in situ* studies is to maintain field-collected fish in clean water for recovery under controlled laboratory conditions. For instance, Minissi *et al.* analyzed the erythrocyte micronucleus frequencies in barbel fish *Barbus plebejus* from two riverine environments (Mignone and Tiber rivers, Italy) and in specimens reared under laboratory conditions for 50 and 100 days after being captured from the same rivers, which served as the control group.[91] They reported that the barbels examined 100 days after capture showed a remarkable decrease in micronucleus frequencies. Similarly, Ergene *et al.* established a control group by transferring *Clarias gariepinus* specimens to laboratory and keeping them in clean water for 60 days after collecting them from the Goksu Delta (Turkey).[58] Furthermore, Galindo and Moreira studied the genotoxic effects of marine pollution on marine fish *Bathygobius soporator* collected at three locations in Salvador, Bahia (Brazil).[63] The authors compared the micronucleus and NA frequencies in erythrocyte and gill cells of fish collected *in situ* with those of fish transferred to the laboratory and kept in clean water for 90 days. After 90 days of recovery under laboratory conditions, significant decreases in the MN and NA frequencies were reported. Thus, it is possible to suggest that a duration of between 2 and 3 months is enough to obtain baseline MN and NA frequencies in fish erythrocytes.

One of the important factors that should be taken into account in *in situ* analyses is the possibility of occurrence of season-dependent variations in genotoxicity results. Although some studies reported no clear seasonal differences,[50,92] a large number of *in situ* studies have reported seasonal fluctuations in the levels of genotoxic damage in fish.[35,47,63,93] Apart from the discharge/activity dynamics, the concentrations of pollutants in water depend on enrichment and dilution phenomena, which are mainly caused by rainfall. Low rainfall and low river flows could have effects through reduced dilution of pollutants, while high rainfall increases pollution through higher leaching of pollutants. On the other hand, a decrease in water quantity in dry seasons could result in increased chemicals concentrations in water, which cause higher levels of genotoxic damage. For instance, Galindo and Moreira reported clear seasonal variations in erythrocyte MN frequencies of frillfin goby (*Bathygobius soporator*) captured from polluted marine environments in Brazil.[63] The authors observed a higher MN frequency in the least rainy month of November (70 mm precipitation), whereas a lower MN frequency was determined in the rainiest month of March

(350 mm precipitation). The authors suggested that precipitation could be a relevant variable in relation to the increase in micronuclei frequency as in the rainier months, surface-water runoff carries chemical drainage into streams or rivers and finally to beaches. Furthermore, Hayashi *et al.* also observed seasonal differences in *Carassius auratus* and *Zacco platypus* collected from the Tomio River (Nara, Japan). The reported micronucleus frequencies in gill and erythrocytes were generally higher in May than in June, August and September.[35] On the other hand, Cavas and Ergene-Gozukara reported higher MN and NA frequencies in erythrocytes of *Mugil cephalus* captured in the summer.[47] Similarly, Wirzinger *et al.* observed higher MN frequencies in erythrocytes of field-collected three-spined sticklebacks (*Gasterosteus aculeatus*) in August.[93] Fish are poikilothermic animals and their metabolism is highly dependent on the environmental temperature. Therefore, the higher MN and NA frequencies observed during the warmer seasons could also be related to increased rates of metabolism and mitotic index caused by the higher seawater temperature.

11.5 Guppy and Peppered Cory as Non-conventional Aquatic Models for Genotoxicity Studies

When using a fish as a model for toxicity biomonitoring, it is essential to consider the sensitivity of the selected model species to different agents. Furthermore, several other factors, such as economics, easy availability, easy laboratory maintenance and breeding, high reproductive capacity, and small but accessible size, are considered as key when selecting and establishing a fish species as a model organism.

The guppy *Poecilia reticulata* (Peters, 1859), of the family Poeciliidae, is one of the world's most widely distributed and most popular freshwater aquarium fish species. This species is native to northwestern South America and is characterized by viviparity. *P. reticulata* shows clear sexual dimorphism. Males have polymorphic color patterns in their dorsal and caudal fins whereas females have a uniform silver color (Figure 11.14). Furthermore, male guppies are smaller than females in size, and they have a modified anal fin called a gonopodium, which is used as an intromittent organ.

Guppies are widely used as a model organism in the fields of toxicology and ecology and *P. reticulata* is one of the eight recommended model organisms in ecotoxicology testing established in the guidelines set by the Organization for Economic Cooperation and Development (OECD).[94] This species is widely used as a model organism in both long- and short-term toxicity studies.[95–98] However, there are only a few studies regarding the use of *P. reticulata* as a model in genetic toxicology studies. For instance Anogwih *et al.*[99] evaluated the genotoxic effects of spinosad on *P. reticulata* using the micronucleus test. Furthermore, Arslan *et al.*[100] reported increased micronucleus frequencies in erythrocytes of *P. reticulata* following exposure to chlorine bleach and dishwashing detergent. *P. reticulata* was also used as

Figure 11.14 Male and female guppy (*Poecilia raticulata*) specimens.

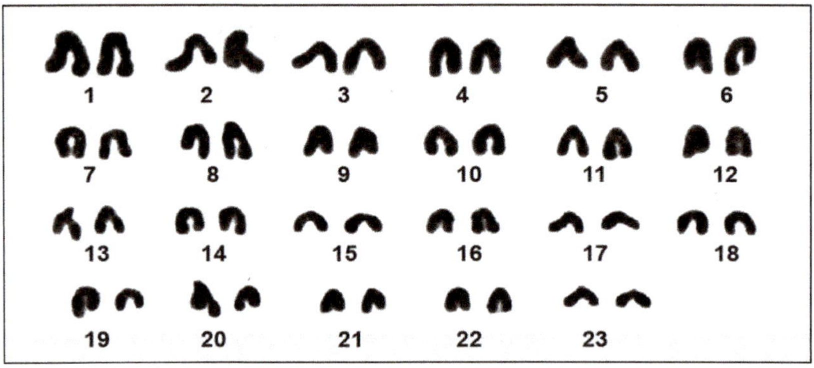

Figure 11.15 Mitotic karyotype of *Poecilia raticulata.*[103]
Reprinted by permission of the publisher (Taylor & Francis Ltd, http://www.tandfonline.com) from E. Lodi, Chromosome complement of the guppy, *Poecilia reticulata* Peters (Pisces, Osteichthyes). *Caryologia*, 1978, **314**, 475–477.

a model organism to evaluate the genotoxic effects of pesticides glyphosate[101] and temephos[102] using the micronucleus test and Comet assay as genotoxicity endpoints.

Karyological analysis of *P. reticulata* revealed a diploid chromosome number of $2n = 46$, consisting of 23 pairs of acrocentric chromosomes (1.3 μm to 2.5 μm) with arm number (NF) 46 (Figure 11.15).[103] The relatively low chromosome number and uniform chromosome structure could allow the use of *P. reticulata* as a model in cytogenetic studies.

The peppered cory *Corydoras paleatus* (Jenyns, 1842), of the family Callichthyidae, is also one of the most common and well-known aquarium fish (Figure 11.16). This species is widely distributed from Southern Brazil, through Uruguay south to Northern Argentina. Females are generally larger than males whereas dorsal fin spine and odontodes are more developed in

Figure 11.16 A peppered cory (*Corydoras paleatus*) specimen.

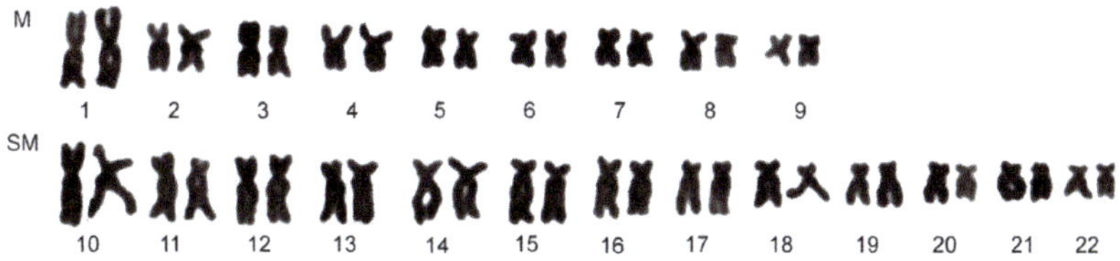

Figure 11.17 Metaphase karyotype of *Corydoras paleatus*. M: Metacentric; SM: submetacentric chromosomes.[108]
Reprinted from M. L. Terancio, M. R. Vicari, M. C. A. Matiello, M. M. Cestari and L. A. C. Bertollo, Cytogenetics of two sympatric Corydoras species (Pisces, Siluriformes, Callichtyidae) of southern Brazil artoni. *Braz. J. Biol.*, 2006, **66**, 191–198.

males. Furthermore, the presence of lanceolate genital papilla is useful to distinguish males from females. Their body shows a yellowish orange ground color with iridescent green coloration.

Although several studies have indicated that *Corydoras paleatus* is a suitable model for environmental toxicity studies,[104,105] this species was first used as a model by one of our study groups in 2005 to evaluate the genotoxic effects of cambium chloride and copper sulfate.[48] Significant increases in micronucleus frequencies were observed in blood gill and liver cells of *C. paleatus* following exposure to both substances. Guiloski *et al.*[106] also used *C. paleatus* as a model organism to evaluate the genotoxicity of the pesticides carbaryl, methyl parathion and deltamethrin. Significantly increased micronucleus frequencies in peripheral erythrocytes of *C. paleatus* exposed to the herbicide roundup were also reported by de Castilhos *et al.*[107]

The karyotype of *C. paleatus* consists of 44 large (~2.5–7 μm) chromosomes (arm number = 88) including nine pairs of metacentric and 13 pairs of submetacentric chromosomes (Figure 11.17).[108] The karyotype composition of this species is also suitable for cytogenetic analysis as it includes a small number of large chromosomes.

Based on the properties mentioned above, it is possible to suggest that both *P. reticulata* and *C. paleatus* possess almost all the characteristics of a suitable aquatic model for genotoxicity studies. They also have further favorable experimental properties, such as small yet accessible size, economic husbandry, and easy breeding and maintaining under laboratory conditions.

References

1. A. Devaux, L. Fiat, C. Gillet and S. Bony, Reproduction impairment following paternal genotoxin exposure in brown trout (*Salmo trutta*) and Arctic charr (*Salvelinus alpinus*), *Aquat. Toxicol.*, 2011, **101**, 405–411.
2. D. R. Dixon and J. T. Wilson, Genetics and marine pollution, *Hydrobiologia*, 2000, **420**, 29–43.
3. A. G. M. Osman, Genotoxicity tests and their contributions in aquatic environmental research, *J. Environ. Prot.*, 2014, **5**, 1391–1399.
4. K. Al-Sabti and C. D. Metcalfe, Fish micronuclei for assessing genotoxicity in water, *Mutat. Res.*, 1995, **343**, 121–135.
5. C. Faßbender and T. Braunbeck, Assessment of genotoxicity in gonads, liver and gills of zebrafish (*Danio rerio*) by use of the comet assay and micronucleus test after in vivo exposure to methyl methanesulfonate, *Bull. Environ. Contam. Toxicol.*, 2013, **91**, 89–95.
6. I. S. De Jesus, M. N. Cestari, M. de, A. Bezerra and P. R. Affonso, Genotoxicity effects in freshwater fish from a Brazilian impacted river, *Bull. Environ. Contam. Toxicol.*, 2016, **96**, 490–495.
7. T. Cavas, Genotoxicity evaluation of metronidazole using the piscine micronucleus test by acridine orange fluorescent staining, *Food. Chem. Toxicol.*, 2011, **49**, 1431–1435.
8. C. G. Olvera-Néstor, E. Morales-Avila, L. M. Gómez-Olivan, M. Galár-Martínez, S. García-Medina and N. Neri-Cruz, Biomarkers of cytotoxic, genotoxic and apoptotic effects in *Cyprinus carpio* exposed to complex mixture of contaminants from hospital effluents, *Bull. Environ. Contam. Toxicol.*, 2016, **96**, 326–332.
9. S. Rodrigues, S. C. Antunes, A. T. Correia and B́. Nunes, Acute and chronic effects of erythromycin exposure on oxidative stress and genotoxicity parameters of *Oncorhynchus mykiss*, *Sci. Total Environ.*, 2016, **545–546**, 591–600.
10. I. Barjhoux, M. Baudrimont, B. Morin, L. Landi, P. González and J. Cachot, Effects of copper and cadmium spiked-sediments on embryonic development of Japanese medaka (*Oryzias latipes*), *Ecotoxicol. Environ. Saf.*, 2012, **79**, 272–278.
11. H. Dabrowska, O. Kopko, A. Góra, I. Waszak and J. Walkusz-Miotk, DNA damage, EROD activity, condition indices, and their linkages with contaminants in female flounder (*Platichthys flesus*) from the southern Baltic Sea, *Sci. Total Environ.*, 2014, **496**, 488–498.

12. I. Ahmad and M. Ahmad, Fresh water fish, *Channa punctatus*, as a model for pendimethalin genotoxicity testing: A new approach toward aquatic environmental contaminants, *Environ. Toxicol.*, 2016, **31**, 1520–1529.
13. W. A. Omar, K. H. Zaghloul, A. A. Abdel-Khalek and S. Abo-Hegab, Genotoxic effects of metal pollution in two fish species, *Oreochromis niloticus* and *Mugil cephalus*, from highly degraded aquatic habitats, *Mutat Res.*, 2012, **746**, 7–14.
14. M. Srut, A. Stambuk, M. Pavlica and G. L. Klobucar, Cage exposure of European sea bass (*Dicentrarchus labrax*) for *in situ* assessment of pollution-related genotoxicity, *Arh. Hig. Rada. Toxicol.*, 2010, **61**, 29–36.
15. A. D. Kligerman, S. E. Bloom and W. M. Howell, *Umbra limi*: a model for the study of chromosome aberration in fishes, *Mutat. Res.*, 1973, **31**, 225–233.
16. C. J. Barker and B. D. Rackham, The induction of sister-chromatide exchanges in cultured fish cells (*Ameca splendens*) by carcinogenic mutagens, *Mutat. Res.*, 1979, **68**, 381–387.
17. R. M. Kocan, L. M. Lantold and K. M. Sabo, Anaphase aberrations: A measure of genotoxicity in mutagen-treated fish cell, *Environ. Mol. Mutagen.*, 1982, **4**, 181–189.
18. K. Al-Sabti, Frequency of chromosomal aberrations in the rainbow trout *Salmo gairdneri* Rich., exposed to detergent and benzene, *Vet. Arch*, 1984, **54**, 83–89.
19. FishBase, ed. R. Froese and D. Pauly, World Wide Web electronic publication. www.fishbase.org, (01/2016), 2016.
20. A. C. Lautrédou, D. D. Hinsinger, C. Gallut, C. H. Cheng, M. Berkani, C. Ozouf-Costaz, C. Cruaud, G. Lecointre and A. Dettai, Phylogenetic footprints of an Antarctic radiation: the Trematominae (Notothenioidei, Teleostei), *Mol. Phylogenet. Evol.*, 2012, **65**, 87–101.
21. A. D. Kligerman, in *Cytogenetic Assays of Environmental Mutagens*, ed. T. C. Hsu, Allanheld, Osmun, 1988, vol. 1, p. 161.
22. A. T. Doherty, The *in vitro* micronucleus assay, *Methods Mol Biol.*, 2012, **817**, 121–141.
23. H. J. Evans, G. J. Nearya and F. S. Williamson, The relative biological efficiency of single doses of fast neutrons and Gamma-rays on *Vicia faba* roots and the effect of oxygen. Part II. Chromosome damage: the production of micronuclei, *Int. J. Radiat. biol.*, 1959, **1**, 216–229.
24. J. A. Heddle, A rapid *in vivo* test for chromosomal damage, *Mutat. Res*, 1973, **18**, 187–190.
25. W. Schmid, The micronucleus test, *Mutat. Res.*, 1975, **31**, 9–15.
26. M. Fenech, The cytokinesis-block micronucleus technique and its application to genotoxicity studies in human population, *Environ. Health Perspect.*, 1993, **101**, 101–107.
27. J. A. Hedle, M. C. Cimino, M. Hayashi, F. Romagna, M. C. Shelby, J. D. Tucker, P. Vanparys and J. T. MacGregor, Micronuclei as an index

of cytogenetic damage: past, present, and future, *Environ. Mol. Mutagen.*, 1991, **18**, 277–291.
28. R. N. Hooftman and W. K. de Raat, Induction of nuclear anomalies (micronuclei) in the peripheral blood erythrocytes of the eastern mudminnow, *Umbra pygmaea* by ethyl methanesulphonate, *Mutat. Res.*, 1982, **104**, 147–152.
29. K. Al-Sabti, Comparative micronucleated erythrocyte cell induction in three cyprinids by five carcinogenic-mutagenic chemicals, *Cytobios*, 1986, **47**, 147–154.
30. G. K. Manna and A. Sadhukhan, Use of cells of gill and kidney of Tilapia fish in micronucleus test (MNT), *Curr. Sci.*, 1986, **55**, 498–501.
31. R. C. Williams and C. D. Metcalfe, Development of an in vivo hepatic micronucleus assay with rainbow trout. Aquatic Toxicology, *Aquat. Toxicol.*, 1992, **23**, 193–202.
32. V. V. Arkhipchuk and N. N. Garanko, Using the nucleolar biomarker and the micronucleus test on *in vivo* fish fin cells, *Ecotoxicol. Environ. Saf.*, 2005, **62**, 42–52.
33. M. Hayashi, T. Sofuni and M. Ishidate Jr., An application of acridine orange fluorescent staining to the micronucleus test, *Mutat. Res.*, 1983, **120**, 241–247.
34. T. Ueada, M. Hayashi, N. Koide, T. Sofuni and J. Kobayashi, A preliminary study of the micronucleus test by acridine orange fluorescent staining compared with chromosomal aberration test using fish erythropoietic and embryonic cells, *Water. Sci. Technol.*, 1992, **25**, 235–240.
35. M. Hayashi, T. Ueda, K. Uyeno, K. Wada, N. Kinae, K. Saotome, N. Tanaka, A. Takai, Y. F. Sasaki, N. Asano, T. Sofuni and Y. Ojima, Development of genotoxicity assay systems that use aquatic organisms, *Mutat. Res.*, 1998, **399**, 125–133.
36. T. Çavaş and S. Ergene-Gozukara, Genotoxicity evaluation of metronidazole using the piscine micronucleus test by acridine orange fluorescent staining, *Environ. Toxicol. Pharmacol.*, 2005, **19**, 107–111.
37. C. Bolognesi, E. Perrone and P. Roffieri, Assessment of micronuclei induction in peripheral erythrocytes of fish exposed to xenobiotics under controlled conditions, *Aquat. Toxicol.*, 2006, **78**, 93.
38. P. M. Costa and M. H. Costa, Genotoxicity assessment in fish peripheral blood: a method for a more efficient analysis of micronuclei, *J. Fish Biol.*, 2007, **71**, 148–151.
39. T. Polarda, S. Jeana, G. Merlinaa, C. Laplanchea, E. Pinellia and L. Gauthiera, Giemsa versus acridine orange staining in the fish micronucleus assay and validation for use in water quality monitoring, *Ecotoxicol. Environ. Saf.*, 2011, **74**, 144–149.
40. Y. Suzuki, Y. Nagae, J. Li, H. Sabaka, K. Mazowa, A. Takahashi and H. Shimuzu, The micronucleus test and erythropoiesis: effects of erythropoietin and a mutagen on the ratio of polychromatic to normochromatic erythrocytes (P/N ratio), *Mutagenesis*, 1989, **4**, 420–424.

41. C. G. Alimba and A. A. Bakare, In vivo micronucleus test in the assessment of cytogenotoxicity of landfill leachates in three animal models from various ecological habitats, *Ecotoxicology*, 2016, **25**, 310–319.
42. S. Pacheco and M. A. Santos, Biotransformation, genotoxic, and histopathological effects of environmental contaminants in European eel (*Anguilla anguilla* L.), *Ecotoxicol. Environ. Saf.*, 2002, **53**, 331–347.
43. T. Çavaş, *In vivo* genotoxicity of mercury chloride and lead acetate: Micronucleus test on acridine orange stained fish cells, *Food. Chem. Toxicol.*, 2008, **46**, 352–358.
44. F. D. Gokalp Muranli and U. Guner, Induction of micronuclei and nuclear abnormalities in erythrocytes of mosquito fish (*Gambusia affinis*) following exposure to the pyrethroid insecticide lambda-cyhalothrin, *Mutat. Res.*, 2011, **726**, 104–108.
45. K. R. Carrasco, K. L. Tilbury and M. S. Mayers, Assessment of the piscine micronuclei test as an *in situ* biological indicator of chemical contaminants effects, *Can. J. Fish. Aquat. Sci.*, 1990, **47**, 2123–2136.
46. T. Çavaş and S. Ergene-Gozukara, Induction of micronuclei and nuclear abnormalities in *Oreochromis niloticus* following exposure to petroleum refinery and chromium processing plant effluents, *Aquat. Toxicol.*, 2005, **74**, 264–271.
47. T. Çavaş and S. Ergene-Gozukara, Micronucleus test in fish cells: a bioassay for *in situ* monitoring of genotoxic pollution in marine environment, *Environ. Mol. Mutagen.*, 2005, **46**, 64–70.
48. T. Çavaş, N. N. Garanko and V. V. Arkhipchuk, Induction of micronuclei and binuclei in blood, gill and liver cells of fishes subchronically exposed to cadmium chlorid and copper sulphate, *Food Chem. Toxicol.*, 2005, **43**, 569–574.
49. I. Strunjak-Petrovic, N. Topic, N. Popovic, R. Coz-Rakovac and M. Jadam, Nuclear abnormalities of marine fish erythrocytes, *J. Fish. Biol.*, 2009, **74**, 2239–2249.
50. J. Carrola, N. Santos, M. J. Rocha, A. Fontainhas-Fernandes, M. A. Pardal, R. A. Monteiro and E. Rocha, Frequency of micronuclei and other nuclear abnormalities in erythrocytes of the grey mullet from the Mondego, Douro and Ave estuaries Portugal, *Environ. Sci. Pollut. Res. Int.*, 2014, **21**, 6057–6058.
51. C. Ruiz de Arcaute, S. Soloneski and L. M. Larramendy, Toxic and genotoxic effects of the 2,4-dichlorophenoxyacetic acid (2,4-D)-based herbicide on the Neotropical fish *Cnesterodon decemmaculatus*, *Ecotoxicol Environ. Saf.*, 2016, **128**, 222–229.
52. C. K. Hemachandra and A. Pathiratne, Combination of physico-chemical analysis, *Allium cepa* test system and *Oreochromis niloticus* erythroc ytebased comet assay/nuclear abnormalities tests for cyto-genotoxicity assessments of treated effluents discharged from textile industries, *Ecotoxicol. Environ. Saf.*, 2016, **131**, 54–64.
53. W. Corredor-Santamaría, M. Serrano-Gómez and Y. M. Velasco-Santamaría, Using genotoxic and haematological biomarkers as an

evidence of environmental contamination in the Ocoa River native fish Villavicencio-Meta Colombia, *Springerplus*, 2016, **5**, 351.
54. P. E. Tolbert, C. M. Shy and J. W. Allen, Micronuclei and other nuclear abnormalities in buccal smears: methods and development, *Mutat. Res.*, 1992, **271**, 69–77.
55. L. Serrano-Garcia and R. Montero-Montoya, Micronuclei and chromatid buds are the result of related genotoxic events, *Environ. Mol. Mutagen.*, 2001, **38**, 38–45.
56. N. Shimizu, N. Itoh, H. Utiyama and G. M. Vahl, Selective entrapment of extrachromosomally DNA by nuclear budding and micronucleation during the S-phase, *J. Cell Biol.*, 1998, **140**, 1307–1320.
57. N. Shimizu, T. Shimura and T. Tanaka, Selective elimination of acentric double minutes from cancer cells through the extrusion of micronuclei, *Mutat. Res.*, 2000, **448**, 81–90.
58. S. Ergene, T. Çavaş, A. Celik, N. Koleli, F. Kaya and A. Karahan, Monitoring of nuclear abnormalities in peripheral erythrocytes of three fish species from the Goksu Delta (Turkey): genotoxic damage in relation to water pollution, *Ecotoxicology*, 2007, **16**, 385–391.
59. V. Rodilla, Origin and evolution of binucleated cells and binucleated cells with micronuclei in cisplatin-treated CHO cultures, *Mutat. Res.*, 1993, **70**, 411–421.
60. L. Xanthopouloua, J. D. A. Delhantya, A. Maniaa, T. Mamasa, P. Serhalb and S. B. Sengupta, The nature and origin of binucleate cells in human preimplantation embryos: relevance to placental mesenchymal dysplasia, *Reprod. BioMed. Online*, 2011, **22**, 362–370.
61. T. Çavaş and S. Ergene-Gozukara, Micronuclei, nuclear lesions and interphase silver-stained nucleolar organizer regions (AgNORs) as cyto-genotoxicity indicators in *Oreochromis niloticus* exposed to textile mill effluent, *Mutat. Res.*, 2003, **538**, 81–91.
62. S. Ergene, T. Çavaş, A. Celik, N. Koleli and C. Aymak, Evaluation of river water genotoxicity using the piscine micronucleus test, *Environ. Mol. Mutagen.*, 2007, **48**, 421–429.
63. T. P. Galindo and L. M. Moreira, Evaluation of genotoxicity using the micronucleus assay and nuclear abnormalities in the tropical sea fish *Bathygobius soporator* (Valenciennes, 1837) (Teleostei, Gobiidae), *Genet. Mol. Biol.*, 2009, **32**, 394–398.
64. F. Ayylon and E. Garcia-Vazques, Induction of micronuclei and other nuclear abnormalities in European minnow *Phoxinus phoxinus* and mollie *Poecilia latipinna*: an assessment of the fish micronucleus test, *Mutat. Res.*, 2000, **467**, 177–186.
65. J. Barsiene, V. Dedonyte, A. Rybakovas, L. Andreikenaite and O. K. Andersen, Investigation of micronuclei and other nuclear abnormalitiesin peripheral blood and kidney of marine fish treated with crude oil, *Aquat Toxicol.*, 2006, **78**, 99–104.
66. O. Ostling and K. J. Johanson, Microelectrophoretic study of radiation-induced DNA damages in individual mammalian cells, *Biochem. Biophys. Res. Commun.*, 1984, **123**, 291–298.

67. N. P. Singh, M. Y. McCoy, R. R. Tice and E. L. Scheider, A simple technique for quantitation of low levels of DNA damage in individual cells, *Exp. Cell. Res.*, 1988, **175**, 184–191.
68. P. L. Olive, J. P. Banath and R. E. Durand, Heterogeneity in radiation-induced DNA damage and repair in tumor and normal cells measured using the "comet" assay, *Radiat. Res.*, 1990, **122**, 86–94.
69. H. Kobayashi, C. Sugiyama, Y. Morikawa, M. Hayashi and T. Sufuni, A comparison between manual microscopic analysis and computerized image analysis in the single cell gel electrophoresis assay, *MMS Commun.*, 1995, **3**, 103–115.
70. N. P. Singh, The comet assay: Reflections on its development, evolution and applications, *Mutat. Res.*, 2016, **767**, 23–30.
71. R. Pandrangi, M. Petras, S. Ralph and M. Vrzoc, Alkaline single cell gel (comet) assay and genotoxicity monitoring using bullheads and carp, *Environ. Mol. Mutagen.*, 1995, **26**, 345–356.
72. K. Deventer, Detection of genotoxic effects on cells of liver and gills of *B. rerio* by means of single cell gel electrophoresis, *Bull. Environ. Contam. Toxicol.*, 1996, **56**, 911–918.
73. C. L. Mitchelmore and J. K. Chipman, Detection of DNA strand breaks in brown trout (*Salmo trutta*) hepatocytes and blood cells using the single cell gel electrophoresis (comet) assay, *Aquat. Toxicol.*, 1998, **41**, 161–182.
74. D. E. Nacci, S. Cayula and E. Jackim, Detection of DNA damage in individual cells from marine organisms using the single cell gel assay, *Aquat. Toxicol.*, 1996, **35**, 197–210.
75. O. Garcia, T. Mandina, A. I. Lamadrid, A. Diaz, A. Remigio, Y. Gonzalez, J. Piloto, J. E. Gonzalez and A. Alvarez, Sensitivity and variability of visual scoring in the comet assay. Results of an inter-laboratory scoring exercise with the use of silver staining, *Mutat. Res*, 2004, **566**, 25–34.
76. D. Anderson, T. W. Yu, B. J. Phillips and P. Schmezer, The effect of various antioxidants and other modifying agents on oxygen-radical-generated DNA damage in human lymphocytes in the COMET assay, *Mutat. Res.*, 1994, **307**, 261–271.
77. J. Palus, E. Dziubaltowska and K. Rydzynski, DNA damage detected by the comet assay in the white blood cells of workers in a wooden furniture plant, *Mutat. Res.*, 1999, **444**, 61–74.
78. M. Pitarque, A. Creus, R. Marcos, J. A. Hughes and D. Anderson, Examination of various biomarkers measuring genotoxic endpoints from Barcelona airport personnel, *Mutat. Res.*, 1999, **440**, 195–204.
79. M. De Boeck, N. Touil, G. De Visscher, P. A. Vande and M. Kirsch-Volders, Validation and implementation of an internal standard in comet assay analysis, *Mutat. Res.*, 2000, **469**, 181–197.
80. M. B. Gyori, G. Venkatachalam, P. S. Thiagarajan, D. Hsu and M. V. Clementa, OpenComet: An automated tool for comet assay image analysis, *Redox Biol.*, 2014, **2**, 457–465.
81. C. Helma and M. Uhl, A public domain image-analysis program for the single-cell gel-electrophoresis (comet) assay, *Mutat. Res.*, 2000, **466**, 9–15.

82. S. S. Sandhu and W. R. Lower, *In situ* assessment of genotoxic hazards of environmental pollution, *Toxicol. Ind. Health*, 1989, **5**, 73–83.
83. W. Pilcher, S. Miles, S. Tang, G. Mayer and A. Whitehead, Genomic and genotoxic responses to controlled weathered oil exposures confirm and extend field studies on impacts of the deepwater horizon oil spill on native killifish, *PloS One*, 2014, **9**, e106351.
84. J. M. Culp, R. B. Lowell and K. J. Cash, Integrating mesocosm experiments with field and laboratory studies to generate weight-of-evidence risk assessments for large rivers, *Environ. Toxicol. Chem.*, 2000, **19**, 1167–1173.
85. K. Hyllanda, B. B. Skeia, G. Brunborgb, T. Langc, M. J. Gubbinsd, J. le Goffe and T. Burgeotf, DNA damage in dab (*Limanda limanda*) and haddock (*Melanogrammus aeglefinus*) from European seas, *Mar. Environ. Res.*, 2017, **124**, 54–60.
86. A. Simonyan, B. Gabrielyan, S. Minasyan, G. Hovhannisyan and R. Aroutiounian, Genotoxicity of water contaminants from the basin of Lake Sevan, Armenia evaluated by the comet assay in Gibel carp (*Carassius auratus gibelio*) and *Tradescantia* bioassays, *Bull. Environ. Contam. Toxicol.*, 2016, **96**, 309–313.
87. J. M. Grizzle, S. Horowitz and D. R. Strength, Caged fish as monitors of pollution: Effects of chlorinated effluent from a seawater treatment plant, *J. Am. Water Res. Assoc.*, 1988, **24**, 951–959.
88. A. Oikari, Caging techniques for field exposures of fish to chemical contaminants, *Aquat. Toxicol.*, 2006, **78**, 370–381.
89. B. Kushwaha, S. Pandey, S. Sharma, R. Srivastava, R. Kumar, N. S. Nagpure, A. Dabas and S. K. Srivastava, *In situ* assessment of genotoxic and mutagenic potential of polluted river water in *Channa punctatus* and *Mystus vittatus*, *Int. Aquat. Res.*, 2012, **4**, 16.
90. G. I. V. Klobucar, A. Stambuk, M. Pavlica, M. Sertic Peric, B. Kutuzovic Hackenberger and K. Hylland, Genotoxicity monitoring of freshwater environments using caged carp (*Cyprinus carpio*), *Ecotoxicology*, 2010, **19**, 77–84.
91. S. Minissi, E. Ciccotti and M. Rizzoni, Micronucleus test in erythrocytes of *Barbus plebejus* (Teleostei, Pisces) from two natural environments: a bioassay for the *in situ* detection of mutagens in freshwater, *Mutat. Res.*, 1996, **367**, 245–251.
92. J. M. Gutiérrez, S. Villar and A. Acuña Plaván, Micronucleus test in fishes as indicators of environmental quality in subestuaries of the Río de la Plata (Uruguay), *Mar Pollut Bull.*, 2015, **91**, 518–523.
93. G. Wirzinger, L. Weltje, J. Gercken and H. Sordyl, Genotoxic damage in field-collected three-spined sticklebacks (*Gasterosteus aculeatus* L.): a suitable biomonitoring tool? *Mutat. Res.*, 2007, **628**, 19–30.
94. OECD (1992), *Test No. 203: Fish, Acute Toxicity Test*, OECD Publishing, Paris.
95. R. Sarikaya, M. Selvi, O. Koçak and F. Erkoç, Investigation of acute toxicity of fenitrothion on guppies *Poecilia reticulata*, *J. Appl. Toxicol.*, 2007, **27**, 318–321.

96. M. Selvi, R. Sarikaya, F. Erkoç and O. Koçak, Investigation of acute toxicity of chlorpyrifos-methyl on guppy *Poecilia reticulata*, *Chemosphere*, 2005, **60**, 93–96.
97. J. A. Anogwih, Toxicity of pirimiphos methyl (Actellic 25EC) on *Anopheles gambiae* s.s., *Culex quinquefasciatus* (Diptera: Culicidae), and potential biocontrol agent, *Poecilia reticulata* (Pisces: Poeciliidae), *J. Econ. Entomol.*, 2014, **107**, 1440–1446.
98. T. L. Rocha, A. P. Santos, A. T. Yamada, C. M. Soares, C. L. Borges, A. M. Bailão and S. M. Sabóia-Morais, Proteomic and histopathological response in the gills of *Poecilia reticulata* exposed to glyphosate-based herbicide, *Environ. Toxicol. Pharmacol.*, 2015, **40**, 175–186.
99. J. A. Anogwih, W. A. Makanjuola and L. O. Chukwu, Spinosad Induced Cytogenotoxic Effects on the mosquito fish, *Poecilia reticulata*, *J. Clinic. Toxicol.*, 2013, **S12**, 001.
100. P. Arslan, M. A. Dalgic, S. Saricakmak, N. Sarigil, S. Ulker and B. Kocak Memni, Investigation of the genotoxic effects of chlorine bleach and dishwashing detergent on guppy (*Poecilia reticulata* Peters, 1859) by using the micronucleus test, *MAKU FEBED*, 2011, **4**, 29–37.
101. J. de Souza Filho, C. D. Sousa, C. D. da Silva, S. M. de Sabóia-Morais and C. K. Grisolia, Mutagenicity and genotoxicity in gill erythrocyte cells of *Poecilia reticulata* exposed to a glyphosate formulation, *Bull. Environ. Contam. Toxicol.*, 2013, **91**, 583–587.
102. B. B. Pereira and E. O. de Campos Júnior, Enzymatic alterations and genotoxic effects produced by sublethal concentrations of organophosphorous Temephos in *Poecilia reticulata*, *J. Toxicol. Environ. Health A*, 2015, **78**, 1033–1037.
103. E. Lodi, Chromosome complement of the guppy, *Poecilia reticulata* Peters (Pisces, Osteichthyes), *Caryologia*, 1978, **314**, 475–477.
104. R. Sarikaya, M. Selvi and F. Erkoç, Investigation of acute toxicity of fenitrothion on peppered corydoras (*Corydoras paleatus*) (Jenyns, 1842), *Chemosphere*, 2004, **56**, 697–700.
105. S. F. Pesce, J. Cazenave, M. V. Monferrán, S. Frede and D. A. Wunderlin, Integrated survey on toxic effects of lindane on neotropical fish: *Corydoras paleatus* and *Jenynsia multidentata*, *Environ. Pollut.*, 2008, **156**, 775–783.
106. I. C. Guiloski, S. C. Rossi, C. A. da Silva and H. C. de Assis, Insecticides biomarker responses on a freshwater fish *Corydoras paleatus* (Pisces: Callichthyidae), *J. Environ. Sci. Health B.*, 2013, **48**, 272–277.
107. N. de Castilhos Ghisi and M. Cestari, Genotoxic effects of the herbicide Roundup® in the fish *Corydoras paleatus* (Jenyns 1842) after short-term, environmentally low concentration exposure, *Environ. Monit. Assess.*, 2013, **185**, 3201–3207.
108. M. L. Terancio, M. R. Vicari, M. C. A. Matiello, M. M. Cestari and L. A. C. Bertollo, Cytogenetics of two sympatric *Corydoras* species (Pisces, Siluriformes, Callichtyidae) of southern Brazil artoni, *Braz. J. Biol.*, 2006, **66**, 191–198.

CHAPTER 12

Genotoxic and Biochemical Responses Triggered by Polycyclic Aromatic Hydrocarbons in Freshwater and Marine Fish: Tambaqui and Seahorse as Bioindicators

ADRIANA REGINA CHIPPARI-GOMES,*[a] FREDERICO AUGUSTO CARIELLO DELUNARDO,[a] HELEN SADAUSKAS-HENRIQUE,[b,c] SUSANA BRAZ MOTA[b] AND VERA MARIA FONSECA DE ALMEIDA-VAL[b]

[a] Universidade Vila Velha – UVV, Laboratório de Ictiologia Aplicada (LAB PEIXE), Rua Comissário José Dantas de Melo, n° 21, Boa Vista, 29102-770, Vila Velha, ES, Brazil, Email: frednativa@yahoo.com.br; [b] Instituto Nacional de Pesquisas da Amazônia – INPA, Laboratório de Ecofisiologia e Evolução Molecular (LEEM), Av. André Araújo, n° 2936, Petrópolis, Manaus, AM, Brazil, Email: susanabmota@gmail.com; helensadauskas@gmail.com; veraval@inpa.gov.br; [c] UNISANTA (Universidade Santa Cecília), Programa de Pós-Graduação em Sustentabilidade de Ecossistemas Costeiros e Marinhos, Rua Oswaldo Cruz, 277 (Boqueirão), 11045-907, Santos, São Paulo, Brazil
*Email: adriana.chippari@uvv.br

Issues in Toxicology No. 33
Ecotoxicology and Genotoxicology: Non-traditional Aquatic Models
Edited by Marcelo L. Larramendy

Published by the Royal Society of Chemistry, www.rsc.org

12.1 Introduction

The aquatic environment comprises various ecosystems, such as rivers, lakes, estuaries, oceans and seas.[1] These ecosystems are used for different purposes, including as water sources or for power generation and irrigation,[2] as well as being the final recipient of domestic and industrial effluents. As a consequence, impacts on the aquatic environment are inevitable. These impacts have occurred and still occur in both marine and freshwater environments, and, depending on the proportions of pollutants, may remain, causing ecological disturbances for long periods.

Among the environmental impacts caused by anthropogenic activities, the increased pollution of the aquatic environment is among the main causes of authorities' concern and attention. Besides causing a decrease in the water quality, pollution also harms aquatic biota. Once pollutants are dumped into the aquatic environment, they are distributed and interact with the biota according to its features and to the conditions of the receiving environment, being subject to chemical, physical and biological transformations.[3,4] Concerning the biota, its structure can be modified by replacing sensitive organisms with those that are resistant to polluted environments, but this consequently changes the native community and reduces the species diversity in that location, which may bring losses to nature and humankind. Before the pollution reaches such a destructive level, the organisms inhabiting the contaminated environments will initially respond at different levels, with alterations at cellular, physiological, biochemical and morphological levels. After these responses, changes in mortality will occur and possible replacement of species may take place. Ongoing pollution events, if they become routine and chronic, will result in such changes in the fish community and aquatic biota. Instead, in the case of acute events, the effects will be more drastic, and significant mortality may occur.

Among the organisms employed as bioindicators, fish have characteristics that make them excellent experimental models for studies on aquatic toxicology and ambient biomonitoring because they are sensitive to chemicals and may indicate the potential for human exposure to genotoxic substances present in the water.[5] Fish accumulate pollutants directly from contaminated water or indirectly through ingestion of contaminated food; fish also react sensitively to environmental changes or stressors in the form of low concentrations of genotoxic substances present in the environment, which may affect the genetic diversity of aquatic populations.[6,7] In addition, these organisms respond to toxic agents in a similar way to higher vertebrates, allowing a better assessment of substances that are potentially teratogenic, mutagenic and carcinogenic to humans.[8]

Among the types of contaminants that affect water bodies and subsequently fish and other aquatic organisms, we can distinguish: metals present in domestic and industrial effluents and released in effluents of mining companies; polycyclic aromatic hydrocarbons (PAHs) from oil and its derivatives; pesticides drained from plantations; so-called emerging contaminants, in which we find medications, drugs, and cosmetics released daily in domestic sewage; and many others.

12.2 Contamination by Polycyclic Aromatic Hydrocarbons (PAHs)

The marine environment has been the scene of many drastic environmental disasters, mostly caused by accidents involving large oil spills all over the world, such as the ones caused by the ships Torry Canyon (1967), Amoco Cadiz (1978), Exxon Valdez (1989), Aegean Nakhodka (1997), and Prestige (2002), and the Deepwater Horizon drilling rig (2010). Similar accidents, however with smaller magnitudes, have also occurred in river basins, such as in Negro River (Brazil, 1999), Barigüi and Iguaçu Rivers (Brazil, 2000), Flúvia River (Spain, 2004), and many others.

These events occurred in marine and freshwater, causing pollution by PAHs, which can be acute, as in the above-mentioned cases, or chronic. Acute cases are usually catastrophic and sporadic events; the effects are quickly detected since in these events we can find high mortalities of fish, birds, and other aquatic organisms due to the direct contact with the oil slick (insoluble oil fraction). These situations are widely broadcast in international media. On the other hand, there are chronic events where small amounts of PAHs are routinely released into the aquatic environment, due to small leaks and routine procedures, *etc.*, which may also cause damage but no massive mortality of organisms; such events are in general barely noticed by the media. In these cases, what happens is that the organisms are routinely exposed to these toxic agents, most often the soluble oil fraction, composed of PAHs, which cause cellular damage as well as biochemical and morphological changes. In the beginning, these changes are not perceptible, but if this event continues, anatomical (morphological) damages can be observed, jeopardizing the survival of the organism, as well as their reproductive capacity. To achieve these levels of changes, the damage that initially was at cellular and structural levels starts to reach higher levels (at the ecological level) as the population structure and the communities are affected, leading to an ecological imbalance. Thus, in the majority of cases, the most serious damage that occurs to aquatic biota is not that caused by major accidents, but by routine and quiet release of pollutants. This situation of acute and chronic pollution events does not occur exclusively when oil and its derivatives are released into the environment, but also when other pollutants, such as metals, pesticides, domestic and industrial sewage and others, are released, causing similar effects on aquatic biota.

The oil, when in contact with water, is divided into insoluble and soluble phases. The insoluble fraction is the part of the oil that remain on the water's surface and forms a layer that prevents gas exchange and light penetration through the water column, reducing photosynthesis and worsening oxygen availability in the affected areas.[9] The fish, when seeking the water surface in search of oxygen, may even impregnate their gills with this insoluble fractions, affecting gas exchange and causing internal hypoxia; this fact can lead to the death of the organism.[9] On the other hand, the soluble fraction is diluted throughout the water column, being almost imperceptible to human vision. However, this is the fraction that contains PAHs, which are considered the primary cause of oil toxicity since these compounds have carcinogenic and mutagenic properties.[10] Once inside the cells, these PAHs are metabolized by enzymes from several mixed-function systems of the Cytochromes P450 super family, leading to the formation of highly reactive metabolites (more toxic than the parent compound) that interact with the DNA molecule, producing a variety of lesions.[11–15]

Among the effects observed in fish caused by exposure to PAHs, several authors include the induction to oxidative stress;[16–25] induction of ROS production;[26] genotoxic, mutagenic and carcinogenic effects.[5,27–33] In addition, there is also the occurrence of morphological damage, as histopathological lesions caused in the gills and liver,[34–40] as well as the damage at most relevant ecological levels, such as losses in breeding success owing to the impact caused during the recruitment of populations, since the PAHs can also act as endocrine disruptors[41–45] or cause changes in gonadal development.[46–49]

12.3 The Effects of Water's Physical and Chemical Parameters on the Toxicity of Contaminants

The physicochemical factors of water regulate the bioavailability of different contaminants that affect fish; these are temperature, pH, dissolved oxygen, transparency, hardness, dissolved organic carbon, and salinity,[50] as well as the chemical form of the metal in the water. These factors may increase or decrease the toxicity of various contaminants in water bodies.

In the case of dissolved organic carbon (DOC), which corresponds to humic substances (humic acid, fulvic acids and humins), there may be a reduction in the bioavailability of metals in the milieu owing to their chemical structure, which exhibits a high affinity to adsorb organic substances, such as pesticides, and to form complexes with metal ions.[51] According to Sunda and Hansen,[52] as the concentration of dissolved organic carbon (DOC) increases in water, the number of binding sites for metal increases, promoting an increase in the formation of the metal–DOC complex, and so reducing the concentration of the bioavailable species of this metal. Thus, the organic matter content in these cases can be used as a protective parameter against the toxicity of metal cations. However, depending on the environmental conditions, metals adsorbed by DOC can be released back into the water column, and thus become bioavailable to the organisms.[53] In fact, studies by Matsuo *et al.*[54] and Sadauskas-Henrique *et al.*[55] showed that the elevation of the level of natural organic matter (NOM) of Negro river (AM-Brazil) can increase the solubility of PAHs, and consequently may increase the toxicity of this substance in the fish body.

Temperature also influences the toxicity of contaminants since increased temperature is directly related to increased metabolic rate in fish, stimulating the absorption of metals[56–59] or other contaminants. In addition, the increase in water temperature leads to a series of consequences: the amount of dissolved oxygen in the water decreases, fish oxygen consumption increases,[60] and the bioavailability of xenobiotics also increases, since the solubility of these xenobiotics is dependent on the water temperature,[61,62] causing increases in the toxicity of contaminants at elevated temperatures. The elevation of temperature also causes increases in the induction of ROS in the organisms,[63–65] resulting in the induction of apoptosis and oxidative stress.[65] These responses can be even more intensified in the presence of contaminants. Currently, this biotic factor is of great importance as a result of the ongoing global climate changes.[66,67]

Several authors[36,68–70] claim that hypoxic events may exacerbate the toxicity caused by PAHs, causing the weakening of the organism's survival. The association between hypoxia exposure and contact with PAHs may result in increased embryo imperfections and cause delays or impairments in embryo development, either due to a change in metabolism to more toxic metabolites, or to an increased half-life of parental toxic PAHs;[68,69] and maximizes the genotoxic and morphological responses in sea horses.[36]

12.4 The Effects of PAHs on Cellular and Genetic Properties and on the Responses of Antioxidant and Detoxification Enzymes

Over the last three decades, a variety of ecotoxicological tools and genotoxic tests became important for generating biological indicators along with the oxidative stress parameters that were widely used in environmental monitoring and assessment.[71] Many environmental pollutants, such as polycyclic aromatic compounds (PAH), induce oxidative stress in aquatic organisms. The oxidative stress results from the appearance of reactive oxygen species (ROS) in the cell, which may cause injuries such as lipid peroxidation, protein oxidation, and DNA breakdown, inducing loss of strength in the organisms.[72,73] Thus, genotoxic damage in aquatic biota has been widely used as an essential tool to detect DNA single-strand breaks in animals acutely exposed to many contaminants.[74]

When ROS generation exceeds the antioxidant capacity of the organism, oxidative stress and oxidative damage occur in the cells, tissues and organs.[75] To neutralize ROS and avoid the oxidative stress, cells increase their antioxidant defense pathways, which include antioxidant enzymes, such as superoxide dismutase (Cu–Zn SOD), catalase (CAT) and glutathione peroxidase (Se-GPx), as well as non-enzymatic antioxidants, such as glutathione tripeptide (GSH). Some PAHs, such as phenanthrene, benzo[*a*]pyrene, and naphthalene, promote oxidative stress in different aquatic organisms.[76,77] These reactions probably occur due to the PAHs' high lipophilicity, *i.e.*, lipid affinity, which causes lipid peroxidation (LPO). Furthermore, the toxicity of PAHs can also be related to the characteristics of the water. In freshwater, some elements, such as natural organic matter (or dissolved organic carbon, DOC), resulting from animal and vegetable residues being released into the environment, can lead to interactions with PAH molecules as well as promote direct effects in tissues and organs such as the gills.[78] Additionally, many types of DNA damage can occur when organisms are exposed to PAH contamination. These damages include DNA single- and double-strand breaks, inter- and intra-strand crosslinks, DNA adducts, and DNA–protein crosslinks.[79]

PAHs may cause genotoxicity, and these damages can be classified into three stages: First is the formation of adducts with toxic molecules. Second is DNA modification, including single- and double-strand breakages, changes in DNA repair, nucleotide base oxidation, and crosslinks in advanced stages.[80] The third, and most severe, damage is the occurrence of cells presenting altered functions, cell proliferation, mutagenesis, and eventually carcinogenesis.[81]

The toxicity of some PAHs depends on many factors, such as the characteristics of the water, as mentioned above. Besides freshwater, the adsorption of PAHs onto the surface of the organisms can also occur in seawater with significantly high levels of PAH. Similarly, the presence of PAHs in marine water can stimulate ROS production, antioxidant defenses, and oxidative damage in the organisms, and such toxicity will depend on some water characteristics, particularly the level of salinity and the presence of phytoplankton and macroalgae.[82,83]

12.5 The Amazonian Fish *Colossoma macropomum* (Osteichthyes, Serrasalmidae) as a Model to Assess Water Pollution

Colossoma macropomum, commonly named "tambaqui" (Figure 12.1), is a Characid (sub-family Serrasalmidae) that is widely distributed in South America, especially occurring in the Amazon and Orinoco River basins.[84] This fish species can reach 1 m in length and 30 kg in weight, and is

Figure 12.1 *Colossoma macropomum* specimen.
Photo credit: Levy de Carvalho Gomes.

considered the biggest species in the order Characiformes in the Amazon basin.[85] When in their natural environment, their feeding behavior changes in accordance with the water levels and seasons, feeding on fruits and seeds in flooded forests during the high water period, and phytoplankton, zooplankton, and periphyton in whitewater rivers and tributaries in low water periods.[86] During these migrations between flooded forests and whitewater rivers and tributaries, tambaqui cope with extreme alterations of water quality parameters, such as: pH; ion content; dissolved and particulate organic carbon; temperature; and dissolved oxygen.[85,87–89]

Several studies have demonstrated that tambaqui developed a wide spectrum of biological advantages (adaptive capacity) to cope with the extreme conditions of their environment.[90–92] The survival of tambaqui in acidic water (pH 4–3.5) is related with the ability to avoid ion disturbances, preventing the inhibition of active ion uptake and the acceleration of diffusive ion losses at the gills.[93] Under conditions of moderate to severe environmental hypoxia, the tambaqui developed the ability to explore the upper part of the water column (water–air interface), which is rich in oxygen. The lower lip of the fish becomes greatly enlarged, facilitating the skimming of the water from the water–air interface.[94,95]

The myriad of adjustments developed by the tambaqui to face the never-ending environmental challenges that naturally occur in their habitats have given this fish species great biological importance, which has encouraged increasing numbers of studies evaluating the physiological mechanisms behind these adjustments. Additional environmental challenges, such as water body contamination with a variety of xenobiotics, are now combined with the natural environmental challenges it faces, pushing the fish to its biological limits.[96] Considering that the tambaqui is a very resilient fish species due its evolutionary history of facing challenging natural environments, this fish species has characteristics that make it a feasible model for ecotoxicological studies. In the last two decades, several studies used the tambaqui as a model for toxicological studies,[54,55,97,98] where this fish species contributed significantly to the understanding of the effects of the pollutants in the Amazon water bodies.

12.5.1 Crude Oil Contamination

The Amazon basin is second biggest producer of gas and oil in Brazil with 1.3 tons of gas and 130 000 barrels of crude oil per day.[99] This economic exploitation of gas and hydrocarbons in the Amazon forest started in 1986 inside one of the most environmentally sensitive areas of the planet. The crude oil is extracted in the Petroliferous Province of Urucu (PPU), located at 280 kilometers from the center of Coari/AM municipality, and then transported in tankers across the Solimões River to Manaus City where it is refined.[100] Although strict safety procedures are currently observed, there is always a risk of an oil spill during the exploitation and transportation of the Urucu crude oil. Moreover, tropical environments have been the recipients of urban and industrial contamination due to uncontrolled population growth and rapid industrialization,[101] which increases the risk of contamination of the water bodies with PAHs from the crude oil. The effects of crude oil contamination are well documented for marine organisms.[102–104] However, our understanding of the effects of crude oil in freshwater habitats has not advanced significantly since Val and Almeida-Val[100] published one of the first considerations addressing this subject. Currently, we have a better understanding, although we are still dealing with several other factors regarding water characteristics and chronic contamination with PAHs in the Amazon basin.

Taking into account the widespread contamination of tropical aquatic environments by petroleum hydrocarbons and the scarcity of relevant bioindicator species in tropical water bodies, an experiment was recently performed to validate a suite of biomarkers for tambaqui.[105] This study demonstrates the potential of biotransformation and antioxidant enzymes, the presence of PAHs metabolites in fish bile and, especially, the genotoxic effects in blood cells of tambaqui as relevant biomarkers in tropical environments. Genotoxic effects caused by benzo[*a*]pyrene in the blood cells of tambaqui presented a dose-dependent response.[105] In the same way, Silva *et al.*[106] also demonstrated increased DNA damage in blood cells of tambaqui intraperitoneally injected with benzo[*a*]pyrene. Genotoxic effects were also found by other studies in tambaqui exposed to crude oil and PAHs. Increases in DNA damage in the blood cells of tambaqui revealed time- and dose-dependent responses when exposed to waterborne naphthalene, benzo-[*a*]pyrene and Urucu crude oil. However, the authors observed erythrocytic abnormalities only in tambaqui exposed to the highest benzo[*a*]pyrene concentration while the Urucu crude oil caused a time- and dose-dependent increase.[107] Similarly, other authors observed DNA damage in blood cells of tambaqui exposed to 2.83 and 3.96 ml L^{-1} (25% and 35% of the LC_{50}-96 h) of waterborne Urucu crude oil for 96 h while the erythrocytic abnormalities increased only at the highest concentration.[108] Both DNA damage and erythrocytic abnormalities can be considered good biomarkers for PAH contamination in Amazon water bodies. However, because of the longer time of exposure and higher dose of PAHs necessary to induce erythrocytic abnormalities, these genotoxic biomarkers should be considered together to avoid misinterpretation of the data.

Biotransformation and antioxidant enzymes in tambaqui are also considered of great importance to evaluate PAH contamination. Glutathione *S*-transferase (GST) plays a fundamental role in Phase II biotransformation, participating in xenobiotic conjugation reactions with non-enzymatic antioxidants such as the tripeptide glutathione in its reduced form (GSH), which acts as the main antioxidant in the cell and is a cofactor for the action of GST, resulting in the efficient elimination of the metabolites from the organism.[109] GST is considered a reliable biomarker in a variety of fish species exposed to PAHs.[108,110–112] The same is true for tambaqui, where GST activity increases in hepatic tissue after 6 h and 48 h of exposure to 1% and 3% waterborne Urucu crude oil, respectively[107] (Table 12.1). Hepatic GST also

Table 12.1 Summary of the bioindicator responses found in tambaqui, *Colossoma macropomum*, exposed to crude oil and PAHs.

Treatments	Doses/time	Bioindicators	Reference
Intraperitoneal injection of BaP	1; 10; 100 and 1000 μmoles BaP kg^{-1}/96 h	↑ PAHs metabolites in bile; ↑ DNA damage in blood cells; ↑ GST, EROD and LPO in liver; ↑ s-SDH in plasma	Sadauskas-Henrique *et al.* (*in press*)[105]
Intraperitoneal injection of BaP	4; 8 and 16 μmolar BaP kg^{-1}/96 h	Overexpression of *Ras* oncogene in liver	Silva *et al.* (*in press*)[106]
Intraperitoneal injection of BaP	8; 16 and 32 μmolar BaP kg^{-1}/96 h	Irreparable histopathological changes in the liver	Silva *et al.* (*in press*)[106]
Waterborne BaP	0.3; 0.9 μmolar/48 h	↑ Erythrocytic abnormalities	Oliveira (2010)[107]
Waterborne BaP and naphthalene	0.1; 0.3; 0.9 μmolar/6; 12; 24; 48 h	↑ DNA damage in blood cells	Oliveira (2010)[107]
Waterborne Urucu crude oil	Oil slick of 0.32 mm/72 h	↑ LDH, CS and CCO in muscle	Duncan (1998)[119]
Waterborne Urucu crude oil	1% and 3% (v/v)/6; 12; 24; 48 h	↑ DNA damage in blood cells	Oliveira (2010)[106]
Waterborne Urucu crude oil	1% and 3% (v/v)/48 h	↑ Erythrocytic abnormalities; ↑ FA	Oliveira (2010)[106]
Waterborne Urucu crude oil	2.83 and 3.96 ml L^{-1}/96 h	↑ DNA damage in blood cells; ↑ Erythrocytic abnormalities; ↑ GST; ↑ FA	Kochhann *et al.* (2013)[108]
Waterborne Urucu crude oil + NOM (Rio Negro, Amazon, Brasil)	5.68 ml L^{-1}/24 h	↑ PAHs metabolites in bile; ↑ DNA damage in blood cells; ↑ GST, EROD and LPO in liver ↑ s-SDH in plasma	Sadauskas-Henrique *et al.* (2016)[55]
Waterborne Urucu crude oil + NOM (Luther Marsh, Ontario, Canada)	5.68 ml L^{-1}/96 h	↑ EROD activity and ↑ CYP1A concentration in liver	Matsuo *et al.* (2006)[54]

increases in tambaqui exposed to waterborne Urucu crude oil at 2.83 and 3.96 ml L^{-1} (25% and 35% of the LC_{50}-96 h, respectively) for 96 h. Similarly, the exposure of tambaqui to three waterborne benzo[*a*]pyrene concentrations (0.1, 0.3, and 0.9 μM) caused increases in hepatic GST after 24 h (Table 12.1). Despite these hepatic GST increases not showing dose- and time-dependence responses, this enzyme plays an important role when analyzed together with other biomarkers.[55] Correlations between the GST activity and the activity of catalase (CAT), an antioxidant enzyme that neutralizes the effects of reactive oxygen species (ROS) in biological membranes, were observed for tambaqui exposed to waterborne Urucu crude oil[108] and intraperitoneally injected with benzo[*a*]pyrene[105] (Table 12.1). This positive correlation suggests their co-involvement in PAH metabolism in tambaqui.

Tambaqui exposed to crude oil and to PAHs can also experience hepatocellular damage. The presence of some enzymes in fish plasma could indicate recent contamination with crude oil and PAHs. Tambaqui exposed to waterborne Urucu crude oil for 48 h and 96 h[112] presented high values of alkaline phosphatase (AF) in plasma (Table 12.1). Tambaqui intraperitoneally injected with 10, 100 and 1000 μmoles benzo[*a*]pyrene kg^{-1} experienced elevated concentrations of the enzyme sorbitol dehydrogenase in the serum (s-SDH), indicating hepatocellular damage. At the same time, the elevations in the lipid peroxidation of the liver at these three highest benzo[*a*]pyrene doses supports a relevant relationship between dysfunctional cellular membrane functions and elevated s-SDH activity in the serum of tambaqui[55] (Table 12.1). In fact, benzo[*a*]pyrene caused irreversible histopathological liver damage (necrosis) in tambaqui intraperitoneally injected with 8, 16 and 32 μM benzo[*a*]pyrene kg^{-1}. At the same time, overexpression of the *Ras* oncogene was observed in the liver of these fishes, which demonstrates the carcinogenic effects of BaP to tambaqui.[106]

Amazon fish species also face challenges related to possible interactions between the physical and chemical characteristics of the Amazon water bodies and crude oil toxicity. For example, natural organic matter (NOM) is a water quality parameter that can interfere with the uptake and bioconcentration of organic pollutants by aquatic organisms. Some studies indicate that NOM might increase the solubility of PAHs[113,114] and may interfere directly in the physiology of aquatic organisms.[115–118] Indeed, it was demonstrated that the Rio Negro NOM induces CYP1A expression and ethoxyresorufin-*O*-deethylase (EROD) activity in the liver of tambaqui, acting as aryl hydrocarbon receptor agonists[54] (Table 12.1). While CYP1A is induced in tambaqui by crude oil, the presence of humic substances (component present in NOM) could elevate the levels of CYP1A and EROD induced by NOM and crude oil. Moreover, interactions between PAH and NOM might increase the solubility of PAHs with a consequent increase in toxicity. Recent studies demonstrated that tambaqui acclimated in natural Rio Negro water (rich in NOM) and acutely exposed to waterborne Urucu crude oil showed a marked increase in the absorption of PAHs, which was associated with significant hepatic (alterations of biotransformation and antioxidant enzymes) and genotoxic alterations (DNA damage in blood cells) in comparison with tambaqui exposed to Urucu crude oil in control water (with no NOM)[55] (Table 12.1). These data suggest that the NOM of the Rio Negro water might cause mild chemical stress responses in fish, which can make them more susceptible to the deleterious effects of the crude oil and PAHs.[55]

Another feature regarding the crude oil contamination in Amazon water bodies is related to the adaptations tambaqui developed to skim the surface and catch the oxygen-rich surface layer of the water. This adaptation maintains the oxygen supply during hypoxic periods but increases the amount of oil pumped across the gills, reducing the gas exchange capacity of the gill tissue and reducing the blood oxygen content.[100] For instance, the exposure of tambaqui to waterborne Urucu crude oil drastically reduced (by around 87%) the concentration of dissolved oxygen in the blood.[100] The consequent decreases of the metabolic rates of tissues involve the regulation of the levels of anaerobic and anaerobic enzymes. Lactate dehydrogenase (LDH), citrate synthase (CS) and cytochrome C oxidase (CCO) were increased in the muscle of tambaqui exposed to a Urucu crude oil slick of 0.32 mm for 72 h[119] (Table 12.1). These metabolic adjustments of tambaqui happen as an adaptive response to the low oxygen availability, increased energetic demand, and decreased oxygen transport in blood as a result of the increased methaemoglobin levels.[100]

Putting these data together, it is possible to visualize the potential of the tambaqui as a bioindicator of freshwater contamination. Moreover, the tambaqui can be considered as a model for toxicological studies owing to its high economic and scientific value in tropical regions, including South America and Asia,[120] due to its widespread availability from aquaculture ventures, and because this fish species is very resistant to environmental challenges.[88,95]

12.6 Seahorse *Hippocampus reidi* (Osteichthyes, Syngnatidae) as a Model to Assess Marine and Estuary Water Pollution

The seahorses, *Hippocampus* spp. (Figure 12.2), belong to the Syngnatidae family, occupying both temperate and tropical ocean regions and usually found in estuarine and coastal environments, such as reefs, bays, seaweed banks, and sea grass.[121,122] Their unique body morphology, along with their curiosities and special traits, including the pregnancy of males, have made these species a charismatic flagship group for many topics in marine conservation.[123–125]

Unlike most fish, seahorses are poor swimmers and rarely venture into the open sea to pursue prey or colonize new areas.[126,127] They have a prehensile tail for grasping holdfasts tightly.[128] This grasping ability helps seahorses anchor to vegetation or other substrates, which allows them to more effectively forage for food and avoid predators.[129] These sedentary characteristics make seahorse populations particularly susceptible to anthropogenic

Figure 12.2 *Hippocampus reidi* specimen.
Photo credit: Laila C. Campos Medeiros.

disturbances and result in increased exposure to toxic substances when pollution events occur.[33] Two species of seahorses are naturally found along the Brazilian coast: the lined seahorse, *Hippocampus erectus* (Perry, 1810), and the long snout seahorse, *Hippocampus reidi* (Ginsburg, 1933).[130] The latter is the most abundant species; therefore, we selected this species as a bioindicator of contamination of marine and estuarine environments by petroleum and its derivatives in our studies.

Concerning the genotoxicity results, Delunardo[33] exposed fish to three concentrations of crude oil (10, 20 and 30 $g\,kg^{-1}$) for 96 h and found DNA breakage, both through Comet assay and micronucleus analysis (MN). After exposure to crude oil, the animals showed a significant relationship between the DNA damage and increased crude oil concentration, indicating a dose-dependent response (Table 12.2), *i.e.*, the amount of DNA damage in *H. reidi* increased with increased availability of this contaminant in the environment. These results are in accordance with Santos *et al.*,[131] who assessed the genotoxicity of the diesel water-soluble fraction (DWSF) in seahorses exposed to three different dilutions (1 : 1000, 1 : 500 and 1 : 100), as well as with previous investigations,[132–135] thus demonstrating that these tests are efficient for rapid *in vitro* evaluation of PAH contamination. Additionally, Negreiros *et al.*[36] observed that the association of hypoxia with the components of petroleum aggravated the genotoxic responses, producing a significant increase in DNA strand breaks in *H. reidi* compared to the control groups (Table 12.2).

Table 12.2 Mean values of DNA damage index (DI) measured by alkaline Comet assay and mean values of micronuclei frequency (‰) (MN) in peripheral erythrocytes of seahorse, *Hippocampus reidi*, exposed to crude oil or diesel oil.

Treatments	DI	MN	Reference
Diesel oil (DWSF) 96 h			
0 (control)	38	0.0	Santos *et al.* (2010)[131]
1:1000	115[a]	2.0[a]	
1:500	153[a]	3.5[a]	
1:100	227[a]	6.0[a]	
Crude oil 96 h			
0 ml L^{-1} (control)	78	1.8	Delunardo *et al.* (2015)[33]
10 ml L^{-1}	253[a]	5.5[a]	
20 ml L^{-1}	260[a]	4.4[a]	
30 ml L^{-1}	280[a]	8.7[a]	
Crude oil			
0 ml L^{-1} (control)	21	0.0	Delunardo *et al.* (2013)[33]
10 ml L^{-1} - 96 h	127[a]	3.8[a]	
10 ml L^{-1} - 168 h	171[a]	2.3[a]	
Recoveryof 168 h	21	0.0	
Crude oil + hypoxia 8 h			
0 ml L^{-1} (control)	6	0.0	Negreiros *et al.* (2011)[36]
18 ml L^{-1}	74[a]	6.2[a]	
1.5 mg O_2 L^{-1}	45[a]	0.3	
18 ml L^{-1} + 1.5 mg O_2 L^{-1}	95[a]	6.7[a]	

[a]Indicates a significant difference ($p \leq 0.05$) compared to the control exposure group (0 mL L^{-1}).

However, the frequencies of nuclear abnormalities (NA), such as lobed nuclei, dumbbell-shaped or segmental nuclei, and kidney-shaped nuclei,[136] indicated the lack of sensitivity of these biological markers as observed in a recent study that evaluated the acute and subchronic exposure of *H. reidi* to the water-soluble fraction of diesel oil (unpublished data), similarly to the study done by Delunardo *et al.*[33] Accordingly, some authors also described that there is not always a direct association between exposure to genotoxic agents and changes in MN and NA frequencies,[137,138] although several studies described the presence of NA in fish cells that were exposed to contaminants.[108,139–143]

Delunardo *et al.*[33] investigated the acute (12, 24, 48 and 96 h) and subchronic (168 and 336 h) responses of *H. reidi* exposed to crude oil (10 ml L^{-1}) and observed that all treatments induced significantly higher DNA damage as compared to their respective control groups, with a peak damage index (DI) reached at 168 h and at 96 h for micronuclei frequency. In addition, these authors observed the ability of fish to recover (168 h without contamination) after 96 h of exposure to the crude oil using the same genotoxic approaches. As already mentioned, in opposition to most chemical compounds, the biotransformation pathway of the PAHs present in crude oil or petroleum derivatives transforms these metabolites into highly reactive toxic intermediates,[14] named reactive oxygen species (ROS), which cause severe oxidative damage to DNA molecules if the speed of production of these free radicals exceeds the capacity of the intra- and extra-cellular defense mechanisms.[143] Thus, the increase in DNA damage in *H. reidi* erythrocytes increases with the time of exposure to the crude oil; this is related to the metabolization of the oil, which generates ROS and causes damage to the DNA of the exposed animals, resulting in the detected lesions.

The recovery experiment demonstrated that a 168 h recovery period following 96 h exposure to crude oil was sufficient to repair the observed genotoxic damages. Considering the presence of potentially genotoxic agents in this pollutant, this recovery could be explained by the repair by excision of the nucleotides of adducts that could have been introduced by some of the petroleum components. This mechanism of repair is known to involve the retrieval of a DNA fragment from around the lesion and constitutes the main repair mechanism for damage induced by PAHs.[144]

Concerning the biochemical responses Delunardo *et al.*[33] observed that GST levels decreased in a dose-dependent manner. Inhibition of GST indicates the inability to efficiently conjugate PAHs that are present in crude oil using the glutathione pathway or an alternative pathway, such as uridine diphosphate glucuronyltransferase (UDP-GT) or sulfotransferase conjugation.[145] Another possible explanation for these findings is that PAHs inhibited the substrate (reduced glutathione, GSH) as this enzyme serves as a substrate for conjugation with electrophilic intermediates under the catalytic action of GST.[146] Variations of GSH levels result in variations of GST levels. Severe oxidative stress may suppress GSH levels (owing to a loss of adaptive mechanisms) and cause oxidation of GSH to GSSG with a consequent decrease in GST activities.[147] These results indicate that care should be taken when using this enzyme as a biomarker in ecosystems that are polluted with PAHs because PAHs might exert opposite effects on GST activity.

Negreiros *et al.*[36] and Delunardo *et al.*[39] also found gill histopathological alterations, as follows: hypertrophy of the pavement cells and proliferation of the lamellar epithelium, vascular congestion, epithelial detachment and fusion of the lamellae (Figure 12.3). The damages observed were classified as defense (inflammatory) or compensatory (cell proliferation) responses to the actions of xenobiotic substances present in the water and not characterized as representing a survival risk to individuals because they were occasional and reversible. However, the individuals showing these alterations become more vulnerable in situations such as escape or food capture because the changes impose respiratory perturbations, which reduce blood oxygenation and cause a hydroelectrolytic imbalance.[148,149]

In summary, the results obtained in these studies show clearly that crude oil compounds and derivatives induce alterations in *H. reidi* at the biochemical, cellular and tissue levels. Hopefully, these results may stimulate future experiments on this subject, which will provide a clearer understanding of the effects of oil spills on marine Neotropical fish species.

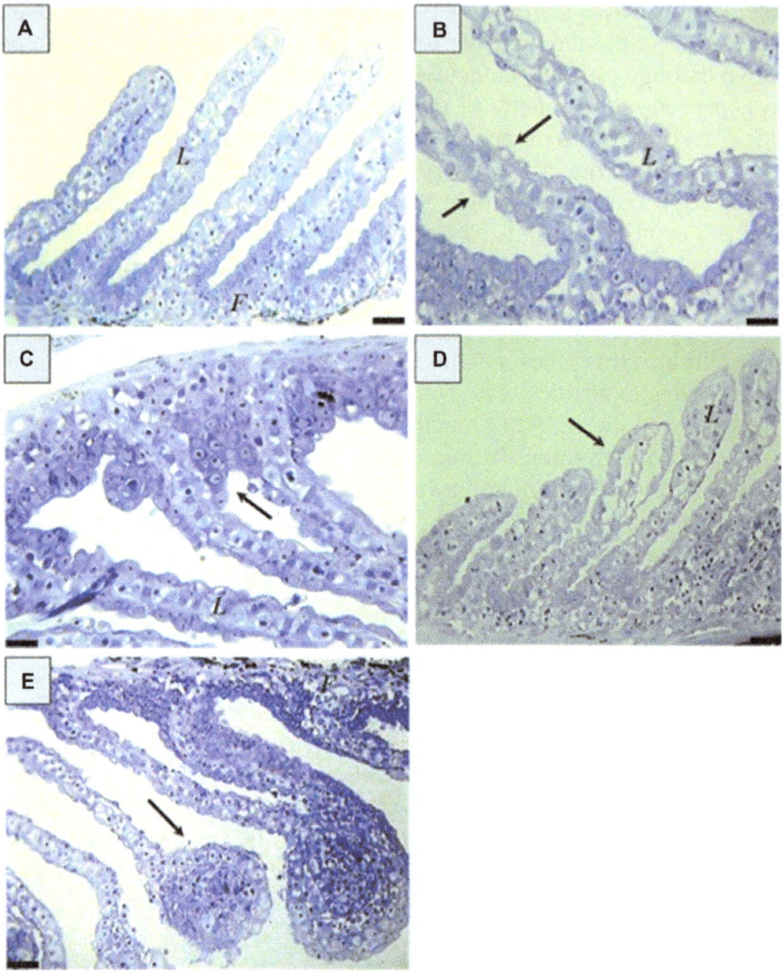

Figure 12.3 Gills of *Hippocampus reidi*. (A) Control fish showing normal structure of the gill. Fish exposed to crude oil showing (B) hypertrophy of the pavement cells (arrow), (C) lamellar epithelial cell hyperplasia (arrow), (D) epithelial detachment (arrow) and (E) vascular congestion (arrow). F: filament; L: lamella. Scale bar = 10 μm.

Besides, it is demonstrated that *H. reidi* might be an excellent bioindicator. This Neotropical fish has low mobility and does not migrate from contaminated areas. Therefore, this fish is likely to be exposed to oil spills. Seahorses in general may have great potential as sentinel species to monitor marine environments.

Acknowledgements

This work was supported by a research grant from FAPES (Proc. #61902861). F. A. C. Delunardo received a PhD fellowship from the Brazilian FAPES.

H. Sadauskas-Henrique received a post PhD fellowship from the Brazilian CNPq. S. B. Mota received a master's fellowship from the Brazilian CAPES. V. M. F. Almeida-Val received a research fellowship from the Brazilian CNPq. Additionally, the authors thank the Brazilian National Research Council (CNPq) and Amazonas State Research Foundation (FAPEAM) to ADAPTA Project (Environmental Adaptations of Aquatic Organisms) for their support of some of the analysis addressed here.

References

1. C. R. Costa, P. Olivi, C. M. R. Botta and E. L. G. Espindola, A toxicidade em ambientes aquáticos: discussão e métodos de avaliação, *Quim. Nova*, 2008, **31**(7), 1820–1830.
2. D. S. L. Moraes and B. Q. Jordão, Degradação de recursos hídricos e seus efeitos sobre a saúde humana, *Rev. Saude Pública*, 2002, **36**, 370–374.
3. L. Bergman and D. M. Pugh, *Environmental Toxicology, Economics and Institutions: the Atrazine Case Study*, Kluwer Academic Publishers, 1994, vol. 8, pp. 1–89.
4. F. Peres, J. C. Moreira and G. S. Dubois, in *Agrotóxicos, saúde e ambiente: uma introdução ao tema*, ed. F. Peres and J. C. Moreira, Editora FIOCRUZ, Rio de Janeiro, 2003, pp. 21–41.
5. T. P. Vanzella, C. B. R. Martinez and I. M. S. Cólus, Genotoxic and mutagenic effects of diesel oil water soluble fraction on a neotropical fish species, *Mutat. Res.*, 2007, **631**, 36–43.
6. T. Çavas and S. Ergene-Gözükara, Induction of micronuclei and nuclear abnormalities in *Oreochromis niloticus* following exposure to petroleum refinery and chromium processing plant effluents, *Aquat. Toxicol.*, 2005, **74**, 264–271.
7. S. T. Matsumoto, M. S. Mantovani, M. I. A. Malaguttii, A. L. Dias, I. C. Fonseca and M. A. Marin-Morales, Genotoxicity and mutagenicity of water contaminated with tannery effluents, as evaluated by the micronucleus test and comet assay using the fish *Oreochromis niloticus* and chromosome aberrations in onion root-tips, *Genet. Mol. Biol.*, 2006, **29**, 148–158.
8. C. Torres, de Lemos, P. M. Rodel, N. G. Terra, N. C. D. Oliveira and B. Erdtmann, River water genotoxicity evaluation using micronucleus assay in fish erythrocytes, *Ecotoxicol. Environ. Saf.*, 2007, **66**, 391–401.
9. A. L. Val, V. M. F. Almeida-Val and A. R. Chippari-Gomes, in *Hypoxia and Petroleum: Extreme Challenges for Fish of the Amazon*, Environmental Protection Agency, Ecosystems Research Division, Athens, Georgia, USA, 2004, vol. 600, pp. 227–241.
10. E. Aas, T. Baussant, L. Balk, B. Liewenborg and O. K. Andersen, PAH metabolites in bile, cytochrome P4501A and DNA adducts as environmental risk parameters for chronic oil exposure: a laboratory experiment with Atlantic cod, *Aquat. Toxicol.*, 2000, **51**, 241–258.

11. A. Tuvikene, Responses of fish to polycyclic aromatic hydrocarbons (PAHs), *Ann. Zool. Fenn.*, 1995, **32**, 295–309.
12. P. Meador, E. Casillas, C. A. Sloan and U. Varanasi, Bioaccumulation of polycyclic aromatic hydrocarbons by marine organisms, *Mar. Ecol.: Prog. Ser.*, 1995, **123**, 107–124.
13. R. F. Lee and S. Steinert, Use of the single cell gel electrophoresis/comet assay for detecting DNA damage in aquatic (marine and freshwater) animals, *Mutat. Res.*, 2003, **544**, 43–64.
14. R. van der Oost, J. Beyer and N. P. E. Vermulen, Bioaccumulation of polycyclic aromatic hydrocarbons by marine organisms, *Environ. Toxicol. Pharmacol.*, 2003, **13**, 57–149.
15. P. Perrichon, F. Akcha, K. Le Menach, M. Goubeau, H. Budzinski, X. Coustin and P. Bustamante, Parental trophic exposure to three aromatic fractions of polycyclic aromatic hydrocarbons in the zebrafish: Consequences for the offspring, *Sci. Total Environ.*, 2015, **524–525**, 52–62.
16. M. Pacheco and M. A. Santos, Biotransformation, endocrine, and genetic responses of *Anguilla anguilla L.* to petroleum distillate products and environmentally contaminated waters, *Environ. Int.*, 2001, **26**, 149–155.
17. J. Zhang, X. Wang, H. Guo, J. Wu and Y. Xue, Biotransformation, endocrine, and genetic responses of *Anguilla anguilla L.* to petroleum distillate products and environmentally contaminated waters, *Ecotoxicol. Environ. Saf.*, 2004, **58**(1), 110–116.
18. E. M. Jönsson, C. Carlsson, R. W. Smith and P. Pärt, Cytochrome P4501A induction in rainbow trout gills and liver following exposure to waterborne indigo, benzo[a]pyrene and 3,3_,4,4_,5-pentachlorobiphenyl, *Aquat. Toxicol.*, 2006, **79**, 78–86.
19. E. M. Jönsson, B. Brunström and I. Brandt, The zebrafish gill model: induction of CYP1A, EROD and PAH adduct formation, *Aquat. Toxicol.*, 2009, **91**, 62–70.
20. Y. Sun, Y. Yin, J. Zhang and Y. Xu, Hydroxylradical generation and oxidative stress in *Carassius auratus* liver, exposed to pyrene, *Ecotoxicol. Environ. Saf.*, 2008, **71**(2), 446–453.
21. C. A. Silva, C. A. Oliveira Ribeiro, A. Katsumiti, M. L. P. Araujo, E. M. Zandona, G. P. Costa Silva, J. Maschio, H. Roche and H. C. Silva de Assis, Evaluation of waterborne exposure to oil spill 5 years after an accident in southern Brazil, *Ecotoxicol. Environ. Saf.*, 2009, **72**, 400–409.
22. A. Trisciani, I. Corsi, C. D. Torre, G. Perra and S. Focardi, Hepatic biotransformation genes and enzymes and PAH metabolites in bile of common sole (*Solea solea*, Linnaeus, 1758) from an oil-contaminated site in the Mediterranean Sea: a field study, *Mar. Pollut. Bull.*, 2011, **62**(4), 806–814.
23. L. G. Luís and L. Guilhermino, Short-term toxic effects of naphthalene and pyrene on the common prawn (*Palaemon serratus*) assessed by a

multi-parameter laboratorial approach: mechanisms of toxicity and impairment of individual fitness, *Biomarkers*, 2012, **17**, 275–285.
24. T. F. Holth, D. P. Eidsvoll, E. Farmen, M. B. Sanders, C. Martínez-Gómez, H. Budzinski, T. Burgeot, L. Guilhermino and K. Hylland, Effects of water accommodated fractions of crude oils and diesel on a suite of biomarkers in Atlantic cod (*Gadus morhua*), *Aquat. Toxicol.*, 2014, **154**, 240–252.
25. F. L. Bettim, G. L. Galvan, M. M. Cestari, C. I. Yamamoto and H. C. S. Assis, Biochemical responses in freshwater fish after exposure to water-soluble fraction of gasoline, *Chemosphere*, 2016, **144**, 1467–1474.
26. M. Afifi, A. Alkaladi, O. A. A. Zinada and M. Couderchet, Alteration in antioxidant genes expression in some fish caught from Jeddah and Yanbu coast as a bio-indicator of oil hydrocarbons pollution, *Saudi J. Biol. Sci.*, 2015, DOI: 10.1016/j.sjbs.2015.06.014.
27. M. Teles, M. Pacheco and M. A. Santos, *Anguilla anguilla* L. liver ethoxyresorufin O-demethylation, glutathione S-transferase, erythrocytic nuclear abnormalities, and endocrine responses to naphthalene and b-naphthoflavone, *Ecotoxicol. Environ. Saf.*, 2003, **55**, 291–298.
28. W. H. L. Siu, J. Cao, R. W. Jack, R. S. S. Wu, B. J. Richardson, L. Xu and P. K. S. Lam, Application of the comet and micronucleus assays to the detection of B[a]P genotoxicity in haemocytes of the green-lipped mussel (*Perna viridis*), *Aquat. Toxicol.*, 2004, **66**, 381–392.
29. IARC-International Agency for Research on Cancer, *Some non-heterocyclic polycyclic aromatic hydrocarbons and some related exposures*, IARC Monographs on the Evaluation of Carcinogenic Risks to Humans, Lyon, France, 2010, **92**.
30. N. Wessel, R. Santos, D. Menard, K. Le Menach, V. Buchet, N. Lebayon, V. Loizeau, T. Burgeot, H. Budzinski and F. Akcha, Relationship between PAH biotransformation as measured by biliary metabolites and EROD activity, and genotoxicity in juveniles of sole (*Solea solea*), *Mar. Environ. Res.*, 2010, **69**, 71–73.
31. N. Wessel, R. Santos, D. Menard, K. Le Menach, V. Buchet, N. Lebayon, V. Loizeau, T. Burgeot, H. Budzinski and F. Akcha, Genotoxic and enzymatic effects of fluoranthene in microsomes and freshly isolated hepatocytes from sole (*Solea solea*), *Mar. Environ. Res.*, 2010, **69**, S71–S73.
32. M. Le Dû-Lacoste, F. Akcha, M. H. Dévier, B. Morin, T. Burgeot and H. Budzinski, Comparative study of different exposure routes on the biotransformation and genotoxicity of PAHs in the flatfish species, *Scophthalmus maximus*, *Environ. Sci. Pollut. Res.*, 2013, **20**, 690–707.
33. F. A. C. Delunardo, L. R. Carvalho, B. F. Silva, M. Galão, A. L. Val and A. R. Chippari-Gomes, Seahorse (*Hippocampus reidi*) as a bioindicator of crude oil exposure, *Ecotoxicol. Environ. Saf.*, 2015, **117**, 28–33.
34. F. M. Akaishi, H. C. Silva de Assis, S. C. G. Jakobi, D. R. Eiras-Stofella, S. D. St-Jean, S. C. Courtenay, E. F. Lima, A. L. R. Wagener, A. L. Scofield and C. A. Ribeiro, Morphological and neurotoxicological findings in

tropical freshwater fish (*Astyanax* sp.) after water waterborne and acute exposure to water soluble fraction (WSF) of crude oil, *Arch. Environ. Contam. Toxicol.*, 2004, **46**, 244–253.
35. J. D. Simonato, C. L. B. Guedes and C. B. R. Martinez, Biochemical, physiological, and histological changes in the neotropical fish *Prochilodus lineatus* exposed to diesel oil, *Ecotoxicol. Environ. Saf.*, 2008, **69**, 112–120.
36. L. A. Negreiros, B. F. Silva, M. G. Paulino, M. N. Fernandes and A. R. Chippari-Gomes, Effects of hypoxia and petroleum on the genotoxic and morphological parameters of *Hippocampus reidi*, *Comp. Biochem. Physiol., Part C: Toxicol. Pharmacol.*, 2011, **153**, 408–414.
37. R. Agamy, Impact of laboratory exposure to light Arabian crude oil, dispersed oil and dispersant on the gills of the juvenile brown spotted grouper (*Epinephelus chlorostigma*): A histopathological study, *Mar. Environ. Res.*, 2013, **86**, 46–55.
38. R. Agamy, Sub chronic exposure to crude oil, dispersed oil and dispersant induces histopathological alterations in the gills of the juvenile rabbit fish (*Siganus canaliculatus*), *Ecotoxicol. Environ. Saf.*, 2013, **92**, 180–190.
39. F. A. C. Delunardo, B. F. Silva, M. G. Paulino, M. N. Fernandes and A. R. Chippari-Gomes, Genotoxic and morphological damage in *Hippocampus reidi* exposed to crude oil, *Ecotoxicol. Environ. Saf.*, 2013, **87**, 1–9.
40. N. J. Brown-Peterson, M. Krasnec, R. Takeshita, C. N. Ryan, K. J. Griffitt, C. Lay, G. D. Mayer, K. M. Bayha, W. E. Hawkins, I. Lipton, J. Morris and R. J. Griffitt, Genotoxic and morphological damage in *Hippocampus reidi* exposed to crude oil, *Aquat. Toxicol.*, 2015, **165**, 197–209.
41. R. L. Cooper and R. J. Kavlock, Endocrine disruptors and reproductive development: a weight-of-evidence overview, *J. Endocrinol.*, 1997, **152**, 159–166.
42. P. R. R. Monteiro, M. A. Reis-Henriques and J. Coimbra, Plasma steroid levels in female flounder (*Platichthys flesus*) after chronic dietary exposure to single polycyclic aromatic hydrocarbons, *Mar. Environ. Res.*, 2000, **49**, 453–467.
43. P. R. R. Monteiro, M. A. Reis-Henriques and J. Coimbra, Polycyclic aromatic hydrocarbons inhibit in vitro ovarian steroidogenesis in the flounder (*Platichthys flesus* L.), *Aquat. Toxicol.*, 2000, **48**, 549–559.
44. C. A. Pollino, E. Georgiades and D. A. Holdway, Physiological changes in reproductively active rainbowfish (*Melanotaenia fluviatilis*) following exposure to naphthalene, *Ecotoxicol. Environ. Saf.*, 2009, **72**, 1265–1270.
45. A. Hawliczek, B. Nota, P. Cenijn, J. Kamstra, B. Pieterse, R. Winter, K. Winkens, H. Hollert, H. Segner and J. Legler, Developmental toxicity and endocrine disrupting potency of 4-azapyrene, benzo[b]fluorene and retene in the zebrafish *Danio rerio*, *Reprod. Toxicol.*, 2012, **33**, 213–223.
46. L. L. Johnson, E. Casillas, T. K. Collier, B. B. McCain and U. Varanasi, Contaminant effects on ovarian development in English sole (*Parophrys*

vetulus) from Puget Sound, Washington, *Can. J. Fish. Aquat. Sci.*, 1988, **45**, 2133–2146.
47. L. L. Johnson, T. K. Collier and J. E. Stein, An analysis in support of sediment quality thresholds for polycyclic aromatic hydrocarbons (PAHs) to protect estuarine fish, *Aquat. Conserv.*, 2002, **12**, 517–538.
48. E. Casillas, D. Misitano, L. L. Johnson, L. D. Rhodes, T. K. Collier, J. E. Stein, B. B. Mccain and U. Varanasi, Inducibility of spawning and reproductive success of female English sole (*Parophrys vetulus*) from urban and nonurban areas of pugetsound, Washington, *Mar. Environ. Res.*, 1991, **31**, 99–122.
49. P. Thomas and L. Budiantara, Reproductive life history stages sensitive to oil and naphthalene in Atlantic croaker, *Mar. Environ. Res.*, 1995, **39**, 147–150.
50. S. A. Buratini and A. Brandelli, in *Bioacumulação*, ed. P. A. Zagatto and E. Bertoletti, 2006, Rima, São Carlos, SP, Brasil, pp. 55–88.
51. W. Stum and J. J. Morgan, *Aquatic Chemistry: An Introduction Emphasizing Chemical equilibria in Natural Waters*, 1981, John Wiley & Sons, 2nd edn.
52. W. G. Sunda and P. J. Hansen, in *Chemical Speciation of Copper in River Water: Effect of Total Copper, pH, Carbonate, and Dissolved Organic Matter*, ed. E. A. Jenne, ACS Symposium Series 93, Washington DC, 1979, pp. 147–180.
53. A. Bianchini, S. E. Martins and M. B. Jorge, *O Modelo do Ligante Biótico e suas Aplicações em Ecotoxicologia*, Apostila, Rio Grande do Sul, 2009.
54. A. Y. O. Matsuo, B. R. Woodin, C. M. Reddy, A. L. Val and J. J. Stegeman, Humic substances and crude oil induce cytochrome P450 1A expression in the Amazonian fish species *Colossoma macropomum* (Tambaqui), *Environ. Sci. Technol.*, 2006, **40**, 2851–2858.
55. H. Sadauskas-Henrique, R. M. Duarte and V. M. F. Almeida-Val, Influence of the natural Rio Negro water on the toxicological effects of a crude oil and its chemical dispersion to the Amazonian fish *Colossoma macropomum*, *Environ. Sci. Pollut. Res.*, 2016, **23**(19), 19764–19775.
56. H. N. Yang and H. C. Chen, Uptake and Elimination of cadmium by Japanese eel, *Anguilla japonica*, at various temperatures, *Bull. Environ. Contam. Toxicol.*, 1996, **56**, 670–676.
57. L. Vergauwen, A. Hagenaars, R. Blust and D. Knapen, Temperature dependence of long-term cadmium toxicity in the zebrafish is not explained by liver oxidative stress: evidence from transcript expression to physiology, *Aquat. Toxicol.*, 2013, **126**, 52–62.
58. M. Abdel-Tawwab and M. Wafeek, Influence of water temperature and waterborne cadmium toxicity on growth performance and metallothionein-cadmium distribution in different organs of Nile tilapia, *Oreochromis niloticus* (L.), *J. Therm. Biol.*, 2014, **45**, 157–162.
59. J. Grasset, E. Ollivier, B. Bougas, G. Yannic, P. G. C. Campbell, L. Bernatchez and P. Couture, Combined effects of temperature

changes and metal contamination at different levels of biological organization in yellow perch, *Aquat. Toxicol.*, 2016, **177**, 324–332.
60. W. R. Barrionuevo and W. W. Burggren, O_2 Consumption and Heart Rate in Developing Zebrafish (*Danio rerio*): Influence of Temperature and Ambient O_2, *Am. J. Physiol.*, 1999, **276**, 505–513.
61. E. H. W. Heugens, A. J. Hendriks, T. Dekker, N. M. Van Straalen and W. Admiraal, A review of the effects of multiple stressors on aquatic organisms and analysis of uncertainty factors for use in risk assessment, *Crit. Rev. Toxicol.*, 2001, **31**, 247–284.
62. R. Osterauer and H. Köhler, Temperature-dependent effects of the pesticides thiacloprid and diazinon on the embryonic development of zebrafish (*Danio rerio*), *Aquat. Toxicol.*, 2008, **86**, 485–494.
63. D. L. Madeira, H. N. Narciso, C. Cabral and M. S. Vinagre, Influence of temperature in thermal and oxidative stress responses in estuarine fish, *Comp. Biochem. Physiol., Part A: Mol. Integr. Physiol.*, 2013, **166**, 237–243.
64. A. J. Li, P. T. Y. Leung, V. W. W. Bao, G. C. S. Lui and K. M. Y. Leung, Temperature-dependent physiological and biochemical responses of the marine medaka *Oryzias melastigma* with consideration of both low and high thermal extremes, *J. Therm. Biol.*, 2014, **36**, 116–123.
65. C. H. Cheng, F. F. Yang, S. A. Liao, Y. T. Miao, C. X. Ye, A. L. Wang, J. W. Tan and X. Y. Chen, High temperature induces apoptosis and oxidative stress in pufferfish (*Takifugu obscurus*) blood cells, *J. Therm. Biol.*, 2015, **53**, 172–179.
66. M. Daufresne, K. Lengfellner and U. Sommer, Global warming benefits the small in aquatic ecosystems, *Proc. Natl. Acad. Sci. U. S. A.*, 2009, **106**, 12788–12793.
67. IPCC, Intergovernmental Panel on Climate Change, *Fifth Assessment Report on Climate Change 2013: The Physical Science Basis*, Final Draft Underlying Scientific-Technical Assessment. Working group 1, Geneva.
68. C. W. Matson, A. R. Timme-Laragy and R. T. Di Giulio, Fluoranthene, but not benzo[a]pyrene, interacts with hypoxia resulting in pericardial effusion and lordosis in developing zebrafish, *Chemosphere*, 2008, **74**, 149–154.
69. C. R. Fleming, S. M. Billiard and R. T. Di Giulio, Hypoxia inhibits induction of aryl hydrocarbon receptor activity in topminnow hepatocarcinoma cells in an ARNTdependent manner, *Comp. Biochem. Physiol., Part C: Toxicol. Pharmacol.*, 2009, **150**, 383–389.
70. S. Dasgupta, R. T. DiGiulio, B. D. Drollette, D. L. Plata, B. J. Brownawell and A. E. McElroy, Hypoxia depresses CYP1A induction and enhances DNA damage, but has minimal effects on antioxidant responses in sheepshead minnow (*Cyprinodon variegatus*) larvae exposed to dispersed crude oil, *Aquat. Toxicol.*, 2016, **177**, 250–260.
71. N. M. Van Straalen, Ecotoxicology becomes stress ecology, *Environ. Sci. Technol.*, 2003, **37**, 324–330.

72. H. Kappus, Oxidative stress in chemical toxicity, *Arch. Toxicol.*, 1987, **60**, 144–149.
73. P. Lemaire, L. Forlin al and D. R. Livingstone, Responses of hepatic biotransformation and antioxidant enzymes to CYP1A-inducers (3-methylcholanthrene,[beta]-naphthoflavone) in sea bass (*Dicentrarchus labrax*), dab (*Limanda limanda*) and rainbow trout (*Oncorhynchus mykiss*), *Aquat. Toxicol.*, 1996, **36**, 141–160.
74. N. P. Singh, M. T. McCoy, R. R. Tice and E. L. Schneider, A simple technique for quantification of low levels of DNA damage in individual cells, *Exp. Cell Res.*, 1988, **175**, 184–191.
75. J. M. Mates, Effects of antioxidant enzymes in the molecular control of reactive oxygen species, *Toxicology*, 2000, **153**, 83–104.
76. Y. Sun, H. Yu, J. Zhang, Y. Yin, H. Shi and X. Wang, Bioaccumulation, depuration and oxidative stress in fish *Carassius auratus* under phenanthrene exposure, *Chemosphere*, 2006, **63**, 1319–1327.
77. J. Wen and L. Pan, Short-term exposure to benzo[a]pyrene causes oxidative damage and affects haemolymph steroid levels in female crab *Portunus trituberculatus*, *Environ. Pollut.*, 2016, **208**, 486–494.
78. R. M. Duarte, D. S. Smith, A. L. Val and C. M. Wood, Dissolved organic carbon from the upper Rio Negro protects zebrafish (*Danio rerio*) against ionoregulatory disturbances caused by low pH exposure, *Sci. Rep.*, 2016, **6**, 20377.
79. R. D. Wood, M. Mitchell, J. Sgouros and T. Lindahl, Human DNA repair genes, *Science*, 001, **291**, 1284–1289.
80. A. S. Fonseca, L. A. G. Magalhães, A. L. Mencalha, M. Geller and F. Paoli, Low intensity infrared laser affects expression of oxidative DNA repair genes in mitochondria and nucleus, *Laser Phys.*, 2014, **24**, 115605.
81. J. M. Monserrat, P. E. Martínez, L. Geracitano, L. L. Amado, C. M. Gaspar Martins, G. L. Leães Pinho, I. S. Chaves, M. Ferreira-Cravo, J. Ventura-Lima and A. Bianchini, Pollution biomarkers in estuarine animals: Critical review and new perspectives, *Comp. Biochem. Physiol., Part C: Toxicol. Pharmacol.*, 2007, **146**, 221–234.
82. R. J. Law, V. J. Dawes, R. J. Woodhead and P. Matthiessen, Polycyclic aromatic hydrocarbons (PAH) in seawater around England and Wales, *Mar. Pollut. Bull.*, 1997, **34**, 306–322.
83. P. Perrichon, K. L. Menach, F. Akcha, J. Cachot, H. Budzink and P. Bustamante, Parental trophic exposure to three aromatic fractions of polycyclic aromatic hydrocarbons in the zebrafish: Consequences for the offspring, *Sci. Total Environ.*, 2016, **568**, 952–966.
84. L. Campos-Baca and C. C. Kohler, Aquaculture of *Colossoma macropomum* and Related Species in Latin America, *Am. Fish. Soc. Symp.*, 2005, **46**, 541–561.
85. C. A. Araújo-Lima and M. Goulding, *Os frutos do tambaqui: ecologia, conservação e cultivo na Amazônia*, Sociedade Civil Mamirauá MCT/CNPq, Tefé-AM, 1998.

86. A. C. B. Oliveira, Seasonality of energy sources of *Colossoma macropomum* in a floodplain lake in the Amazon – lake Camaleão, Amazonas, Brasil, *Fish. Manage. Ecol.*, 2006, **13**, 135–142.
87. K. Furch, in *The Amazon. Limnology and Landscape Ecology of a Mighty Tropical River and Its Basin*, eds H. Sioli and W. Junk, Dordrecht, 1984, pp. 167–199.
88. A. L. Val and V. M. F. Almeida-Val, *Fishes of the Amazon and Their Environment, Zoophysiology*, Springer Berlin Heidelberg, Berlin, Heidelberg, 1995.
89. I. Walker and P. A. Henderson, in *Physiology and Biochemistry of Fishes of the Amazon*, ed. A. L. Val, V. M. F. Almeida-Val and D. J. Randall, Instituto Nacional de Pesquisas da Amazônia, Manaus, 1996, pp. 7–22.
90. R. Gonzalez and R. W. Wilson, Patterns of ion regulation in acidophilic fish native to the ion-poor, acidic Rio Negro, *J. Fish Biol.*, 2001, **58**, 1680–1690.
91. R. J. Gonzalez, C. M. Wood, M. L. Patrick and A. L. Val, Diverse Strategies for Ion Regulation in Fish Collected from the Ion-Poor, Acidic Rio Negro, *Physiol. Biochem. Zool.*, 2002, **75**, 37–47.
92. C. M. Wood, R. W. Wilson, R. J. Gonzalez, M. L. Patrick, H. L. Bergman, A. Narahara and A. L. Val, Responses of an Amazonian Teleost, the Tambaqui (*Colossoma macropomum*), to Low pH in Extremely Soft Water, *Physiol. Zool.*, 2012, **71**, 658–670.
93. R. W. Wilson, C. M. Wood, R. J. Gonzalez, M. L. Patrick, H. L. Bergman, A. Narahara, A. L. Val and S. Diego, Ion and Acid-Base Balance in Three Species of Amazonian Fish during Gradual Acidification of Extremely Soft Water, *Physiol. Biochem. Zool.*, 1999, **72**, 277–285.
94. E. Braum and W. J. Junk, Morphological adaptation of two Amazonian characoids (Pisces) for surviving in oxygen efficient waters, *Int. Rev. Gesamten Hydrobiol.*, 1982, **67**, 869–886.
95. A. L. Val, V. M. F. Almeida-Val and D. J. Randall, in *Fish Physiology*, ed. A. L. Val, V. M. F. Almeida-Val and D. J. Randall, Academic Press, London,, 2005, p. 634.
96. A. L. Val, P. M. Fearnside and V. M. F. Almeida-Val, Environmental disturbances and fishes in the Amazon,*J. Fish Biol.*, 2016, **89**, 192–193.
97. R. M. Duarte, R. T. Honda and A. L. Val, Acute effects of chemically dispersed crude oil on gill ion regulation, plasma ion levels and haematological parameters in tambaqui (*Colossoma macropomum*), *Aquat. Toxicol.*, 2010, **97**, 134–141.
98. S. Braz-Mota, L. M. L. Fé, F. A. C. Delunardo, H. Sadauskas-Henrique, V. M. F. Almeida-Val and A. L. Val, Roundup® exposure promotes gills and liver impairments, DNA damage and inhibition of brain cholinergic activity in the Amazon teleost fish *Colossoma macropomum*, *Chemosphere*, 2015, **135**, 53–60.
99. W. S Almeida and N. M Souza, Coari: petróleo e sustentabilidade – um exemplo amazônico, *Desenvolv. Meio Ambient.*, 2008, **17**, 69–92.

100. A. L. Val and V. M. F. Almeida-Val, in *Biology of Tropical Fishes*, ed. A. L. Val and V. M. F. Almeida-Val, INPA, Manaus, 1999, p. 460.
101. United Nations Documents, 1987. Report of the World Commission on Environment and Development: Our Common Future.
102. C. Martínez-Gómez, J. A. Campillo, J. Benedicto, B. Fernández, J. Valdés, I. García and F. Sánchez, Monitoring biomarkers in fish (*Lepidorhombus boscii* and *Callionymus lyra*) from the northern Iberian shelf after the Prestige oil spill, *Mar. Pollut. Bull.*, 2006, **53**, 305–314.
103. J. Jung, M. Kim, U. H. Yim, S. Y. Ha, J. G. An, J. H. Won, G. M. Han, N. S. Kim, R. F. Addison and W. J. Shim, Biomarker responses in pelagic and benthic fish over 1 year following the Hebei Spirit oil spill (Taean, Korea), *Mar. Pollut. Bull.*, 2011, **62**, 1859–1866.
104. K. W. Nkpaa, M. O. Wegwu and E. B. Essien, Assessment of polycyclic aromatic hydrocarbons (PAHs) levels in two commercially important fish species from crude oil polluted waters of Ogonil and their carcinogenic health risks, *J. Environ. Earth Sci.*, 2013, **3**, 128–138.
105. H. Sadauskas-Henrique, R. M. Duarte, M. M. Gagnon and V. M. F. Almeida-Val, Validation of a suite of biomarkers of fish health in the tropical bioindicator Tambaqui (*Colossoma macropomum*), *Ecol. Indic.*, 2017, **73**, 443–451.
106. G. S. Silva, L. M. L. Fé, M. N. P. Silva and V. M. F. Almeida-Val, *Ras* oncogene and Hypoxia-inducible factor-1 alpha (hif-1α) expression in Amazon fish *Colossoma macropomum* (Cuvier, 1818) exposed to benzo[a]pyrene, *Genet. Mol. Biol.*, in press.
107. C. P. F. Oliveira, Biomarcadores moleculares, genotóxicos e enzimáticos de efeitos do petróleo em tambaqui (*Colossoma macropomum*): Subsídios para o monitoramento ambiental na Amazônia, PhD thesis, Universidade Federal do Amazonas – UFAM, 2010.
108. D. Kochhann, S. M. A. Brust, F. X. V. Domingos and A. L. Val, Linking hematological, biochemical, genotoxic, and behavioral responses to crude oil in the Amazon fish *Colossoma macropomum* (Cuvier, 1816), *Arch. Environ. Contam. Toxicol.*, 2013, **65**, 266–275.
109. R. Rinaldi, E. Eliasson, S. Swedmark and R. Morgenstern, Reactive intermediates and the dynamics of glutathione transferases, *Drug Metab. Dispos.*, 2002, **30**, 1053–1058.
110. A. L. S. Tim-Tim, F. Morgado, S. Moreira, R. Rangel, A. J. A. Nogueira, A. M. V. M. Soares and L. Guilhermino, Cholinesterase and glutathione S-transferase activities of three mollusc species from the NW Portuguese coast in relation to the "Prestige" oil spill, *Chemosphere*, 2009, **77**, 1465–1475.
111. J. D. Simonato, M. N. Fernandes and C. B. R. Martinez, Gasoline effects on biotransformation and antioxidant defenses of the freshwater fish *Prochilodus lineatus*, *Ecotoxicology*, 2011, **20**, 1400–1410.
112. D. Kochhann, M. M. Jardim, F. X. V. Domingos and A. L. Val, Biochemical and behavioral responses of the Amazonian fish *Colossoma*

macropomum to crude oil: The effect of oil layer on water surface, *Ecotoxicol. Environ. Saf.*, 2015, **111**, 32–41.

113. H. H. Cho, J. Choi, M. N. Goltz and J. W. Park, Combined effect of natural organic matter and surfactants on the apparent solubility of polycyclic aromatic hydrocarbons, *J. Environ. Qual.*, 2002, **31**, 275–280.
114. H. Lippold, U. Gottschalch and H. Kupsch, Joint influence of surfactants and humic matter on PAH solubility. Are mixed micelles formed? *Chemosphere*, 2008, **70**, 1979–1986.
115. C. E. W. Steinberg, A. Paul, S. Pflugmacher, T. Meinelt, R. Klöcking and C. Wiegand, Pure humic substances have the potential to act as xenobiotic chemicals - A review, *Fresenius Environ. Bull.*, 2003, **12**, 391–401.
116. C. E. W. Steinberg, S. Kamara, V. Y. Prokhotskaya, L. Manusadžianas, T. A. Karasyova, M. A. Timofeyev, Z. Jie, A. Paul, T. Meinelt, V. F. Farjalla, A. Y. O. Matsuo, B. K. Burnison and R. Menzel, Dissolved humic substances - Ecological driving forces from the individual to the ecosystem level? *Freshwater Biol.*, 2006, **51**, 1189–1210.
117. M. A. Timofeyev, Z. M. Shatilina, A. V. Kolesnichenko, D. S. Bedulina, V. V. Kolesnichenko, S. Pflugmacher and C. E. W. Steinberg, Natural organic matter (NOM) induces oxidative stress in freshwater amphipods *Gammarus lacustris sars* and *Gammarus tigrinus* (Sexton), *Sci. Total Environ.*, 2006, **366**, 673–681.
118. C. E. W. Steinberg, in *Advances in the Physicochemical Characterization of Dissolved Organic Matter: Impact on Natural and Engineered Systems*, ed. F. L. Rosario-Ortiz, 2014, pp. 1–28.
119. W. P. Duncan, Estresse metabólico e dano celular em *Colossoma macropomum* e *Hoplosternum littorale* expostos ao petróleo, PhD thesis, Universidade Federal do Amazonas – UFAM, 1998.
120. Food and Agriculture Organization of the United Nations – FAO, 2016. Introduced species fact sheet [www Document]. URL http://fao.org/fishery/introsp/3885/en. (accessed 5.20.16).
121. P. R. Teske and L. B. Beheregaray, Evolution of seahorses' upright posture was linked to Oligocene expansion of seagrass habitats, *Biol. Lett.*, 2009, **5**, 521–523.
122. L. Willadino, L. P. Souza-Santos, R. C. S. Mélo, A. P. Brito, N. C. S. Barros, C. M. V. Araújo-Castro, D. B. Galvão, A. Gouveia, C. G. Regis and R. O. Cavalli, Ingestion rate, survival and growth of newly released seahorse *Hippocampus reidi* fed exclusively on cultured live food items, *Aquaculture*, 2012, **360–361**, 10–16.
123. S. A Lourie, A. C. J Vincent and H. J Hall, *Seahorses: An Identification Guide to the World's Species and Their Conservation*, Project Seahorse, London, 1999, p. 214.
124. N. Zhang, X. Bin, C. Y. Mou, W. L. Yang, J. W. Wei, L. Liang, J. J. Zhu, J. C. Du, X. K. Wu, L. T. Ye, Z. Y. Fu, Y. Lu, J. H. Lin, Z. Z. Sun, J. Su, M. L. Dong and A. L. Xu, Molecular profile of the unique species of

traditional Chinese medicine, Chinese seahorse (*Hippocampus kuda* Bleeker), *FEBS Lett.*, 2003, **550**, 124–134.

125. H. J. Koldewey and K. M. Martin-Smith, A global review of seahorse aquaculture, *Aquaculture*, 2010, **302**, 131–152.
126. S. J. Foster and A. C. J. Vincent, Life history and ecology of seahorses: implications for conservation and management, *J. Fish Biol.*, 2004, **65**, 1–61.
127. I. L. Rosa, R. R. N. Alves, K. M. Bonifácio, J. S. Mourão, F. M. Osório, T. P. R. Oliveira and M. C. Nottingham, Fishers' knowledge and seahorse conservation in Brazil, *J. Ethnobiol. Ethnomed.*, 2005, **1**, 1–15.
128. M. M. Porter, E. Novitskaya, A. B. Castro-Cesena, M. A. Meyers and J. McKittrick, Highly deformable bones: unusual deformation mechanisms of seahorse armor, *Acta Biomater.*, 2013, **9**, 6763–6770.
129. D. Kleiber, L. K. Blight, I. R. Caldwell and A. C. J. Vincent, The importance of seahorses and pipefishes in the diet of marine animals, *Rev. Fish Biol. Fish.*, 2011, **21**, 205–223.
130. I. L. Rosa, T. L. Dias and J. K. Baum, Threatened fishes of the world: *Hippocampus reidi* Ginsburg, 1933 (Syngnathidae), *Environ. Biol. Fishes*, 2002, **64**, 378.
131. C. A. Santos, L. S. Novaes and L. C. Gomes, Genotoxic effects of the diesel water- soluble fraction on the seahorse *Hippocampus reidi* (Teleostei: Syngnathidae) during acute exposure, *Zoologia*, 2010, **27**, 956–960.
132. C. Gravato and M. A. Santos, Genotoxicity biomarkers' association with B(a)P biotransformation in *Dicentrarchus labrax* L, *Ecotoxicol. Environ. Saf.*, 2003, **55**, 352–358.
133. M. Teles, M. Pacheco and M. A. Santos, *Anguilla anguilla* L. liver ethoxyresorufin O-deethylation, glutathione S-transferase, erythrocytic nuclear abnormalities, and endocrine responses to naphthalene and beta-naphthoflavone, *Ecotoxicol. Environ. Saf.*, 2003, **55**, 291–298.
134. V. M. Andrade, T. R. O. Freitas and J. Silva, Comet assay using mullet (*Mugil* sp.) and sea catfish (*Netuma* sp.) erythrocytes for the detection of genotoxic pollutants in aquatic environment, *Mutat. Res.*, 2004, **560**, 57–67.
135. R. R. Otter, J. Meier, K. M. Kubach, J. M. Lazorchak and S. J. Klaine, The effects of urbanization on *Lepomis macrochirus* using the comet assay, *Ecotoxicol. Environ. Saf.*, 2012, **84**, 299–303.
136. M. G. Pacheco and M. A. Santos, Induction of micronuclei and nuclear abnormalities in the erythrocytes of *Anguilla anguilla* L. exposed either to cyclophosphamide or to bleached kraft pulp mill effluent, *Fresenius Environ. Bull.*, 1996, **5**, 746–751.
137. V. Bombail, D. Aw, E. Gordon and J. Batty, Applications of the comet and micronucleus assays to butterfish (*Pholis gunnellus*) erythrocytes from the Firth of Forth, Scotland, *Chemosphere*, 2005, **44**, 383–392.
138. D. Cavalcante, C. Martinez and S. Sofia, Genotoxic effects of Roundups on the fish *Prochilodus lineatus*, *Mutat. Res.*, 2008, **655**, 41–46.

139. M. Pacheco and M. A. Santos, Induction of EROD activity and genotoxic effects by Polycyclic aromatic hydrocarbons and resin acids on the juvenile eel (*Anguilla anguilla* L.), *Ecotoxicol. Environ. Saf.*, 1997, **38**, 252–259.
140. L. Serrano-Garcia and L. Montero-Montoya, Micronuclei and chromatine buds are related genotoxic events, *Environ. Mol. Mutagen.*, 2001, **38**, 38–45.
141. T. Çavaş and S. Ergene-Gözükara, Induction of micronuclei and nuclear abnormalities in *Oreochromis niloticus* following exposure to petroleum refinery and chromium processing plant effluents, *Aquat. Toxicol.*, 2005, **74**, 264–271.
142. S. Ergene, T. Çavaş, A. Celik, N. Koleli, F. Kaya and A. Karahan, Monitoring of nuclear abnormalities in peripheral erythrocytes of three fish species from the Goksu Delta (Turkey): genotoxic damage in relation to water pollution, *Ecotoxicology*, 2007, **16**, 385–391.
143. J. Cadet, T. Douki, D. Gasparutto and J. Luc Ravanat, Oxidative damage to DNA: formation, measurement and biochemical features, *Mutat. Res.*, 2003, **531**, 5–23.
144. D. A. Scicchitiano, Transcription past DNA adducts derived from polycyclic aromatic hydrocarbons, *Mutat. Res.*, 2005, **577**, 146–154.
145. C. Silva, C. Oliveira, C. Gravato and J. R. Almeida, Behavior and biomarkers as tools to assess the acute toxicity of benzo(a)pyrene in the common prawn *Palaemon serratus*, *Mar. Environ. Res.*, 2013, **90**, 39–46.
146. W. J. Habig, M. J. Pabst and W. B. Jacoby, Glutathione S-transferase, the first enzymatic step in mercapturic acid formation, *J. Biol. Chem.*, 1974, **249**, 7130–7139.
147. M. Oliva, L. M. Gonzalez De Canales, C. Gravato, L. Guilhermino and J. A. Perales, Biochemical effects and polycyclic aromatic hydrocarbons (PAHs) insenegal sole (*Solea senegalensis*) from a Huelva estuary (SW Spain), *Ecotoxicol. Environ. Saf.*, 2010, **73**, 1843–1851.
148. M. N. Fernandes and A. F. Mazon, in *Fish Adaptation*, ed. A. L. Val and B. G. Kapoor, Science Publishers, Enfield, 2003, p. 418.
149. J. D. Simonato, A. C. Albinati and C. B. R. Martinez, Effects of the water woluble fraction of diesel fuel oil on some functional parameters of the Neotropical freshwater fish *Prochilodus lineatus* Valenciennes, *Bull. Environ. Contam. Toxicol.*, 2006, **76**, 505–511.

CHAPTER 13

Blenniidae *and* Syngnathidae: *Partially Unexplored Reservoirs of Sentinel Species for Environmental Monitoring Studies*

M. M. SANTOS*[a,b] AND N. M. MONTEIRO*[c,d]

[a] CIMAR/CIIMAR, LA – Interdisciplinary Centre of Marine and Environmental Research, Group of Endocrine Disruptors and Emerging Contaminants, University of Porto, Av. General Norton de Matos, s/n, 4450 Matosinhos, Portugal; [b] FCUP – Department of Biology, Faculty of Sciences, University of Porto, Rua do Campo Alegre, 4169-007 Porto, Portugal; [c] CIBIO/InBIO – Research Centre in Biodiversity and Genetic Resources,University of Porto, rua Padre Armando Quintas, 4485-661 Vairão, Portugal; [d] CEBIMED – Faculty of Health Sciences, University Fernando Pessoa, rua Carlos da Maia 296, 4200-150 Porto, Portugal
*Email: santos@ciimar.up.pt; nm.monteiro@icloud.com

13.1 Introduction

In order to assure the good ecological and environmental status (GES) of its inland and marine waters, the European Union implemented, during the past decade, an ecosystem-based management approach. Two of the main implemented legal EU instruments are the Water Framework Directive

Issues in Toxicology No. 33
Ecotoxicology and Genotoxicology: Non-traditional Aquatic Models
Edited by Marcelo L. Larramendy

Published by the Royal Society of Chemistry, www.rsc.org

2000/60/EC (WFD) and the Marine Strategy Framework Directive 2008/56/EC (MSFD).[1] These documents aimed for a good ecological (WFD) and environmental status (MSFD) of all European water bodies (*i.e.*, inland surface, transitional, coastal and ground waters) and marine waters to be rapidly achieved during this decade, either by 2015 (WFD) or 2020 (MSFD). This narrow time frame forced EU member states to swiftly perform in-depth assessments and establish capable monitoring programs for European aquatic ecosystems. These directives, which rely on good knowledge of ecosystem functioning under very specific environmental conditions, highlighted the importance of monitoring both the chemical and ecological/environmental status of water bodies and marine waters to evaluate progress towards the established ambitious objectives.

Although the WFD and MSFD monitoring programs involve the combined use of both chemical and biological parameters, the use of biological responses, namely biomarkers and bioassays, is expected to contribute to the integration of chemical and biological data, providing an inclusive insight into the quality of a water body.[1–3] Understandably, particular concern is devoted to the group of stressors formed by a myriad of environmental chemicals. A wide range of contaminants is continuously introduced into the aquatic environment, mostly associated with industrial, agricultural and domestic waste run-offs.[4] These contaminants can cause a variety of sub-lethal effects on living organisms that are difficult to anticipate based on chemical analyses alone.[5,6] Therefore, during the last decade, the use of biological responses as early warning signals of ecosystem health has been successfully tested as an adequate methodology to study the sub-lethal effects of pollutants.[1,6] Biomarkers can provide important information to establish cause–effect relationships in ecological quality assessment, both within WFD and MSFD,[7,8] simultaneously allowing the unravel of the underlying mechanisms of disruption.[2,9,10]

Response to environmental stress within a biological system initially involves changes at the molecular/biochemical level that may subsequently lead to ecosystem impacts.[11,12] Ecological endpoints are sensitive to environmental stress only when deleterious effects altering the community structure have already taken place.[1] Hence, it is essential to use endpoints that rely on sub-lethal effects that are capable of working as early-warning signals.

Although mussels have been extensively used as sentinel species in a large number of monitoring programs, a single taxon cannot be representative of the overall diversity present in aquatic ecosystems. Therefore, sentinel organisms with distinct genetic make-ups, spanning numerous trophic levels, should be used and integrated within a holistic approach.[12] Hence, especially in marine areas, there is a tangible need for new sentinel species that can be used in the assessment of environmental stress, including anthropogenic pollution. Teleost fish are particularly interesting given the available knowledge on their genetic background, which shows vast homologies to that of mammals. Additionally, good knowledge on the biology and ecology of several teleosts is already available.

Here, we review the potentially useful information already available on the distribution, habitat, life history and reproduction of two families of teleost fish, *Blenniidae* and *Syngnathidae*, particularly focusing on marine and estuarine species that have already been shown to adequately respond to xenobiotics. Special emphasis is given to literature that presents evidence that supports the use of both families as promising sources of sentinel species for pollution monitoring. Gaps in current knowledge are also identified and potential research priorities established.

13.2 *Blenniidae*

The selection of appropriate sentinel species is a key aspect in the establishment of monitoring programs. A good sentinel species should display specific characteristics, such as: (i) abundant and easy to sample, (ii) a wide geographical distribution and restricted home range, and (iii) sensitivity to the stressors addressed in the monitoring study. Most species from the family *Blenniidae* seem to effortlessly meet these criteria. Among most of the teleost fish families whose species inhabit rocky intertidal and subtidal areas, blennids are probably the most exhaustively studied from a behavioural point of view. This family is highly abundant in temperate and tropical habitats, being present both in marine, estuarine and freshwater ecosystems. The *Blenniidae* family is divided into 57 genera, encompassing 387 species. Within blennids from temperate habitats, the shanny, *Lipophrys pholis* (Figure 13.1) is one of the most well studied species in north-eastern Atlantic intertidal ecosystems, and many works have been published dealing with its ecology, ontogeny, reproductive biology and behaviour.[13–16]

13.2.1 Habitat and Distribution

The shanny displays several of the characteristics required in a key sentinel species. Namely, it is an intertidal resident, easily found in rock-pools and

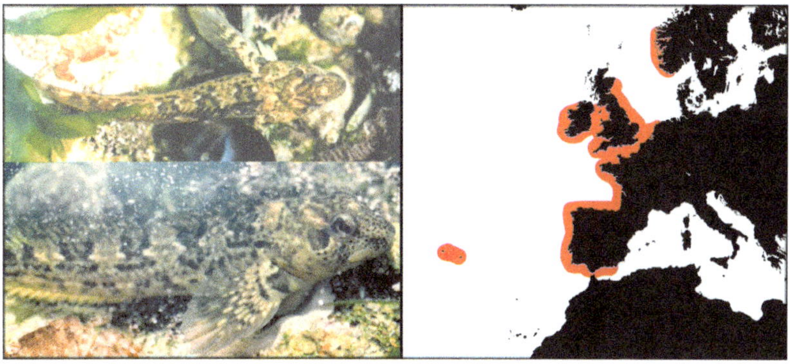

Figure 13.1 The biogeographical distribution of the shanny, *Lipophrys pholis* (adapted from ref. 17; images courtesy of Nuno Monteiro).

crevices, from Mauritania to Norway,[17] including the Azores Islands and parts of the Mediterranean (Figure 13.1), being locally abundant and extremely easy to catch. Its ecology and behaviour have been intensively studied as the shanny is a key species in the dynamics of the intertidal area in European rocky shore ecosystems.[13–15] After recruitment to the intertidal zone (approximately 3 months after hatching), the shanny displays a restricted home range during the rest of its life-cycle,[15,18,19] a fact that makes the blennid a worthy representative of local environmental conditions, including exposure to pollutants.

13.2.2 Reproduction

In recent years, a large set of data on the reproductive season, spermatogenesis, oogenesis and genital morphology of male and females was put forward in order to validate the use of the shanny as a sentinel in monitoring studies.[20–22]

The breeding season of *L. pholis* near the southern limit of distribution of the species (Portugal) occurs during cold-water periods, between November and May. In contrast, in more septentrional latitudes, the shanny experiences a contracted reproductive period, occurring mostly during summer months.[17] The basic pattern of oogenesis in *L. pholis* is similar to that already described for other blennids[23–25] as well as other teleosts.[26,27] Histological observations of the ovaries showed that all stages of ovarian germ cell development are represented during the entire year, at least along the Portuguese coast, suggesting that the shanny is an asynchronous spawner, with eggs being recruited in several batches during the breeding season. *L. pholis* ovarian development is also asynchronous, and seven ovarian germ cells have been described (oogonia, early and late perinuclear oocytes, cortical-alveolar oocytes, early vitellogenic oocytes, vitellogenic oocytes and spawning oocytes).[20] Along the Portuguese coast, early oogenesis occurs in May, mid-oogenesis in September, and spawning takes place especially from November to January. Gonad development of both males and females, described based on stereologic analysis, correlates well with the gonadosomatic index (GSI). In fact, this index seems to provide a good indication of the reproductive status of *L. pholis*, coinciding with the already described breeding season in Portuguese waters.[13,28]

Paralleling our observations on ovarian development, *L. pholis* testes were also observed to be asynchronously arranged, with various cell types observable during all seasons, although in different proportions.[21] According to the frequency of male gonad cell types, three stages of gonadal maturation were defined for populations along the Portuguese coast. Early spermatogenesis (May): the seminiferous tubules are mainly formed by primary spermatogonia cysts, with type A and B spermatogonial cells more evident during this stage. Mid spermatogenesis (September): testes in this stage gradually increased in size and the walls of the seminiferous tubules became thinner, while still showing germinal cysts containing all stages of cell development. The proportions of primary and secondary spermatocytes increase in comparison with type A and B spermatogonial cells. The spermatids and spermatozoa are predominantly in the late developing stages and spermatozoa accumulate in the central part of the lobules. Although the spermatogenic cell frequency indicates that the reproductive period of *L. pholis* extends from November to May, spermatogonia are seen within the seminiferous tubules throughout the entire year along the Portuguese coast.[21] Given the recurrent observation of spermatogonia, namely during the breeding season, it seems that *L. pholis* males are capable of multiple spawning episodes, as already observed for other *Blenniidae*.[21] Interestingly, Shackley and King[29] observed that *L. pholis* females are also capable of spawning several times during the breeding season. In *L. pholis* populations from higher latitudes, authors[23,29] described additional 'spent' and 'resting' periods. Since these stages are not observed in southern populations, it can be hypothesized that the extended breeding period of *L. pholis* near the species' southern limit of distribution prevents the occurrence of such stages since the gonads prepare for the next developmental phase right after the closure of the previous one. It can be hypothesised that this species presents reproductive investment strategies that vary according to latitude: a higher immediate investment in higher latitudes, where the breeding season is shorter, thus explaining the observed 'spent' and 'resting' periods, and a less pronounced, but more expanded, investment in the south.

Information on the genital morphology of male and female *L. pholis* was recently provided,[22] and it is now possible to distinguish males and females directly in the field, based on the genital morphology. The use of non-invasive sex determination procedures may be extremely useful not only for ecological studies but also as a proxy for the detection of environmental exposure to endocrine disruptive chemicals. In fact, in *Gobiidae*, a family of teleost fish also showing dimorphic genital papillae, the feminization of male urogenital papilla was observed in the sand goby *Pomatoschistus minutus* in areas with higher levels of environmental estrogens.[30]

The available detailed knowledge on the shanny's reproductive cycle, time of spawning, and gonad development constitutes relevant baseline information that can be used to detect the impact of both natural and anthropogenic stress. This information can, thus, potentiate the integration of *L. pholis* as a sentinel species in monitoring programs designed to evaluate chemical pollution effects in European marine ecosystems.

13.2.3 Use as Sentinel Species in Monitoring Studies

The blenny *L. pholis* was initially used as a sentinel species to monitor hydrocarbon pollution associated with the *Sea Empress* oil spill.[31–33] More recently, several studies validated the use of this blenny for pollution monitoring associated with hydrocarbon contamination and xenoestrogens. In fact, there is now evidence that the shanny is also highly responsive to a wide variety of pollutants, from those present in oil spills,[31–33] to neurotoxic compounds,[34] organic contaminants[35] or estrogenic chemicals.[36,37]

In order to assess the species' sensitivity to organic contaminants under field conditions, *L. pholis* responses were evaluated along the Portuguese coast at six sites reflecting different degrees of anthropogenic contamination.[35] The induction of two biomarkers extensively validated in the assessment of PAHs contamination, ethoxyresorufin-*O*-deethylase activity (EROD) and fluorescent aromatic compounds (FACs), was examined. In parallel, mussels were also collected at the same locations and levels of 16 PAHs and selected heavy metals were determined. The animals collected in the urban areas showed a significant induction of EROD and FACs (up to six-fold induction) when compared with the reference sites. Additionally, a positive correlation was observed between the biomarkers and PAH levels in mussel tissues. These results show that *L. pholis* is responsive to organic contaminants such as PAHs. The six-fold EROD induction between contaminated and reference sites is above the induction levels reported for other field studies using teleost fish, thus supporting the sensitivity of this species to PAH exposure.[33] The levels of FACs in *L. pholis* bile showed a positive correlation with EROD activity and PAH levels in mussel tissues, supporting the combined used of EROD and FACs in the assessment of organic contaminant pollution, particularly PAHs. In the scope of the same project, other authors[34] characterised acetylcholine (AChE), butyrylcholine (BChE) and propionylcholine (PrChE) esterases in *L. pholis* muscle, showing that AChE was predominant. The use of eserine sulphate and BW284c51 (0.64–800 μM), and iso-OMPA (0.08–16 mM), confirmed the presence of true cholinesterases (ChEs) as well as the presence of pseudocholinesterases. The field application of these markers in *L. pholis*, sampled in seven locations along the Portuguese coast, revealed that fish were likely to be affected by neurotoxic compounds. A significant depletion of AChE in animals collected at urban and industrialised sites, compared with those from reference locations, was observed.

In another study,[33] aimed at validating the use of *L. pholis* in pollution monitoring associated with petrogenic hydrocarbon contamination, a multi-biomarker approach study was carried out one week after a moderate oil spill from the waste treatment plant (WTP) of a major Portuguese refinery in the north of Portugal. Fish collected at 2 km from the accident displayed a significant induction of ethoxyresorufin-*O*-deethylase activity (EROD) and fluorescent aromatic compounds (FACs) in bile (up to a five-fold induction) in comparison with the pre-spill scenario. The findings correlated well with PAH accumulation in mussel tissues, which further validated the responsiveness of *L. pholis* to petrogenic hydrocarbon contamination.

13.2.4 Response to Genotoxic Compounds

The blenny *L. pholis* was used in two distinct geographic areas to monitor the genotoxic effects of oil spills. Following the *Sea Empress* oil spill in

Pembrokeshire (Wales), higher levels of DNA adducts were recorded in specimens collected from a site affected by the oil spill, in comparison with reference sites.[31,32] More recently, following a small-scale spill of WWTP in an oil refinery in northern Portugal, genotoxic effects were evaluated one week after the spill in fish collected in impacted and references site, using the erythrocytic nuclear abnormalities (ENA) assay. An induction of approximately 15% in ENA was observed in animals collected in the vicinity of the incident. In contrast, no significant differences were recorded at the reference site. Furthermore, the overall levels of ENA were at in the impacted site in comparison with the non-impacted area. For most marine species inhabiting relatively clean areas, the normal background ENA levels are in the range of 3–25‰,[38,39] which is similar to the observed levels in *L. pholis* erythrocytes along the Portuguese coast (25‰). The background level of ENA in the impacted site before the spill was already twice that recorded at the reference site, thus indicating that in the pre-spill period fish were already exposed to chronic levels of genotoxic agents. After the small-scale spill, a further 15% increase in ENA was observed in fish collected from the impacted area (Cabo do Mundo). In an attempt to validate, under laboratorial conditions, the ENA induction observed in the field,[33] adult *L. pholis* were exposed for eight days to the model PAH B[*a*]P. Although B[*a*]P has been described as a model carcinogen, *L. pholis* exposure did not increase ENA above the background levels of control animals. In fact, the absolute ENA level observed for the highest B[*a*]P concentration was approximately half of that recorded in field animals from the oil spill-impacted area in Portugal.[33] The authors suggested that this lack of ENA induction could be related to the exposure period, although laboratory data from other fish species indicate that ENA induction after a genotoxic insult usually takes place between 3 and 7 days.[40] Further studies should evaluate the dynamics of ENA induction in *L. pholis* following the insult of additional genotoxic chemicals under longer exposure periods.

13.2.5 Response to Estrogenic Chemicals

Natural and synthetic compounds mimicking endogenous estrogens have been reported over the past two decades to act as one of the main groups of endocrine-disrupting chemicals in aquatic ecosystems. Therefore, in addition to inland and transitional waters, the validation of sentinel species to monitor the impact of such chemicals in marine habits is still required.[30] In order to evaluate, in the field, *L. pholis* sensitivity to estrogenic chemicals,[36] males were collected from two sites reflecting different degrees of anthropogenic contamination. The vitellogenin II gene (VTGII) was initially isolated and its liver expression evaluated by RT-PCR from field samples. A significant induction of gene expression was observed in the specimens collected in the urban area, when compared to the reference site, suggesting exposure to ECs, despite the lack of testis-ova in the animals showing elevated copies of VTGII mRNA. In a follow-up study,[37] a sensitive real-time

PCR-based method was validated to study the dynamic response of VTG I and VTG II upon exposure to the model EC 17a-ethinylestradiol (EE2). The findings of the laboratory study indicate that *L. pholis* VTG I proved to be not only more inducible but also more sensitive to EE2 exposure than VTG II, within the same range of concentrations. VTG I gene induction was 475-fold higher than VTG II at 15 ng L^{-1} EE2, and 13-fold at 5 ng L^{-1} EE2. These levels of VTG I induction, following EE2 exposure, are comparable to those observed in other sentinel species as well as the model teleost zebra fish (*Danio rerio*). Combined, these studies indicate that, in the field, expression of VTG I in *L. pholis* should be preferentially used in the screening of EC exposure due to its higher induction.

13.3 Syngnathids

The term 'syngnathid' stems from a particular aspect of the cranial osteology of these small teleosts, namely the fusion of the jaws forming the characteristic tube-like snout ending in a reduced mouth aperture that severely constrains prey size.[41] Apart from feeding structures, syngnathids also characteristically display an armoured body, covered by bony plates arranged in a series of rings, that allows for increased protection against predation. Ironically, this protective body armour has long contributed to the commercial exploitation of syngnathids, namely seahorses, as morbid souvenirs.[42] Still, amongst the several peculiarities that allow for the quick identification of syngnathids, probably the most distinctive characteristic is the evolutionary result of a series of adaptations for extreme male parental care that ultimately resulted in the unique phenomenon of male pregnancy.

13.3.1 Habitat and Distribution

The family *Syngnathidae*, which includes seahorses (subfamily *Hippocampinae*), seadragons, pipehorses and pipefishes (subfamily *Syngnathinae*), currently comprises approximately 300 species.[43,44] Syngnathids are present in most shallow tropical and warm-temperate waters (Figure 13.2), presenting a higher diversity in the Indo-West Pacific.[45] While seahorses are almost exclusively marine (*Hippocampus capensis*, present in brackish estuarine waters, is a notable exception), there are several pipefish species that can be found in brackish water (*e.g. Syngnathus abaster*) or freshwater (*e.g. Microphis deocata*). The coastal distribution of many syngnathids and/or their regular presence in estuaries and coastal lagoons, especially during the breeding season, is particularly noteworthy from an ecotoxicological standpoint. While these specific environments perform a relevant ecological function, serving as spawning and nursery areas for numerous fish species,[46] they are also intensively impacted by chemical contamination.[47,48] For instance, seagrass meadows, a preferential habitat for several syngnathid species, have been repeatedly reported as directly impacted by exposure to endocrine disrupting chemicals.[49,50]

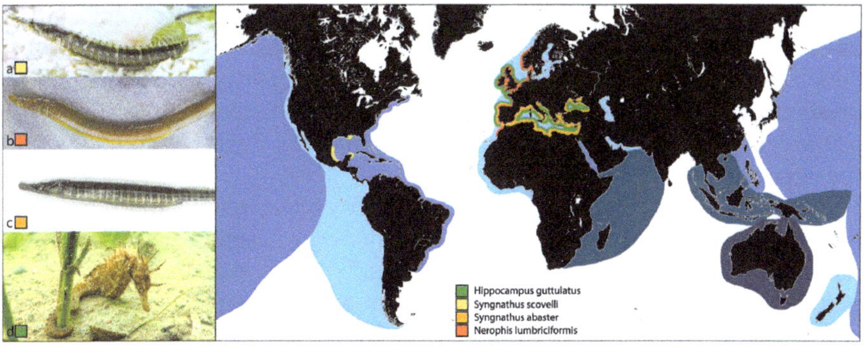

Figure 13.2 A simplified representation of marine syngnathid genera distribution (adapted from ref. 45, 52 and 91), with increasingly darker shades of blue indicating higher diversity areas. Solid coastal contour lines indicate the geographical distribution of the four species already used in ecotoxicological studies [yellow (a): *Syngnathus scovelli*, image courtesy of Emily Rose; red (b): *Nerophis lumbriciformis*, image courtesy of Nuno Monteiro; orange (c): *Syngnathus abaster*, image courtesy of Mário Cunha; green (d): *Hippocampus guttulatus*, image courtesy of the Hydro-ecology and Fisheries Biology Group at CCMAR).

13.3.2 Reproduction

In syngnathids, the female deposits oocytes into a specialized brooding structure, located either in the abdomen or in the tail of the male.[51] The phylogenetic relationships within syngnathids have been addressed using either the position and closure method of the incubating area[51] or, more recently, using mitochondrial DNA.[52,53] Somehow surprisingly, both approaches agree that the evolutionary radiation of this group was accompanied by a diversification of structures involved in parental care, from an ancestral pipefish, that probably presented a rather simple brooding structure similar to the simplest incubating area of the *Nerophinae*, to the sealed pouch of the *Hippocampinae*. A comparative anatomical study of three different brooding structures has been presented in ref. 54, describing the morphology and ultrastructure of the incubating surface of *Nerophis ophidion, S. abaster* and *H. hippocampus*. Despite all the immediately observable differences in the complexity of the brooding structures, implying distinct functions related to different reproductive strategies,[55] the common presence of a vascularized dermis hints at a close connection between fathers and developing young. The simple egg attachment in a male pipefish ancestor, that carried the eggs away from nest predators and the threat of sneaker fertilizations, was probably the starting point to a new set of evolutionary possibilities and constraints within the *Syngnathidae* family.[55] In marsupium-lacking pipefish, such as those from the genus *Nerophis* or *Entelurus*, the open pouch functions primarily as an adequate substratum for egg development, even though some modifications in the brood area epithelium (*e.g.*, convoluted microridges) are visible, suggesting extended contact between father and offspring. Other complex morphological and ultrastructural modifications are visible in species that present a brood pouch, such as those from the genus *Syngnathus* or *Hippocampus*, namely the presence of mitochondria-rich cells that are associated with an important osmoregulatory role,[56,57] and a more vascularized dermis contacting with the eggs.

Given the highly developed form of viviparity observed in eutherians, there has been a historical hesitancy in using the term pregnancy for the unique male viviparity observed in syngnathids.[58] The relative importance of patrotrophy during embryonic development remains partially undetermined as most of the energetic reserves derive from the yolk-rich oocytes produced by females. Nevertheless, research suggests that active nutrient supplementation by the father does occur.[59–61] Our own preliminary observations confirm the existence of nutrient supplementation by showing that, in the pipefish *S. typhle*, the developing brood gains weight from the onset to the end of pregnancy, an unexpected result if embryos depended solely on oocyte encapsulated resources (lecithotrophy).

Today, there seems to be a general consensus that the male's marsupium is much more than a passive structure that holds and camouflages the developing young and is functionally equivalent to the amniote (mammal and reptile) uterus. Evidence of osmoregulation,[57] waste transport and gas exchange,[62] as well as immune protection prior to the development of the innate immune system of the embryos,[63] confirms the large spectrum of activities performed by the male incubating structures during each pregnancy event. Unsurprisingly, the transcripts of the seahorse brood pouch revealed homology with genes of reproductive function in pregnant mammals, reptiles, and other live-bearing fish.[62]

Recently, a darker aspect of male pregnancy was uncovered with the observation of post-copulatory sexual selection mechanisms acting within the male's brood pouch.[64] In the Gulf pipefish, *S. scovelli*, males increase rates of offspring abortion in pregnancies from unattractive mothers, possibly as a way of saving valuable resources for future reproductive events. Our own research corroborates these findings, in a different pipefish species (*S. abaster*), by showing that males negatively impact ongoing pregnancies when exposed to a sexier female, by increasing the number of aborted embryos or by giving birth to significantly smaller offspring (Monteiro, unpublished data). Curiously, in the same pipefish, we were also able to show that, in normal conditions, males are able to decrease the brood's developmental heterogeneity.

Owing to their unique morphology, distribution, home range, behavior and mode of reproduction, syngnathids have become flagship species for the conservation of resident inshore fish communities.[65] Even though there is still little information on endocrine disruption effects on syngnathids, this family can be viewed as an interesting model for the assessment of ecosystem health due to several factors: (i) considerable life span, (ii) preferential preference for coastal habitats so more prone to anthropogenic pressure (*e.g.* pollutants), and (iii) a large body of literature addressing issues related with development,[66–68] reproductive behavior[69–73] and mating systems.[74–76]

13.3.3 Use as Sentinel Species in Monitoring Studies

As researchers working on non-model organisms increasingly depend on reliable complete genome sequences of the target species, the lack of available syngnathid genomes seemed to be the major obstacle to the wider usage of pipefish and seahorses as relevant ecotoxicological models. The public release of syngnathid genomes is near and researchers will soon have a wider set of genetic tools to explore.[77] Nevertheless, with the available tools and still using a reduced number of species, several authors have already started showing the potential importance of syngnathids in ecotoxicology.

Due to their geographic distribution (*e.g. S. abaster* and *N. lumbriciformis*, see Figure 13.2), low mobility, small home range and diet, many syngnathid species are prone to bioaccumulate xenobiotics present in the surrounding environment and thus serve as valuable sentinels for chemical pollution monitoring. A recent study[78] showed that the long-snouted seahorse, *H. guttulatus*, despite the low body lipid contents, bioaccumulates organochlorine compounds (OCPs), as well as heavy metals and polycyclic aromatic hydrocarbons (PAHs), mirroring pollutant concentrations in the sampled locations, thus reinforcing the potential value of seahorses, as well as other syngnathids with similar life histories, as sentinels for monitoring programs.

Over the last few decades, industrial development and urbanization along marine coasts contributed to an alarming accumulation of organic pollutants, some of them exhibiting endocrine disrupting activity. These compounds are known to potentially interfere with the internal homeostasis and hormone-controlled physiological processes of several different groups of organisms, namely fish.[79] A major fraction of the endocrine disrupting chemicals that enter the environment are chemicals that mimic natural estrogens, such as the pharmaceutical 17α-ethinylestradiol (EE2), which is relatively stable and binds to estrogen receptors with high affinity. Prolonged contact with EE2 is known to severely impact fish reproduction and, ultimately, population sustainability. A classic study, using the fathead minnow (*Pimephales promelas*), reported sudden reproductive shutdown after chronic exposure to low levels of EE2.[80] The deleterious impact of EE2 in syngnathids has also been repeatedly investigated, and confirmed in at least three pipefish (*S. abaster, S. scovelli* and *N. lumbriciformis*; see Figure 13.2). Nevertheless, the overwhelming interest in EE2 demonstrated by researchers working with syngnathids, when compared with other xenobiotics, is not always only linked to conservational concerns. In fact, EE2 effects have been especially tested in a restricted group of pipefish exhibiting a common characteristic: sex-role reversal. In sex-role-reversed species, where sexual selection acts more strongly on females than on males, females tend to evolve elaborate secondary sexual traits while males are generally the choosier sex. These species expectedly became exceptional models in the study of the evolution of sex differences and sex roles. Although sex-role-reversed mating systems are infrequent, there are examples scattered among many taxonomical groups.[81] Exceptionally, there are several sex-role-reversed species within the *Syngnathidae* family.[73,82] Given that the proximate effects of sex-role reversal remain poorly understood, the response of sex-role-reversed syngnathids to EE2 (*e.g.*, morphology, behavior, and reproductive physiology) became an interesting path worth exploring.

Some of the most bizarre and elaborate traits observed in nature, like the peacock's tail, are usually the product of sexual selection. The expression of these secondary sexual traits has profound repercussions in individual fitness, allowing for access to mates and successful fertilizations. Since the expression of secondary sexual traits is typically regulated by sex steroids, endocrine disruptors such as EE2 can ultimately affect mating dynamics. Partridge *et al.*,[83] using the sex-role-reversed Gulf pipefish as a model, performed short-term exposures to EE2 and showed that adult male pipefish quickly developed female-like secondary sexual traits, at concentrations of 100 ng L^{-1}. These results were consistent with previous observations conducted by Ueda *et al.*[84] Once exposed to EE2, males started to develop persistent iridescent stripes (the secondary sex trait normally expressed by females; see Figure 13.2a) that negatively impacted their fitness through diminished attractiveness. Exposed males showed altered hepatosomatic and gonadosomatic indexes, accompanied by histological alterations in the liver and gonads, as well as amplified VTG levels.[84] These results showed that endocrine disruptors have the potential to imbalance pre-copulatory mate choice mechanisms.

More recently, Rose *et al.*[85] addressed the effects of EE2 exposure (2 and 5 ng L^{-1}) on pre- and post-mating episodes of selection in the sex-role-reversed Gulf pipefish. Unlike what has been observed in the sand goby (*Pomatoschistus minutus*), a species with orthodox sex roles, where sexual selection was impacted by the feminization of males,[86,85] found no significant changes in the opportunity for selection in both male and female Gulf pipefish. At the lowest EE2 concentration, females that mated with multiple males experienced increased fitness (they produced more eggs and had higher embryo survivorship) while males seemed largely unaffected. Nevertheless, a rise in EE2 concentration to 5 ng L^{-1} (an ecologically relevant concentration) resulted in a potentially dramatic populational impact. While females were largely unaffected, the males' reproductive potential was seriously compromised as they became unable to mate and carry pregnancies to term, showing abnormal brood pouch morphology. Rose *et al.*[85] predicted that EE2 concentrations approaching 5 ng L^{-1} or higher are prone to reduce the reproductive potential of Gulf pipefish populations, with the effect being mediated by the impact on males rather than on females. These results are largely concordant with those from Sarria *et al.*,[87] using two other pipefish, *N. lumbriciformis* and *S. abaster*, which differ in the degree of sex-role reversal. Here, the authors also concluded that courtship behavior resisted estrogenic exposure (and might not be the most sensitive indicator of contamination). While behavior seemed largely unaffected, a complete absence of pregnant males was registered, in both species, at the highest EE2 concentration (18 ng L^{-1}). At intermediate concentrations (3 and 9 ng L^{-1}), while pregnancies were still observed, the percentage of lost eggs was higher in EE2 exposed groups. The authors hypothesized that the absence of pregnancies in the highest EE2 concentration was caused by a disruption occurring prior to the moment of copulation. This hypothesis was based on the fact that (i) oocyte volume, in both species, decreased as the contaminant concentration rose and (ii) increased transcription of VTG I was observed in males of both species when exposed to EE2. Curiously, Sarria *et al.*[87] noted that males of the most sex-role-reversed species (*N. lumbriciformis*) seemed more sensitive to VTG induction, an observation that can open new insights into the molecular basis of the mechanisms behind sex-role reversal as VTG expression is well known to be regulated by estrogens.

In a recent study investigating gene expression patterns in the liver of the sex-role-reversed Gul pipefish, using next-generation RNA-sequencing technology,[88] detected sexually dimorphic expression patterns. The obtained sex-specific transcriptomes were comparable to those of fishes with conventional sex-roles. Nevertheless, when females, pregnant and non-pregnant males were exposed to 5 ng L^{-1} of EE2, feminization of the male liver transcriptome was observed (upregulated genes included classic estrogen-response genes, such as vitellogenin, choriogenin, and zona pellucida). These results suggest that the ancestral state of estrogen-regulated female reproductive physiology is conserved in sex-role-reversed vertebrates. Interestingly, these results are again coherent with the observations of Sarria *et al.*,[87] using two pipefish species with distinct degrees of sex-role reversal. Independently from the degree of sex-role reversal, imbalances in the oogenesis process, induction of vitellogenin in males and the absence of pregnancies in the highest EE2 concentrations suggest that reproductive physiology is independent of the degree of sex-role reversal.

Relatively low levels of EE2 have been shown to have drastic effects on gene expression levels, reproductive development, and behavior in syngnathids. EE2 impact, however, is not circumscribed to mature and reproducing individuals. EE2 insult during early life has the potential to modulate population structure, either directly through increased mortality or by causing inappropriate aggregation events, thus affecting the number of young that will reach adulthood. Given the lack of information on the impact of endocrine disruptors on fish early life dispersal patterns, Sárria *et al.*[89] analyzed the vertical distribution of newborn pipefish. In *S. abaster*, despite no significant differences in overall mortality, newborn tended to shift their vertical distribution towards the surface in a dose-dependent manner (EE2 concentrations of 8, 12 and 36 ng L^{-1}). Normally, immediately upon exiting the marsupium, pipefish express a clear preference for a distribution close to the substratum.[67] Previous studies by Sárria *et al.*[90] also showed that during these early development phases, juvenile pipefish exposed to EE2 are more vulnerable to predation since they tend to spend less time in secluded areas. Thus, it can be hypothesized that EE2-exposed juveniles become particularly vulnerable to predators, given their risky behaviors and unorthodox vertical distribution. A follow-up of the continuously exposed pipefish confirmed that EE2 effects were also noticeable upon sexual maturity, namely by the alteration of several primary and secondary sexual characters (males: significant delay in the expression of a developed marsupium and darker body pigmentation; females: delay in gonad maturation). Given the immediate implications for population dynamics, most studies on endocrine disruptors almost invariably focus on very specific life history stages, such as embryonic development or reproduction, thus creating a gap in our knowledge of what happens in between. During this intermediate phase, newborns and juveniles face numerous challenges, whose outcome may impair reproduction or even survival. Thus, when addressing the effects of contaminants, fish larvae behavior should be considered as a valuable endpoint, given the obvious implications on individual fitness and population stability and persistence.

Despite the ubiquitous coexistence of several classes of endocrine disruptors in most aquatic ecosystems, there is still limited information regarding their combined effects. In an attempt to expand our understanding of the combined action of xenobiotics acting through dissimilar modes of action, Sarria *et al.*[90] conducted a 7 day exposure experiment with the estrogenic chemical ethinylestradiol (EE2; 3 and 9 ng L^{-1}) and tributyltin (TBT; 10 and 50 ng Sn L^{-1}), singly and in binary mixtures, using pipefish newborns (*S. abaster*). Even though the obtained patterns were generally difficult to interpret, when addressing newborn size, a general trend suggests that while EE2 seems to depress growth, TBT produces the opposite results. Growth was diminished when juveniles were exposed to a mixture of the highest concentrations of both TBT and EE2, suggesting that TBT effects in juvenile growth are counteracted when in the presence of the highest EE2 concentrations. When considering the effects of the selected EDCs on the surface of the fish pupil, when adjusted for global eye surface, an increase in EE2 or TBT concentration was translated into larger pupil surfaces. Nevertheless, when both contaminants were combined, an opposite trend emerged. When analyzing juvenile behavior (number of movement bursts and time spent on the aquarium's most secluded area, away from a potential predator), EE2 increased both the number of bursts and time spent on the protected area. TBT alone favored an increase in time spent in the secluded area, but did not alter movement bursts. When in combination, TBT depressed EE2 effects on the number of bursts in the protected area, whereas EE2, at 9 ng L^{-1}, significantly decreased the time spent in the protected area at the highest TBT exposure level. Difficult to interpret? Absolutely. Unsurprisingly, this seems to be part of the take-away message conveyed by the authors. While single exposure to the tested xenobiotics produced interpretable effects, with concentration additive effects that impacted growth, vision and behavior, their combined effects are largely unpredictable, apart from the fact that mixtures of TBT and EE2 have the potential to severely impact endpoints known to affect pipefish survival.

Syngnathids, like many other teleosts with similar life histories (*e.g.* blennids), are prone to serve as valuable sentinel species for ecotoxicological studies. Available data (*e.g.*, physiology, behaviour, gene expression) show that several syngnathids are able to correctly signal sub-lethal pollutant levels throughout distinct ontogenic stages. Nevertheless, what makes syngnathids truly exceptional is definitively their unique mode of reproduction, male pregnancy. As such, this family of truly remarkable fish provides a promising comparative model for researchers working on other forms of viviparity.

13.4 Conclusion and Future Perspectives

Most of the available data dealing with the biological responses of the shanny and several syngnathids when exposed to pollutants revealed a generally consistent pattern of high sensitivity. It becomes evident that these two taxonomical groups show great, but still unharvested, potential to be incorporated into the risk assessment of pollution monitoring programs, such as those projected under the European Water Policy legislation.

Although this chapter focuses mostly on ecosystem management of European waters, both families are ubiquitously distributed throughout the world, with species dispersed through many distinct ecosystems. Hence, we believe that most of the advantages here identified for the few species that were actually tested are likely to be valid for close relatives, in different regions and contrasting habitats.

Despite the initial research focus aimed towards priority and estrogenic chemicals, there is today general concern about the potential impact of the so-called ‘emerging pollutants’ such as pharmaceuticals, personal care products, nanoparticles and microplastics. Although no data are yet available on the impact of these stressor in blennids and syngnathids, we believe that given their established record of sensitivity, they are likely to be impacted by emerging pollutants. Future studies should address this issue, now grounded on a solid, and rapidly growing, body of information available for these taxa.

Finally, it has been clearly stressed by several environmental agencies that understanding the mode of action (MOA) is a central aim in risk assessment of environmental pollutants.[12] The advances in omics technologies now offer the possibility of unparalleled insights into pollutants’ MOA. In fact, massive sequencing is nowadays possible at reduced costs, thanks to the development of different NGS platforms that allow for an entire genome to be sequenced in a short period of time. The current unavailability of complete blennid and syngnathid genomes is still a major drawback for the speedier integration of these promising groups into cutting-edge ecotoxicological studies. While, undeservingly, blennids do not currently seem to gather sufficient attention to press researchers into unveiling a full genome, syngnathids have been receiving much more visibility, primarily due to male pregnancy.Thus, we may expect, in the near future, for a more consensual integration of these non-model species into ecotoxicological studies, especially when addressing the impact of emerging pollutants in viviparity.

References

1. M. Martinez-Haro, R. Beiras, J. Bellas, R. Capela, J. P. Coelho, I. Lopes, M. Moreira-Santos, A. M. Reis-Henriques, R. Ribeiro, M. M. Santos and J. C. Marques, A review on the ecological quality status assessment in aquatic systems using community based indicators and ecotoxicological tools: what might be the added value of their combination? *Ecol Indic*, 2015, **48**, 8–16.
2. I. J. Allan, B. Vrana, R. Greenwood, G. A. Mills, B. Roig and C. Gonzalez, A "toolbox" for biological and chemical monitoring requirements for the European Union's Water Framework Directive, *Talanta*, 2006, **69**, 302–322.
3. J. A. Hagger, M. B. Jones, D. R. P. Leonard, R. Owen and T. S. Galloway, Biomarkers and integrated environmental risk assessment: Are there more questions than answers? *Integr. Environ. Assess. Manage.*, 2006, **2**, 312–329.
4. D. Fuentes-Rios, R. Orrego, A. Rudolph, G. Mendoza, J. F. Gavilan and R. Barra, EROD activity and biliary fluorescence in *Schroederichthys chilensis* (Guichenot 1848): Biomarkers of PAH exposure in coastal environments of the South Pacific Ocean, *Chemosphere*, 2005, **61**, 192–199.
5. R. van der Oost, J. Beyer and N. P. E. Vermeulen, Fish bioaccumulation and biomarkers in environmental risk assessment: a review, *Environ. Toxicol. Pharmacol.*, 2003, **13**, 57–149.
6. D. Webb, M. M. Gagnon and T. Rose, Interseasonal variability in biomarkers of exposure in fish inhabiting a southwestern Australian estuary, *Environ. Toxicol.*, 2005, **20**, 522–532.
7. European-Commission, *Common Implementation Strategy for the Water Framework Directive (2000/60/EC): Guidance Document No. 19 Guidance on surface water chemical monitoring under the Water Framework Directive*, Luxembourg, 2009.
8. European-Commission, *Common Implementation Strategy for the Water Framework Directive (2000/60/EC): Guidance Document No. 25 Guidance on chemical monitoring of sediment and biota under the Water Framework Directive*, Luxembourg, 2010.
9. A. Collins, D. G. Ohandja, D. Hoare and N. Voulvoulis, Implementing the Water Framework Directive: a transition from established monitoring networks in England and Wales, *Environ. Sci. Policy*, 2012, **17**, 49–61.
10. V. N. de Jonge, M. Elliott and V. S. Brauer, Marine monitoring: Its shortcomings and mismatch with the EU water framework directive's objectives, *Mar. Pollut. Bull.*, 2006, **53**, 5–19.

11. L. F. Castro and M. M. Santos, To bind or not to bind: the taxonomic scope of nuclear receptor mediated endocrine disruption in invertebrate phyla, *Environ. Sci. Technol.*, 2014, **48**, 5361–5363.
12. M. M. Santos, R. Ruivo, M. Lopes-Marques, T. Torres, C. B. de los Santos, L. F. C. Castro and T. Neuparth, Statins: An undesirable class of aquatic contaminants? *Aquat. Toxicol.*, 2016, **174**, 1–9.
13. C. Faria, V. Almada and E. Goncalves, Juvenile recruitment, growth and maturation of *Lipophrys pholis* (Pisces: Blenniidae), from the west coast of Portugal, *J. Fish Biol.*, 1996, **49**, 727–730.
14. J. Dodd, R. N. Gibson and R. N. Hughes, Use of cues by *Lipophrys pholis* L. (Teleostei, Blenniidae) in learning the position of a refuge, *Behav. Processess.*, 2000, **49**, 69–75.
15. N. M. Monteiro, S. M. Quinteira, K. Silva, M. N. Vieira and V. C. Almada, Diet preference reflects the ontogenetic shift in microhabitat use in Lipophrys pholis, *J. Fish Biol.*, 2005, **67**, 102–113.
16. C. Faria, R. Borges, F. Gil, V. Almada and E. Gonçalves, Embryonic and larval development of *Lipophrys pholis* (Pisces: Blenniidae), *Sci. Mar.*, 2002, **66**, 21–26.
17. C. D. Zander, in *Fishes of the North-eastern Atlantic and the Mediterranean*, ed. P. J. P. Whitehead, M. L. Bauchot, J. C. Hureau, J. Nielsen and E. Tortonese, UNESCO, Paris, 1986, vol. III, pp. 1096–1112.
18. C. Faria and V. Almada, Microhabitat segregation in three rocky intertidal fish species in Portugal: does it reflect interspecific competition? *J. Fish Biol.*, 2001, **58**, 145–159.
19. C. Faria, V. Almada and M. Nunes, Patterns of agonistic behaviour, shelter occupation and habitat preference in juvenile *Lipophrys pholis*, *Coryphoblennius galerita* and *Gobius cobitis*, *J. Fish Biol.*, 1998, **53**, 1263–1273.
20. F. Ferreira, M. Santos, M. Reis-Henriques, M. Vieira and N. Monteiro, The annual cycle of oogenesis in the shanny, *Lipophrys pholis* (Pisces: Blenniidae), *Sci. Mar.*, 2011, **76**, 273–280.
21. F. Ferreira, M. M. Santos, M. A. Reis-Henriques, M. N. Vieira and N. M. Monteiro, The annual cycle of spermatogenesis in *Lipophrys pholis* (Blenniidae), a recently proposed sentinel species for pollution monitoring, *Ichthyol. Res.*, 2011, **58**, 360–365.
22. F. Ferreira, M. M. Santos, M. A. Reis-Henriques, N. M. Vieira and N. M. Monteiro, Sexing blennies using genital papilla morphology or ano-genital distance, *J. Fish Biol.*, 2010, **77**, 1432–1438.
23. S. Z. Qasim, The biology of *Blennius pholis* L. (Teleostei), *Proc. Zool. Soc. London*, 1957, **128**, 161–208.
24. J. Dunne, Littoral and Benthic Investigations on the West-Coast of Ireland - VII. (Faunistic and Ecological-Studies) - The biology of the shanny, *Blennius pholis* L. (Pisces) at Carna, Connemara, *Proc. R. Ir. Acad., Sect. B*, 1977, **77**, 207–226.
25. M. Carrassón and M. Bau, Reproduction and gonad histology of *Aidablennius sphynx* (Pisces: Blenniidae) of the Catalan Sea (Northwestern Mediterranean), *Sci. Mar.*, 2003, **67**, 461–469.

26. C. R. Tyler and J. P. Sumpter, Oocyte growth and development in teleosts, *Rev.: Methods Technol. Fish Biol. Fish.*, 1996, **6**, 287–318.
27. E. Lubzens, G. Young, J. Bobe and J. Cerda, Oogenesis in teleosts: How fish eggs are formed, *Gen. Comp. Endocrinol.*, 2010, **165**, 367–389.
28. V. C. Almada, E. N. Barata, E. J. Goncalves and R. F. Deoliveira, On the Breeding-Season of *Lipophrys pholis* (Pisces, Blenniidae) at Arrabida, Portugal, *J. Mar. Biol. Assoc. U. K.*, 1990, **70**, 913–916.
29. S. E. Shackley and P. E. King, The reproductive cycle and its control: frequency of spawning and fecundity in *Blennius pholis* L, *J. Exp. Mar. Biol. Ecol.*, 1977, **30**, 73–83.
30. P. Rodrigues, M. A. Reis-Henriques, J. Campos and M. M. Santos, Urogenital papilla feminization in male *Pomatoschistus minutus* from two estuaries in northwestern Iberian Peninsula, *Mar. Environ. Res.*, 2006, **62**, S258–S262.
31. B. Lyons, J. Harvey and J. Parry, An initial assessment of the genotoxic impact of the Sea Empress oil spill by the measurement of DNA adduct levels in the intertidal teleost *Lipophrys pholis*, *Mutat Res., Genet. Toxicol. Environ. Mutagen.*, 1997, **390**, 263–268.
32. J. S. Harvey, B. P. Lyons, T. S. Page, C. Stewart and J. M. Parry, An assessment of the genotoxic impact of the Sea Empress oil spill by the measurement of DNA adduct levels in selected invertebrate and vertebrate species, *Mutat Res., Genet. Toxicol. Environ. Mutagen.*, 1999, **441**, 103–114.
33. M. M. Santos, M. Sole, D. Lima, B. Hambach, A. M. Ferreira and M. A. Reis-Henriques, Validating a multi-biomarker approach with the shanny *Lipophrys pholis* to monitor oil spills in European marine ecosystems, *Chemosphere*, 2010, **81**, 685–691.
34. M. Sole, G. Lobera, D. Lima, M. A. Reis-Henriques and M. M. Santos, Esterases activities and lipid peroxidation levels in muscle tissue of the shanny *Lipophrys pholis* along several sites from the Portuguese Coast, *Mar. Pollut. Bull.*, 2008, **56**, 999–1007.
35. D. Lima, M. M. Santos, A. M. Ferreira, C. Micaelo and M. A. Reis-Henriques, The use of the shanny *Lipophrys pholis* for pollution monitoring: A new sentinel species for the northwestern European marine ecosystems, *Environ. Int.*, 2008, **34**, 94–101.
36. F. Ferreira, M. M. Santos, L. F. C. Castro, M. A. Reis-Henriques, D. Lima, M. N. Vieira and N. M. Monteiro, Vitellogenin gene expression in the intertidal blenny *Lipophrys pholis*: a new sentinel species for estrogenic chemical pollution monitoring in the European Atlantic coast? *Comp. Biochem. Physiol., Part C: Toxicol. Pharmacol.*, 2009, **149**, 58–64.
37. F. Ferreira, N. M. Monteiro, M. N. Vieira, M. A. Reis-Henriques, L. F. Castro and M. M. Santos, A real-time PCR assay for differential expression of vitellogenin I and II genes in the liver of the sentinel fish species *Lipophrys pholis*, *Toxicol. Mech. Methods*, 2013, **23**, 591–597.
38. M. Oliveira, M. Pacheco and M. A. Santos, Cytochrome P4501A, genotoxic and stress responses in golden grey mullet (Liza aurata) following short-term exposure to phenanthrene, *Chemosphere*, 2007, **66**, 1284–1291.

39. P. Van Ngan, V. Gomes, M. J. A. C. R. Passos, K. A. Ussami, D. Y. F. Campos and A. J. D. Rocha, Biomonitoring of the genotoxic potential (micronucleus and erythrocyte nuclear abnormalities assay) of the Admiralty Bay water surrounding the Brazilian Antarctic Research Station "Comandante Ferraz," King George Island, *Polar Biol.*, 2007, **30**, 209–217.
40. T. Cavas and S. Ergene-Gozukara, Induction of micronuclei and nuclear abnormalities in *Oreochromis niloticus* following exposure to petroleum refinery and chromium processing plant effluents, *Aquat. Toxicol.*, 2005, **74**, 264–271.
41. H. Leysen, E. R. Dumont, L. Brabant, L. Van Hoorebeke and D. Adriaens, Modelling stress in the feeding apparatus of seahorses and pipefishes (Teleostei: Syngnathidae), *Biol. J. Linn. Soc.*, 2011, **104**, 680–691.
42. S. Foster, S. Wiswedel and A. Vincent, Opportunities and challenges for analysis of wildlife trade using CITES data - seahorses as a case study, *Aquat. Conserv. Mar. Freshwater Ecosyst.*, 2016, **26**, 154–172.
43. W. N. Eschmeyer and J. D. Fong, Species by family/subfamily, http://researcharchive.calacademy.org/research/ichthyology/catalog/SpeciesByFamily.asp, (accessed 10/7/2016, 2016).
44. R. Froese and D. Pauly, FishBase, www.fishbase. org).
45. C. E. Dawson, *Indo-Pacific pipefishes (Red Sea to the Americas)*, Ocean Springs, MS (USA) Gulf Coast Research Lab., 1985.
46. A. Franco, P. Franzoi, S. Malavasi, F. Riccato, P. Torricelli and D. Mainardi, Use of shallow water habitats by fish assemblages in a Mediterranean coastal lagoon, *Estuarine, Coastal Shelf Sci.*, 2006, **66**, 67–83.
47. J. Ridgway and G. Shimmield, Estuaries as repositories of historical contamination and their impact on shelf seas, *Estuarine, Coastal Shelf Sci.*, 2002, **55**, 903–928.
48. G. Pojana, A. Gomiero, N. Jonkers and A. Marcomini, Natural and synthetic endocrine disrupting compounds (EDCs) in water, sediment and biota of a coastal lagoon, *Environ. Int.*, 2007, **33**, 929–936.
49. R. Francois, F. T. Short and J. H. Weber, Accumulation and Persistence of Tributyltin in Eelgrass (Zostera-Marina L) Tissue, *Environ. Sci. Technol.*, 1989, **23**, 191–196.
50. J. C. Chesworth, M. E. Donkin and M. T. Brown, The interactive effects of the antifouling herbicides Irgarol 1051 and Diuron on the seagrass Zostera marina (L.), *Aquat. Toxicol.*, 2004, **66**, 293–305.
51. E. S. Herald, From pipefish to seahorse: a study of phylogenetic relationships, *Proc. Natl. Acad. Sci. U. S. A.*, 1959, **29**, 465–473.
52. A. B. Wilson, A. Vincent, I. Ahnesjö and A. Meyer, Male pregnancy in seahorses and pipefishes (family Syngnathidae): rapid diversification of paternal brood pouch morphology inferred from a molecular phylogeny, *J. Hered.*, 2001, **92**, 159–166.
53. A. Wilson, I. Ahnesjö, A. Vincent and A. Meyer, The dynamics of male brooding, mating patterns, and sex roles in pipefishes and seahorses (family Syngnathidae), *Evolution*, 2003, **57**, 1374–1386.

54. M. Carcupino, Functional significance of the male brood pouch in the reproductive strategies of pipefishes and seahorses: a morphological and ultrastructural comparative study on three anatomically different pouches, *J. Fish Biol.*, 2002, **61**, 1465–1480.
55. N. Monteiro, V. Almada and M. Vieira, Implications of different brood pouch structures in syngnathid reproduction, *J. Mar. Biol. Assoc. U. K.*, 2005, **85**, 1235–1241.
56. S. Watanabe, T. Kaneko and Y. Watanabe, Immunocytochemical detection of mitochondria-rich cells in the brood pouch epithelium of the pipefish, *Syngnathus schlegeli*: structural comparison with mitochondria-rich cells in the gills and larval epidermis, *J. Cell Tissue Res.*, 1999, **295**, 141–149.
57. J. Ripley, Osmoregulatory role of the paternal brood pouch for two *Syngnathus* species, *Comp. Biochem. Physiol.*, 2009, **154**, 98–104.
58. K. Stolting and A. Wilson, Male pregnancy in seahorses and pipefish: beyond the mammalian model, *BioEssays*, 2007, **29**, 884–896.
59. T. W. Haresign and S. E. Shumway, Permeability of the marsupium of the pipefish Syngnathus fuscus to [14C]-alpha amino isobutyric Acid, *Comp. Biochem. Physiol., Part A: Mol. Integr. Physiol.*, 1981, **69**, 603–604.
60. J. L. Ripley and C. M. Foran, Direct evidence for embryonic uptake of paternally-derived nutrients in two pipefishes (Syngnathidae: *Syngnathus* spp.), *J.Comp. Physiol. B*, 2009, **179**, 325–333.
61. C. Kvarnemo, K. B. Mobley, C. Partridge, A. G. Jones and I. Ahnesjo, Evidence of paternal nutrient provisioning to embryos in broad-nosed pipefish Syngnathus typhle, *J. Fish Biol.*, 2011, **78**, 1725–1737.
62. C. M. Whittington, O. W. Griffith, W. Qi, M. B. Thompson and A. B. Wilson, Seahorse brood pouch transcriptome reveals common genes associated with vertebrate pregnancy, *Mol. Biol. Evol.*, 2015, **32**, 3114–3131.
63. P. Melamed, Y. Xue, J. F. D. Poon, Q. Wu, H. Xie, J. Yeo, T. W. J. Foo and H. K. Chua, The male seahorse synthesizes and secretes a novel C-type lectin into the brood pouch during early pregnancy, *FEBS J.*, 2005, **272**, 1221–1235.
64. K. A. Paczolt and A. G. Jones, Post-copulatory sexual selection and sexual conflict in the evolution of male pregnancy, *Nature*, 2010, **464**, 401–404.
65. M. Yasue, A. Nellas and A. C. J. Vincent, Seahorses helped drive creation of marine protected areas, so what did these protected areas do for the seahorses? *Environ. Conserv.*, 2012, **39**, 183–193.
66. N. Monteiro, V. Almada and M. Vieira, Early life history of the pipefish *Nerophis lumbriciformis* (Pisces: Syngnathidae), *J. Mar. Biol. Assoc. U. K.*, 2003, **83**, 1179–1182.
67. K. Silva, N. Monteiro, V. Almada and M. Vieira, Early life history of *Syngnathus abaster*, *J. Fish Biol.*, 2006, **68**, 80–86.
68. S. Sommer, C. M. Whittington and A. B. Wilson, Standardised classification of pre-release development in male-brooding pipefish, seahorses, and seadragons (Family Syngnathidae), *BMC Dev. Biol.*, 2012, **12**, 39.

69. K. Silva, V. C. Almada, M. N. Vieira and N. M. Monteiro, Female reproductive tactics in a sex-role reversed pipefish: scanning for male quality and number, *Behav. Ecol.*, 2009, **20**, 768–772.
70. K. Silva, M. N. Vieira, V. C. Almada and N. M. Monteiro, The effect of temperature on mate preferences and female-female interactions in Syngnathus abaster, *Anim. Behav.*, 2007, **74**, 1525–1533.
71. N. Monteiro, M. Vieira and V. Almada, The courtship behaviour of the pipefish Nerophis lumbriciformis: reflections of an adaptation to intertidal life, *Acta Ethol.*, 2002, **4**, 109–111.
72. A. Berglund, Risky Sex - Male pipefishes mate at random in the presence of a predator, *Anim. Behav.*, 1993, **46**, 169–175.
73. A. Berglund and G. Rosenqvist, Sex role reversal in pipefish, *Adv. Study Behav.*, 2003, **32**, 131–167.
74. A. Jones and J. Avise, Microsatellite analysis of maternity and the mating system in the Gulf pipefish *Syngnathus scovelli*, a species with male pregnancy and sex-role reversal, *Mol. Ecol.*, 1997, **6**, 203–213.
75. A. Jones and J. Avise, Polygynandry in the dusky pipefish *Syngnathus floridae* revealed by microsatellite DNA markers, *Evolution*, 1997, **51**, 1611–1622.
76. E. E. McCoy, A. G. Jones and J. C. Avise, The genetic mating system and tests for cuckoldry in a pipefish species in which males fertilize eggs and brood offspring externally, *Mol. Ecol.*, 2001, **10**, 1793–1800.
77. K. Mobley, C. Small and A. Jones, The genetics and genomics of Syngnathidae: pipefishes, seahorses and seadragons, *J. Fish Biol.*, 2011, **78**, 1624–1646.
78. M. I. Nenciu, V. Coatu, A. Oros, D. Rosioru, D. Tiganus and N. Rosoiu, Pollutant Bioaccumulation in the long-snouted seahorse at the Romanian coast, *J Environ Prot Ecol*, 2014, **15**, 1650–1659.
79. J. P. Nash, D. E. Kime, L. T. M. van der Ven, P. W. Wester, F. Brion, G. Maack, P. Stahlschmidt-Allner and C. R. Tyler, Long-term exposure to environmental concentrations of the pharmaceutical ethynylestradiol causes reproductive failure in fish, *Environ. Health Perspect.*, 2004, **112**, 1725–1733.
80. K. A. Kidd, P. J. Blanchfield, K. H. Mills, V. P. Palace, R. E. Evans, J. M. Lazorchak and R. W. Flick, Collapse of a fish population after exposure to a synthetic estrogen, *Proc. Natl. Acad. Sci. U. S. A.*, 2007, **104**, 8897–8901.
81. M. Eens and R. Pinxten, Sex-role reversal in vertebrates: behavioural and endocrinological accounts, *Behav. Processes.*, 2000, **51**, 135–147.
82. G. Rosenqvist and A. Berglund, Sexual signals and mating patterns in Syngnathidae, *J. Fish Biol.*, 2011, **78**, 1647–1661.
83. C. Partridge, A. Boettcher and A. G. Jones, Short-term exposure to a synthetic estrogen disrupts mating dynamics in a pipefish, *Horm. Behav.*, 2010, **58**, 800–807.
84. N. Ueda, C. Partridge, J. Bolland and J. Hemming, Effects of an environmental estrogen on male Gulf pipefish, *Syngnathus scovelli*

(Evermann and Kendall), a male brooding Teleost, *Bull. Environ. Contam. Toxicol.*, 2005, **74**, 1207–1212.

85. E. Rose, K. A. Paczolt and A. G. Jones, The effects of synthetic estrogen exposure on premating and postmating episodes of selection in sex-role-reversed Gulf pipefish, *Evol Appl*, 2013, **6**, 1160–1170.
86. M. Saaristo, J. A. Craft, K. K. Lehtonen and K. Lindstrom, Sand goby (*Pomatoschistus minutus*) males exposed to an endocrine disrupting chemical fail in nest and mate competition, *Horm. Behav.*, 2009, **56**, 315–321.
87. M. P. Sárria, M. M. Santos, L. F. Castro, N. M. Vieira and N. M. Monteiro, Estrogenic chemical effects are independent from the degree of sex role reversal in pipefish, *J. Hazard. Mater.*, 2013, **263**(Pt 2), 746–753.
88. E. Rose, S. P. Flanagan and A. G. Jones, The effects of synthetic estrogen exposure on the sexually dimorphic liver transcriptome of the sex-role-reversed Gulf pipefish, *PloS One*, 2015, **10**, e0139401.
89. M. P. Sárria, M. M. Santos, M. A. Reis-Henriques, N. M. Vieira and N. M. Monteiro, Drifting towards the surface: A shift in newborn pipefish's vertical distribution when exposed to the synthetic steroid ethinylestradiol, *Chemosphere*, 2011, **84**, 618–624.
90. M. P. Sárria, M. M. Santos, M. A. Reis-Henriques, N. M. Vieira and N. M. Monteiro, The unpredictable effects of mixtures of androgenic and estrogenic chemicals on fish early life, *Environ. Int.*, 2011, **37**, 418–424.
91. C. E. Dawson, in *Fishes of the North-eastern Atlantic and the Mediterranean*, ed. M. L. B. P. J. P. Whitehead, J. C. Hureau, J. Nielson and E. Tortonese, United Nations Educational, Scientific and Cultural Organization Paris, 1986, vol. 2, pp. 628–639.

CHAPTER 14

The Use of the Ten Spotted Live-bearer Fish Cnesterodon decemmaculatus *(Jenyns, 1842) (Pisces, Poeciliidae) in the Genotoxic Evaluation of Environmental Pollutants*

S. SOLONESKI*[a,b] AND M. L. LARRAMENDY[a,b]

[a] Cátedra de Citología, Facultad de Ciencias Naturales y Museo, Universidad Nacional de La Plata, Calle 64 N° 3, B1904AMA La Plata, Argentina; [b] Consejo Nacional de Investigaciones Científicas y Técnicas (CONICET), Buenos Aires, Argentina
*Email: ssoloneski@yahoo.com.ar

14.1 Introduction

Fish are the most numerous and phylogenetically varied vertebrate group, and they represent a key animal paraphyletic group for understanding vertebrate evolution, physiology, immunology, and developmental disorders as well as several mechanisms related to many disease processes.[1] The term "fish" is defined negatively and excludes the tetrapods (*i.e.*, amphibians, reptiles, birds, and mammals), which descend from the same origin, and thus is not considered a proper grouping in systematic biology. The

Issues in Toxicology No. 33
Ecotoxicology and Genotoxicology: Non-traditional Aquatic Models
Edited by Marcelo L. Larramendy

Published by the Royal Society of Chemistry, www.rsc.org

traditional term "*pisces*" (also "*ichthyes*") is considered a typological but not a phylogenetic classification. Fish are important in broad terms of ecology and food production. Fish possess most of the tissue types of mammals except for certain tissue types, like breast, prostate, and lung tissues, because those organs are specific to mammals.[1] However, fish respond to toxicants in a similar manner to higher vertebrates.[2] They have a great ability to metabolize many xenobiotics and accumulate different types of pollutants.[3] Together with amphibians and reptiles, fish belong to a group called poikilotherm organisms, where the body temperature varies over time according to environmental thermal conditions. They are distributed in practically all zones of the aquatic habitat, have great commercial and recreational value, and play an important role in monitoring aquatic pollution as they respond with great sensitivity to changes in their environment.[4] They have been employed as useful sentinel organisms to detect many environmental pollutants as well as in the detection of several kinds of contamination sources.[1] In addition, fish have an exclusive xenobiotic metabolism as well as a low rate of repair mechanism compared to mammals.[5] Another characteristic of fish is that they have been shown to be more sensitive to the induction of DNA damage,[6] and several reports have shown DNA repair to be slower in fish than in mammals.[7]

For many years, fish have been extensively used to detect environmental toxicity in both acute and chronic bioassays to generate mandated worldwide water quality criteria. Regulatory agencies, the private sector, and academia combine their expertise to determine the toxicity levels of many pollutants and demand sufficient high quality information to serve as the basis for establishing the safe exposure levels of many xenobiotics. Worldwide regulatory institutions, *e.g.*, the United States Environmental Protection Agency (EPA, www.epa.gov), the International Agency for Research on Cancer (IARC, www.iarc.fr), the World Health Organization (WHO, www.who.int), and the European Chemicals Agency (ECHA, www.echa.auropa.eu), among others, have contributed to the regulation and control of the many pollutants.

Aquatic ecosystems are continuously being contaminated with pollutant derivatives from anthropogenic activities, including heavy metals, microbial toxins, pesticides, and polycyclic aromatic hydrocarbons, among many others, and these are the most influential genotoxicants for fish. Among animal species, fish are the inhabitants that cannot run away from the detrimental effects of these pollutants.[8] These pollutants can affect different stages of the aquatic food chain, which may lead to the disturbance of the whole ecosystem. Fish can accumulate these toxic compounds in their body from the dissolved water phase, but also from sediments and food.[8]

Teleostei represent more than 50% of all known fish species and have gained considerable attention from researchers with applied scientific interests. Teleostei, such as the zebrafish (*Danio rerio*), the medakafish (*Oryzias latipes*), the rainbow trout (*Oncorhynchus mykiss*), the guppy (*Poecilia reticulata*), and the pufferfish (*Fugu rubripes*), among several others, are increasingly being used as vertebrate model systems in various fields of biology, especially as research models for studies in genetics and developmental biology. The special attention to these fish species for a wide range of fundamental and applied research purposes has earned them the status of model fish species.

Poeciliidae is a family of freshwater and brackish water fish belonging to the order Cyprinodontiformes, which comprises mostly small freshwater fish, including well-known live-bearing aquarium fish such as guppies, mollies, platies, and swordtails. The original distribution of the family was from the southeastern United States to north of the La Plata river in Argentina as well as the continent of Africa, including Madagascar. In addition, some genera like *Poecilia* and *Gambusia* have been identified in hot spring pools.[9] However, as a result of the release of aquarium specimens and the widespread use of species of the genera *Poecilia* and *Gambusia* for mosquito control, they are now found in all tropical and subtropical areas worldwide.[10] These genera were expected to eat mosquito larvae and help in the control of malaria, but in many cases these teleosts had a negative impact on native fish populations.[11] In general, the maximum length for the family is about 100 mm, and in most species of this fish the females are larger than the males. The subfamily *Poeciliinae* (Garman, 1895), containing 27 genera, are live-bearing fish because they are fish that retain the eggs inside the body and then give birth to live free-swimming young (http://www.itis.gov). Live-bearing fish are nearly all members of the Poeciliinae, including mollies (*Poecilia vivipara*), guppies (*P. reticulata*), platies (*Xiphophorus maculatus*), and swordtails (*Xiphophorus* sp.), among others, and many of this species are very popular in the aquarium trade. In recent years, numerous species kept in aquaria have been released into the wild, often with adverse effects on native fish (http://www.itis.gov).

So far, seven species of *Cnesterodon* sp. have been reported in this regard: *C. carnegiei* (Haseman, 1911); *C. brevirostratus* (Rosa and Costa, 1993); *C. septentrionalis* (Rosa and Costa, 1993); *C. hypselurus* (Lucinda and Garavello, 2001); *C. omorgmatos* (Lucinda and Garavello, 2001); *C. raddai* (Meyer and Etzel, 2001) and *C. decemmaculatus* (Jenyns, 1842) (http://www.itis.gov).

The use of the ten spotted live-bearer fish, *C. decemmaculatus*, as a model organism can be accounted for by several attributes, including the relative ease of rearing and breeding in captivity, high fecundity, rapid development, and short generation time.[12] There are several main reasons why this fish could be used as a vertebrate model in modern biology: their easy portability, their relatively easy use in toxicity testing, and their easy maintenance in laboratory conditions, as well as their potential for *in situ* field monitoring.[13]

Whereas *D. rerio* and *O. latipes* and their features as experimental model systems are well known among researchers, little is known about the applications or advantages of research on *C. decemmaculatus*, which represents an ideal teleost model for both toxicity and environmental screening. The use of *C. decemmaculatus* in research laboratories for toxicology screening requires a solid knowledge of the husbandry of this organism to ensure efficient reproduction and maintenance of healthy specimens.[13] Investigation of *C. decemmaculatus* reproduction in both the wild and in laboratory conditions is of high importance for husbandry under optimal conditions.[13] The understanding of *C. decemmaculatus* reproductive biology and behaviour has recently been increased by several investigations groups.[13]

The ten spotted live-bearer fish, *C. decemmaculatus*, is a small neotropical cyprinoid fish originally from South America with extensive distribution in Neotropical America, including Laguna dos Patos in Uruguay, the Negro and Salado River basins and western drainages of Argentina, and small coastal drainages of Uruguay and Argentina.[14] The species attains high population densities in a wide variety of water bodies. *C. decemmaculatus* is a small fish and viviparous, with a marker for sexual differentiation (Figure 14.1).

It is microomnivorous, benthic-pelagic, and a nonmigratory fish. Usually the maximum length of the female is bigger than the maximum length of the male, with females and males reaching sizes of 45 mm and 36 mm, respectively.[12] Their body shape is laterally compressed (flattened, side-to-side) with a straight back and a high caudal peduncle.[13] The mouth is terminal and oblique, and the head is small, with the presence of prominent eyes.[13] The body pattern presents between seven and ten vertical spots on each side of the body, and juveniles also have thick dark vertical spots. The background body colour is brown-silver, silvery, or light olive, grading to cream or white on the belly. The vertical spots are mostly dark brown or black. The body shapes of males and females of the ten spotted live-bearer fish are very different and distinctive. Males and females are easily distinguished by a clearly sexual dimorphism. Male possess anterior rays of an anal fin modified into an elongated intromittent genitallium[12] (Figure 14.2).

Females grow much bigger than males and they have much heavier-looking bodies. The bodies of females are much more rounded compared to

Figure 14.1 Photograph showing a *Cnesterodon decemmaculatus* population kept in laboratory aquaria conditions.

Figure 14.2 Adult male of *Cnesterodon decemmaculatus*. Males are smaller and thinner than females, and are characterized by an elongated gonopodium. Bar represents 1 cm.

Figure 14.3 Adult female of *Cnesterodon decemmaculatus*. Females are larger than males and often display a distended abdomen due to pregnancy. Bar represents 1 cm.

those of males. When females are pregnant, the body may look stuffed and very full (Figure 14.3).

C. decemmaculatus is highly tolerant to extreme environmental conditions, is easy to handle and breed under laboratory conditions, similar to guppies (*P. reticulata*) or medakas (*O. latipes*),[15] and additionally has comparatively large ranges of tolerance to many environmental conditions, *e.g.*, temperature, salinity, and pH, all preferred conditions for toxicological studies.[15] Moreover, their maintenance under laboratory conditions is inexpensive and they are easy to care for in large numbers under the appropriate photoperiod conditions.[15] Their high fecundity and the short generation time make the ten spotted live-bearer fish an excellent vertebrate model system for several biological studies, especially in toxicological screening.

14.2 Fish as a Model for Aquatic Genotoxicology

Anthropogenic activities, including industrial, agricultural, and urban wastes, are a major cause of concern for human health and they also represent a serious threat to aquatic biota.[16] The biological effects of pollutants differ significantly between different organisms, reflecting variability in the pattern of exposure, uptake routes, metabolism following uptake, rates of accumulation, and sensitivity of the target organs.[16] It has been suggested that genotoxic exposures could have long-term consequences for living species, and hence ecosystem quality.[17] Thus, understanding the effects of pollutants on wildlife and ecosystems will require detailed study of many different species, representing a wide range of taxa. Aquatic animals, especially fish, are recognized as excellent indicators of genotoxic, mutagenic, and carcinogenic effects of indoor and outdoor environmental pollutants, usually agricultural wastes, chemicals, polycyclic aromatic hydrocarbons, organic endocrine-disrupting compounds, heavy metals, *etc.* Fish have been employed as valid models for understanding mechanisms of human cancer and its prevention, as wild organism indicators of environmental contamination, as indicators of potential human exposure to carcinogens in contaminated aquatic sites, and as potential organisms for *in situ* field monitoring of environmental contamination in groundwater in the vicinity of toxic waste sites.[18,19] Fish could be selected as models in genotoxicological studies since they are very sensitive bioindicators of water quality and can highlight the potential danger of new emerging pollutants introduced in the aquatic environment, *e.g.*, nanomaterials, biopesticides, and pesticides mixtures.[20]

Genotoxic pollution of aquatic ecosystems comprises the introduction of contaminants with mutagenic, teratogenic, and carcinogenic potential into their principal media and genomes of the resident organisms.[21,22]

Genotoxicity is a deleterious action that affects a cell's genetic material, disturbing its integrity.[5,23] It involves the detection of compounds capable of causing DNA damage with the aim of understanding the potential biological consequences and molecular mechanisms of genetic material.[24] The chromosomal aberration and the micronucleus tests are the most commonly used and well validated *in vitro* and *in vivo* for detecting DNA damage at the chromosome level.[25,26] The purpose of the chromosome aberration test is to identify xenobiotics that cause either chromosome or chromatid type structural aberrations.[27] With the majority of mutagens, induced aberrations are of the chromatid type, but chromosome-type aberrations also occur. An increase in polyploidy may indicate that a chemical has the potential to induce numerical aberrations.[28] Micronuclei originate from chromosomal (or chromatid) fragments that have not been incorporated into the daughter nucleus following mitosis due to the chromosome breaking (clastogenic event) or mitotic spindle dysfunction (aneugenic event).[25,29] The assay has proved to be useful for measuring DNA damage in *in vivo* and *in situ* systems in the presence of genotoxic compounds.[30–33] Furthermore, regulatory agencies consider that the micronucleus test has emerged as one of the preferred methods for the assessment of cytogenetic damage and has virtually replaced the use of chromosome aberration tests because the micronucleus test is faster, considerably less expensive, and requires minimal training.[27] In addition to micronuclei, several studies have described the presence of other nuclear abnormalities as indicators of DNA instability, namely, binucleated cells, blebbed nuclei, lobed nuclei, and notched nuclei in cells exposed to genotoxic substances.[33–35] When microscopically analyzing micronuclei, one can see they are morphologically identical to but smaller than the main nuclei. The criteria used in the identification of micronuclei are as follows: (1) the diameter of the micronucleus should be less than one-third to one-sixth that of the main nucleus; (2) the micronucleus should be round or oval in shape; (3) the micronucleus should be separated from or marginally overlap with the main nucleus as long as there is clear identification of the nucleus boundary; (4) the micronucleus should be the same staining intensity as or a lighter staining intensity than the main nucleus and nonrefractable, and therefore be readily distinguished from artefacts such as staining particles; (5) the micronucleus should be not linked or connected to the main nucleus; and (6) no more than four or five micronuclei should be associated with the principal nucleus.[25,29] Similarly, other nuclear abnormalities could be analyzed as indicators of DNA instability: (1) blebbed nuclei, where cells present one nucleus with a relatively small evagination of the nuclear membrane containing euchromatin; (2) lobed nuclei, where cells present a nucleus with small evaginations of the nuclear membrane larger than blebs; and (3) notched nuclei, where cells present a nucleus with vacuoles and appreciable depth into a nucleus without containing nuclear material, among others[34,36] (Figure 14.4).

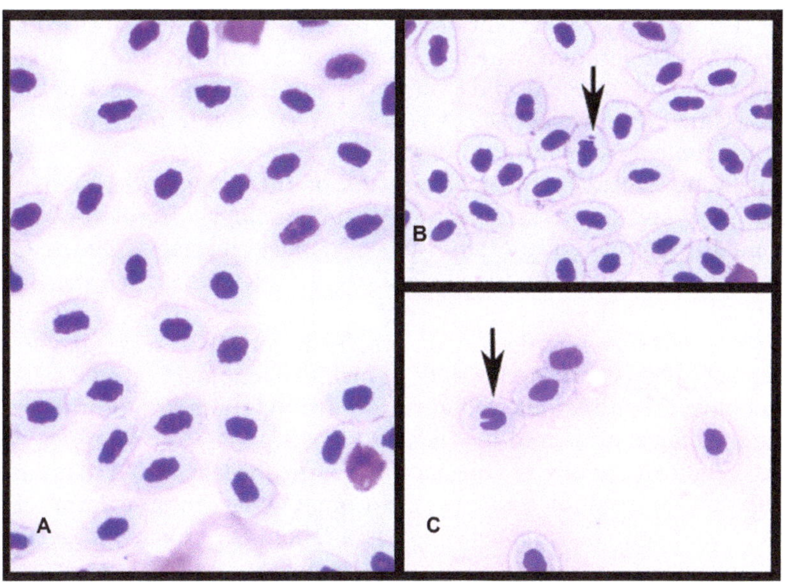

Figure 14.4 Photomicrographs from a blood smear of *C. decemmaculatus* after *in vivo* treatment with the glyphosate-based formulation DMA® showing mature erythrocytes exhibiting micronuclei (arrow). Cells were stained with 5% Giemsa and captured with a light microscope. Magnification, 1000×. Cells are approximately 10 µm along the long axis. Reprinted from *Ecotoxicol. Environ. Saf.*, **89**, J. Vera-Candioti, S. Soloneski and M. L. Larramendy, Evaluation of the genotoxic and cytotoxic effects of glyphosate-based herbicides in the ten spotted live-bearer fish *Cnesterodon decemmaculatus* (Jenyns, 1842), 166–173. Copyright (2013), with permission from Elsevier.

Aquatic ecosystems can receive a number of potentially toxic substances, among which pesticides released from agronomic, industrial, and other human activities are especially significant due to their toxicity, their bioaccumulation potential, and their ability to induce damage in DNA.

The significance of environmental pollution in the genotoxic effects in aquatic organisms is still important for detecting damage exerted by many xenobiotics. Knowledge of the acute and chronic toxicity of a xenobiotic helps in predicting and preventing several types of damage to the aquatic organisms in receiving waters, and this information is then useful in regulating toxic waste discharges.[37]

In recent years, the Comet assay has emerged as a potential tool for genotoxicity testing and regulatory submissions.[38,39] The Comet assay is a simple, sensitive, and fast procedure that can detect primary DNA lesions and repair in any eukaryotic cell type after xenobiotic exposure, and can be used to identify and quantify short-lived DNA damage.[40] In this technique, a small number of cells are resuspended in a microscope slide containing a thin agarose gel, and the cells are then lysed, electrophoresed, and stained with a fluorescent DNA-binding dye.[40] Cells with increased DNA damage display increased migration of nuclear DNA, which resembles the shape of a comet (Figure 14.5). The extent of DNA damage is further quantified by the length of DNA migration, which is visually or automatically determined in 50–100 randomly selected and nonoverlapping nucleoids. DNA damage is then classified into five classes (0–I, undamaged; II, minimum damage; III, medium damage; IV, maximum damage).[36,40]

Actually, to elucidate the mechanisms of the damage detected in the Comet assay and create a more robust methodology, several studies have attempted to correlate the extent of DNA damage measured by the Comet and micronuclei bioassays in cells evaluated under the same experimental conditions.[41]

Using *C. decemmaculatus* as an organism model, several genotoxicity studies have been conducted with this small fish employing the Comet assay methodology on different cellular systems. A summary of the results reported so far is presented in Table 14.1.

The genotoxic effect of the herbicide glyphosate, as a commercial formulation, was studied using the Comet assay in circulating blood erythrocytes from *C. decemmaculatus* as a laboratory-exposed target fish.[42] The acute genotoxic effects of glyphosate in experiments performed for 2 and 4 days of exposure under laboratory conditions were estimated after exposure to two agrochemical commercial formulations containing 48% glyphosate. Significant increases in DNA single-strand breaks were observed in erythrocytes of fish exposed to 3.90 $mg\,L^{-1}$ Panzer® and 22.90 $mg\,L^{-1}$ Credit® for both exposure times. When analyzing the genetic DNA damage index (GDI) values in fish exposed to Panzer® and Credit®, a significant increase of the GDI was observed for both commercial formulations due to an enhancement in the frequency of type II–IV nucleoids and a concomitant decrease of type 0–I nucleoids over negative control values (Figure 14.5).[42]

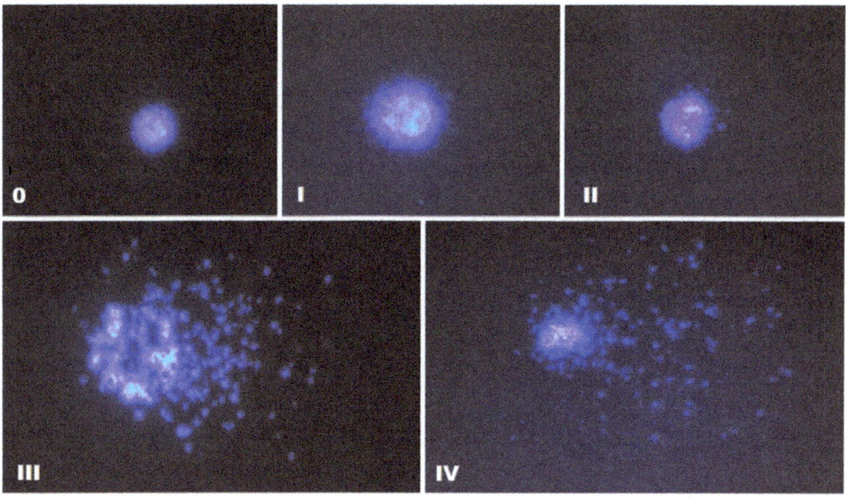

Figure 14.5 Digitized comet images showing undamaged (0–I) and damaged nucleoids (II–IV) of circulating blood cells from *Cnesterodon decemmaculatus* after *in vivo* exposure. They represent classes 0–IV as used for visual scoring (0–I: undamaged, II: minimum damage, III: medium damage, IV: maximum damage). Nucleoids were stained with 4′,6-diamidino-2-phenylindole (DAPI) and captured with a fluorescent microscope. Magnification, 1000×.
Reprinted from *Ecotoxicol. Environ. Saf.*, **128**, C. Ruiz de Arcaute, S. Soloneski and M. L. Larramendy, Toxic and genotoxic effects of the 2,4-dichlorophenoxyacetic acid (2,4-D)-based herbicide on the Neotropical fish *Cnesterodon decemmaculatus*, 222–229. Copyright (2016), with permission from Elsevier.

Similar results were also reported in *C. decemmaculatus* exposed to two 48% chlorpyrifos-based insecticides, namely, Lorsban* 48E® and CPF Zamba®.[42] The experiments were conducted for 2 and 4 days under laboratory conditions, and the results showed a significant increase in the DNA single-strand breaks of fish circulating erythrocytes after exposure to 0.008 mg L^{-1} Lorsban* 48E® and 0.052 mg L^{-1} CPF Zamba®. The GDI values in those fish exposed to Lorsban* 48E® and CPF Zamba® showed a significant increase after exposure to both commercial formulations due to an enhancement, over negative control values, in the frequency of type II–IV nucleoids and a concomitant decrease of type 0–I nucleoids (Figure 14.5).[42]

The genotoxic effect of the carbamic insecticide pirimicarb was studied using the piscine Comet assay in peripheral blood erythrocytes from *C. decemmaculatus* exposed to two 50% pirimicarb-based insecticides, namely Aficida® and Patton Flow®.[42] Both insecticide commercial formulations exerted a significant increase in DNA single-strand breaks in erythrocytes of fish exposed to 25 mg L^{-1} Aficida® and 22 mg L^{-1} Patton Flow® for both exposure times. The GDI values in those fish exposed to Aficida® and Patton Flow® showed a significant increase for both commercial

Table 14.1 Evaluation of genotoxic effects employing several biomarkers on *Cnesterodon decemmaculatus* exposed to different xenobiotics.

Biomarker	Test	Target tissue/ organ	Pollutant/chemical analyzed[a]	Exposure time (days)	Agent, concentration (LEC)[b]	Results	References
DNA damage	Comet assay	Blood	Glyphosate (Panzer®)	2, 4	3.90 mg L^{-1}	+[c]	42
			Glyphosate (Credit®)	2, 4	22.90 mg L^{-1}	+[c]	42
			Chlorpyrifos (Lorsban*48E®)	2, 4	0.008 mg L^{-1}	+[c]	42
			Chlorpyrifos (CPF Zamba®)	2, 4	0.052 mg L^{-1}	+[c]	42
			Pirimicarb (Aficida®)	2, 4	25 mg L^{-1}	+[c]	42
			Pirimicarb (Patton Flow®)	2, 4	22 mg L^{-1}	+[c]	42
			Dicamba (Banvel®)	2, 4	410 mg L^{-1}	+[c]	43
			2,4-D (2,4-D DMA)	2, 4	252 mg L^{-1}	+[c]	33
Chromosomal damage	Frequency of micronuclei	Blood	Pirimicarb (Aficida®)	2, 4	50 mg L^{-1}	+[c]	44
			Pirimicarb (Patton Flow®)	2, 4	22 mg L^{-1}	+[c]	45
			Glyphosate (Panzer®)	2, 4	3.90 mg L^{-1}	+[c]	46
			Glyphosate (Credit®)	2, 4	22.90 mg L^{-1}	+[c]	46
			Chlorpyrifos (Lorsban*48E®)	2, 4	0.008 mg L^{-1}	+[c]	32
			Chlorpyrifos (CPF Zamba®)	2, 4	0.052 mg L^{-1}L		32
			Dicamba (Banvel®)	2, 4	1229 mg L^{-1}	+[c]	43
			2,4-D (2,4-D DMA®)	2, 4	252 mg L^{-1}	+[c]	33
	Other nuclear abnormalities	Blood	2,4-D (2,4-D DMA®)	2, 4	504 mg L^{-1}	+[c]	33
			Dicamba (Banvel®)	2, 4	410 mg L^{-1}	+[c]	43

[a]Between parentheses () trade name of commercial formulation employed in the study.
[b]LEC, lowest effective concentration.
[c]+, Positive.

formulations due to an enhancement, over negative control values, in the frequency of type II–IV nucleoids and a concomitant decrease of type 0–I nucleoids (Figure 14.5).[42]

Induction of DNA single-strand breaks estimated by the alkaline Comet assay was also evaluated in *in vivo* circulating erythrocytes of *C. decemmaculatus* exposed to a 57.71% dicamba-based herbicide, namely, Banvel®.[43] *C. decemmaculatus* fish exposed to 410, 820, and 1229 mg L^{-1} concentrations of the dicamba-based formulation for different exposure times (2 and 4 days) demonstrated a significant induction of DNA single-strand breaks in blood circulating cells for all concentrations assayed and both exposure times. When analyzing the GDI values in fish exposed to Banvel®, a significant increase in the GDI was observed after exposure to the commercial formulation due to an enhancement, over negative control values, in the frequency of type II–IV nucleoids and a concomitant decrease of type 0–I nucleoids (Figure 14.5).[43]

Finally, the Comet assay methodology was used to analyze the DNA damage exerted by the herbicide 2,4-D in *C. decemmaculatus* exposed to the 54.8% 2,4-D-based commercial herbicide DMA®.[33] The authors revealed acute genotoxicity caused by the herbicide 2,4-D when 256, 504, and 756 mg L^{-1} concentrations of the 2,4-D-based commercial formulation were used. They observed a significant increase in DNA damage in blood peripheral erythrocytes following either 2 or 4 days of exposure. When analyzing the GDI values in fish exposed to DMA®, a significant increase was observed due to an increase, over negative control values, in the frequency of type II–IV nucleoids and a concomitant decrease of type 0–I nucleoids (Figure 14.5).[33]

When the erythrocyte micronucleus test was employed in *C. decemmaculatus* to monitor aquatic pollutants displaying mutagenic features, several results were reported (Table 14.1). Micronucleus induction was reported in *C. decemmaculatus* exposed to two 50% pirimicarb-based insecticides, namely, Aficida® and Patton Flow®.[44,45] Both insecticide commercial formulations exerted a significant increase in DNA single-strand breaks in erythrocytes of fish exposed to 25 mg L^{-1} Aficida® and 22 mg L^{-1} Patton Flow® after 2 and 4 days of exposure. Similarly, *C. decemmaculatus* demonstrated a significant induction of micronuclei in circulating erythrocytes after treatment with 3.9 mg L^{-1} Panzer® and 22.90 mg L^{-1} Credit®.[46]

Similar results were also reported in *C. decemmaculatus* exposed to two 48% chlorpyrifos-based insecticides, Lorsban* 48E® and CPF Zamba®.[32] The results showed a significant increase in the frequency of micronuclei in circulating erythrocytes after exposure to 0.008 mg L^{-1} Lorsban* 48E® and 0.052 mg L^{-1} CPF Zamba®.[32] Similarly, the frequency of micronuclei was estimated in circulating erythrocytes of *C. decemmaculatus* exposed to the 57.71% dicamba-based herbicide Banvel®.[43] *C. decemmaculatus* exposed to 410, 820, and 1229 mg L^{-1} concentrations of the dicamba-based formulation for different exposure times (2 and 4 days) demonstrated a significant induction of micronuclei in circulating blood cells after exposure to the higher concentration (1229 mg L^{-1}) assayed after both exposure times. When the micronucleus methodology was employed to analyze the DNA damage exerted by the herbicide 2,4-D in *C. decemmaculatus* exposed to the 54.8% 2,4-D-based commercial herbicide DMA®, a significant increase in micronucleus frequency was reported (Figure 14.4). The authors revealed acute genotoxicity caused by the herbicide 2,4-D when 256, 504, and 756 mg L^{-1} concentrations of the commercial formulation were employed after 2 and 4 days of exposure.[33] Nuclear abnormalities other than micronuclei were evaluated on erythrocytes to assess genetic damage. These studies demonstrated that both dicamba- and 2,4-D-based commercial formulations were genotoxic in *C. decemmaculatus* in fish exposed for 2 or 4 days. The results further indicated that nuclear abnormalities other than micronuclei, such as blebbed nuclei and notched nuclei, were significantly induced (Figure 14.4).[33,43]

Cytotoxicity biomarkers have proved most useful as tools to clarify the biochemical or metabolic alterations involved in the mechanisms of toxic action of several xenobiotics at the cellular level, such as free radicals, irritants, and/or genotoxic compounds.[47]

Using the teleostei *C. decemmaculatus* as an organism model, several cytotoxicity studies have been conducted on different target tissues/organs. A summary of the results reported so far is presented in Table 14.2.

The cytotoxic effect of the carbamic insecticide pirimicarb was studied using the erythrocyte/erythroblast ratio as a biomarker in peripheral blood erythrocytes from *C. decemmaculatus* exposed to two 50% pirimicarb-based insecticides, namely, Aficida® and Patton Flow®.[44,45] Both insecticide commercial formulations exerted a significant increase in cellular cytotoxicity, revealed by a decreased proportion of circulating erythrocytes and an enhancement of erythroblasts found after 2 and 4 days of exposure to 50 mg L^{-1} Aficida® and 22 mg L^{-1} Patton Flow®, respectively.[44,45] Similar results were also reported in *C. decemmaculatus* exposed to two 48% chlorpyrifos-based insecticides, namely, Lorsban* 48E® and CPF Zamba®.[32] The experiments were conducted for 2 and 4 days under laboratory conditions, and the results showed a significant increase in cellular cytotoxicity estimated by the circulating erythrocyte/erythroblast ratio after exposure to 0.008 mg L^{-1} Lorsban* 48E® and 0.052 mg L^{-1} CPF Zamba®.[32] Finally, the cytotoxic effect of the herbicide glyphosate, as a commercial formulation containing 48% glyphosate, was studied using the erythrocyte/erythroblast ratio in circulating blood erythrocytes from *C. decemmaculatus* as a laboratory-exposed target fish.[46] The acute cytotoxic effect of glyphosate in experiments performed for 2 and 4 days of exposure under laboratory conditions was estimated after fish were exposed to two agrochemical commercial formulations of glyphosate. No cellular cytotoxicity was found after Panzer® and Credit® treatment, regardless of both the concentration and the sampling time. Analysis of cytotoxicity by estimation of the

Table 14.2 Evaluation of cytotoxic effects employing several biomarkers on *Cnesterodon decemmaculatus* exposed to different xenobiotics.

Biomarker	Target tissue/organ	Pollutant/chemical analyzed[a]	Exposure time (days)	Agent, concentration (LEC or HIC)[b]	Results	References
Erythrocyte/erythroblast ratio	Blood	Pirimicarb (Aficida®)	2, 4	50 mg L^{-1}	+[c]	44
		Pirimicarb (Patton Flow®)	2, 4	22 mg L^{-1}	+[c]	45
		Chlorpyrifos (Lorsban*48E®)	2, 4	0.008 mg L^{-1}	+[c]	32
		Chlorpyrifos (CPF Zamba®)	2, 4	0.052 mg L^{-1}	+[c]	32
		Glyphosate (Panzer®)	2, 4	11.80 mg L^{-1}	−[d]	46
		Glyphosate (Credit®)	2, 4	68.80 mg L^{-1}	−[d]	46
Hepatosomatic index	Liver	Chlorpyrifos (Clorfox®)	42	0.0084 μl L^{-1}	+[c]	48
		Glyphosate (Roundup Max®)	42	0.2 mg L^{-1}	−[d]	48
Enzymatic activity						
Acetylcholinesterase	Brain	Chlorpyrifos (Clorfox®)	42	0.0084 μl L^{-1}	+[c]	48
		Glyphosate (Roundup Max®)	42	2 mg L^{-1}	−[d]	
		Paraquat (a.i.)	4	20 mg L^{-1}	−[d]	51
		Paraquat (Osaquat®)	4	20 mg L^{-1}	−[d]	50
		Environmental water sample without anthropogenic activity			+[c]	52
		Highly polluted peri-urban water			+[c]	54
		Highly polluted peri-urban water			+[c]	53
Acetylcholinesterase	Muscle	Chlorpyrifos (Clorfox®)	42	0.0084 μl L^{-1}	−[d]	48
		Glyphosate (Roundup Max®)	42	2 mg L^{-1}	−[d]	
		Environmental water sample without anthropogenic activity			+[c]	52
	Liver	Environmental water sample without anthropogenic activity			+[c]	52
Acetylcholinesterase	Whole body	Glyphosate (a.i.)	4	35 mg L^{-1}	−[d]	55
		Glyphosate formulation	4	1 mg L^{-1}	+[c]	55

Table 14.2 (*Continued*)

Biomarker	Target tissue/organ	Pollutant/chemical analyzed[a]	Exposure time (days)	Agent, concentration (LEC or HIC)[b]	Results	References
Catalase	Muscle	Chlorpyrifos (Clorfox®)	42	0.0084 μl	–[d]	48
		Glyphosate (Roundup Max®)	42	L^{-1}2 mg L^{-1}	–[d]	
	Liver	Chlorpyrifos (Clorfox®)	42	0.0084 μl L^{-1}	+[c]	48
		Glyphosate (Roundup Max®)	42	2 mg L^{-1}	–[d]	
Glutathione-*S*-transferase	Brain	Chlorpyrifos (Clorfox®)	42	0.0084 μl L^{-1}	–[d]	48
		Glyphosate (Roundup Max®)	42	2 mg L^{-1}	–[d]	
	Muscle	Chlorpyrifos (Clorfox®)	42	0.0084 μl L^{-1}	–[d]	48
		Glyphosate (Roundup Max®)	42	2 mg L^{-1}	–[d]	
	Liver	Chlorpyrifos (Clorfox®)	42	0.0084 μl L^{-1}	+[c]	48
		Glyphosate (Roundup Max®)	42	2 mg L^{-1}	–[d]	
	Gills	Chlorpyrifos (Clorfox®)	42	0.0084 μl L^{-1}	–[d]	48
		Glyphosate (Roundup Max®)	42	2 mg L^{-1}	–[d]	
Aspartate aminotransferase/alanine aminotransferase ratio	Liver	Chlorpyrifos (Clorfox®)	42	0.0084 μl L^{-1}	+[c]	48
		Glyphosate (Roundup Max®)	42	2 mg L^{-1}	+[c]	
Alkaline phosphatase	Liver	Chlorpyrifos (Clorfox®)	42	0.0084 μl L^{-1}	–[d]	48
		Glyphosate (Roundup Max®)	42	2 mg L^{-1}	+[c]	48
Total protein content	Brain Muscle Liver	Environmental water sample without anthropogenic activity			+[c]	52
ATP phosphohydrolase	Gill	Highly polluted peri-urban water			+[c]	54
Aminotransferase	Liver	Highly polluted peri-urban water			+[c]	54

[a] Between parentheses () trade name of commercial formulation employed in the study; a.i., active ingredient.
[b] LEC, lowest effective concentration; HIC, highest ineffective concentration.
[c] +, Positive.
[d] –, Negative.

hepatosomatic index in liver of *C. decemmaculatus* exposed to the 48% chlorpyrifos-based insecticide Clorfox® showed a significant increase in the somatic index in fish exposed for 42 days to an environmentally relevant pesticide concentration of 0.0084 $\mu l\,L^{-1}$ Clorfox®.[48] On the other hand, this index did not present significant differences for fish treated with the glyphosate-based herbicide Roundup Max®.[48]

Antioxidant enzymes and nonenzymatic systems are essential for the conversion of reactive oxygen species into harmless metabolites, and they are implicated in the protection and restoration of normal metabolism and cellular function.[49] Bonifacio *et al.*[48] showed experimentally that the chlorpyrifos-based insecticide Clorfox® can induce oxidative stress in *C. decemmaculatus* and impair their health due to its redox potential.[48] They investigated brain tissue of *C. decemmaculatus* exposed to different concentrations of chlorpyrifos and glyphosate following a 42 day static renewal bioassay. The results of this study clearly showed that the fish experienced oxidative stress as characterized by significant modulation of the activity of antioxidants such as acetylcholinesterase in the brain after exposure to the 0.0084 $\mu l\,L^{-1}$ chlorpyrifos-based insecticide, but not to the 2 $mg\,L^{-1}$ glyphosate-based commercial herbicide formulation.[48] Oxidative stress analyzed by estimation of the levels of acetylcholinesterase in brains of *C. decemmaculatus* exposed to the pure herbicide paraquat as well as the paraquat-based herbicide Osaquat® showed negative results after fish were exposed to 20 $mg\,L^{-1}$ concentrations of both compounds for 4 days.[50,51] Menéndez-Helman *et al.*[52] reported that environmental water samples without anthropogenic activity can affect the acetylcholinesterase activity in the brain in relation to circannual rhythms. Finally, a significant modulation of acetylcholinesterase activity in brains of *C. decemmaculatus* after exposure to highly polluted periurban water was reported.[53,54] When acetylcholinesterase activity was measured in the muscle of *C. decemmaculatus* exposed to different concentrations of chlorpyrifos- and glyphosate-based formulations, negative results were observed after 42 day static renewal bioassays.[48] Contrarily, positive results were observed in acetylcholinesterase activity of liver and muscle of *C. decemmaculatus* after exposure to a water sample without anthropogenic activity in relation to circannual rhythms.[52] Finally, negative results were observed when acetylcholinesterase activity was estimated in the whole body of *C. decemmaculatus* exposed to 35 $mg\,L^{-1}$ pure glyphosate, but positive findings were reported after exposure to the 1 $mg\,L^{-1}$ glyphosate-based commercial formulation for 4 days.[55] Bonifacio *et al.*[48] used catalase activity estimation in fish to analyze the oxidative status after chlorpyrifos-based insecticide (Clorfox®) and glyphosate-based herbicide (Roundup Max®) exposure. In both experiments, the results were negative after a 42 day exposure period, when the muscle tissue were examined, as well as in liver of fish exposed to glyphosate, but positive results were reported in liver of *C. decemmaculatus* exposed to the chlorpyrifos-based insecticide.[48] When glutathione-*S*-transferase activity was measured in *C. decemmaculatus* after exposure to several compounds for

42 days, different findings were observed according to the tissue analyzed. In the brain, muscle, and gills, the glutathione-*S*-transferase activity after treatment with both the chlorpyrifos-based insecticide and the glyphosate-based herbicide was not significantly different from that of the control group.[48] In the liver of *C. decemmaculatus*, the glutathione-*S*-transferase activity was significantly increased after exposure to 0.0084 $\mu l\,L^{-1}$ Clorfox®,[48] but not when the fish were exposed to a 2 $mg\,L^{-1}$ glyphosate-based formulation.[48] When the aspartate aminotransferase/alanine aminotransferase ratio was analyzed in livers of *C. decemmaculatus,* positive results were observed after exposure to 0.0084 $\mu l\,L^{-1}$ Clorfox® and 2 $mg\,L^{-1}$ Roundup Max® commercial formulations.[48] Estimation of alkaline phosphatase activity in livers of *C. decemmaculatus* gave positive results after glyphosate-based formulation exposure but negative results after chlorpyrifos-based insecticide exposure.[48] When the total protein content was measured in brain, muscle, and liver of *C. decemmaculatus* exposed to an environmental water sample without anthropogenic activity, positive results were observed in relation to circannual rhythms.[52] Finally, positive results were demonstrated when the activities of ATP phosphohydrolase in gills and aminotransferase in liver were analyzed in fish exposed to samples of highly polluted periurban water.[54]

14.3 Perspectives

The increasing focus on the use of nontraditional fish species for ecotoxicology and genotoxicology research not only lies in the ability to reveal genotoxic and cytotoxic effects induced by many potential environmental xenobiotics, but also helps researchers understand the real performance of such ecosystems. *C. decemmaculatus* is proving itself as a useful animal model system for studying environmental toxicity, developmental biology, genetics, and genotoxicology. The increased attention to the use of nontraditional species of fish in ecotoxicology and genotoxicology research lies in their ability not only to reveal the genotoxic and cytotoxic effects of many potential environmental xenobiotics, but also to help researchers understand the performance of real ecosystems. As a versatile tool *in vivo* and *in situ*, *C. decemmaculatus* has a number of features that make it attractive as a laboratory model organism for toxicity studies. Thus, *C. decemmaculatus* may be considered as a sentinel organism due to its role in the trophic chain.

Acknowledgements

The support from the National University of La Plata (Grants 11/N699, 11/N746 and 11/N817) and the National Council for Scientific and Technological Research (CONICET, PIP N° 0344) from Argentina are acknowledged.

References

1. D. A. Powers, Fish as model systems, *Science*, 1989, **246**, 352–358.
2. K. Al-Sabti and C. D. Metcalfe, Fish micronuclei for assessing genotoxicity in water, *Mutat. Res.*, 1995, **343**, 121–135.
3. C. K. Grisolia and C. M. T. Corderio, Variability in micronucleus induction with different mutagens applied to several species of fish, *Gene. T Nol. Bios.*, 2000, **23**, 235–239.
4. J. F. L. G. Solbe, in *Handbook of Ecotoxicology*, ed. P. Calow, Blackwell Scientific Publication, Oxford, England, 1993, vol. I, pp. 66–82.
5. P. A. White and J. Rasmussen, The genotoxic hazards of domestic wastes in surface water, *Mutat. Res.*, 2001, **410**, 223–236.
6. S. Raisuddin and A. N. Jha, Relative sensitivity of fish and mammalian cells to sodium arsenate and arsenite as determined by alkaline single-cell gel electrophoresis and cytokinesis-block micronucleus assay, *Environ. Mol. Mutagen.*, 2004, **44**, 83–89.
7. S. Espina, C. S. Vanegas, S. Bottlo and Villarnieva, Acute toxicity and synergism of cadmium and zinc in white shrimp, *Penaeus setiferus*, *Bull. Environ. Contam. Toxicol.*, 1997, **58**, 87–92.
8. F. E. Olaifa, A. K. Olaifa, A. A. Adelaja and A. G. Owolabi, Heavy metal contamination of *Clarias gariepinus* from lake and fish farm in Ibadan, Nigeria, *Afr. J. Biomed. Res.*, 2004, **7**, 145–148.
9. D. E. McAllister, Introduction of tropical fishes into a hot spring near Banff, Alberta, Canada, *Can. Field Nat.*, 1969, **83**, 31–35.
10. R. Kant, S. Haq, H. C. Srivastava and V. P. Sharma, Review of the bioenvironmental methods for malaria control with special reference to the use of larvivorous fishes and composite fish culture in central Gujarat, India, *J. Vector Borne Dis.*, 2013, **50**, 1–12.
11. R. Froese and D. Pauly, *Cyprinidae. FishBase. World Wide Web electronic publication*, 2016.
12. R. Menni, *Peces y ambientes en la Argentina continental*, Monografías del Museo Argentino de Ciencias Naturales. ISSN 1515-7652, Buenos Aires, Argentina, 2004.
13. A. Somma, M. Mastrángelo and L. Ferrari, *Manual de producción de Cnesterodon decemmaculatus en laboratorio*, Editorial Utopías, Buenos Aires, Argentina, 2011.
14. P. H. F. Lucinda, in *Checklist of the Freshwater Fishes of South and Central America*, ed. R. E. Reis, S. O. Kullander and C. J. Ferraris Jr, EDIPUCRS, Porto Alegre, 2003, vol. 1, pp. 555–581.
15. R. C. Menni, S. E. Gómez and F. López Armengol, Subtle relationships: freshwater fishes and the chemistry of water in southern South America, *Hydrobiologia*, 1996, **328**, 173–197.
16. S. M. Rhind, Anthropogenic pollutants: a threat to ecosystem sustainability? *Philos. Trans. R. Soc., B*, 2009, **364**, 3391–3401.
17. A. N. Jha, Ecotoxicological applications and significance of the comet assay, *Mutagenesis*, 2008, **23**, 207–221.

18. Y. Ishikawa, Medakafish as a model system for vertebrate developmental genetics, *Bioessays*, 2000, **22**, 487–495.
19. B. A. Barut and L. I. Zon, Realizing the potential of zebrafish as a model for human disease, *Physiol. Genomics*, 2000, **2**, 49–51.
20. R. Meffe and I. de Bustamante, Emerging organic contaminants in surface water and groundwater: A first overview of the situation in Italy, *Sci. Total Environ.*, 2014, **481**, 280–295.
21. M. T. Smith, The mechanism of benzene-induced leukemia: a hypothesis and speculations on the causes of leukemia, *Environ. Health Perspect.*, 1996, **104**, 1219–1225.
22. A. Fagr, A. M. El-Shehawi and M. A. Seehy, Micronucleas test in fish genome: A sensitive monitor for aquatic pollution, *Afr. J. Biotechnol.*, 2008, **7**, 606–612.
23. T. Cavaş, *In vivo* genotoxicity evaluation of atrazine and atrazine-based herbicide on fish *Carassius auratus* using the micronucleus test and the comet assay, *Food Chem. Toxicol.*, 2013, **49**, 1431–1435.
24. P. K. M. Nagarathna, M. Johnson Wesley, P. Sriram Reddy and K. Reena, Review on genotoxicity, its molecular mechanisms and prevention, *Int. J. Pharm. Sci. Rev. Res.*, 2013, **22**, 236–243.
25. M. Fenech, Cytokinesis-block micronucleus cytome assay, *Nat. Protoc.*, 2007, **2**, 1084–1104.
26. T. Cavaş and S. Ergene-Gözükara, Micronuclei, nuclear lesions and interphase silver-stained nucleolar organizer regions (AgNORs) as cyto-genotoxicity indicators in *Oreochromis niloticus* exposed to textile mill effluent, *Mutat. Res.*, 2003, **538**, 81–91.
27. C. Winder and N. H. Stacey, *Occupational Toxicology*, CRC Press, New York, 2nd edn, 2004.
28. OECD, *In Vitro* mammalian chromosome aberration test, *OECD Guidelines for the Testing of Chemicals*, Organization for Economic Co-operation and Development, 1997, vol. 473, pp. 1–48.
29. M. Fenech, The micronucleus assay determination of chromosomal level DNA damage, *Methods Mol. Biol.*, 2008, **410**, 185–216.
30. D. Ali, N. S. Nagpure, S. Kumar, R. Kumar, B. Kushwaha and W. S. Lakra, Assessment of genotoxic and mutagenic effects of chlorpyrifos in freshwater fish *Channa punctatus* (Bloch) using micronucleus assay and alkaline single-cell gel electrophoresis, *Food Chem. Toxicol.*, 2009, **47**, 650–656.
31. N. Nikoloff, G. S. Natale, D. Marino, S. Soloneski and M. Larramendy, Flurochloridone-based herbicides induced genotoxicity effects on *Rhinella arenarum* tadpoles (Anura: Bufonidae), *Ecotoxicol. Environ. Saf.*, 2014, **100**, 275–281.
32. J. Vera-Candioti, S. Soloneski and M. L. Larramendy, Chlorpyrifos-based insecticides induced genotoxic and cytotoxic effects in the ten spotted live-bearer fish, *Cnesterodon decemmaculatus* (Jenyns, 1842), *Environ. Toxicol.*, 2013, **29**, 1390–1398.
33. C. Ruiz de Arcaute, S. Soloneski and M. L. Larramendy, Toxic and genotoxic effects of the 2,4-dichlorophenoxyacetic acid (2,4-D)-based

herbicide on the Neotropical fish *Cnesterodon decemmaculatus*, *Ecotoxicol. Environ. Saf.*, 2016, **128**, 222–229.
34. T. Cavaş and S. Ergene-Gözükara, Induction of micronuclei and nuclear abnormalities in *Oreochromis niloticus* following exposure to petroleum refinery and chromium processing plant effluents, *Aquat. Toxicol.*, 2005, **74**, 264–271.
35. F. D. Gökalp Muranli and U. Güner, Induction of micronuclei and nuclear abnormalities in erythrocytes of mosquito fish (*Gambusia affinis*) following exposure to the pyrethroid insecticide lambda-cyhalothrin, *Mutat. Res.*, 2011, **726**, 104–108.
36. T. Cavaş and S. Könen, Detection of cytogenetic and DNA damage in peripheral erythrocytes of goldfish (*Carassius auratus*) exposed to a glyphosate formulation using the micronucleus test and the comet assay, *Mutagenesis*, 2007, **22**, 263–268.
37. USEPA, *Methods for Measuring the Acute Toxicity of Effluents and Receiving Waters to Freshwater and Marine Organisms*, USEPA, 5th edn, 2002, **821-R-02-012.**
38. R. R. Tice, in *Biomonitoring and Biomarkers as Indicators of Environmental Change*, ed. F. M. Butterworth, L. D. Corkum and J. Guzman-Rincon, Plenum Press, New York, 1995, pp. 69–79.
39. R. R. Tice, P. W. Andrews and N. P. Singh, in *DNA Damage and Repair in Human Tissues*, ed. B. M. Sutherland and A. D. Woodhead, Plenum Press, New York, 1990, p. 291.
40. A. R. Collins, Workshop on single cell gel electrophoresis (the comet assay). Meeting Report, *Mutagenesis*, 1992, **7**, 357–358.
41. J. L. He, W. L. Chen, L. F. Jin and H. Y. Jin, Comparative evaluation of the *in vitro* micronucleus test and the comet assay for the detection of genotoxic effects of X-ray radiation, *Mutat. Res.*, 2000, **469**, 223–231.
42. J. Vera-Candioti, S. Soloneski and M. L. Larramendy, Single-cell gel electrophoresis assay in the ten spotted live-bearer fish, *Cnesterodon decemmaculatus* (Jenyns, 1842), as bioassay for agrochemical-induced genotoxicity, *Ecotoxicol. Environ. Saf.*, 2013, **98**, 368–373.
43. C. Ruiz de Arcaute, S. Soloneski and M. L. Larramendy, Evaluation of the genotoxicity of a herbicide formulation containing 3,6-dichloro-2-metoxybenzoic acid (dicamba) in circulating blood cells of the tropical fish *Cnesterodon decemmaculatus*, *Mutat. Res.*, 2014, **773**, 1–8.
44. J. Vera-Candioti, S. Soloneski and M. L. Larramendy, Genotoxic and cytotoxic effects of the formulated insecticide Aficida® on *Cnesterodon decemmaculatus* (Jenyns, 1842) (Pisces: Poeciliidae), *Mutat. Res.*, 2010, **703**, 180–186.
45. J. Vera-Candioti, S. Soloneski and M. L. Larramendy, Pirimicarb-based formulation-induced genotoxicity and cytotoxicity on the fresh water fish *Cnesterodon decemmaculatus* (Jenyns, 1842) (Pisces, Poeciliidae), *Toxicol. Ind. Health*, 2015, **31**, 1051–1060.
46. J. Vera-Candioti, S. Soloneski and M. L. Larramendy, Evaluation of the genotoxic and cytotoxic effects of glyphosate-based herbicides in the ten

spotted live-bearer fish *Cnesterodon decemmaculatus* (Jenyns, 1842), *Ecotoxicol. Environ. Saf.*, 2013, **89**, 166–173.
47. F. Mazzotti, E. Sabbioni, M. Ghiani, B. Cocco, R. Ceccatelli and S. Fortaner, *In vitro* assessment of cytotoxicity and carcinogenic potential of chemicals: evaluation of the cytotoxicity induced by 58 metal compounds in the Balb/3T3 cell line, *ATLA, Altern. Lab. Anim.*, 2001, **29**, 601–611.
48. A. F. Bonifacio, J. Cazenave, C. Bacchetta, M. L. Ballesteros, M. A. Bistoni, M. V. Amé, L. Bertrand and A. C. Hued, Alterations in the general condition, biochemical parameters and locomotor activity in *Cnesterodon decemmaculatus* exposed to commercial formulations of chlorpyrifos, glyphosate and their mixtures, *Ecol. Indic.*, 2016, **67**, 88–97.
49. M. Selvi, T. Cavaş, A. Cağlan Karasu Benli, B. Koçak Memmi, N. Çinkiliç, A. S. Dinçel, O. Vatan, D. Yilmaz, R. Sarikaya, T. Zorlu and F. Erkoç, Sublethal toxicity of esbiothrin relationship with total antioxidant status and *in vivo* genotoxicity assessment in fish (*Cyprinus carpio* L., 1758) using the micronucleus test and comet assay, *Environ. Toxicol.*, 2013, **28**, 644–651.
50. W. D. Di Marzio, J. L. Alberdi, M. E. Sáenz and M. D. C. Tortorelli, Effects of paraquat (Osaquat® formulation) on survival and total cholinesterase activity in male and female adults of *Cnesterodon decemmaculatus* (Pisces, Poeciliidae), *Environ. Toxicol. Water Qual.*, 1998, **13**, 55–59.
51. W. D. Di Marzio and M. C. Tortorelli, Effects of paraquat on survival and total cholinesterase activity in fry of *Cnesterodon decemmaculatus* (Pisces, Poeciliidae), *Bull. Environ. Contam. Toxicol.*, 1994, **52**, 274–278.
52. R. J. Menéndez-Helman, G. V. Ferreyroa, M. dos Santos, Afonso and A. Salibián, Circannual rhythms of acetylcholinesterase (AChE) activity in the freshwater fish *Cnesterodon decemmaculatus*, *Ecotoxicol. Environ. Saf.*, 2015, **111**, 236–241.
53. F. R. de la Torre, L. Ferrari and A. Salibián, Freshwater pollution biomarker: response of brain acetylcholinesterase activity in two fish species, *Comp. Biochem. Physiol., Part C: Toxicol. Pharmacol.*, 2002, **131**, 271–280.
54. F. R. de la Torre, A. Salibián and L. Ferrari, Assessment of the pollution impact on biomarkers of effect of a freshwater fish, *Chemosphere*, 2007, **68**, 1582–1590.
55. R. J. Menendez-Helman, G. V. Ferreyroa, M. D. S. Alfonso and A. Salibian, Glyphosate as an acetylcholinesterase inhibitor in *Cnesterodon decemmaculatus*, *Bull. Environ. Contam. Toxicol.*, 2012, **88**, 6–9.

CHAPTER 15

Genotoxicity in Urodele Amphibians Pleurodeles waltl *and* Ambystoma mexicanum *(*Lissamphibia, Caudata*) Exposed to Freshwater Pollutants: A Historical View*

L. GAUTHIER* AND F. MOUCHET

EcoLab, Université de Toulouse, CNRS, INPT, UPS, Toulouse, France
*Email: laury.gauthier@univ-tlse3.fr

15.1 Introduction

Among the many different tools developed to evaluate the biological hazards of environmental genotoxic pollutants, those addressing entire organisms in water are of particular interest for the monitoring and prediction of aquatic ecosystems.[1,2] In aquatic vertebrates, cytogenetic damage was first envisaged to evaluate the cellular impact of genotoxic substances. From the previous observations of Evans[3] on mammals, who suggested evaluating cytogenetic damage by counting cells with micronuclei under the microscope, numerous aquatic organisms (mainly fish) were tested to help the detection of genotoxic substances in water using the induction of micronuclei, such as the eastern mudminnow,[4] the mudminnow and the brown bullhead,[5] the white

Issues in Toxicology No. 33
Ecotoxicology and Genotoxicology: Non-traditional Aquatic Models
Edited by Marcelo L. Larramendy

Published by the Royal Society of Chemistry, www.rsc.org

croaker[6] or the fossil cat,[7] for the first of them. Other vertebrate test organisms have been suggested to investigate the potential mutagenic risk, like the anuran amphibians *Rana catesbeiana*[8] and *Caudiverbera caudiverbera*[9] after exposure to gamma radiation. Before these studies on anuran amphibians, Jaylet and his co-workers developed a pioneering work to measure the micronuclei induction in red blood cells in the urodele amphibian (*Amphibia, Caudata, Salamandridae*) *Pleurodeles waltl.*, the sharp-ribbed-salamander or Spanish ribbed-newt, in their laboratory in Toulouse, France.

In this chapter, we will retrace the scientific adventure of the discovery and design of this new method using particularly interesting non-conventional organisms to evaluate the potential mutagenic risk in the aquatic medium: the use of the urodele amphibians *Pleurodeles waltl* and *Ambystoma mexicanum.*

15.2 Origins

Most of the work performed in this field of research was initiated in the early 1980s by the team of Professor André Jaylet in the laboratory of general biology, Toulouse, France (currently Center for Developmental Biology) at the University of Toulouse. Professor Jaylet was a cytogeneticist in the service of Professor J. C. Beetschen (a French embryologist) and worked on the experimental remodeling of chromosomal assortment in the amphibian *Pleurodeles waltl* (Michahelles, 1830) and its applications in developmental biology. He was the first to establish the somatic karyotype of *P. waltl*[10] and of other amphibians: *Pelodytes punctatus*,[11] *Euproctus asper*[12] and *Salamandra salamandra.*[13] Jaylet's research then focused on the study of the experimental remodeling of the chromosomes of *P. waltl* after irradiation of adults and of its offspring.

During his numerous works, A. Jaylet was brought to work on the embryonic origin of blood cells in the larvae of *Pleurodeles*. He applied his findings on the experimental polyploidy to develop interesting cellular markers for the study of the embryonic origin of the blood strain cells in *P. waltl* embryos and larvae. Together with P. Deparis, immunologist at the Center for Developmental Biology, they demonstrated the truncal origin and the role of the endoderm in the blood cell ontogeny in pleurodeles embryos and larvae.[14] During their work, A. Jaylet had the opportunity to observe the formation of small masses of chromatin in the cytoplasm of red blood cells of pleurodeles larvae exposed to clastogenic substances in the rearing medium. From this observation, the idea of a new experimental tool for the detection of potential mutagenic substances using amphibians was born.

In the general biology laboratory, several animal species were reared for embryology studies. The neotenic urodele *Ambystoma mexicanum* was mainly used as a model organism for developmental and cell regenerative work. Because of the resemblance between the young larval stages of both species, A. Jaylet had the idea of using larvae of *Ambystoma mexicanum* to develop a similar micronuclei assay to that for the pleurodeles.

15.3 Biology and Status of *Pleurodeles waltl* and *Ambystoma mexicanum*

Named "*waltlii*" (abbreviation: *waltl*) in 1830 by its first descriptor, the German physician and zoologist Karl Michahelles (1807–1834), who dedicated the name to the German physician and entomologist Joseph Waltl (1805–1888), the ribbed newt frequents the quiet and clean waters (ponds, lakes, ditches and slow-moving streams with vegetation cover) of the Iberian peninsula (Spain, Portugal) and Morocco. Adults are mostly found under stones or in mud in their aquatic habitats. This is the largest newt of the European fauna (about 25–30 cm long). It has been classified as "near threatened" (NT) by IUCN since 2006, but because this species is probably in significant decline in its natural range, due to agricultural pollution, increasing urbanization and introduced predators, it is likely to be ranked VU "vulnerable" in the future.[15] In laboratory experiments, pleurodeles has been used to study the effects of chromosomal and hormonal sexual differentiation and more recently in regenerative, reproduction and developmental experiments in space in microgravity conditions.[16]

Ambystoma mexicanum, the Mexican salamander or Mexican walking fish, more commonly known as "axolotl", is a well-known neotenic salamander. It originates from high altitude lakes underlying Mexico City. Since 2010, IUCN has listed this species as critically endangered (CR) because of the very small area of occupancy of the wild type (less than 10 km^2) and the continuous decline of the natural population due to the urbanization of Mexico city, the consequent water pollution and the high abundance of introduced fish impacting the axolotl populations through competition and predation.[17] Axolotls are also extensively used as model organisms in laboratories to study vertebrate embryology and development, neoteny and metamorphosis, as well as the vertebrate regenerative system. Indeed, the axolotl does not heal by scarring and is capable of regeneration of entire lost appendages as well as accepting transplants from other individuals and restoring their full functionality. It is also a popular exotic pet so that it is assumed now that all international trade is in captive-bred animals.

15.4 Genotoxicity Testing in Urodeles Using the Micronucleus Assay

15.4.1 From the Idea to the Standardized Protocol

The initial work of A. Jaylet on *Pleurodeles waltl*[18] helped to clarify the effects of X-rays on the chromosomes of this species provided with large chromosomes ($2n = 24$) and with a particularly high DNA content.[19] As with radiation, clastogenic substances lead to chromosomal damage (inversions, translocations, deletions) that can be revealed by examining the karyotypes. These lesions, qualified as "radiomimetics", require a laborious consideration to carefully quantify the consequences on chromosomes of the exposed organism. The observation by A. Jaylet of the presence of micronuclei in the red blood cells of pleurodeles larvae exposed to clastogenic substances or radiation allowed a more convenient evaluation of the genotoxic effects than the classical karyotype analysis. Based on this observation, in the early 1980s Jaylet's experimental mutagenesis team in the Center of Developmental Biology in Toulouse designed a new method to measure the potential genotoxic effects of pure substances or radiation after urodelian larvae exposure. Siboulet and collaborators[20] evaluated the micronucleated red blood cells induction level in pleurodeles larvae exposed to X-ray irradiation for 6 days and established the dose–effect curve from 6 to 1200 rad exposure doses. Six potentially clastogenic substances were also evaluated [benzo[*a*]pyrene (BaP), carbaryl, *N*-nitrosocarbaryl, aziridine, caffeine and formaldehyde]. The data obtained were compared to other test methods more classically applied in toxicology. Grinfeld and collaborators[21] exposed pleurodeles larvae to different concentrations of BaP for various lengths of time and the micronuclei frequencies in circulating blood were plotted at different times after the end of the treatment. In this work, the uptake and release of the substance was also studied using tritiated BaP. Jaylet and collaborators[22] determined the most suitable larval stage for testing chemicals using pleurodeles larvae reared in water containing one substance (BaP, ethyl methane sulfonate, diethylsulfate or *N*-ethyl-*N*′-nitro-*N*-nitrosoguanidine). Response curves as a function of treatment duration over a period of 16 days were plotted for three concentrations of each compound in order to optimize the micronucleus test conditions. At the same time, because of the similarity of the young larval stages of pleurodeles with those of axolotls, Jaylet and collaborators[23] determined the optimal larval stage to be used for the testing of substances in water using axolotl larvae. In this work, the results obtained with BaP and ethyl methane sulfonate exposures show that axolotl also holds promise as an *in vivo* test system for the detection of clastogenic substances in water. Another advantage of using the axolotl is the complementarity of its breeding period, under our latitudes, with those of pleurodeles. Indeed, while the pleurodeles reproduction period is from autumn to spring in the best case, that of the axolotl runs from winter to late summer. Thus, the combined use of both species makes spawning available almost all year without requiring artificial fertilization.

These last three publications are the essential first step in the development of a larval exposure protocol for the two amphibian species to evaluate the genotoxicity of pure substances added to the rearing water of the animals or to water samples taken from the wild, for which we try to assess the genotoxicity. From the beginning of the definition of the assays, Jaylet's team had the idea of trying to develop a standardized protocol for this method to assess the water genotoxicity. After contact with the French Association for Standardization (AFNOR), *via* the Applied Chemistry Research Institute (IRCHA), which at that time in France carried out the standardization of the *Daphnia magna* acute toxicity test, the first version of a standardized protocol of the test was deposited in 1987 as a specification document with the AFNOR.[24] This document is the only one referring simultaneously to the micronucleus test procedure using both pleurodeles and ambystoma amphibians. Later, the specification document evolved; the first French standard related to the evaluation of the genotoxicity of water samples using pleurodeles larvae was published in May 1992.[25] Compared to the original 1987 document, the standard was restricted to only one urodelian species, *Pleurodeles waltl*, and the scope was extended to all kinds of water samples (surface water, groundwater, industrial and urban effluents) and not just pure substances added to the rearing medium of the larvae. Additional references supported this extended domain of application. Eight years later, in September 2000, a new standardized procedure was published by AFNOR taking into account substantial modifications of the original test procedure, mainly due to a change in the amphibian test organism developed in the laboratory.[26] Indeed, thanks to the work of Catherine Zoll-Moreux and under the supervision of Dr Vincent Ferrier in the Center of Developmental Biology,[27,28] the micronucleus test was transferred to the model amphibian *Xenopus laevis* considering some real advantages of this easy-to-use laboratory anuran species.[29] However, considering the cytogenetic advantage of urodeles compared to anuran, the use of the larvae of pleurodeles was maintained in the new standard and appears in Annex A of the new protocol. Other fundamental modifications have been brought to the previous standard of 1992, including the extension of the scope to solid matter (soils, sludge, wastes, *etc.*) through organic or aqueous extraction before testing, precisions on the rearing advantages of the new test organism and adaptations to the new genotoxicity test procedures, and the addition of references supporting the new standard.[26] Finally, in August 2006, due to the lack of data on the same subject at the international level, the amphibian micronucleus test was brought in by the International Organization for Standardization (ISO), which published the current ISO reference on the subject, including the specific procedure for the use of pleurodeles larvae.[30] In November of the same year, the International ISO standard was transferred to a French NF ISO standard in the AFNOR catalogue.

15.4.2 Protocol and Practicability of the Newt Micronucleus Test

The detailed protocol is given in Annex A of the standardized procedure ISO 21427-1 for pleurodeles[29] and in the AFNOR fascicule T90-325 for pleurodeles and axolotl (AFNOR, 1987).[26] No hormonal induction is needed for both urodeles because of their natural laying period. Sexually mature males and females are placed together in normal tap water filtered through active charcoal at 20 ± 1 °C for pleurodeles and 18 ± 1 °C for axolotl. A few days later, viable eggs are maintained in a bowl containing 5–6 L of normal tap water filtered through active charcoal at 18 °C until the swimming stage. They are then transferred to an aquarium containing the same type of water at 20 ± 2 °C for pleurodeles and 18 ± 2 °C for axolotl until they reached the appropriate development stage for experimentation (about 6 weeks after laying). The young urodele larvae are fed every day on freshly hatched artemia (*Artemia* spec.), then the food is switched to thawed *Chironomus* larvae. Each day, one third of the water is renewed before feeding. The water to which the urodeles larvae are exposed is reconstituted water (water filtered through active charcoal, to which was added nutritive salts [294 $mg L^{-1}$ $CaCl_2 \cdot 2H_2O$, 123.25 $mg L^{-1}$ $MgSO_4 \cdot 7H_2O$, 64.75 $mg L^{-1}$ $NaHCO_3$, 5.75 $mg L^{-1}$ KCl]) at 20 °C. Exposure begins at stage 53 of the development table[31] for pleurodeles larvae, when the hind limbs present four well-formed digits with an outline of the fifth, and with 30 mm long axolotl larvae when the hind limb buds of the larvae exhibit slight indentations (onset of the formation of the two first digits). The larvae are taken from the same hatch to reduce inter-animal variability. They are exposed in groups of 20 (100 mL $larvae^{-1}$) in 5 L glass flasks containing either a control medium—negative and positive control—or the test medium (different concentrations of pure substances or environmental/industrial effluents). The negative control is reconstituted water. Positive controls are systematically performed in each experiment to check both the responsiveness of the larvae and the correct operation of the experiment. The positive control is Cyclophosphamide ([6055-19-2]), as recommended in the standardized procedure at 20 $mg L^{-1}$.[26] Urodeles larvae are submitted to a 12 h light and 12 h dark cycle. They are fed every day on thawed chironomus larvae. The exposure flasks are maintained their temperature at 20 °C ± 1 °C.

At the end of the 12 days of exposure, a single smear of blood is prepared from each anaesthetized larva (tricaine methane sulfonate) after cardiac puncture with heparinized micropipettes (20% solution at 5000 IU mL^{-1}). After fixing in methanol and staining with hematoxylin, the smears are screened under a microscope (oil immersion lens, 1500×). The number of erythrocytes that contain one micronucleus or more (micronucleated erythrocytes, MNE) is determined in a sample of 1000 erythrocytes per larva. For each group of animals, the results (number of micronucleated erythrocytes per thousand, MNE‰) obtained for the individual larvae were arranged in increasing order of magnitude. The medians and quartiles were then calculated. The statistical method used to compare the medians was based on the recommendations of McGill *et al.*[32] and consisted of determining the theoretical medians of samples of size n (where $n \geq 7$) and their 95% confidence limits expressed by:

$$M \pm 1.57 \times \frac{IQR}{\sqrt{n}} \tag{15.1}$$

where M is the median and IQR is the inter-quartile range (upper quartile–lower quartile). Under these conditions, the difference between the theoretical medians of the test groups and the theoretical median of the negative control group is significant to within 95% certainty if there is no overlap. In this case, the induction of micronucleus in exposed larvae is considered as a significant genotoxic response. It may be noted that genotoxicity is assessed up to the concentrations at which signs of acute toxicity begin to appear in the exposed larvae such as mortality.

Since 2008, acute (mortality) and chronic (growth) toxicities are determined under the same standardized exposure procedures on the same larval group. At the end of exposure, acute toxicity is examined by counting dead animals and is expressed as a percentage. Chronic toxicity (growth inhibition) is evaluated by measuring the total body length of each larva at the beginning of the exposure and at the end of the exposure. Measurements were performed using Mesurim image analysis software.[33] Statistical analyses were performed on the organisms' size using a Kruskall–Wallis test, followed by Dunn's test to analyze differences between groups. For the sake of clarity, graphic representation is based on growth rates (GR), calculated for each group as follows:

$$\mathrm{GR} = \left(\frac{(\mathrm{AG}) - (\text{negative control AG})}{\text{negative control AG}}\right) \times 100 \tag{15.2}$$

with AG, the average growth within a treatment group, determined as the difference between the average size at the end and the average size at the beginning of the experiment.

15.4.3 Genotoxicity Testing of Chemical Substances and (Nano)particles

Pleurodeles larvae were used to evaluate the potential genotoxicity of 19 organic substances in water, most of them being known or suspected mutagenic or carcinogenic substances in mammals. The results were compared with published data from other tests used to detect the clastogenic or mutagenic properties of chemicals. Finally, it was concluded that pleurodeles larvae represent a valuable model for the detection of waterborne carcinogens.[34] This was confirmed by Lecurieux and collaborators,[35] who studied the effects of three carcinogenic substances in water. In their publication of 1993, Fernandez and co-workers[36] compiled most of the preliminary data obtained in the laboratory using the amphibian micronucleus assay, after exposure of pleurodeles, axolotl and also xenopus larvae to 47 different chemicals and X-rays. For comparative purposes, literature data have been collected on other short-term genotoxicity tests and on long-term carcinogenicity assays in rodents. The results obtained with the newt micronucleus test, using pleurodeles larvae, were found to correlate better with those of the Salmonella Ames test than with those of the rodent micronucleus assay. Similarly, in a comparative study using newt larvae and the bacterial Ames test, Godet and Vasseur[37] demonstrated the superiority of the newt micronucleus assay over the Ames test for detecting the genotoxicity of metals and for revealing the additive effect of chromium on the genotoxicity of iron (undetected with the Ames test). Fernandez and L'haridon[38] compared the genotoxicity in the newt larvae of seven polycyclic aromatic hydrocarbons (PAHs) and evaluated the influence of 24 hours of UVA irradiation of the media containing the substance to be tested. Using the same test system, L'haridon *et al.*[39] evaluated the genotoxicity of amines and/or potential nitrosating agents *in vivo*. They demonstrated the genotoxic effects of *N*-nitrosoatrazine and *N*-nitrosodiethanolamine at high concentrations but concluded that it was probably impossible to detect their effects in natural conditions. Djomo *et al.*[40] confirmed the strong genotoxic potential of BaP and the weak effect of naphthalene in newt larvae exposed to these PAHs as part of the major fraction of hydrocarbons found in crude oil. Zoll *et al.*[41] observed chromosomal aberrations and abnormal cell division, as well as induction of micronuclei in pleurodeles larvae exposed to low concentrations of mercuric chloride (MC) and methyl mercuric chloride (MMC) in water. The authors measured bioconcentration factors of 600 and 1200 in larvae exposed for 12 days to MC and MMC, respectively. The newt micronucleus test has also been applied by Ferrier *et al.*[28] to evaluate representative environmental pollutants at contaminated sites in the south west of France, like carbendazim, bis[tributyltin]oxide, potassium dichromate, cadmium chloride and pyrene. More recently, new contaminants, also called emerging contaminants, such as manufactured nanoparticles (NPs), have emerged in the ecotoxicology area. Increasingly present in the daily life of our developed society, they are used in various fields, such as medicine, electronics and industry. Nano(eco)genotoxicology was thus born since NPs are expected to enter the environment during some stages of their life cycle, especially in the aquatic compartment, which may concentrate all kinds of pollution. Urodeles larvae were used for the first time for the evaluation of NP genotoxicity after 12 days of exposure of *Ambystoma mexicanum* to carbon nanotubes,[42] and of *Pleurodeles waltl* to NPs of cerium dioxide (CeO_2).[43] Carbon nanotubes were not genotoxic to *A. mexicanum* larvae at any concentration tested (from 0.1 mg L^{-1} to 1 g L) in water, although black masses of carbon nanotubes were observed inside the gut.[42] Dose-dependent genotoxic effects were measured in *P. waltl* larvae exposed to 1 and 10 mg L^{-1} of CeO_2 nanoparticles in water.[43]

15.4.4 Contribution to Human Public Health and Water Quality Assessment

15.4.4.1 *Drinking Water Disinfection Treatment Processes*

Since the preliminary works of Gauthier,[44] who demonstrated the high sensitivity of the newt micronucleus assay, pleurodeles larvae have been widely used for the detection of the genotoxic potential of natural freshwater samples taken from contaminated rivers or from waters from different treatment processes. Thus, using pleurodeles larvae, Jaylet *et al.*[45] evaluated the mutagenic activity in drinking water taken directly from the tap supplying the laboratory. Groups of larvae were reared in tap water, while control animals were reared in tap water filtered over sand and activated charcoal to remove micropollutants. Separate tests carried out in samples of tap water taken at different times throughout the year gave positive responses depending on the time of year. The authors concluded that this test was able to detect clastogens in normal drinking water and that it could be used for quality control of drinking water during various stages in the treatment of raw water. Genotoxic micropollutants in drinking water can come from a variety of origins. They can comprise: (i) residual chlorine from the chlorination processes used for water disinfection, (ii) substances produced by the action of chlorine on organic matter forming halogenated organic compounds, or (iii) actual substances present in raw water.

Gauthier *et al.*[46] demonstrated the clastogenic potential of free chlorine and of monochloramine, used in various disinfection processes in water treatment plants, at concentrations found in water distribution networks. Using chlorinated water samples fully dechlorinated with sodium hyposulfite in increasing concentrations, the same authors reported the elimination of the genotoxic effects initially observed in pleurodeles larvae, thus demonstrating the important role of the presence of chlorinated substances in the mutagenic risk in tap water. In the same way, disinfection processes using ozone were studied by Jaylet *et al.*[47] Indeed, due to its high oxidizing, bleaching and deodorizing properties, ozone treatment is widely applied to water treatment plants for disinfection processes. Pleurodeles larvae were exposed to various ozonized river water samples. Paradoxically, genotoxic effects were observed only in larvae exposed to the less ozonized water samples, corresponding to a quarter of the initial ozone water demand of the river water, whereas the application of strong ozone concentrations led to non-genotoxic effects in the exposed larvae.

Because the presence of potential mutagenic micropollutants in treated water samples may be produced by the direct or indirect action of chlorine on organic matter in the raw water, Lecurieux *et al.*[48,49] studied the genotoxic potential of different trihalomethanes and halogenated acetonitriles commonly identified in chlorinated water samples using pleurodeles larvae and bacterial test systems. From their results, they deducted some particular structure–activity relationships. In the case of trihalomethanes, the authors noticed that the presence of bromine substituents generally led to significant genotoxic effects and that the genotoxicity of haloacetonitriles with bromine substituents generally appeared to be higher than that of the corresponding chlorinated acetonitriles. Moreover, the genotoxicity of the chlorinated acetonitriles seems to increase with the number of chlorine substituents in the molecule. Among the other organohalogenated compounds formed during water chlorination of humic and amino acid-rich water samples, the genotoxicity of chloral hydrate and chloropicrin was tested using both bacterial test-systems and the newt micronucleus assay.[50] In this work, only chloral caused a significant increase in the frequency of micronucleated red blood cells in the exposed pleurodeles larvae but at non-realistic treated water concentration (200 $\mu g\,mL^{-1}$).

These key studies led to an important collaborative program with industrial water treatment managers in France in the early 1990s. As a result, some findings were able to trigger water treatment process modifications, which led to enhanced water quality production in drinking water treatment plants.

15.4.4.2 Evaluation of Wastewater Treatment Processes and Monitoring of the Water Quality

Until the early 1990s, few attempts have been made to incorporate *in vivo* genotoxicity assays in the quality assessment of water effluents. A first comparison between different approaches was made by Van Der Gaag *et al.*[51] who compared the genotoxicity results obtained with *in vivo* (sister chromatid exchange in the fish *Nothobranchius rachowi*, micronuclei induction in the mussel *Mytilus edulis* and in the newt *Pleurodeles waltl*), and *in vitro* test systems (Ames test and SOS Chromotest) exposed to the effluent of a waste water treatment unit that receives effluents from various petrochemical industries. The impact of the sterilization methods required for *in vitro* test systems was evaluated using the *in vivo* test procedures applied to the microfilters and the XAD column extracts with flow-through at pH 2 and 7, and physical-chemical analysis of the samples was conducted to characterize the different steps of the sample preparation before biological testing. In this pilot study, the authors demonstrated that the analytical techniques used to monitor waste waters only cover some of the organic components that can be a potential risk to health and that the sterilization techniques used, which contribute to lowering the genotoxicity measured with the *in vivo* test systems, may explain the failure of bacterial test systems to detect genotoxic effects in this study. Gauthier *et al.*[52] published the results of a first environmental study using only pleurodeles larvae to detect the potential genotoxicity of different waste waters and industrial effluents (tannery effluents from wool and leather industries, oil refinery wastes, and effluents from the petrochemical industry). The authors concluded that the experiments carried out with the newt micronucleus assay on industrial effluents and wastes of various origins underlined the ability of the method to detect the anthropogenic genotoxicity of polluted natural waters and to directly evaluate the impact of aquatic contamination on the exposed ecosystems. The simplicity of the test method could allow its use in routine monitoring of the contamination of a river and in the evaluation of programs intended to reduce pollution.

The newt micronucleus test was therefore considered as a welcome additional tool to help in decision making by the organizations responsible for supervising the discharge of waste into watercourses. Godet and Vasseur[37] compared the genotoxic potential of eight effluents from five industrial sites in France (an urban water treatment plant, a surface treatment plant, a polystyrene production factory, a paper pulp factory and a plant for the production of chlorinated derivatives), using the *in vivo* newt micronucleus assay and the *in vitro* bacterial Ames test. The authors obtained positive responses in the assays with six industrial effluents from various origins. Globally, the Ames test was found to be less sensitive than the newt micronucleus assay, which was preferably used to focus on the study of a metallurgy effluent to explain the combined action of metal contaminants present in the industrial effluent. The authors suggested the use of the sensitive newt micronucleus assay to reveal the possible synergistic effects of pollutant mixtures in water. They also proposed a strategy combining the use of both test systems (Ames and micronucleus test) to reveal the potential genotoxic effects of water samples and effluents. In their comparative amphibian micronucleus work of 1995, Zoll-Moreux and Ferrier[27] studied five waste waters of various origins using both pleurodeles and xenopus larvae. The authors reported the results obtained after exposure of pleurodeles larvae to a coke works effluent before and after biological treatment as well as the river water downstream of the discharge point. Dose–response relationships were observed for both the untreated and treated water samples. The same type of result was obtained after exposure of pleurodeles larvae to an urban effluent taken from the outflow of a waste water treatment plant outside a French town of around 200 000 inhabitants. Clear dose–response relationships were also evidenced even on a diluted effluent. In the same way, Ferrier *et al.*[28] published supplementary results obtained with both amphibian species on waste water from a chemical plant, a steel factory and a paper mill and concluded that both species had similar sensitivity to reveal the genotoxic potential of waste water samples using the micronucleus assay, with a slight advantage to xenopus. Other types of effluents have been studied, such as roadway rainfall effluents, as reported by Gauthier and Ferrier.[53] In the framework of an environmental study on the impact of such effluents on the surrounding environment, the genotoxicity of six rainfall samples was evaluated using the newt micronucleus assay during different rainfall periods over a year. The authors observed positive responses in the larvae exposed to water sampled in a water retention tank receiving the rainfall effluents of a characterized roadway section in the south of France. Three effluents produced clear dose–effect responses in the exposed amphibian larvae and physical-chemical analysis of the same samples indicated that part of the genotoxic effects measured in these experiments could probably be attributed to the high levels of PAHs found in the three positive effluents (up to 188 ng L^{-1}).

15.5 Genotoxicity in Pleurodeles Larvae: Investigations on the Mechanisms of Action

15.5.1 Use of Organic and Inorganic Contaminants as Model Molecules

Among the substances studied in the laboratory in Toulouse, the organic contaminants PAHs played a major role understanding the mechanisms of action involved in the formation of the micronuclei in red blood cells of pleurodeles larvae exposed to contaminated waters. So, in 1987, Fernandez and Jaylet[54] observed reduced clastogenic effects in larvae exposed to BaP in presence of 2(3)-*tert*-butyl-(4)hydroxyanisole (BHA), compared to BaP-exposed larvae. In this paper, the authors assumed that in the newt, BHA, an antioxidant currently used as a food additive (E 320), influences various stages in the metabolic transformation and/or detoxification mechanisms of BaP, so that BHA could have, for example, lowered the formation of reactive intermediates (like reactive oxygen species) or prevent them from reaching their target on DNA or on the mitotic apparatus. Marty *et al.*[55] studied the effects of the same substance (BaP), involving the action of the hepatic cytochrome P450-dependent monooxygenases to explain the mechanisms controlling the formation of hydroxylated metabolites in newt larvae. The authors demonstrated that the metabolic pathways identified in the newt larvae were not fundamentally different from those of mammals or fish and respond to classical inducers of the 3-methylcholanthrene (3-MC) type. Marty *et al.*[56] characterized the enzymatic activities cytochromes P450 IA and IIB in newt larvae pre-induced with 3-MC or phenobarbital. They observed very poor sensitivity of the enzymatic activity after induction with phenobarbital and season- and sex-dependence of the measured 3-MC-induced activities, comparable to that observed in rat liver. Using primary brain cell cultures from *Pleurodeles waltl* larvae, Calevro *et al.*[57] investigated the potential of Cd(II) to induce oxidative DNA damage by measuring the frequencies of DNA strand breaks and DNA base modifications recognized by the bacterial formamidopyrimidine-DNA glycosylase (Fpg protein). They concluded that Cd was not able to induce oxidative DNA base modifications in larval brain cells directly, but it was capable of generating DNA strand breaks and interfering with the repair of oxidative DNA damage, explaining the early life stage neurotoxicity of this metal in *P. waltl.*

15.5.2 DNA Lesions Revealed by the Alkaline Comet Assay, Another Method for Genotoxicity Testing

DNA damage can also be revealed using the Comet assay as primary DNA damage. The Comet assay, as first described by Östling and Johanson,[58] and further developed by Singh *et al.*[59] and Olive *et al.*,[60] is a sensitive method widely used for detecting DNA damage in individual cells (double- and single-strand breaks, alkali-labile sites, excision repair sites and crosslinks), induced by a variety of genotoxic agents, such as industrial chemicals, biocides, agrochemicals, and pharmaceuticals.[61] Numerous studies have demonstrated its capacity to detect low levels of DNA damage, its requirement for few cells, its low cost, and its speed of execution and analysis for assessing the genotoxicity of chemicals toward numerous aquatic species and, notably, on amphibians.[62–68] This assay was then adapted, for the first time, for use on *Pleurodeles waltl* larvae by Mouchet.[69,70] Cells are mixed with low melting agarose, placed on microscope slides and lysed by an alkaline buffer with ionic detergents. The slides are then transferred to a horizontal electrophoresis tank containing alkaline buffer in order to allow the DNA to unwind. Cells with increased DNA damage display increased migration from the nuclear region towards the anode. Slide analysis is performed using fluorescence microscopy, after staining the slides with ethidium bromide solution to measure DNA damage parameters, such as the tail DNA (percentage of DNA in the tail), and/or tail length (TL, length of the tail, distance between the head and the last DNA fragment).

The experimental exposure conditions are basically the same for the micronucleus test and the Comet assay and allow the comparison of different kinds of DNA damage. Both genetic endpoints are different. The micronucleus assay detects chromosomal and/or genomic mutations (chromosomal damage and/or alteration of mitotic spindles), whereas the alkaline Comet assay detects primary DNA damage, expressed as single- and double-strand breaks, alkali-labile sites that are expressed as single-strand breaks and single-strand breaks associated with incomplete repaired excision (sites present at the time of cell lysis). Given that the induction of DNA damage measured in the two bioassays stems from different mechanisms and that they have different response times, the Comet assay could be proposed for use with amphibians as a relevant complementary method for the assessment of the genotoxicity of chemicals. For example, the results of the Comet assay and micronucleus test were compared in *P. waltl* larvae for their ability to detect DNA damage with concentrations of captan from 12.25 $\mu g\,L^{-1}$ to 2 $mg\,L^{-1}$[69] and of cadmium from 0.25 to 2 $mg\,L^{-1}$[70] and with varying exposure times. The Comet assay showed captan and cadmium to be genotoxic from the first day of exposure while the micronucleus test showed they were not genotoxic to *P. waltl* at all concentrations. The Comet assay was concluded to be more sensitive than the micronucleus test for detecting genotoxicity of captan and cadmium.

15.6 *P. waltl* and *A. mexicanum* as Interesting Tools for Developmental and Molecular Toxicology

Urodelian were among the first organisms used to investigate the developmental effects in embryology and in experimental biology, since the 20th century. Because they were also found to be sensitive and effective organisms in environmental quality studies, some authors observed the biological impacts of contaminants on the development of both species.

For example, Calevro *et al.*[71] evaluated the effects of *in vivo* exposure to environmentally realistic and supra-environmental concentrations of Cr, Al and Cd in laboratory conditions, by analyzing the incidence of abnormalities and lethality on the development of *P. waltl*, from the early blastula stage until stage 32/33. The authors showed that Al (0.15, 0.75 and 1.5 mmol L^{-1}) and Cr (0.75 and 1.5 mmol L^{-1}) were toxic to *P. waltl* development and that Cd was highly toxic to embryos at concentrations ranging from 0.18 to 50 μmol L^{-1}. Calevro *et al.*[72] observed malformations *in vivo* in embryos exposed to this same range of Cd concentration, as well as delay and arrest of development in a dose-dependent manner. Flament *et al.*[73] analyzed the effects of Cd on *P. waltl* gonadogenesis from stage 42 to late stage 54 and demonstrated the lack of a direct effect of cadmium on sex determination/differentiation but a strong inhibitory effect on metamorphosis. Ortiz *et al.*[74] assessed the effects of ammonium nitrate fertilizer on embryos and larvae of a few European amphibians including *P. waltl*, which were collected from ponds in Spain. The authors showed that *P. waltl* did not suffer lethal effects but suffered abnormalities or decreased growth at the highest fertilizer concentration of 200 mg NO_3^{-} L^{-1} for 15 days.

Using axolotl, Robles-Mendoza *et al.*[75] evaluated the toxicity of chlorpyrifos (CPF) and malathion (MLT) on *A. mexicanum* embryos and larvae exposed 96 h at stage 44 and 54 from 0.5 to 30 mg MLT L^{-1} and on embryos maintained for 9 days without pesticide in order to identify possible recuperation. Results showed that embryos were less sensitive to acute exposure to persistent organic pollutants than axolotl larvae and that the toxicity of CPF in larval stages was greater than that of MLT. The same team also evaluated the effect of CPF concentrations on esterase activity (acetylcholinesterase and carboxylesterase) and bioconcentration of CPF, in relation to the motor activity of *A. mexicanum* juveniles exposed for 48 h to 0.05 and 0.1 mg CPF L^{-1}.[76] CPF is finally bioconcentrated into axolotls. The internal CPF concentrations are related to the observed inhibition activity of AChE and CbE. Important alterations to axolotls were identified even with short exposure times and low concentrations of CPF. Ortiz-Ordoñez *et al.*[77] investigated the level of lipid peroxidation, the cellular ultrastructure and the liver histology in *A. mexicanum* juveniles to evaluate alterations after exposure to sediment elutriates loaded with heavy metals (Pb, Cd and Hg). Axolotl is also an excellent model for molecular toxicology and tissue regeneration studies. Shoots *et al.*[78] investigated 2,3,7,8-tetrachlorodibenzo-*p*-dioxin (TCDD) toxicity with an isolated aryl hydrocarbon receptor (AHR) from whole *Ambystoma mexicanum*. The authors predicted that *A. mexicanum* lacks sensitivity to TCDD toxicity. From their part, Ponomareva *et al.* (2015) showed complete regeneration of amputated tails of *A. mexicanum* embryos after 7 days and identified signaling pathways associated with tissue regeneration.

15.7 Eco(geno)toxicity Testing with Pleurodeles in Environmentally Relevant Approaches: A Contribution to Ecosystem Health

15.7.1 Use of Micro- and Mesocosms for the Environmental Evaluation of Nanoparticles

Genotoxicity tests, as described above, which are conducted under controlled laboratory conditions on single species, such as pleurodeles larvae, are essential to determine whether the contaminant is (geno)toxic towards pleurodeles and to understand its (geno)toxicity mechanisms. A standardized assay necessarily constitutes the first step of an ecotoxicological evaluation. Nevertheless, this test does not account for the complex environmental parameters, such as trophic relations (competition and predation for example), and which would interfere with the genotoxicity expression. Contrarily to most anuran amphibians, pleurodeles larvae present the advantage of being carnivorous, placing them at a higher level in the aquatic trophic chain than primary producers or primary consumers. To better understand the effects of nanoparticles on freshwater ecosystems, a recent study intended to reproduce relevant exposure scenarios *via* the use of more environmentally relevant systems: in microcosm conditions in which experimental trophic chains have been exposed to CeO_2 NPs.[79] Previously, Bour *et al.*[80] reviewed different studies using methods that allow for integrated ecotoxicological assessments of NPs and described their characteristics and associated endpoints. The systems reviewed allow for studies of toxicity in single species and/or *via* the trophic route as well as studies at the community level. Bour *et al.*[43] also studied the potential toxicity of two types of CeO_2 NPs (small citrate-coated spheres and large uncoated plates) for four organisms (diatoms, chironomids and two amphibian species), using single species test systems, including *Pleurodeles waltl* larvae. Various toxicities were observed in the different organisms and, notably, an amphibian species-dependent response. In this study, pleurodeles larvae exhibited clear growth rate inhibition after exposure to the highest concentration ($10\ mg\,L^{-1}$) of the non-coated CeO_2 NP and a dose-dependent genotoxic response, different from that of the anuran *Xenopus laevis* (no genotoxic effect) in this study. Because pleurodeles was found to be quite sensitive in this work, the authors suggested possible drastic consequences in the case of biomagnification using pleurodeles larvae in real-world exposure conditions using an experimental trophic chain. This work was done by Bour and her co-workers.[79] In this paper, the authors reported severe toxicity on pleurodeles larvae used as top consumers in a three-level experimental trophic chain used for the toxicity assessment of CeO_2 NPs. Different species at three trophic levels, namely (i) bacteria, fungi, diatoms, (ii) *Chironomus riparius* larvae, and (iii) carnivorous pleurodeles larvae; were put together in a battery of nine dynamic microcosms over a 6 week period in the presence of CeO_2 NPs. No clear effects were reported on the primary compartment, except for changes in the bacterial community from the third week of contamination. No toxic effects were observed on chironomids, despite significant NP accumulation in larvae. In contrast, significant mortality was recorded on pleurodeles larvae exposed in the microcosms whereas no toxicity was observed in the same larvae directly exposed to the NPs in laboratory test conditions or *via* the trophic route. The authors suggested that indirect effects with the bacterial compartment or partial CeO_2 NPs dissolution in the microcosm may have induced the pleurodeles mortality in these experiments. Anyway, this work demonstrated the feasibility of the use of urodelian amphibian larvae as a top predator in a carnivorous trophic chain to study the environmental effects of CeO_2 NP exposure and validated the importance of studying the biological effects in microcosm conditions in order to detect biological effects that could not be observed with the classical standardized assays.

15.7.2 Ecogenotoxicity Monitoring with Urodeles in Field Settings

Instead of performing genotoxicity evaluations in controlled laboratory settings, levels of genotoxicity can be also evaluated in free-living populations of urodeles. Ecogenotoxicity monitoring studies on urodeles are certainly less abundant than laboratory studies. To our knowledge, to date no work refers to the use of pleurodeles or axolotl to highlight ecotoxicological effects in environmental conditions directly in the wild. However, a few other urodelian species have been studied with this objective. Davis and Floyd[81] determined the baseline levels of genotoxic stress, in terms of frequency of MNE, in a free-living population of *Cryptobranchus alleganiensis alleganiensis*, also known as the giant hellbenders salamander, which lives in northern Georgia, USA. MNE were detected in 1.16% of erythrocytes and authors obtained a negative relationship between snout–vent length and MNE frequency, indicating an age-related increase in clearance of cells with nuclear damage. In this study, the subjects differed in age, in contrast with the majority of micronucleus studies, which are performed at specific developmental stages. Udroiu *et al.*[82] brought adults and larvae of the Italian crested newt (*Triturus carnifex*) from natural ponds, which differ in their radon content, to the laboratory to measure micronuclei in adult shed skin cells and larval gill cells. The micronucleus frequencies were higher in *Triturus carnifex* from the radon-rich site compared to those from the radon-free site. The larval gills seemed to be more sensitive than the adult tissues. This method is concluded as a relevant application of the micronucleus assay in red blood cells of amphibians.

15.8 Conclusion

The casual observation of the presence of micronuclei in red blood cells of pleurodeles larvae exposed to mutagenic or carcinogenic substances in water led to the development of a particularly effective tool for the detection of genotoxic effects in contaminated waters. Various factors contribute to the high sensitivity of the urodelian model for the detection of genotoxic compounds. It has been demonstrated that the larvae strongly accumulate pollutants from the surrounding medium.[21,41,55] Dose–effect relationships have often been demonstrated with various complex mixtures taken from the wild or sampled before and after treatments in real industrial conditions, so that the use of the newt micronucleus test has been recommended for water quality treatment controls in many occasions.[27,83] The sensitivity of the method allows direct testing without the requirement for micropollutant concentration techniques, leading to extracts (required for *in vitro* test systems), which are usually not fully representatives of the initial water sample. Thus, using pleurodeles larvae, drinking water samples have been tested directly from the tap or taken at different treatment steps in water treatment plants and were found to be genotoxic, whereas other test systems probably would not have revealed any mutagenic activity. Relatively few studies have been devoted to the use of *Ambystoma mexicanum* for environmental quality assessment, while the axolotl is rather used as a model species in other scientific disciplines, such as molecular biology, genetic or developmental biochemistry and enzymology.

Compared to anurans, the main advantages of urodelians are the fact that their cells contain a relatively small number of chromosomes with a large amount of DNA, which greatly facilitates their study from the cytogenetic point of view. Combined to the particular physiology of the larval erythropoiesis (unlike most of the anurans, larval erythropoiesis occurs predominantly in circulating blood and not in other hematopoietic organs), these biological characteristics makes the urodelian particularly sensitive and efficient biomarkers for the detection of genotoxic effects in water using erythrocytes cellular models. Despite these advantages, today, the majority of the work devoted to the use of amphibians for the detection of the genotoxic effects in water is related to anurans. Zoll[84] demonstrated the main advantages of the anuran from the technical point of view of the rearing of the animals using the amphibian *Xenopus laevis*, a laboratory model species whose breeding is long established. The ISO amphibian micronucleus assay is mainly devoted to the use of xenopus tadpoles for the evaluation of genotoxins in water due to its ease of use. Anyway, amphibians have long been the organisms of choice for monitoring the quality of aquatic environments due to their particular physiology, their permeability to pollution and natural sensitivity of young stages of development. The works reported in this chapter have mainly contributed to the recognition of the use of amphibians as particularly sensitive genotoxicity bioindicators and among them, the pioneer role of the urodelian *P. waltl* and *A. mexicanum* must be recognized.

However, the interest in the use of these non-model amphibian species in ecotoxicology lies not only in their ability to reveal genotoxic effects of environmental contaminants, but also to better understand the ecological mechanisms of the functioning of the ecosystems. For example, our recent use of the carnivorous pleurodeles larvae[79] allowed us to make assumptions about the impacts of aquatic contamination with NPs along a reconstituted trophic chain in microcosms. In this case, numerous ecotoxicological end-points were measured on the various organisms of each trophic level and considering the pleurodeles larvae, acute, chronic and geno-toxicity end-points were analyzed to consider the potential direct and indirect effects of the contaminant on the different compartments in the microcosm.

Mostly from the original work of Professor Jaylet and his team at the University of Toulouse, the use of the urodele amphibians *P. waltl* and *A. mexicanum* finally gave birth, nearly three decades later, to a remarkable quality assessment tool for the aquatic ecosystems. The sensitive ISO standardized procedure provides a solid basis on which managers of the quality of the environment are now able to rely to effectively assess the effects of pollutants at a reasonable cost. It is also a development tool of choice for the understanding of the functional mechanisms of the disturbances that occur in contaminated ecosystems.

Acknowledgements

This chapter is dedicated to the memory of Professor André Jaylet, deceased on May 26th, 1989 at the age of 53 years.

References

1. I. Chouroulinkov and A. Jaylet, Contamination of aquatic systems and genetic effects, in *Aquatic Ecotoxicology Fundamental Concepts and Methodologies*, ed. A. Boudou and F. S. Ribeyres, CRC Press Inc., Boca Raton, Florida, 1989, Part 4, pp. 211–235.
2. A. Jaylet and C. Zoll, Tests for detection of genotoxins in fresh-water, *Aquat. Sci.*, 1990, 2(2), 151–166.
3. H. J. Evans, G. J. Neary and F. S. Williamson, The relative biological efficiency of single doses of fast neutrons and gamma rays on *Vicia faba* roots and the effect of oxygen. II Chromosome damage, the production of micronuclei, *Int. J. Radiat. Biol.*, 1959, **1**, 216–229.
4. R. N. Hooftman and W. K. De Raat, Induction of nuclear anomalies (micronuclei) in the peripheral blood erythrocytes of the eastern mudminnow *Umbra pygmaea* by ethyl methane sulphonate, *Mutat. Res.*, 1982, **123**, 147–152.
5. C. D. Metcalfe, Induction of micronuclei and nuclear abnormalities in the erythrocytes of mudminnows (*Umbra limi*) and brown bullheads (*Ictalurus nebulosus*), *Bull. Environ. Contam. Toxicol.*, 1988, **40**, 489–495.
6. K. R. Carrasco, K. L. Tilbury and M. S. Myers, Assessment of the piscine micronucleus test as an *in situ* indicator of chemical contaminant effects, *Can. J. Fish. Aquat. Sci.*, 1990, **47**(11), 2123–2136.

7. R. K. Das and N. K. Nanda, Induction of micronuclei in peripheral erythrocytes of fisch *Heteropneustes fossilis* by mitomycin C and paper mill effluent, *Mutat. Res.*, 1986, **75**, 65–71.
8. P. W. Krauter, S. L. Anderson and F. L. Harisson, Radiation-induced micronuclei in peripheral erythrocytes of *Rana catesbeiana*: an aquatic animal model for *in vivo* genotoxicity studies, *Environ. Mol. Mutagen.*, 1987, **10**, 285–291.
9. W. Venegas, I. Hermosilla, J. F. Gavilan, R. Naveas and P. Carasco, Larval stages of the anuran amphibian *Caudiverbera caudiverbera*: a biological model for studies for genotoxic agents, *Biol. Soc. Biol. Concept.*, 1987, **58**, 171–179.
10. J. C. Beetschen and A. Jaylet, Le caryotype somatique de l'Amphibien Urodèle *Pleurodeles waltlii Michah*, *C. R. Acad. Sci.*, 1961, **253**, 3055–3057.
11. A. Jaylet, Les chromosomes somatiques de l'amphibien anoure *Pelodytes punctatus* (Daudin), *C. R. Acad. Sci.*, 1964, **258**, 5523–5526.
12. A. Jaylet, Le caryotype de l'amphibien urodèle Euproctus asper (Dugés), *Chromosoma*, 1966, **18**, 79–87.
13. A. Kessous, H. Caussinus, A. Jaylet and J. C. Beetschen, Essai d'analyse biométrique du caryotype de l'Amphibien Urodèle *Salamandra salamadra L*, *Chromosoma*, 1968, **23**, 324–332.
14. P. Deparis and A. Jaylet, The role of endoderm in blood cell ontogeny in the newt *Pleurodeles waltl*, *J. Embryol. Exp. Morphol.*, 1984, **81**, 37–47.
15. P. Beja, J. Bosch, M. Tejedo, P. Edgar, D. Donaire-Barroso, M. Lizana, I. Martinez-Solano, A. Salvador, M. Garcia-Paris, E. R. Gil, T. Slimani, Mouden El Hassan El, P. Geniez and T. Slimani, *Pleurodeles waltl. The IUCN Red List of Threatened Species*, 2009, p. 9, DOI: 10.2305/IUCN.UK.2009.RLTS.T59463A11926338.en.
16. L. Gualandris-Parisot, D. Husson, F. Foulquier, P. Kan, J. Davet, C. Aimar, C. Dournon and A. M. Duprat, *Pleurodeles waltl*, amphibian, Urodele, is a suitable biological model for embryological and physiological space experiments on a vertebrate, *Adv. Space Res.*, 2001, **28**(4), 569–578.
17. L. Zambrano, P. M. Reidl, J. McKay, R. Griffiths, B. Shaffer, O. Flores-Villela, G. Parra-Olea and D. Wake, Ambystoma mexicanum, 2010, The IUCN Red List of Threatened Species 2010: e.T1095A3229615. http://www.iucnredlist.org/details/full/1095/0, (accessed June 2016).
18. A. Jaylet and C. Baquier, Accidents chromosomiques obtenus à l'état hétérozygote dans la descendance viable de mâles irradiés, chez le triton *Pleurodeles waltlii Michah*, *Cytogenetics*, 1967, **6**, 390–401.
19. A. Lima-de-Faria, *Handbook of Molecular Cytology*, Elsevier, Amsterdam, 1969, p. 1509.
20. R. Siboulet, S. Grinfeld, P. Deparis and A. Jaylet, Micronuclei in red blood cells of the newt *Pleurodeles waltl Michah*: induction with X-rays and chemicals, *Mutat. Res.*, 1984, **125**, 275–281.
21. S. Grinfeld, A. Jaylet, R. Siboulet, D. Deparis and I. Chouroulinkow, Micronuclei in red blood cells of the newt *Pleurodeles waltl* after

treatment with benzo(a)pyrene dependence on dose, length of exposure, posttreatment time and uptake for the drug, *Environ. Mutagenen.*, 1986, **8**, 41–51.
22. A. Jaylet, P. Deparis and D. Gaschinard, Induction of micronuclei in peripheral erythrocytes of axolotl larvae following *in vivo* exposure to mutagenic agents, *Mutagenesis*, 1986, **1**, 211–215.
23. A. Jaylet, P. Deparis, V. Ferrier, S. Grinfeld and R. Siboulet, A new micronucleus test using peripheral blood erythrocytes of the newt *Pleurodeles waltl* to detect mutagen in fresh water pollution, *Mutat. Res.*, 1986, **164**, 245–257.
24. AFNOR Association Française de Normalisation (the French National Organization for quality regulations), Détection en milieu aquatique de la génotoxicité d'une substance vis-à-vis de larves de batraciens (*Pleurodeles waltl et Ambystoma mexicanum*) – essai des micronoyaux, 1987, Paris, p. 12.
25. AFNOR Association Française de Normalisation (the French National Organization for quality regulations), Essai des eaux – Evaluation de la génotoxicité au moyen de larves de batraciens (*Pleurodeles waltl.*) – NF T 90–325, 1992, Paris, p. 15.
26. AFNOR Association Française de Normalisation (the French National Organization for quality regulations), Qualité de l'Eau. Evaluation de la génotoxicité au moyen de larves d'amphibien (*Xenopus laevis, Pleurodeles waltl*) – NF T 90–325, 2000, ICS: 13.020.40; 13.060.70., Paris, p. 17.
27. C. Zoll-Moreux and V. Ferrier, Etude comparative entre le test Jaylet et le test micronoyau Xénope, http://www.eau-seine-normandie.fr/fileadmin/mediatheque/Fiches_Etudes/fiches/ecol/94ecol3.htm, (accessed June 2016), 1995, 44, p. 317.
28. V. Ferrier, L. Gauthier and C. Zoll-Moreux, Genotoxicity tests in amphibians – A review, in *Microscale Testing in Aquatic Toxicology: Advances, Techniques and Practice*, ed. P. Wells, K. Lee and C. Blaise, 1998, p. 507.
29. P. Van Hummelen, C. Zoll, J. Paulussen, M. Kirsh-Volders and A. Jaylet, The micronucleus test in xenopus: a new and simple in vivo technique for detection of mutagens in fresh water, *Mutagenesis*, 1989, **4**(1), 12–16.
30. ISO International Standard, Water quality – Evaluation of genotoxicity by measurement of the induction of micronuclei – Part 1: evaluation of genotoxicity using amphibian larvae. ISO 21427–1, ICS: 13.060.70, Genova CH, 2006, p. 15.
31. L. Galien and M. Durocher, Tables de développement chez *Pleurodeles waltl*, *Bull. Biol. Fr. Belg.*, 1957, **91**, 97–117.
32. R. McGill, J. W. Tukey and W. A. Larsen, Variations of Box Plots, *Am. Stat.*, 1978, **32**(1), 12–16.
33. J. F. Madre, 2006. Software Mesurim. Académie d'Amiens, http://acces.ens-lyon.fr/acces/logiciels/mesurim/telechargement, (accessed June 2016).
34. M. Fernandez, L. Gauthier and A. Jaylet, Use of newt for *in vivo* genotoxicity testing of water: results on 19 compounds evaluated by the micronucleus test, *Mutagenesis*, 1989, **4**, 17–26.

35. F. Lecurieux, D. Marzin and F. Erb, Genotoxic activity of three carcinogens in peripheral blood erythrocytes of the newt *Pleurodeles waltl*, *Mutat Res.*, 1992, **283**(3), 157–160.
36. M. Fernandez, J. L'Haridon, L. Gauthier and C. Zoll-Moreux, Amphibian micronucleus test(s): a simple and reliable method for evaluating *in vivo* genotoxic effects of freshwater pollutants and radiations. Initial assessment, *Mutat. Res.*, 1993, **292**(1), 83–99.
37. F. Godet and P. Vasseur, Evaluation de la genotoxicité des effluents : étude comparative des tests d'Ames et micronoyaux Triton, 1994, http://www.km-dev.com/eaufrance/francais/etudes/pdf/etude_29.pdf, (accessed June 2016).
38. M. Fernandez and J. L'haridon, Influence of lighting conditions on toxicity and genotoxicity of various PAH in the newt *in vivo*, *Mutat. Res.*, 1992, **298**, 31–41.
39. J. L'haridon, M. Fernandez, V. Ferrier and J. Bellan, Evaluation of the genotoxicity of N-nitrosoatrazine, N-nitrosodiethanolamine and their precursors *in vivo* using the newt micronucleus test, *Water Res.*, 1993, **27**(5), 855–862.
40. J. E. Djomo, V. Ferrier, L. Gauthier, C. Zoll-Moreaux and J. Marty, Amphibian micronucleus test in vivo: evaluation of the genotoxicity of some major polycyclic aromatic hydrocarbons found in a crude oil, *Mutagenesis*, 1995, **10**, 223–226.
41. C. Zoll, E. Saouter, A. Boudou, F. Ribeyre and A. Jaylet, Genotoxicity and bioaccumulation of methyl mercury chloride *in vivo* in the newt *Pleurodeles waltl*, *Mutagenesis*, 1988, **3**, 337–343.
42. F. Mouchet, P. Landois, E. Flahaut, P. Pinelli and L. Gauthier, Assessment of the potential *in vivo* ecotoxicity of Double-Walled Carbon Nanotubes (DWNTs) in water, using the amphibian *Ambystoma mexicanum*, *Nanotoxicology*, 2007, **1**(2), 149–156.
43. A. Bour, F. Mouchet, L. Verneuil, L. Evariste, J. Silvestre, E. Pinelli and L. Gauthier, Toxicity of CeO2 nanoparticles at different trophic levels - Effects on diatoms, chironomids and amphibians, *Chemosphere*, 2015, **120**, 230–236.
44. L. Gauthier, Etude du pouvoir génotoxique des eaux de surface, potables ou en cours de traitement, par la formation de micronoyaux chez le triton: *Pleurodeles waltl*, PhD thesis, University Paul Sabatier France, 1989, p. 108.
45. A. Jaylet, L. Gauthier and M. Fernandez, Detection of mutagenicity in drinking water using a micronucleus test in newt larvae (*Pleurodeles waltl*), *Mutagenesis*, 1987, **2**, 211–214.
46. L. Gauthier, Y. Levy and A. Jaylet, Evaluation of the clastogenicity of water treated with sodium hypochlorite or monochloramine using a micronucleus test in the newt larvae (*Pleurodeles waltl*), *Mutagenesis*, 1989, **4**, 170–173.
47. A. Jaylet, L. Gauthier and Y. Levy, Detection of genotoxicity in chlorinated of ozonated drinking water using an amphibian micronucleus test,

in *Genetic Toxicology of Complex Mixtures: Short-term Bioassays in the Analysis of Complex Environmental Mixtures*, ed. M. D. Waters *et al.*, Plenum Press, New-York, 1990, vol. VI, pp. 71–80.
48. F. Lecurieux, L. Gauthier, F. Erb and D. Marzin, Use of the SOS chromotest, the Ames-fluctuation test and the newt micronucleus test to study the genotoxicity of four trihalomethanes, *Mutagenesis*, 1995, **10**(4), 333–341.
49. F. Lecurieux, S. Giller, L. Gauthier, F. Erb and D. Marzin, Study of the genotoxic activity of six halogenated acetonitriles, using the SOS chromotest, the Ames-fluctuation test and the newt micronucleus test, *Mutat. Res.*, 1995, **341**(4), 289–302.
50. S. Giller, F. Lecurieux, L. Gauthier, F. Erb and D. Marzin, Genotoxicity assay of chloral hydrate and chloropicrine, *Mutat. Res.*, 1995, **348**, 147–152.
51. M. A. Van Der Gaag, L. Gauthier, A. Noordslj, Y. Levi and M. N. Wrisberg, Methods to measure genotoxins in waste water: evaluation with *in vivo* and *in vitro* tests, in *Genetic Toxicology of Complex Mixtures: Short-term Bioassays in the Analysis of Complex Environmental Mixtures*, ed. M. D. Waters *et al.*, Plenum Press, New-York, 1990, vol. VI, pp. 215–232.
52. L. Gauthier, M. A. Van der Gaag, J. L'Haridon, V. Ferrier and M. Fernandez, *In vivo* detection of waste and industrial effluent: use of the newt micronucleus test (Jaylet test), *Sci. Total Environ.*, 1993, **138**, 249–269.
53. L. Gauthier and V. Ferrier, Effluents autoroutiers et tests de toxicité, in *Autoroutes et aménagements, Interactions avec l'environnement*, ed. Presses Polytechniques et Universitaires Romandes, 2004, p. 328.
54. M. Fernandez and A. Jaylet, An antioxidant protects against the clastogenic effects of benzo(a) pyrene in the newt *in vivo*, *Mutagenesis*, 1987, **2**, 293–296.
55. J. Marty, P. Lesca, A. Jaylet, C. Ardourel and J. L. Rivière, *In vivo* and *in vitro* metabolism of benzo(a)pyrene by the larva of the newt, *Pleurodeles waltl*, *Comp. Biochem. Physiol., C.*, 1989, **93**(2), 213–219.
56. J. Marty, J. L. Rivière, M. J. Guinaudy, P. Kremers and P. Lesca, Induction and characterization of cytochromes P450IA and -IIB in the newt, *Pleurodeles waltl*, *Ecotoxicol. Environ. Saf.*, 1992, **24**, 144–154.
57. F. Calevro, D. Beyersmann and A. Hartwig, Effect of cadmium (II) on the extent of oxidative DNA damage in primary brain cell cultures from Pleurodeles larvae, *Toxicol. Lett.*, 1998, **94**(3), 217–225.
58. O. Östling and K. J. Johanson, Microelectrophoretic study of radiation-induced DNA damages in individual mammalian cells, *Biochem. Biophys. Res. Commun.*, 1984, **123**(1), 291–298.
59. N. P. Singh, M. T. McCoy, R. R. Tice and L. E. Schneider, A simple technique for quantitation of low levels of DNA damage in individual cells, *Exp. Cell. Res.*, 1988, **175**, 184–191.
60. P. L. Olive, J. P. Banath and R. E. Durand, Heterogeneity in radiation-induced DNA damage and repair in tumor and normal cells measured using the "comet" assay, *Radiat. Res.*, 1990, **122**, 86–94.
61. D. W. Fairbairn, P. L. Olive and K. L. O'Neill, The comet assay: a comprehensive review, *Mutat. Res.*, 1995, **339**, 37–59.

62. S. Ralph, M. Petras, R. Pandrangi and M. Vrzoc, Alkaline single cell gel (comet) assay and genotoxicity monitoring using two species of tadpoles, *Environ. Mol. Mutagen.*, 1996, **28**, 112–120.
63. C. Clements, S. Ralph and M. Petras, Genotoxicity of select herbicides in Rana catesbeiana tadpoles using the alkaline single-cell gel DNA electrophoresis (comet) assay, *Environ. Mol. Mutagen.*, 1997, **29**, 277–288.
64. S. Ralph and M. Petras, Caged amphibian tadpoles and in situ genotoxicity monitoring of aquatic environments with the alkaline single cell gel electrophoresis (comet) assay, *Mutat. Res.*, 1998, **413**(3), 235–250.
65. S. Ralph and M. Petras, Comparison of sensitivity to methyl methane sulphonate among tadpole developmental stages using the alkaline single-cell gel electrophoresis (comet) assay, *Environ. Mol. Mutagen.*, 1998, **31**(4), 374–382.
66. F. Mouchet, L. Gauthier, C. Mailhes, V. Ferrier and A. Devaux, Comparative study of the comet assay and the micronucleus test in amphibian larvae (*Xenopus laevis*) using benzo(a)pyrene, ethyl methane sulfonate, and methyl methane sulfonate: Establishment of a positive control in the amphibian comet assay, *Environ. Toxicol.*, 2005, **20**, 74–84.
67. F. Mouchet, C. Mailhes, L. Gauthier, V. Ferrier, M. J. Jourdain and A. Devaux, Biomonotoring of the genotoxic potential of draining water from dredged sediments, using the comet and micronucleus tests on amphibian (*Xenopus laevis*) and bacterial assays (Mutatox and Ames tests), *J. Toxicol. Environ. Health, Part A*, 2005, **68**, 811–832.
68. F. Mouchet, G. Gauthier, C. Mailhes, V. Ferrier, M. J. Jourdain, G. Triffault and A. Devaux, Biomonitoring of the genotoxic potential of aqueous extracts of soils and bottom ash resulting from municipal solid waste incineration, using the comet and micronucleus tests on amphibian (*Xenopus laevis*) larvae and bacterial assays (Mutatox and Ames tests), *Sci. Total Environ.*, 2006, **355**, 232–246.
69. F. Mouchet, L. Gauthier, C. Mailhes, V. Ferrier and A. Devaux, Comparative evaluation of the genotoxicity of captan in amphibian larvae (*Xenopus laevis* and *Pleurodeles waltl*) using the comet assay and the micronucleus test, *Environ. Toxicol.*, 2006, **21**(3), 264–277.
70. F. Mouchet, L. Gauthier, M. Baudrimont, P. Gonzalez, C. Mailhes, V. Ferrier and A. Devaux, Comparative evaluation of the toxicity and genotoxicity of cadmium in amphibian larvae (*Xenopus laevis* and *Pleurodeles waltl*) using the comet assay and the micronucleus test, *Environ. Toxicol.*, 2007, **22**(4), 422–435.
71. F. Calevro, S. Campani, C. Filippi, R. Bastitoni, P. Deri, S. Bucci, M. Ragghianti and G. Mancino, Bioassays for testing effects of Al, Cr and Cd using development in the amphibian Pleurodeles waltl and regeneration in the planarian Dugesia etrusca, *Aquat. Ecosyst. Health Manage.*, 1999, **2**(3), 281–288.
72. F. Calevro, S. Campani, M. Ragghianti, S. Bucci and G. Mancino, Tests of toxicity and teratogenicity in biphasic vertebrates treated with heavy metals ($Cr3+$, $Al3+$, $Cd2+$), *Chemosphere*, 1998, **37**(14–15), 3011–3017.

73. S. Flament, S. Kuntz, A. Chesnel, I. Grillier-Vuissoz, C. Tankozic, M. Penrad-Mobayed, G. Auque, P. Shirali, H. Schroeder and D. Chardard, Effect of cadmium on gonadogenesis and metamorphosis in *Pleurodeles waltl.* (urodele amphibian), *Aquat. Toxicol.*, 2003, **64**(2), 143–153.
74. M. E. Ortiz, A. Marco, N. Saiz and M. Lizana, Impact of ammonium nitrate on growth and survival of six European amphibians, *Arch. Environ. Contam. Toxicol.*, 2004, **47**, 234–239.
75. C. Robles-Mendoza, C. García-Basilio, S. Cram-Heydrich, M. Hernández-Quiroz and C. Vanegas-Pérez, *Chemosphere*, 2009, **74**(5), 703.
76. C. Robles-Mendoza, S. R. Zúñiga-Lagunes, C. A. Ponce de León-Hill, J. Hernández-Soto and C. Vanegas-Pérez, Esterases activity in the axolotl Ambystoma mexicanum exposed to chlorpyrifos and its implication to motor activity, *Aquat. Toxicol.*, 2011, **105**(3–4), 728–734.
77. E. Ortiz-Ordoñez, E. López-López, J. E. Sedeño-Díaz, E. Uría, I. A. Morales, M. E. Pérez and M. Hibayama, Liver histological changes and lipid peroxidation in the amphibian Ambystoma mexicanum induced by sediment elutriates from the Lake Xochimilco, *J. Environ. Sci.*, 2016, **46**, 156–164.
78. J. Shoots, D. Fraccalvieri, D. G. Francks, M. S. Denison, M. E. Hahn, L. Bonati and W. H. Powell, An Aryl Hydrocarbon Receptor from the Salamander *Ambystoma mexicanum* Exhibits Low Sensitivity to 2,3,7,8-Tetrachlorodibenzo-*p*-dioxin, *Environ. Sci. Technol.*, 2015, **49**(11), 6993–7001.
79. A. Bour, F. Mouchet, S. Cadarsi, J. Silvestre, L. Verneuil, D. Baque, E. Chauvet, J. M. Bonzom, C. Pagnout, H. Clivot, I. Fourquaux, M. Tella, M. Auffan, J. Silvestre, L. Gauthier and E. Pinelli, Toxicity of CeO_2 nanoparticles on a freshwater experimental trophic chain: a study in environmentally relevant condition through the use of mesocosms, *Nanotoxicology*, 2015, **10**(2), 245–255.
80. A. Bour, F. Mouchet, J. Silvestre, L. Gauthier and E. Pinelli, Environmentally relevant approaches to assess nanoparticles ecotoxicity: a review, *J. Hasard. Mater.*, 2015, **283**, 764–777.
81. A. K. Davis and T. M. Floyd, Evaluating levels of genotoxic stress in eastern hellbenders (*Cryptobranchus alleganiensis alleganiensis*) using the erythrocyte micronucleus assay, *Comp. Clin. Pathol*, 2013, **23**(5), 1189–1193.
82. I. Udroiu, A. Sgura, L. Vignoli, M. A. Bologna, M. D'Amen, D. Salvi, A. Ruzza, A. Antoccia and C. Tanzarella, Micronucleus test on Triturus carnifex as a tool for environmental biomonitoring. Environ, *Environ. Mol. Mutagen.*, 2015, **56**(4), 412–417.
83. F. Godet, P. Vasseur and M. Babut, Essais de génotoxicité *in vitro* et *in vivo* applicables à l'environnement hydrique, *Rev. Sci. Eau*, 1993, **6**(3), 285–314.
84. C. Zoll-Moreux, Conséquence de la contamination du milieu hydrique par des sels de mercure, du benzo(a)pyrène ou des pesticides organochlorés, chez deux amphibiens *Pleurodeles waltl* et Xenopus laevis, PhD thesis, University Paul Sabatier France, 1991.

CHAPTER 16

The Use of Aquatic Life-stages of European Amphibians in Toxicological Studies

NORMAN WAGNER* AND BRUNO VIERTEL

Trier University, Department of Biogeography, Universitätsring 15, 54296 Trier, Germany
*Email: wagnern@uni-trier.de

16.1 Introduction

Worldwide, amphibians are the most threatened vertebrate group. One-third of the nearly 7500 described species are threatened with extinction.[1–4] There are multiple causes for the global amphibian decline, which can also act in synergistic and cumulative ways.[5] Furthermore, factors are regionally specific, for instance, the emerging infectious disease chytridiomycosis—caused by the amphibian chytrid fungus *Batrachochytrium dendrobatidis*—is supposed to be the main driver of catastrophic species and population declines from mountain and remote areas in the Tropics.[6,7] However, the causes for area-wide amphibian population declines are mainly of anthropogenic origin and environmental contamination—especially the use of large quantities of pesticides—is one of six suggested main factors.[8–10] Pesticide use is increasing to satisfy the overwhelming food demand of the growing human population on limited areas of productive agricultural land.[11] Increasing use of pesticides is negatively correlated with farmland biodiversity.[12] Many amphibian populations are also persisting in cultivated landscapes, which have been largely transformed for agriculture.[13]

Issues in Toxicology No. 33
Ecotoxicology and Genotoxicology: Non-traditional Aquatic Models
Edited by Marcelo L. Larramendy

Published by the Royal Society of Chemistry, www.rsc.org

Environmental contamination of aquatic amphibian habitats with pesticides can occur by way of direct over-spraying,[14] wind drift,[15] run-off and drainage.[16,17] Wagner *et al.*[18] showed that, even within protected areas, strictly protected European amphibian species can be at high risk of pesticide exposure. Smalling *et al.*[19,20] observed that higher pesticide amounts accumulated in frog tissues compared to water and sediment samples. Although clear evidence of causal relationships between population declines and increasing use of agrochemicals is widely lacking,[21] Mann *et al.*[13] concluded that the increasing use of agrochemicals and the resulting impacts on anuran health have the potential to be significant. In several laboratory and field studies, negative effects of environmentally relevant pesticide concentrations at the individual level have been observed, which included in addition to acute toxicity, mainly chronic effects and delayed effects, such as reduced body indices and prolonged or shortened metamorphosis.[22–26] In particular, such—more subtle—effects are difficult to assess for amphibians in the field. For example, mortality of larvae during pond drying can be remarkably increased due to prolonged metamorphosis (induced by sublethal pesticide concentrations).[27] It becomes even harder when other biotic and abiotic factors are considered, like the presence of predators,[28,29] competition between larvae,[30] different water chemistry[31] or contamination with pesticide mixtures.[32,33] Furthermore, exposure risks and adverse effects of pesticides were found to be species- and formulation-specific.[22,34,35] The latter can be mainly caused by differences in the added substances and not different active ingredients.[21] Additionally, reproductive toxicity of pesticides is specific for particular developmental processes.[36–38]

There are different objectives for toxicity testing using aquatic life-stages of amphibians:

(1) To achieve basic knowledge on the impairment of developmental steps under dosages high enough to provoke toxicity. Traditionally, the African clawed frog (*Xenopus laevis*, all species according to Frost[4]) is implemented in the Frog Embryo Teratogenesis Assay-*Xenopus* (FETAX) to detect the teratogenic potential of chemicals.[39] The metamorphosis assay[40] is recommended to study effects on the endocrine system in terms of endocrine disruption. *X. laevis* and other anuran clades are proposed by the guidelines as laboratory species (developmental toxicity studies).[41]
(2) To study effects on development after exposure to relatively low dosages that are realistic in the agricultural landscape (ecotoxicological developmental studies).
(3) In one experimental trial, both objectives can be co-located in respect of appropriate dosages. Dose-range finding studies are recommended to ensure the validity of the study.

In this chapter, larval amphibians distributed in Europe are discussed in terms of use as laboratory species. One main reason is the detection of the above-mentioned species-specific effects of pesticides on development. For practical reasons not all European species can be investigated. The representativity of their reproductive biology on the one hand and easy breeding and husbandry on the other are criteria for species selection. Out of the three lissamphibian orders, the salamanders and newts (urodeles) and the frogs (anurans) are candidates for toxicity testing. The third order, the non-European gymnophiones, is not under discussion here.

16.2 Aquatic Life-stages of Amphibians as Test Organisms in Toxicological Studies

Amphibians are the only tetrapods (four-legged and terrestrial vertebrates) that develop free living aquatic larvae. Their basic type is represented by the predatory larvae of the urodeles (salamanders and newts), resembling in outlines the larva of the ancestor of the recent amphibians (lissamphibia), which are in all probability the carboniferous temnospondyl dissorophoids.[42] Larval traits in common are the broad head clearly distinct from the long trunk and a long tail. Three pairs of outer bushy gills are laterally positioned at the posterior margin of the head in the level of a ventral cutaneous fold. The lower and upper jaws are tooth bearing. In some taxa tooth-like structures appear on the branchial arches (visceral or pharyngeal skeleton) forming gill rakers to prevent the prey from escaping through the gill slits. Salamander larvae are sit-and-wait predators. The prey is ingested by suction feeding supported by the broad mouth and the pharyngeal skeleton and the robust musculature.[43] The gills are ventilated only by temporal movement or shivering. Most larvae have lungs.[44] At an early stage salamander larvae develop extremities, in many taxa first the fore limbs and slightly later the hind limbs. Balancers occur at the cheeks. Metamorphosis proceeds in a long time axis. The larvae accomplish the habit of the adults step-by-step, indicated by the differentiation of hands and feet, change of the shapes of head and trunk, loss of gills and tail fins, increase of pigment density and change of pigment pattern.[45,46] Incorporation of xenobiotics during food ingestion or *via* the skin and the gills are conceivable. When catching the prey, amounts of water get in close contact to the oral cavity and may be swallowed together with the prey and incorporated *via* the buccal epithelium and the digestive tract. The role of the amphibian skin and the gills in the transport of chemicals is outlined below.

There is less experience with salamander larvae in toxicity studies than in anurans. This is caused by the small clutches of different age in most species and the carnivorous mode of nutrition. Both are not an advantage for toxicity studies. A sufficient number of eggs of the same age is necessary to achieve an appropriate sample size. Availability of animal food with contents stated by analysis is not always guaranteed. However, information on food properties is a must in toxicity studies (see below). As already mentioned, species selection should take into account the practicability of the respective species in the studies. Spotted salamander (*Ambystoma maculatum*) and tiger salamander (*A. tigrinum*)—distributed in North America—are proposed by the OECD guidelines.[41] Their reproductive biology resembles the wide-spread terrestrial European newt species smooth newt (*Lissotriton vulgaris*), palmate newt (*Lissotriton helveticus*), alpine newt (*Ichthyosaura alpestris*) and northern crested newt (*Triturus cristatus*). In spring they migrate to pools and ditches and, after inner fertilization *via* spermatophor, glue single eggs to the submerged plants. The normal tables of development characterize developmental steps by outer morpholgical traits (see below). The normal table of *A. maculatum* is also applicable for the European newt species.[46,47] *Lissotriton vulgaris*, mainly distributed in the lowland, and *Lissotriton helveticus*, in the colline zone, may function as surrogate species for urodeles.

The phlylogenetic origin of anuran amphibians (frogs and toads) remains unclear. The fossil record does not provide evidence of transitory forms from the temnospondyl occurrence to the recent anuran morphology.[42,48,49] Anuran larvae are unique within vertebrates.[50,51] They differ in many traits from salamander larvae. Most of their features that are also in the focus of toxicological risk assessment are the result of dramatic change from the ancestral amphibian morphology. The presumably pre-existing temnospondyl physiognomy is phylogenetically transformed into a short and roundish head-trunk region without separation of the two portions. The tail is longer than the body and is not the ancestral tail but a new structure, as demonstrated by vertebral ontogeny.[52] Limb buds occur early during premetamorphosis. However, differentiation starts not before the onset of prometamorphosis, which is late in anuran life.[53,54] The embryonic gills are gradually covered by a lateral and ventral cutaneous fold, the operculum, which encloses a peribranchial chamber but leaving open one or two orifices, the spiraculum/spiracula. The formerly outer embryonic gills are reduced and at their base new gills develop, remaining until metamorphosis. The gills are ventilated by permanent fish-like movements of the oral cavity and the pharynx (buccopharynx). The oral orifice is small. Teeth are missing and are replaced in functionality by a horny beak and horny teeth on a field around the mouth, the oral disc (except in pipids and microhylids). The most thrilling novelty is the formation of a filter apparatus posterior of the buccal cavity by means of conversion of parts of the esophagus, the pharynx and the posterior region of the buccal cavity.[55] A mucous entrapment on an increased inner surface interacting with the water stream of ventilation enables suspension feeding.[56–59] Particles suspended and transported by the water are entangled by the mucus and are transported by ciliary cells of the enlarged esophagus into the digestive tract.[60,61] The peribranchial chamber drains the water leading outwardly *via* the spiracula, which function as egestion openings. A completely new feeding strategy within amphibians arises. Anuran larvae are facultative or permanent suspension-feeders. Metamorphosis is a short period of around five days with condensed processes of complete or partial reduction and remodeling of organ systems, exceeding the transition in salamanders by far due to the transition of an aquatic omnivorous suspension feeder into a terrestrial carnivore.[53,54] Suspension feeding implicates easy husbandry of anuran larvae. They should be nourished with powdered commercial fish food fulfilling the nutritional requirements. The ingredients should be stated by analysis to be on the safe side so as not to supply food with biologically inadequate nourishment and not to induce effects of malnutrition, which could be mixed up with possible signs of intoxication.

In toxicological studies, two different phylogenetic lines of anuran larvae are employed. In general *Xenopus laevis*, a pipid from southern Africa, became the classic laboratory frog due to its easy handling and breeding, which can be initiated by injection of Human Chorionic Gonadotropin (HCG). The clawed toad was employed in the Frog Embryo Teratogenesis Assay-*Xenopus* (FETAX)[39] and soon became a reference species for subsequent toxicity studies with non-pipid anurans (Figure 16.1 a, b).[25–29,62–64]

Four larval types or morphotypes are described considering external traits, such as the number and position of the spriraculum, or the presence of horny structures and an oral disc,[65] and were attributed to anuran families and phylogenetic lineages.[66] They comprise larval type I of the pipidae (*Xenopus*) and rhinophrynidae (two pairs of latero-ventrally positioned spiracula, horny structures and oral disc missing), larval type III of the Ascaphidae and Alytidae (also known as Discoglossidae, but the older name Alytidae has priority) (*Alytes*, *Latonia* and *Discoglossus*; a single short ventro-median spiraculum, horny beaks and an oral disc with horny teeth) and larval type IV of all remaining taxa (80% of all recent anurans; a single left spiraculum, horny beaks and an oral disc with horny teeth). Analysis of morphological traits in connection with data from molecular evolution of anurans and their larvae demonstrates that larval type III is the basic type and larval type IV is derived with comparably minor alterations. Larval type I occurs in phylogenetic isolation and is understood as the result of extensive apomorphous alterations of larval type III.[4,67–69] (The microhylid larval type II [a single ventro-median long spiraculum, horny beaks and oral disc with horny teeth missing] is derived from the larval type IV and is here not under discussion.) The different apomorphous morphological, and in consequence functional, traits of larval types may lead to different exposure to toxic agents. Two contact zones with xenobiotics are conceivable. The inner surface of the buccopharynx and the outer body surface. Plenty of water is in close contact with the epithelia of the oral cavity, the filter apparatus and the gills during ventilation.[70–73] The amount of water pumped through the buccopharynx differs by species. The highest rate is achieved by *X. laevis*, followed by the larval type IV species *Epidalea calamita* (natterjack toad). Common toad (*Bufo bufo*) larvae lay at the opposite end of the scale and common frog (*Rana temporaria*) between *E. calamita* and *B. bufo*.[58,59] This is a distinct difference to salamander larvae. The body surface of the aquatic life-stages of both groups plays a crucial role in gas exchange.[74,75] The permeability of the skin, including the gills, to xenobiotics seems necessary to assume. It has to be suggested that different ontogenetic stages are sensitive

Figure 16.1 Anura. The evaluation of the early larva of *Xenopus laevis* (larval type I), lateral view in (a) and ventral view in (b), is part of the classic Frog Embryo Teratogenesis Assay-*Xenopus* (FETAX). Due to its wide distribution, *Rana temporaria* is often introduced into studies of toxicity. The images show newly hatched embryos (c), late embryos with external gills (d) before transition into the larva (e). A late embryo (f), just hatched, an early larva (g) and a larva during middle prematamorphosis (h) of *Bombina bombina* demonstrate the developmental changes in the transition of the embryonic phase to the free living larva (larval type III). The early *Discoglossus pictus* larva (i) represents an alternative species for toxicity testing of amphibians (larval type III). Scale bars equal 1 mm in (a), (b), (f), (g) and (i), 2 mm in (c), 3 mm in (d) and 4 mm in (e) and (h). Photographs (a), (b) and (i) taken by Bruno Viertel, photographs (c), (d), (e), (f), (g) and (h) from Burkhard Thiesmeier; reprinted with permission of Laurenti-Verlag.

in different ways to xenobiotics. Fundamental early ontogenetic steps, such as fertilization (formation of the body axis), cleavage (cell proliferation), gastrulation (formation of the organizer and induction of organ anlagen) and neurulation (early formation of neural organs and neural crest), proceed to some extent in isolation from the environment.[76] During late embryonic organogenesis the jelly coat becomes macerated. Contact with the environment increases. Studies on the transfer of xenobiotics through the jelly coat and egg membrane demonstrate that the early embryonic stages are not completely protected against external influences.[77–79]

After hatching development continues and the premetamorphic larva arises, demonstrated by closure of the operculum, formation of the filter apparatus and tail driven locomotory activity due to the differentiation of the musculature and nervous system (Table 16.1). Premetamorphic larvae are feeding and growing stages, especially the early larvae as observed during filtering experiments.[58,59] Marked exposure to xenobiotics is strongly suggested. Continuous steps of organogenesis include the development of the endocrine organs.[55] During metamorphosis, the larvae are shrinking and lose biomass. Organs contributing to metabolism, such as the digestive tract including the larval stomach, liver and pancreas, are dramatically subjected to histological and functional change. Already at early prometamorphosis food ingestion decreases,[58,59] lowering the exposure to xenobiotics.

Experimental design has to be adjusted to the steps of embryonic development, larval growth and differentiation phase and the deep cuts during metamorphosis (Table 16.1). The genuine FETAX[39,83,84] experiment is also the basis for embryonic development testing in non-pipids (Table 16.1). The necessity of testing chemicals for effects on endocrine function was the reason for the implementation of the metamorphosis assay (Table 16.1).[40,41] Variations of these protocols based on the demands of the present guidelines or upcoming guidelines are conceivable. According to the objective of the study, distinct steps or short phases of development may be selected for treatment. Furthermore, variations of the methods are permissive if the rationale is feasible. For instance removal of the jelly coat and the egg membrane is under discussion. There are two lines. Concerns that L-cysteine, applied to remove the jelly coat chemically, could harm the embryos and to provide a more natural condition of exposure.[85] However, dejellying may be adequate for special purposes, such as special teratogenicity studies to directly expose the embryos to the chemical. This is also true for toxicity studies with salamander embryos. Mechanical dejellying may be suitable to minimize harm and changes producing artefacts.

Development of anurans (and urodeles) enables valuable possibilities to adjust the study design to the goal of the experiment. Furthermore, complete life cycle testing, including post-metamorphic development until reproduction, is also conceivable. For all experiments, one or two control groups are a must to get better a handle on the biological variability of the species and to increase the amount of background data for further studies.

The normal tables of development are helpful (*X. laevis:* Nieuwkoop and Faber,[81] larval type III and IV: Gosner[80]), indicating developmental steps by external morphological traits. However, the outer morphological development does not proceed in the same time line in all anuran groups and external development does not exactly mirror the steps of inner developmental processes in a comparative way. This is especially true for embryonic organogenesis after neurulation, the onset of metamorphic climax, the time scale of limb development, such as the time point of protrusion of the fore

Table 16.1 Comparison of normal tables of development of European anurans (according to Gosner[80]) with *X. laevis* as reference species (according to Nieuwkoop and Faber[81]). Onset of climax of metamorphosis, comparison according to: + = McDiarmid and Altig[82] or ++ = OECD.[41] Treatment periods according to the guidelines: T = teratogenicity testing (ASTM[39]), M = metamorphosis assay.

Gosner	Nieuwkoop and Faber		Treatment periods
European larval types III and IV	Larval type I *X. laevis*	External traits	
Embryonic development			
1	1	Fertilization	
2	1	Axis formation	
3–8	2–8	Cleavage	
8/9	8/9	Mid-late blastula	
9	9	Late blastula	T
10	10	Early gastrula	T
12	12	Late gastrula	T
13	13	Early neurula	T
15	17	Late neurula	T
16	18	Neural groove	T
16	19	Initial neural tube	T
17	20	Neural folds fused, still suture	T
18	26	Tail bud	T
19–23	27–43	Continuing organogenesis	T
24	44	Late embryo	T
Premetamorphosis (larval development)			
25	46	Early larva, closed opercula	T
27–33	47–53	Continuous larval organogenesis and growth	
29	51		M1,[a] M[b]
34/35	54	Late larva	M1,[a] M[b]
Prometamorphosis			
36–38	54/55	Onset of hind limbs growth	M1,[a] M2,[a] M4,[a] M[b]
39/40+	56+	Onset of foot differentiation	M1,[a] M2,[a] M4,[a] M[b]
39/40++	57++		M1,[a] M2,[a] M4[a]
Climax of metamorphosis			
41+	58 +	Onset of tail resorption	M1,[a] M2,[a] M3[a]
41++	60–61++	Onset of tail resorption	M1,[a] M2,[a] M3[a]
42	58	Protrusion of the fore limbs	M1,[a] M2,[a] M3[a]
42–45	60–65	Onset of gradual change of the facies, differentiation of hands and feet, gradual change of colour pattern, decrease of food intake, change of behavior	M1,[a] M2,[a] M3[a]
46	66	A frog	

[a] = according to OECD.[41]
[b] = according to = OECD.[40]

limbs and differentiation of the feet, and the change of head morphology (Table 16.1). Different results are achieved according to the morphological criteria the authors select[41,81] (Table 16.1) with the result of a lack of conceptual clarity. Treatment periods should be designed with overlaps of the respective developmental stages to compensate for this weakness. The only normal table combining external and inner development in an extensive way was designed for *X. laevis*.[81]

The choice of species is a crucial prerequisite for the reasonableness of chemical testing. *X. laevis* became the classic laboratory frog because of its easy handling and breeding and was the initial species introduced into chemical testing (FETAX[39]). A huge mass of valuable background data is available, including results from molecular developmental biology research. Several non-pipid anuran species are proposed by the OECD[41] to be implemented into the metamorphosis assay. They comprise the non-European species *Lithobates pipiens* (previously *Rana pipiens*) and *Lithobates catesbeianus* (previously *Rana catesbeiana*), both from northern America (ranidae, larval type IV), and *Glandirana rugosa* (previously *Rana rugosa*, ranidae, Japan, Korea, China, south Siberia, introduced into the United States, larval type IV). The weakness of some species is the availability of eggs, which can only be obtained outside the regular breeding season by artificial fertilization. The pros and cons of species selection are under discussion.[41]

For testing of European species, *Rana temporaria* (Figure 16.1c–e) seems to be suitable (see also ref. 41). The eggs can be sampled in the field. As a result of the wide-spread distribution and the appearance in the valleys up to the mountains, the breeding season is stretched and the fertilized eggs can be collected easily for 3–4 weeks in early spring, though the species is an explosive breeder (Table 16.2). A phylogenetically more basic species is the European fire-bellied toad, *Bombina bombina* (Figure 16.1, f–h, Table 16.2). The species was implemented by Larsen and colleagues[86,87] for pesticide testing. The Mediterranean painted frog (*Discoglossus pictus*, Figure 16.1, i) was introduced from Northern Africa to Southern Europe, but there are also other species in Northern Africa (*D. scovazzi*) and Southern Europe (*D. galganoi, D. jeanneae, D. montalenti, D. sardus*). The benefits of these species, especially *D. scovazzi* and *D. pictus*, are easy handling, spontaneous permanent breeding and fast development (Table 16.2). These species are understood as a practicable surrogate species for toxicity testing in respect of European anuran larvae. An additional advantage is the phylogenetically basic position of the Alytidae and Bombinatoridae, which may include more ancestral physiological functions. For all these 'new' test species the amount of background data is clearly smaller than in *X. laevis*, but this is growing with the implementation of these species in toxicity studies (see also the next paragraph of this chapter).

The endpoints can be divided into two groups. (1) General endpoints are mortality, clinical signs, growth, development and external morphology in respect of malformations. They should be documented independent of the rationale of the experiment. Mortality is measured by counting surviving

Table 16.2 European anuran species proposed for toxicity testing in comparison with *X. laevis*.[39–41] PMG = Pregnant Mare Gonadotropin, HCG = Human Chorionic Gonadotropin.

Distribution	Breeding	Developmental period	Sexual maturity	References
Discoglossus spec., Alytidae, larval type III				
Southern Europe and Northern Africa	Permanent, spontaneous	1–2 months	6–8 months	24
Bombina bombina, Bombinatoridae, larval type III				
East of Central Europe, Eastern and South-Eastern Europe	For oocyte maturation injection of PMG and HCG for breeding	2.5 months	2–3 years	87
Rana temporaria, Ranidae, larval type IV				
Western, Central, West, Eastern Europe and South-Eastern Europe	Explosive, spontaneous	3–4 months in the field and around 2 months in the laboratory	Sexual maturity is achieved after 3 years	41
Xenopus laevis, Pipidae, larval type I				
Southern Africa	Injection of HCG	2 months	1 year	81

larvae. Criteria for life functions are provided by heartbeat and blood circulation in the gills. Clinical signs comprise any change of behavior, including activity and narcosis. Growth should be documented by measurement of body length, tail length (or total length, head–tail length) and, if necessary in the case of deformation, also maximum body width, maximum body height, internarial distance, interorbital distance, maximum tail height, length of anal tube, tail muscle height and tail muscle width.[40,89] Application of a measuring eyepiece in a stereomicroscope or measurement by photodigitalization are recommended. Body weight as an endpoint is problematic for living aquatic animals from the methodological point of view due to the variable amounts of water the larvae may carry to the balance. Developmental delay should be detected with the help of the respective table of normal development. Hind limb length and development is a valuable endpoint for premetamorphic development, onset of prometamorphosis and metamorphic climax. Change of external morphology should be documented with the help of the 'Atlas of Abnormalities: a Guide for the Performance of FETAX'[83] and with the respective normal table. (2) Special endpoints are dependent on the rationale of the experiment. In the metamorphosis assay,[41] steps of organogenesis and still more of the functioning of the endocrine system are monitored. Besides limb development, as outlined above, tail resorption, skin and thyroid development are markers of possible endocrine disturbance. The properties of the skin change during metamorphosis and provide valuable markers. Increased pigment density can be especially observed in the initially transparent premetamorphic larvae of *X. laevis*. This is less the case in the European species due to their early larval pigmentation. However, in the species proposed for toxicity testing (Table 16.2) gradual change during prometamorphosis is noticeable. Staining with eosin or hematoxylin/eosin (HE) may be useful. In addition, immunohistochemical techniques can be applied to detect proteins specific for the skin of the adults. A crucial endpoint in the metamorphosis assay is thyroid development. The size of the thyroid can be measured through the skin (measuring eyepiece or measurement by photodigitalization are recommended). Histopathological examination after HE staining is suitable. Scanning electron microscopy (SEM) and transmission electron microscopy (TEM) are helpful to detect changes on the cell and organelle level. Functional analysis includes measurement of the thyroxin prohormone (T4) and triiodothyronine (T3), and if possible also monoiodotyrosine (MIT) and diiodotyrosine (DIT). Iodothyronine deionase catalyzes conversion of T4 to T3. The other isoform of the enzyme inactivates T4 and T3 due to their conversion to T2 (diiodothyronine) and reverse T3. Measurement of these biochemical markers, including the transport proteins of thyroid hormones and parameters from thyroid hormone receptor biology (*e.g.* receptor binding assays), provides valuable information on the activity of the thyroid. Gene expression assays are conceivable for special purposes.[41]

16.3 Short Review of Toxicological Studies Using Aquatic Life-stages of European Amphibians

One prime question remains: "What is known about the effects of pesticides on the aquatic life-stages of European amphibians?" For this purpose, we conducted a literature and database review and searched the Web of Knowledge and Google Scholar for toxicological studies on the impact of pesticides on European amphibians using the scientific names of all European species + pesticide* or agrochemical* as keywords. Furthermore, we employed the database ECOTOX (US Environmental Protection Agency [USEPA] http://cfpub.epa.gov/ecotox/) and the EFSA report by Fryday and Thompson.[90] We examined the references of the retrieved publications for further information.

16.3.1 European Newts and Salamanders

Only eight studies on the aquatic toxicity of pesticides on European urodeles could be found; only five of them used embryos or larvae (Table 16.3).

The first study using a European urodele species was conducted in the 1970s and investigated lethal effects and clinical signs of DDT exposure in smooth newts (*Lissotriton vulgaris*) (Figure 16.2a,b).[91] Zaffaroni *et al.*[92]

Table 16.3 Overview of studies on the aquatic toxicity of pesticides using European urodele species as test organisms (ordered chronologically).

CAS	Pesticide name	Pesticide type	Formulation	Order	Species	Life-stage	96 h LC_{50} (mg a.i. L^{-1})	Considered endpoints	References	Remarks
50-29-3	DDT	Insecticide	No	Urodela	*Lissotriton vulgaris*	Larvae	Not stated	Survival, clinical signs	91	
12427-38-2	Maneb	Fungicide	No	Urodela	*Triturus carnifex*	Adult	Not stated	Survival, effects on peripheral blood cells	92	*T. carnifex* was previously recognized as a subspecies of *T. cristatus*
1066-45-1	Trimethyltin chloride	Insecticide	No	Urodela	*Triturus carnifex*	Larvae	Not stated	Survival, neurotoxicity	93	Not only aquatic exposure but another trial with intraperitoneal injection
133-06-2	Captan	Fungicide	No	Urodela	*Pleurodeles waltl*	Larvae	Not stated	Genotoxicity, survival, clinical signs (diminished food intake), growth inhibition	96	Genotoxicity investigated by micronucleus test and comet assay
61-82-5	Amitrole	Herbicide	No	Urodela	*Salamandra salamandra*	Larvae	Not stated	Effects on predatory behavior	97	

1071-83-6	Glyphosate	Herbicide	Yes	Urodela	*Chioglossa lusitanica*	Embryos	Not stated	Survival, growth inhibition, hatching time, developmental stage at hatching	98	Additional treatments in combination with ammonium nitrate
1071-83-6	Glyphosate	Herbicide	No	Urodela	*Ichthyosaura alpestris*	Adult	Not stated	Avoidance of contaminated water	99	
1071-83-6	Glyphosate	Herbicide	Yes	Urodela	*Ichthyosaura alpestris*	Adult	Not stated	Avoidance of contaminated water	99	
1071-83-6	Glyphosate	Herbicide	No	Urodela	*Lissotriton helveticus*	Adult	Not stated	Avoidance of contaminated water	99	
1071-83-6	Glyphosate	Herbicide	Yes	Urodela	*Lissotriton helveticus*	Adult	Not stated	Avoidance of contaminated water	99	
1071-83-6	Glyphosate	Herbicide	Yes –	Urodela	*Lissotriton vulgaris*	Larvae	Not stated	Survival, behavior, body mass change, predatory activity	100	

Figure 16.2 Urodela. Image (a) shows an embryo of *Lissotriton vulgaris* in dorsal view just before hatching and (b) a fully grown larva of the same species in lateral view. *Salamandra salamandra* gives birth to fully developed larvae, as demonstrated in the lateral view in image (c). Scale bars equal 1 mm in (a), 4 mm in (b) and 5 mm in (c).
All photographs taken by Burkhard Thiesmeier, reprinted with permission of Laurenti-Verlag.

exposed adult Italian crested newts (*Triturus carnifex*) to the fungicide maneb and studied effects on survival and peripheral blood cells. Larval *T. carnifex* were treated intraperitoneally with the insecticide trimethyltin chloride. While the highest test concentration led to total mortality, lower doses induced neuropathological damage.[93] Since the 1980s, larval Iberian ribbed newts (*Pleurodeles waltl*) have been used in different studies aiming at the detection of genotoxic effects of test substances and waste water.[94,95] However, the genotoxic effects of a fungicide (caplan) on this urodele were investigated only in one study, using the micronucleus test and the Comet assay.[96] More recently, Mandrillon and Saglio[97] used larval fire salamanders (*Salamandra salamandra*, Figure 16.2c) in an experiment on the predator–prey relationship with common frog larvae (*Rana temporaria*). Sub-lethal concentrations of the herbicide amitrole had a detrimental effect on the predatory behavior.[49] Ortiz-Santaliestra *et al.*[98] exposed eggs of the golden-striped salamander (*Chioglossa lusitanica*) to a glyphosate-based herbicide. Treated embryos hatched at a larger size compared to controls. Additional exposure to ammonium nitrate increased this effect. However, mortality was not significantly enhanced in any treatment.[50] Wagner and Lötters[99] did not find effects of water contamination with glyphosate and a glyphosate-based herbicide on site selection of palmate newts (*Lissotriton helveticus*) and alpine newts (*Ichthyosaura alpestris*). Likewise, Ujszegi *et al.*[100] did not find any effects at low concentrations of a glyphosate-based herbicide on survival, behavior, body mass change or predatory activity of smooth newts (*Lissotriton vulgaris*). For other European urodele species, we only found limited information on the effects of other environmental contaminants (*e.g.*, PCB accumulation in olm *Proteus anguinus*[101]).

Taken together, there is an urgent need for more information on the aquatic toxicity of pesticides on European urodeles. Interestingly, the pesticide exposure risk is calculated to be very high for the strictly protected European urodeles *Triturus dobrogicus, T. cristatus, T. carnifex* and *T. karelinii*;[18] for three out of these four.

16.3.2 European Frogs and Toads

There are more data on the effects of pesticides on the aquatic development of European anuran species compared to urodele species (*cf.* Tables 16.3 and 16.4). Cooke[102,103] used caged tadpoles of *Rana temporaria* and *Bufo bufo* as indicators of the contamination of small agricultural water bodies with agrochemicals (in one study[102] the survival of smooth newts, which colonized artificial ponds, was also monitored). The largest amount of data comprised pesticide studies using embryos and larvae of *R. temporaria* and *Bombina bombina* (Table 16.4). As already mentioned, Larsen and colleagues[86–88] established *B. bombina* as an anuran test organism in Denmark. The possibility of breeding fire-bellied toads by injection with HCG is a benefit for laboratory work. However, the relatively low number of eggs is counterproductive.

Usually, survival was considered as a general endpoint. However, as pointed out below, the amount of data is too poor to draw any conclusions from these comparisons. Some studies used realistic exposure scenarios in the field and/or studied sublethal effects of environmentally relevant concentrations (*e.g.*, ref. 88, 102, 103). For example, the studies by Greulich and colleagues[77–79] showed that isoproturon and different cypermethrin formulations passed the jelly coat of eggs from yellow-bellied toads (*Bombina variegata*) and moor frogs (*Rana arvalis*). Wagner *et al.*[38] found effects of two herbicide formulations on *R. temporaria* specific to both the developmental stages and the respective population. For the latter finding, evolved tolerances in individuals from agricultural land were discussed. Furthermore, chronic exposure of larval Perez's frog (*Pelophylax perezi*) to an insecticide and a herbicide commonly used in rice cultivation (where the species is abundant) had no effect on completion of metamorphosis, but a mixture of both pesticides dramatically increased mortality during metamorphosis.[104]

Beside specific pesticide studies, like for European urodele species, for European anurans additional information is also available for other environmental contaminants (*e.g.*, ammonium nitrate fertilizers[131,132]).

Table 16.4 Overview of studies on the effects of pesticides on the aquatic development of European anuran species (order after species name).

CAS	Pesticide name	Pesticide type	Formulation	Order	Species	Life-stage	96 h LC_{50} (mg a.i. L^{-1})	Considered endpoints	References	Remarks
66230-04-4	Esfenvalerate	Insecticide	Yes	Anura	*Bombina bombina*	Embryos	>0.3	Survival, malformations, growth inhibition, lag effects until completion of metamorphosis	86,87	87 = Field study
67747-09-5	Prochloraz	Fungicide	Yes	Anura	*Bombina bombina*	Embryos	>10	Survival, malformations, growth inhibition, lag effects until completion of metamorphosis	86	
52918-63-5	Deltamethrin	Insecticide	Yes	Anura	*Bombina bombina*	Embryos	Not stated	Survival, malformations, growth inhibition	87	Field study
39515-41-8	Fenpropathrin	Insecticide	Yes	Anura	*Bombina bombina*	Embryos	Not stated	Survival, malformations, growth inhibition	87	Field study
52645-53-1	Permethrin	Insecticide	No	Anura	*Bombina bombina*	Embryos	Not stated	Survival, malformations, growth inhibition	87	Field study
1582-09-8	Trifluralin	Herbicide	Yes	Anura	*Bombina bombina*	Embryos	Not stated	Survival, malformations, clinical signs, growth inhibition	105	Included in the EFSA report by Fryday and Thompson[90]
1582-09-8	Trifluralin	Herbicide	Yes	Anura	*Bombina bombina*	Larvae	19.6	Survival, malformations, clinical signs, growth inhibition	105	Included in the EFSA report by Fryday and Thompson[90]
34123-59-6	Isoproturon	Herbicide	Yes	Anura	*Bombina bombina*	Larvae	Not stated	Uptake, survival, malformations, clinical signs, GST activity	77	^{14}C-labeled test substance

34123-59-6	Isoproturon	Herbicide	Yes		Anura	*Bombina bombina*	Eggs	Not stated	Uptake through the jelly coat	77	^{14}C-labeled test substance
34123-59-6	Isoproturon	Herbicide	Yes		Anura	*Bombina variegata*	Larvae	Not stated	Uptake, survival, malformations, clinical signs, GST activity	77	^{14}C-labeled test substance
34123-59-6	Isoproturon	Herbicide	Yes		Anura	*Bombina variegata*	Eggs	Not stated	Uptake through the jelly coat	77	^{14}C-labeled test substance
52315-07-8	Cypermethrin	Insecticide	No		Anura	*Bombina variegata*	Eggs	Not stated	Uptake through the jelly coat	79	^{14}C-labeled test substance
52315-07-8	Cypermethrin	Insecticide	No		Anura	*Bombina variegata*	Larvae	Not stated	Uptake, survival, malformations, clinical signs, GST activity	79	^{14}C-labeled test substance
115-29-7	Endosulfan	Insecticide	No		Anura	*Bufo bufo*	Larvae	0.43	Survival, malformations, clinical signs, effects on gill apparatus morphology and ultrastructure, survival, malformations, morpho-functional features of the epidermis	79,106–108	Only[79] included in the EFSA report by Fryday and Thompson[90]
50-29-3	DDT	Insecticide	No		Anura	*Bufo bufo*	Larvae	Not stated	Survival, clinical signs	91	
60-57-1	Dieldrin	Insecticide	No		Anura	*Bufo bufo*	Larvae	Not stated	Survival, clinical signs	91	
85-00-7	Diquat	Herbicide	No		Anura	*Bufo bufo*	Larvae	Not stated	Survival, clinical signs, developmental rate, growth inhibition	102	Field study
1194-65-6	Dichlobenil	Herbicide	No		Anura	*Bufo bufo*	Larvae	Not stated	Survival, clinical signs, developmental rate, growth inhibition	102	Field study

Table 16.4 (*Continued*)

CAS	Pesticide name	Pesticide type	Formulation	Order	Species	Life-stage	96 h LC_{50} (mg a.i. L^{-1})	Considered endpoints	References	Remarks
7758-98-7	Copper sulfate	Fungicide	No	Anura	*Bufo bufo*	Embryos and larvae	0.08–0.09	Survival, growth inhibition	109	LC_{50} values only presented in a figure, included in the EFSA report by Fryday and Thompson[90]
12427-38-2	Maneb	Fungicide	No	Anura	*Bufo bufo*	Larvae	1.966 (120 h exposure)	Survival, malformations	110	
7758-98-7	Copper sulfate	Fungicide	No	Anura	*Bufotes viridis*	Larvae	0.058 (120 h exposure)	Survival, malformations, clinical signs, developmental rate	111	
1071-83-6	Glyphosate	Herbicide	Yes	Anura	*Bufotes viridis*	Larvae	19.6–19.8	Survival, effects on enzymatic activity	112,113	*X. laevis* more sensitive compared to *B. viridis*; sublethal concentrations in ref. 113 also included pesticide mixtures
12427-38-2	Maneb	Fungicide	No	Anura	*Bufotes viridis*	Larvae	0.332 (120 h exposure)	Survival, malformations	110	
950-37-8	Methidathion	Insecticide and acaricide	No	Anura	*Bufotes viridis*	Larvae	10–25.7	Survival, effects on enzymatic activity	112,113	*X. laevis* more sensitive compared to *B. viridis*; sublethal concentrations in ref. 113 also included pesticide mixtures
7758-98-7	Copper sulfate	Fungicide	No	Anura	*Discoglossus jeanneae*	Embryos and larvae	0.08–0.10	Survival, growth inhibition	109	LC_{50} values only presented in a figure, included in the EFSA report by Fryday and Thompson[90]

7758-98-7	Copper sulfate	Fungicide	No		Anura	*Epidalea calamita*	Embryos	0.08–0.22	Survival, clinical signs (escape behavior)	114	Included in the EFSA report by Fryday and Thompson[90]
7758-98-7	Copper sulfate	Fungicide	No		Anura	*Epidalea calamita*	Larvae	0.11	Survival, clinical signs (escape behavior)	109,113	Included in the EFSA report by Fryday and Thompson[90]
60-51-5	Dimethoate	Insecticide	Yes		Anura	*Hyla arborea*	Larvae	20.27–37.37	Survival, malformations, clinical signs, growth inhibition	115	Fryday and Thompson[90] stated wrong species (*Hyla versicolor* instead of *H. arborea*)
107534-96-3	Tebuconazole	Fungicide	No		Anura	*Hyla intermedia*	Larvae	Not stated	Survival, clinical signs, developmental rate	116	
107534-96-3	Tebuconazole	Fungicide	No		Anura	*Hyla intermedia*	Larvae	Not stated	Survival, clinical signs, developmental rate	116	
7758-98-7	Copper sulfate	Fungicide	No		Anura	*Pelobates cultripes*	Larvae	0.22	Survival, growth inhibition	109	LC_{50} values only presented in a figure, included in the EFSA report by Fryday and Thompson[90]
76-44-8	Heptachlor	Insecticide	No		Anura	*Pelophylax* kl. *esculentus*	Larvae	Not stated	Survival, bioaccumulation, toxic effects on the ventral epidermis	117	Investigation of toxic effects on the ventral epidermis of both larvae and adults
7758-98-7	Copper sulfate	Fungicide	No		Anura	*Pelophylax perezi*	Larvae	0.36–0.57	Survival, growth inhibition	109	LC_{50} values only presented in a figure, included in the EFSA report by Fryday and Thompson[90]

Table 16.4 *(Continued)*

CAS	Pesticide name	Pesticide type	Formulation	Order	Species	Life-stage	96 h LC_{50} (mg a.i. L^{-1})	Considered endpoints	References	Remarks
67375-30-8	Alpha-cypermethrin	Insecticide	No	Anura	*Pelophylax perezi*	Larvae	Not stated	Survival (acute toxicity and completing of metamorphosis)	104	Also combinations of both pesticides
19666-30-9	Oxadiazon	Herbicide	No	Anura	*Pelophylax perezi*	Eggs	Not stated	Hatching success	104	Also combinations of both pesticides
1071-83-6	Glyphosate	Herbicide	No	Anura	*Pelophylax ridibundus*	Larvae	18.3–22.7	Survival, effects on enzymatic activity	112,113	*X. laevis* more sensitive compared to *P. ridibundus;* sublethal concentrations in ref. 113 also included pesticide mixtures
950-37-8	Methidathion	Insecticide and acaricide	No	Anura	*Pelophylax ridibundus*	Larvae	7.34–27.4	Survival, effects on enzymatic activity	112,113	*X. laevis* more sensitive compared to *P. ridibundus*; sublethal concentrations in ref. 113 also included pesticide mixtures
121-75-5	Malathion	Insecticide	Yes	Anura	*Pelophylax ridibundus*	Larvae	29–38	Survival, malformations, clinical signs, growth inhibition	118	Included in the EFSA report by Fryday and Thompson[90]
52315-07-8	Cypermethrin	Insecticide	No	Anura	*Rana arvalis*	Eggs	Not stated	Uptake through the jelly coat	79	^{14}C-labeled test substance
52315-07-8	Cypermethrin	Insecticide	No	Anura	*Rana arvalis*	Larvae	Not stated	Uptake, survival, malformations, clinical signs, GST activity	79	^{14}C-labeled test substance

67375-30-8	Alpha-cypermethrin	Insecticide	No		Anura	*Rana arvalis*	Larvae	Not stated	Survival, malformations, clinical signs, GST activity	79	
67375-30-8	Alpha-cypermethrin	Insecticide	No		Anura	*Rana arvalis*	Eggs, embryos and larvae	Not stated	Survival, malformations, clinical signs, growth inhibiton, metamorphosis	78	
61-82-5	Amitrole	Herbicide	No		Anura	*Rana dalmatina*	Larvae	Not stated	Clinical signs (escape behavior, refuge use)	97	
115-29-7	Endosulfan	Insecticide	No		Anura	*Rana dalmatina*	Larvae	0.074	Survival, malformations, clinical signs, growth inhibition, developmental rates	119	
2921-88-2	Chlorpyrifos	Insecticide	No		Anura	*Rana dalmatina*	Larvae	5.17	Survival (acute toxicity and completing of metamorphosis), malformations (including effects on gill morphology, ultrastructure and gonadal differentiation), clinical signs, growth inhibition	120,121	Included in the EFSA report by Fryday and Thompson[90]
94-75-7	2,4-D	Herbicide	No		Anura	*Rana temporaria*	Larvae	Not stated	Survival, clinical signs	91	
67375-30-8	Alpha-cypermethrin	Insecticide	Yes		Anura	*Rana temporaria*	Larvae	Not stated	Genotoxicity	122	Genotoxicity investigated by micronucleus test, *R. temporaria* larvae more sensitive compared to *Xenopus laevis*

Table 16.4 (*Continued*)

CAS	Pesticide name	Pesticide type	Formulation	Order	Species	Life-stage	96 h LC_{50} (mg a.i. L^{-1})	Considered endpoints	References	Remarks
1066-51-9	Aminomethyl-phosphonic acid	Main metabolite	No	Anura	*Rana temporaria*	Adult	Not stated	Avoidance of contaminated water	99	
131860-33-8	Azoxystrobin	Fungicide	No	Anura	*Rana temporaria*	Eggs, larvae	Not stated	Survival, developmental rate, growth	123	Chronic exposure started with eggs
21725-46-2	Cyanazine	Herbicide	No	Anura	*Rana temporaria*	Eggs, larvae	Not stated	Survival, developmental rate, growth	123	Chronic exposure started with eggs
101205-02-1	Cycloxydim	Herbicide	Yes	Anura	*Rana temporaria*	Larvae	1.3–2.0	Survival, malformations, clinical signs, growth inhibition, time to metamorphosis	28	
50-29-3	DDT	Insecticide	No	Anura	*Rana temporaria*	Embryos, larvae	Not stated	Survival, malformations, clinical signs, developmental rate	91,124–129	In one study, increased predation by *T. cristatus* of hyperactive tadpoles due to DDT exposure was studied
60-57-1	Dieldrin	Insecticide	No	Anura	*Rana temporaria*	Larvae	Not stated	Survival, clinical signs	91	
85-00-7	Diquat	Herbicide	No	Anura	*Rana temporaria*	Larvae	Not stated	Survival, clinical signs, developmental rate, growth inhibition	102	Field study

1194-65-6	Dichlobenil	Herbicide	No	Anura	*Rana temporaria*	Larvae	Not stated	Survival, clinical signs, developmental rate, growth inhibition	102	Field study
115-29-7	Endosulfan	Insecticide	No	Anura	*Rana temporaria*	Larvae	Not stated	Survival, clinical signs (including feeding behavior)	130	Also lag-effects of environmentally relevant concentrations studied
66230-04-4	Esfenvalerate	Insecticide	No	Anura	*Rana temporaria*	Larvae	Not stated	Survival, developmental rate, growth	123	
1071-83-6	Glyphosate	Herbicide	Yes	Anura	*Rana temporaria*	Larvae	10.4–12.2	Survival, malformations, clinical signs, growth inhibition, time to metamorphosis	28	
94-74-6	MCPA	Herbicide	No	Anura	*Rana temporaria*	Larvae	Not stated	Survival, developmental rate, growth	123	
52645-53-1	Permethrin	Insecticide	No	Anura	*Rana temporaria*	Eggs, larvae	Not stated	Survival, developmental rate, growth	123	Chronic exposure started with eggs
23103-98-2	Pirimicarb	Insecticide	No	Anura	*Rana temporaria*	Larvae	Not stated	Survival, developmental rate, growth	123	

16.4 Fish as Standard Test Organisms Used in Pesticide Approval

Comparison to the fish model leads to considerations dependent on the different nature of the test species. Teleost fish are usually implemented as surrogates for acute toxic effects.[133] Their body surface is completely covered by dermal bones, *i.e.*, scales and bone plates, functioning as a barrier to the environment. In adults and in fish larvae with a gradually more permeable skin, the gills play a crucial role in gas and ion exchange[134–137] and, in close connection with this function, also in absorption of xenobiotics.[138,139] Due to ventilation, an amount of water transporting xenobiotics is passing through the oral cavity and the gill bearing pharynx. It is suggested that aquatic life-stages of anurans are more exposed to their aquatic environment and as a consequence to xenobiotics than most teleosts, such as the rainbow trout (*Oncorhynchus mykiss*), a common standard test organism. Special physiological properties, such as resorption *via* the intestinal tract, detoxification mechanisms, and sensitivity to toxic agents and their metabolites, are possibly different in the clades. The results from toxicity studies confirm this assumption (Table 16.3). Furthermore, in terrestrial-life stages (*i.e.*, metamorphs, juveniles and adults) dermal absorption of xenobiotics is much higher than for other vertebrates due to their permeable skin, which is physiologically involved in gas, water and electrolyte exchange with the environment.[140]

In an industry-funded study, Weltje *et al.*[141] concluded that fish are proper surrogates for amphibian larvae. This statement is dependent on simple correlation of 96 h LC_{50} values (about 100 data points) with those of fish. Furthermore, and most disputable, this conclusion is based on comparison (again correlations) of NOEC values from amphibians and fish, which are highly assorted (as pointed out above). This is demonstrated by lumping together results under the header 'development', instead of a comparison of equal developmental steps, *i.e.*, treatment periods, and clearly defined identic endpoints. Together with the fact that fish and amphibian larvae dramatically differ in their reproductive biology (see above) and ecology (in particular, most amphibians use small, often temporary pools in cultivated landscape for reproduction—see below), no general conclusions on the suitability of fish as surrogate species can be made.

16.5 Environmentally Relevant Pesticide Concentrations

In European pesticide approval, modelled concentrations of the active ingredients are employed to calculate 'Toxicity to Exposure Ratios' (TER = concerning acute toxicity, LC_{50} values of test organisms, and concerning chronic effects, EC_{50} values divided by the expected environmental concentrations [and considering safety factors]). However, aquatic amphibian habitats like temporary ponds or vernal pools are usually not protected by no-spray buffer zones.[142] Furthermore, for registration in the European Union, only active ingredients are assessed, but not the formulations used in the field; added substances can dramatically increase toxicity and for such adjuvants environmental concentrations are widely lacking.[21] In general, there is nearly no information on environmental concentrations of pesticides in aquatic amphibian habitats in Europe, especially small ponds next to agricultural land. For example, in Germany, such small ditches, pools and ponds are not seen as 'water bodies' by the water protection act; however, they are crucial for most autochthonous amphibian species.[143] In the German agricultural landscape, Greulich *et al.*[144] found low concentrations of the insecticide alphacypermetrine and peak concentrations up to 22 $\mu g L^{-1}$ of the herbicide isoproturon in a reproduction pond used by nine different amphibian species, especially *Rana arvalis* and common spadefoot toad (*Pelobates fuscus*). Glyphosate was detected in all three breeding ponds of *R. temporaria* populations studied by Wagner *et al.*,[38] also used for reproduction by *B. bufo*, *I. alpestris* and *L. helveticus* (unpublished data). In Denmark, Briggs and Damm[88] measured peak concentrations of $>10\ \mu g L^{-1}$ of different herbicides, but also detected fungicides and insecticides in artificial ponds created for *B. bombina*. The latter authors concluded that buffer zones of 10 m are not sufficient to protect sensitive environments like amphibian reproduction ponds from the spray drift of pesticides.

16.6 Conclusions

1. Anuran development is most suitable to test the effects of pesticides in the laboratory owing to the unique developmental biology of this group. In several studies, European species were successfully implemented. In studies of amphibian developmental toxicity, predominantly aquatic life-stages of European anurans were employed compared to urodeles. There is at least one study on the aquatic toxicity of pesticides for only seven out of 36 European urodeles (19.4%) and for only 15 out of 52 European anurans (28.9%). This is relevant with regard to observed species-specific effects of pesticides. Most background data is available for the aquatic development of *R. temporaria* and *B. bombina*, but also larvae of *Discoglossus* species are suitable owing to their fast development and the possibility to receive large amounts of egg from adults in the laboratory. The goal is to replace *X. laevis* as a non-European species with a reproductive biology different from the more basic European species to achieve more realistic results covering the demands of protection of these species.
2. Properties of urodele reproductive biology increase experimental efforts. Therefore, salamanders should only be implemented if specific problems of pesticide toxicity affecting this group are the goal of the study.
3. Especially due to their dramatically different ecology and reproduction biology as well as the lack of data dealing with environmental

contamination of aquatic amphibian habitats, there is no evidence that fish species are appropriate surrogates to assess toxic effects of pesticides on aquatic amphibian development.

Acknowledgements

We are thankful to Prof. Marcelo L. Larramendy for his invitation to contribute to the present book and Dr Burkhard Thiesmeier for providing photographs of larval amphibians.

References

1. R. A. Alford and S. J. Richards, Global amphibian declines: a problem in applied ecology, *Annu. Rev. Ecol. Evol. Syst.*, 1999, **30**, 133–165.
2. S. N. Stuart, M. Hoffmann, J. S. Chanson, N. A. Cox, R. J. Berridge, P. Ramadi and B. E. Young, *Threatened Amphibians of the World*, Lynx Editions, Barcelona, 2008.
3. D. B. Wake and V. T. Vredenburg, Are we in the midst of the sixth mass extinction? A view from the world of amphibians, *Proc. Natl. Acad. Sci. U. S. A.*, 2008, **105**, 11466–11473.
4. D. R. Frost, Amphibian Species of the World: an Online Reference. Version 6.0. Electronic Database accessible at http://research.amnh.org/herpetology/amphibia/index.html. (last accessed July 2016) American Museum of Natural History, New York, USA. 2016.
5. E. H. C. Grant, D. A. W. Miller, B. R. Schmidt, M. J. Adams, S. M. Amburgey, T. Chambert, S. S. Cruickshank, R. N. Fisher, D. M. Green, B. R. Hossack, P. T. J. Johnson, M. B. Joseph, T. A. G. Rittenhouse, M. E. Ryan, J. H. Waddle, S. C. Walls, L. L. Bailey, G. M. Fellers, T. A. Gorman, A. M. Ray, D. S. Pilliod, S. J. Price, D. Saenz, W. Sadinski and E. Muths, Quantitative evidence for the effects of multiple drivers on continental-scale amphibian declines, *Sci. Rep.*, 2016, **6**, 25625.
6. E. La Marca, K. R. Lips, S. Lötters, R. Puschendorf, R. Ibáñez, J. V. Rueda-Almonacid, R. Schulte, C. Marty, F. Castro, J. Manzanilla-Puppo, J. E. García-Pérez, F. Bolaño, G. Chavez, J. A. Pounds, E. Toral and B. E. Young, Catastrophic population declines and extinctions in Neotropical Harlequin frogs (Bufonidae: *Atelopus*), *Biotropica*, 2005, **37**, 190–201.
7. K. R. Lips, F. Brem, R. Brenes, J. D. Reeve, R. A. Alford, J. C. Voyles, C. Carey, L. Livo, A. P. Pessier and J. P. Collins, Emerging infectious disease and the loss of biodiversity in a Neotropical amphibian community, *Proc. Natl. Acad. Sci. U. S. A.*, 2006, **102**, 3165–3170.
8. J. P. Collins and A. Storfer, Global amphibian declines: sorting the hypotheses, *Divers. Distrib.*, 2003, **9**, 89–98.

9. T. B. Hayes, P. Case, S. Chul, D. Chung, C. Haeffele, K. Haston, M. Lee, V. P. Mai, Y. Marijoua, J. Parker and M. Tsui, Pesticide mixtures, endocrine disruption, and amphibian declines: are we underestimating the impact? *Environ. Health Perspect.*, 2006, **114**, 40–50.
10. W. A. Battaglin, K. L. Smalling, C. Anderson, D. Calhoun, T. Chestnut and E. Muths, Potential interactions among disease, pesticides, water quality and adjacent land cover in amphibian habitats in the United States, *Sci. Total Environ.*, 2016, **566–567**, 320–332.
11. H.-R. Köhler and R. Triebskorn, Wildlife ecotoxicology of pesticides: can we track effects to the population level and beyond? *Science*, 2013, **341**, 759–765.
12. F. Geiger, J. Bengtsson, F. Berendse, W. W. Weisser, M. Emmerson, M. B. Morales, P. Ceryngier, J. Liira, T. Tscharntke, C. Winqvist, S. Eggers, R. Bommarco, T. Part, V. Bretagnolle, M. Plantegenest, L. W. Clement, C. Dennis, C. Palmer, J. J. Onate, I. Guerrero, V. Hawro, T. Aavik, C. Thies, A. Flohre, S. Hänke, C. Fischer, P. W. Goedhart and P. Inchausti, Persistent negative effects of pesticides on biodiversity and biological control potential on European farmland, *Basic Appl. Ecol.*, 2010, **11**, 97–105.
13. R. M. Mann, R. V. Hyne, C. B. Choung and S. P. Wilson, Amphibians and agricultural chemicals: review of the risks in a complex environment, *Environ. Pollut.*, 2009, **157**, 2903–2927.
14. D. G. Thompson, B. F. Wojtaszek, B. Staznik, D. T. Chartrand and G. R. Stephenson, Chemical and biomonitoring to assess potential acute effects of Vision herbicide on native amphibian larvae in forest wetlands, *Environ. Toxicol. Chem.*, 2004, **23**, 843–849.
15. C. Davidson, H. B. Shaffer and M. R. Jennings, Spatial tests of the pesticide drift, habitat destruction, UV-B, and climate-change hypotheses for California amphibian declines, *Conserv. Biol.*, 2002, **16**, 1588–1601.
16. W. M. Edwards, G. B. Triplett Jr. and R. M. Kramer, A watershed study of glyphosate transport in runoff, *J. Environ. Qual.*, 1980, **9**, 661–665.
17. C. D. Brown and W. van Beinum, Pesticide transport via sub-surface drains in Europe, *Environ. Pollut.*, 2009, **157**, 3314–3324.
18. N. Wagner, D. Rödder, C. A. Brühl, M. Veith, P. P. Lenhardt and S. Lötters, Evaluating the risk of pesticide exposure for amphibian species listed in Annex II of the European Union Habitats Directive, *Biol. Conserv.*, 2014, **176**, 64–70.
19. K. L. Smalling, G. M. Fellers, P. M. Kleeman and K. M. Kuivila, Accumulation of pesticides in pacific chorus frogs (*Pseudacris regilla*) from California's Sierra Nevada Mountains, USA, *Environ. Toxicol. Chem.*, 2013, **32**, 2026–2034.
20. K. L. Smalling, R. Reeves, E. Muths, M. Vandever, W. A. Battaglin, M. L. Hladik and C. L. Pierce, Pesticide concentrations in frog tissue

and wetland habitats in a landscape dominated by agriculture, *Sci. Total Environ.*, 2015, **502**, 80–90.
21. N. Wagner, W. Reichenbecher, H. Teichmann, B. Tappeser and S. Lötters, Questions concerning the potential impact of glyphosate-based herbicides on amphibians, *Environ. Toxicol. Chem.*, 2013, **32**, 1688–1700.
22. C. M. Howe, M. Berrill, B. D. Pauli, C. C. Helbing, K. Werry and N. Veldhoen, Toxicity of glyphosate-based pesticides to four North American frog species, *Environ. Toxicol. Chem.*, 2004, **23**, 1928–1938.
23. B. K. Williams and R. D. Semlitsch, Larval responses of three Midwestern anurans to chronic, low-dose exposures of four herbicides, *Arch. Environ. Contam. Toxicol.*, 2010, **58**, 819–827.
24. N. Wagner, S. Lötters, M. Veith and B. Viertel, Effects of an environmentally relevant temporal application scheme of low herbicide concentrations on larvae of two anuran species, *Chemosphere*, 2015, **135**, 175–181.
25. C. Lanctôt, C. Robertson, L. Navarro-Martín, C. Edge, S. D. Melvin, J. Houlahan and V. L. Trudeau, Effects of the glyphosate-based herbicide Roundup WeatherMax® on metamorphosis of wood frogs (*Lithobates sylvaticus*) in natural wetlands, *Aquat. Toxicol.*, 2013, **140–141**, 48–57.
26. D. K. Jones, J. I. Hammond and R. A. Relyea, Very highly toxic effects of endosulfan across nine species of tadpoles: Lag effects and family-level sensitivity, *Environ. Toxicol. Chem.*, 2009, **28**, 1939–1945.
27. R. A. Relyea and N. Diecks, An unforeseen chain of events: lethal effects of pesticides on frogs at sublethal concentrations, *Ecol. Appl.*, 2008, **18**, 1728–1742.
28. R. A. Reylea and N. Mills, Predator-induced stress makes the pesticide carbaryl more deadly to gray treefrog tadpoles (*Hyla versicolor*), *Proc. Natl. Acad. Sci. U. S. A.*, 2001, **98**, 2491–2496.
29. R. A. Relyea, New effects of Roundup® on amphibians: Predators reduce herbicide mortality while herbicides induce anti-predator morphology, *Ecol. Appl.*, 2012, **22**, 634–647.
30. D. K. Jones, J. R. Hammond and R. A. Relyea, Competitive stress can make the herbicide Roundup® more deadly to larval amphibians, *Environ. Toxicol. Chem.*, 2011, **30**, 446–454.
31. A. N. Edginton, P. M. Sheridan, G. R. Stephenson, D. G. Thompson and H. J. Boermans, Comparative effects of pH and Vision herbicide on two life stages of four anuran amphibian species, *Environ. Toxicol. Chem.*, 2004, **23**, 81–822.
32. R. A. Relyea, A cocktail of contaminants: how mixtures of pesticides at low concentrations affect aquatic communities, *Oecologia*, 2009, **159**, 363–376.
33. J. C. Brodeur, M. B. Poliserpi, M. F. D'Andrea and M. Sánchez, Synergy between glyphosate- and cypermethrin-based pesticides during acute exposures in tadpoles of the common South American Toad *Rhinella arenarum*, *Chemosphere*, 2014, **112**, 70–76.

34. R. M. Mann and J. R. Bidwell, The toxicity of glyphosate and several glyphosate formulations to four species of southwestern Australian frogs, *Arch. Environ. Contam. Toxicol.*, 1999, **36**, 193–199.
35. R. A. Relyea and D. K. Jones, The toxicity of Roundup Original MAX® to 13 species of larval amphibians, *Environ. Toxicol. Chem.*, 2009, **28**, 2004–2008.
36. C. M. Bridges, Long-term effects of pesticide exposure at various life stages of the southern leopard frog (*Rana sphenocephala*), *Arch. Environ. Contam. Toxicol.*, 2000, **39**, 91–96.
37. L. M. Biga and A. R. Blaustein, Variations in lethal and sublethal effects of cypermethrin among aquatic stages and species of anuran amphibians, *Environ. Toxicol. Chem.*, 2013, **32**, 2855–2860.
38. N. Wagner, M. Veith, S. Lötters and B. Viertel, Population and life-stage specific effects of two herbicide formulations on the aquatic development of European common frogs (*Rana temporaria*), *Environ. Toxicol. Chem.*, 2017, **36**, 190–200.
39. ASTM (American Society for Testing and Materials), Standard Guide for Conducting the Frog Embryo Teratogenesis Assay-*Xenopus* (FETAX). E1439-98, ASTM International, West Conshohocken, 1998.
40. OECD (Organisation for Economic Co-operation and Development), Test No. 231: Amphibian Metamorphosis Assay, in *OECD Guidelines for Testing of Chemicals. Section 2: Effects on Biotic Systems*, OECD Publishing, Paris, 2009.
41. OECD (Organisation for Economic Co-operation and Development), *Series on Testing and Assessment. Detailed Review Paper on Amphibian Metamorphosis Assay for the Detection of Thyroid Active Substances, No. 46*, OECD Publishing, Paris, 2004.
42. R. Carroll, *The Rise of Amphibians*, The John Hopkins University Press, Baltimore, 2009.
43. S. M. Deban and D. B. Wake, Aquatic feeding in salamanders in *Feeding-form, Function, and Evolution in Tetrapod Vertebrates*, ed. K. Schwenk, Academic Press, San Diego, San Francisco, New York, Boston, London, Sydney, Tokyo, 2000, pp. 65–94.
44. L. Goniakowska-Witalinska, Ultrastructural and morphometric changes in the lung of newt, *Triturus cristatus carnifex* Laur. during ontogeny, *J. Anat.*, 1980, **130**, 571–583.
45. R. Rugh, *Experimental Embryology: Techniques and Procedures*, Burgess, Minneapolis, 1962.
46. H. H. Epperlein and M. Junginger, The normal development of the newt, *Triturus alpestris* (Daudin), *Amphib.-Reptil.*, 1981, **2**, 295–308.
47. R. G. Harrison, *Harrison stages and description of the normal development of the spotted salamander,* Ambystoma punctatum *(Linn.)*, in *Organization and Development of the Embryo*, ed. R. G. Harrison, Yale University Press, New Haven, 1969, pp. 44–66.

48. Z. Roček and J. C. Rage, Anatomical transformations in transition from temnospondyl to proanuran stages, in *Amphibian Biology – Palaeontology*, ed. H. Heatwole and R. L. Carroll, Surrey Beatty & Sons, Chipping Norton, Australia, 2000, vol. 4, pp. 1274–1282.
49. Z. Roček and J. C. Rage, *Proanuran stages (*Triadobatrachus, Czatkobatrachus*)*, in *Amphibian Biology – Palaeontology*, ed. H. Heatwole and R. L. Carroll, Surrey Beatty & Sons, Chipping Norton, Australia, 2000, vol. 4, pp. 1283–1294.
50. R. J. Wassersug, Evolution of anuran life cycles, *Science*, 1974, **185**, 377–378.
51. R. J. Wassersug, The adaptive significance of the tadpole stage with comments on the maintenance of complex life cycles in anurans, *Am. Zool.*, 1975, **15**, 405–417.
52. G. R. Handrigan and R. J. Wassersug, The anuran bauplan: a review of the adaptive, developmental, and genetic underpinnings of frog and tadpole morphology, *Biol. Rev.*, 2007, **82**, 1–25.
53. H. Fox, *Amphibian Morphogenesis*, Humana Press, Clifton, New Jersey, 1984.
54. P. Alberch, Development and evolution of amphibian metamorphosis, in *Trends in Vertebrate Morphology. Proceedings of the 2nd International Symposium on Vertebrate Morphology in Vienna*, ed. H. Splechtna and H. Hilgers, Gustav Fischer, Stuttgart, New York, 1989, pp. 163–173.
55. B. Viertel and S. Richter, Anatomy: viscera and endocrines, in *Tadpoles, the Biology of Anuran Larvae*, ed. R. Altig and R. W. McDiarmid, The University of Chicago Press, Chicago and London, 1999, pp. 92–148.
56. D. B. Seale and R. J. Wassersug, Suspension feeding dynamics of anuran larvae related to their functional morphology, *Oecologia*, 1979, **39**, 259–272.
57. D. B. Seale and H. Beckvar, The competition ability of anuran larvae (genera: *Hyla, Bufo*, and *Rana*) to ingest suspended blue-green algae, *Copeia*, 1980, **1980**, 495–503.
58. B. Viertel, Suspension feeding of anuran larvae at low concentrations of *Chlorella* algae (Amphibia, Anura), *Oecologia*, 1990, **85**, 167–177.
59. B. Viertel, Functional response of suspension feeding anuran larvae to different particle sizes at low concentrations, *Hydrobiologia*, 1992, **234**, 151–173.
60. R. Wassersug, The role of ultraplanktonic entrapment in anuran larvae, *J. Morphol.*, 1972, **1972**, 279–288.
61. B. Viertel, The filter apparatus of anuran larvae. Aspects of the filtering mechanism, in *Trends in Vertebrate Morphology. Proceedings of the 2nd International Symposium on Vertebrate Morphology in Vienna*, ed. H. Splechtna and H. Hilgers, Gustav Fischer, Stuttgart, New York, 1989, pp. 526–533.
62. R. D. Semlitsch, C. M. Bridges and A. M. Welch, Genetic variation and a fitness tradeoff in the tolerance of gray treefrog (*Hyla versicolor*) tadpoles to the insecticide carbaryl, *Oecologia*, 2000, **125**, 179–185.

63. C. M. Bridges and R. D. Semlitsch, Variations in pesticide tolerance of tadpoles among and within species of Ranidae and patterns of amphibian decline, *Conserv. Biol.*, 2000, **14**, 1490–1499.
64. L. Navarro-Martín, C. Lanctôt, P. Jackman, B. J. Park, K. Doe, B. D. Pauli and V. L. Trudeau, Effects of glyphosate-based herbicides on survival, development, growth and sex ratios of wood frogs (*Lithobates sylvaticus*) tadpoles. I: Chronic laboratory exposures to VisionMax®, *Aquat. Toxicol.*, 2014, **154**, 278–290.
65. G. L. Orton, The systematics of vertebrate larvae, *Syst. Zool.*, 1953, **2**, 63–75.
66. P. H. Starrett, Evolutionary patterns in larval morphology, in *Evolutionary Biology of the Anurans*, ed. J. L. Vial, University of Missouri Press, Columbia, 1973, pp. 251–271.
67. L. S. Ford and D. C. Cannatella, The major clades of frogs, *Herpetol. Monogr.*, 1993, **7**, 94–117.
68. A. Haas, Phylogeny of frogs as interfered from primarily larval characters (Amphibia: Anura), *Cladistics*, 2003, **19**, 23–89.
69. K. Roelants, A. Haas and F. Bossuyt, Anuran radiations and the evolution of tadpole morphospace, *Proc. Natl. Acad. Sci. U. S. A.*, 2011, **108**, 8731–8736.
70. N. Gradwell, Comments on gill irrigation in *Rana fuscigula*, *Herpetology*, 1972, **28**, 123–125.
71. N. Gradwell, Gill irrigation in *Rana catesbeiana*, Part I. On the anatomical basis, *Can. J. Zool.*, 1972, **50**, 481–499.
72. R. Wassersug and K. Hoff, A comparative study of the buccal pumping mechanism of tadpoles, *Biol. J. Linn. Soc.*, 1979, **12**, 225–259.
73. R. Wassersug and K. Hoff, Developmental changes in the orientation of the anuran jaw suspension, *Evol. Biol.*, 1982, **15**, 223–246.
74. R. G. Boutilier, D. F. Stiffler and D. P. Toews, Exchange of respiratory gases, ions, and water in amphibious and aquatic amphibians, in *Environmental Physiology of the Amphibians*, ed. M. E. Feder and W. W. Burggren, The University of Chicago Press, Chicago and London, 1992, pp. 81–124.
75. G. R. Ultsch, D. F. Bradford and J. Freda, Physiology, coping with the environment, in *Tadpoles, the Biology of Anuran Larvae*, ed. R. Altig and R. W. McDiarmid, The University of Chicago Press, Chicago and London, 1999, pp. 189–214.
76. S. F. Gilbert, *Developmental Biology*, Sinauer Associates, Inc., Sunderland, Massachusetts, 2010, vol. 9.
77. K. Greulich, E. Hogue and S. Pflugmacher, Uptake, metabolism, and effects on detoxication enzymes of isoproturon in spawn and tadpoles of amphibians, *Ecotoxicol. Environ. Saf.*, 2002, **52**, 256–266.
78. K. Greulich and S. Pflugmacher, Differences in susceptibility of various life stages of amphibians to pesticide exposure, *Aquat. Toxicol.*, 2003, **65**, 329–336.

79. K. Greulich and S. Pflugmacher, Uptake and effects on detoxication enzymes of cypermethrin in embryos and tadpoles of amphibians, *Arch. Environ. Contam. Toxicol.*, 2004, **47**, 489–495.
80. K. L. Gosner, A simple table for staging anuran embryos and larvae with notes on identification, *Herpetology*, 1960, **16**, 183–190.
81. P. D. Nieuwkoop and J. Faber, Normal Table of *Xenopus laevis* (Daudin) - A Systematical and Chronological Survey of the Development from the Fertilized Egg till the End of Metamorphosis, North-Holland Publishing Company, Amsterdam, 1956.
82. R. W. McDiarmid and R. Altig, Research-materials and technique, in *Tadpoles, the Biology of Anuran Larvae*, ed. R. Altig and R. W. McDiarmid, The University of Chicago Press, Chicago and London, 1999, pp. 7–23.
83. J. A. Bantle, J. N. Dumont, R. A. Finch, G. Linder and D. J. Fort, *Atlas of Abnormalities: a Guide for the Performance of FETAX*, Oklahoma State University Press, Stillwater, 1998.
84. J. A. Bantle, R. A. Finch, D. J. Fort, E. L. Stover, M. Hull, M. Kumsher-King and A. M. Gaudet-Hull, Phase III interlaboratory study of FETAX. Part 3. FETAX validation using 12 compounds with and without an exogenous metabolic activation system, *J. Appl. Toxicol.*, 1999, **19**, 447–472.
85. S. Yu, M. R. Wakes, Q. Cai, J. D. Maul and B. P. Cobb, Lethal and sublethal effects of three insecticides on two developmental stages of *Xenopus laevis* and comparison with other amphibians, *Environ. Toxicol. Chem.*, 2013, **32**, 2056–2064.
86. J. Larsen and I. Sorensen, The Effect of Esfenvelerate and Prochloraz on Amphibians with Reference to *Xenopus laevis* and *Bombina bombina*, Ministry of Environment, Copenhagen, 2004.
87. J. Larsen, I. Sorensen and K. Gustafson, The Effects of Selected Pyrethroids on Embryos of *Bombina bombina* during Different Culture and Semi-field Conditions, Ministry of Environment, Danish Environmental Agency, Copenhagen, 2004.
88. L. Briggs and N. Damm, Effects of Pesticides on *Bombina bombina* in Natural Pond Ecosystems, Ministry of Environment, Danish Environmental Agency, Copenhagen, 2004.
89. R. Altig and R. W. McDiarmid, Body plan, development and morphology, in *Tadpoles, the Biology of Anuran Larvae*, ed. R. Altig and R. W. McDiarmid, The University of Chicago Press, Chicago and London, 1999, pp. 24–51.
90. S. Fryday and H. Thompson, *Toxicity of pesticides to aquatic and terrestrial life stages of amphibians and occurrence, habitat use, and exposure of amphibian species in agricultural environments. Question No EFSA-Q-2011-0079 0*, Food and Environment Research Agency, Sand Hutton, York, UK, 2012.
91. A. S. Cooke, The effects of DDT, dieldrin and 2,4-D on amphibian spawn and tadpoles, *Environ. Pollut.*, 1972, **3**, 51–68.

92. N. P. Zaffaroni, T. Zavanella and E. Arias, Peripheral blood cells in the crested newt after long-term exposure to the fungicide manganese ethylenebisdithiocarbamate (Maneb), *Bull. Environ. Contam. Toxicol.*, 1979, **23**, 587–591.
93. S. Gozzo, G. Perretta, U. Andreozzi, V. Monaco and E. Rossiello, Neuropathology induced by trimethyltin in the central nervous system of the urodele *Triturus carnifex*, *Aquat. Toxicol.*, 1994, **30**, 1–11.
94. A. Jaylet, P. Deparis, V. Ferrier, S. Grinfeld and R. Siboulet, A new micronucleus test using peripheral blood erythrocytes of the newt *Pleurodeles waltl* to detect mutagens in fresh-water pollution, *Mutat. Res., Environ. Mutagen. Relat. Subj.*, 1986, **164**, 245–257.
95. A. Jaylet, L. Gauthier and M. Fernandez, Detection of mutagenicity in drinking water using a micronucleus test in newt larvae (*Pleurodeles waltl*), *Mutagenesis*, 1987, **2**, 211–214.
96. F. Mouchet, L. Gauthier, C. Mailhes, V. Ferrier and A. Devaux, Comparative evaluation of genotoxicity of captan in amphibian larvae (*Xenopus laevis* and *Pleurodeles waltl*) using the comet assay and the micronucleus test, *Environ. Toxicol.*, 2006, **21**, 264–277.
97. A. Mandrillon and P. Saglio, Waterborne amitrole affects the predator–prey relationship between common frog tadpoles (*Rana temporaria*) and larval spotted salamander (*Salamandra salamandra*), *Arch. Environ. Contam. Toxicol.*, 2007, **53**, 233–240.
98. M. E. Ortiz-Santaliestra, M. J. Fernández-Beneítez, M. Lizana and A. Marco, Influence of a combination of agricultural chemicals on embryos of the endangered gold-striped salamander (*Chioglossa lusitanica*), *Arch. Environ. Contam. Toxicol.*, 2011, **60**, 672–680.
99. N. Wagner and S. Lötters, Effects of water contamination on site selection by amphibians: experiences from an arena approach with European frogs and newts, *Arch. Environ. Contam. Toxicol.*, 2013, **65**, 98–104.
100. J. Ujszegi, Z. Gál, Z. Mikó and A. Hettyey, No observable effect of a glyphosate-based herbicide on two top predators of temporal water bodies, *Environ. Toxicol. Chem.*, 2015, **34**, 307–313.
101. M. Pezdirc, E. Heath, L. B. Mali and B. Bulog, PCB accumulation and tissue distribution in cave salamander (*Proteus anguinus anguinus*, Amphibia, Urodela) in the polluted karstic hinterland of the Krupa River, Slovenia, *Chemosphere*, 2011, **84**, 987–993.
102. A. S. Cooke, Effects of field applications of the herbicides diquat and dichlobenil on amphibians, *Environ. Pollut.*, 1977, **12**, 43–50.
103. A. S. Cooke, Tadpoles as indicators of harmful levels of pollution in the field, *Environ. Pollut.*, 1981, **25**, 123–133.
104. F. Mesléard, M. Gauthier-Clerc and F. Lambret, Impact of the insecticide alphacypermetrine and herbicide oxadiazon, used singly or in combination, on the most abundant frog in French rice fields, *Pelophylax perezi*, *Aquat. Toxicol.*, 2016, **176**, 24–29.
105. F. Sayim, Toxicity of trifluralin on the embryos and larvae of the red-bellied toad, *Bombina bombina*, *Turk. J. Zool.*, 2010, **34**, 479–486.

106. E. Brunelli, I. Bernabò, C. Berg, K. Lundstedt-Enkel, A. Bonacci and S. Tripepi, Environmentally relevant concentrations of endosulfan impair development, metamorphosis and behaviour in *Bufo bufo* tadpoles, *Aquat. Toxicol.*, 2009, **91**, 135–142.
107. I. Bernabò, A. Guardia, D. La Russa, G. Madeo, S. Tripepi and E. Brunelli, Exposure and post-exposure effects of endosulfan on *Bufo bufo* tadpoles: Morpho-histological and ultrastructural study on epidermis and iNOS localization, *Aquat. Toxicol.*, 2013, **142**, 164–175.
108. E. Brunelli, I. Bernabò, E. Sperone and S. Tripepi, Gill alterations as biomarkers of chronic exposure to endosulfan in *Bufo bufo* tadpoles, *Histol. Histopathol.*, 2010, **25**, 1519–1529.
109. E. Garcia-Munoz, F. Guerrero and G. Parra, Intraspecific and interspecific tolerance to copper sulphate in five Iberian amphibian species at two developmental stages, *Arch. Environ. Contam. Toxicol.*, 2010, **59**, 312–321.
110. M. Gürkan and S. Hayretdağ, Acute toxicity of maneb in the tadpoles of common and green toad, *Arh. Hig. Rada Toksikol.*, 2015, **66**, 189–195.
111. M. Gürkan and S. Hayretdağ, Morphological and histological effects of copper sulfate on the larval development of green toad, *Bufo viridis*, *Turk. J. Zool.*, 2012, **36**, 231–240.
112. A. Güngördü, Comparative toxicity of methidathion and glyphosate on early life stages of three amphibian species: *Pelophylax ridibundus, Pseudepidalea viridis*, and *Xenopus laevis*, *Aquat. Toxicol.*, 2013, **140–141**, 220–228.
113. A. Güngördü, M. Uçkun and Yologlu, Integrated assessment of biochemical markers in premetamorphic tadpoles of three amphibian species exposed to glyphosate- and methidathion-based pesticides in single and combination forms, *Chemosphere*, 2016, **144**, 2024–2035.
114. E. Garcia-Munoz, F. Guerrero and G. Parra, Effects of copper sulfate on growth, development, and escape behavior in *Epidalea calamita* embryos and larvae, *Arch. Environ. Contam. Toxicol.*, 2009, **56**, 557–565.
115. F. Sayim and U. Kaya, Effects of dimethoate on tree frog (*Hyla arborea*) larvae, *Turk. J. Zool.*, 2006, **30**, 261–266.
116. I. Bernabò, A. Guardia, R. Macirella, S. Sesti, A. Crescente and E. Brunelli, Effects of long-term exposure to two fungicides, pyrimethanil and tebuconazole, on survival and life history traits of Italian tree frog (*Hyla intermedia*), *Aquat. Toxicol.*, 2016, **172**, 56–66.
117. C. Fenoglio, A. Grosso, E. Boncompagni, C. Gandini, G. Milanes and S. Barni, Exposure to heptachlor: evaluation of the effects on the larval and adult epidermis of *Rana* kl. *esculenta*, *Aquat. Toxicol.*, 2009, **91**, 151–160.
118. F. Sayim and F. Acute, toxic effects of malathion on the 21st stage larvae of the marsh frog, *Turk. J. Zool.*, 2008, **32**, 99–106.
119. M. Lavorato, I. Bernabò, A. Crescente, M. Denoël, S. Tripepi and E. Brunelli, Endosulfan effects on *Rana dalmatina* tadpoles: quantitative developmental and behavioural analysis, *Arch. Environ. Contam. Toxicol.*, 2013, **64**, 253–262.

120. I. Bernabò, L. Gallo, E. Sperone, S. Tripepi and E. Brunelli, Survival, development, and gonadal differentiation in *Rana dalmatina* chronically exposed to chlorpyrifos, *J. Exp. Zool., Part A*, 2011, **315**, 314–327.
121. I. Bernabò, E. Sperone, S. Tripepi and E. Brunelli, Toxicity of chlorpyrifos to larval *Rana dalmatina*: acute and chronic effects on survival, development, growth and gill apparatus, *Arch. Environ. Contam. Toxicol.*, 2011, **61**, 704–718.
122. Z. Rudek and M. Rożek, Induction of micronuclei in tadpoles of *Rana temporaria* and *Xenopus laevis* by the pyrethroid Fastac 10 EC, *Mutat. Res., Genet. Toxicol. Environ. Mutagen.*, 1992, **298**, 25–29.
123. M. Johansson, H. Piha, H. Kylin and J. Merilä, Toxicity of six pesticides to common frog (*Rana temporaria*) tadpoles, *Environ. Toxicol. Chem.*, 2006, **25**, 3164–3170.
124. A. S. Cooke, The effect of pp′-DDT on tadpoles of the common frog (*Rana temporaria*), *Environ. Pollut.*, 1970, **1**, 57–71.
125. A. S. Cooke, Response of *Rana temporaria* tadpoles to chronic doses of pp′-DDT, *Copeia*, 1973, **1973**, 647–652.
126. A. S. Cooke, The effects of DDT, when used as a mosquito larvicide, on tadpoles of the frog *Rana temporaria*, *Environ. Pollut.*, 1970, **5**, 259–273.
127. D. Osborn, A. S. Cooke and S. Freestone, Histology of a teratogenic effect of DDT on *Rana temporaria* tadpoles, *Environ. Pollut., Ser. A*, 1981, **25**, 305–319.
128. A. S. Cooke, The influence of rearing density on the subsequent response to DDT dosing for tadpoles of the frog *Rana temporaria*, *Bull. Environ. Contam. Toxicol.*, 1979, **21**, 837–841.
129. A. S. Cooke, Selective predation by newts on frog tadpoles treated with DDT, *Nature*, 1971, **22**, 275–276.
130. M. Denoël, B. D'Hooghe, G. F. Ficetola, C. Brasseur, E. De Pauw, J. P. Thomé and P. Kestemont, Using sets of behavioral biomarkers to assess short-term effects of pesticide: a study case with endosulfan on frog tadpoles, *Ecotoxicology*, 2012, **21**, 1240–1250.
131. M. E. Ortiz-Santaliestra, A. Marco, M. J. Fernández and M. Lizana, Influence of developmental stage on sensitivity to ammonium nitrate of aquatic stages of amphibians, *Environ. Toxicol. Chem.*, 2006, **25**, 105–111.
132. Q. Xu and R. S. Oldham, Lethal and sublethal effects of nitrogen fertilizer ammonium nitrate on common toad (*Bufo bufo*) tadpoles, *Arch. Environ. Contam. Toxicol.*, 1997, **32**, 298–303.
133. ASTM (American Society for Testing and Materials), *Standard Guide for Conducting Acute Toxicity Tests on Test Materials with Fishes, Macroinvertebrates, and Amphibians. E 729-96*, ASTM International, West Conshohocken, 2002.
134. M. E. Feder and W. W. Burggren, Cutaneous gas exchange in vertebrates: design, patterns, control, and implications, *Biol. Rev.*, 1985, **60**, 1–45.

135. J. C. Fenwick, Calcium exchange across fish gills, in *Vertebrate Endocrinology: Fundamentals and Biomedical Implications*, ed. P. K. T. Pang and M. P. Schreibman, Academic Press, San Diego, 1989, pp. 319–338.
136. D. H. Evans, P. M. Piermarini and K. P. Choe, The multifunctional fish gill: dominant site of gas exchange, osmoregulation, acid-base regulation and excretion of nitrogeneous waste, *Physiol. Rev.*, 2006, **85**, 97–177.
137. D. H. Evans, P. M. Piermarini and W. T. W. Potts, Ionic transport in the fish gill epithelium, *J. Exp. Zool.*, 1999, **283**, 641–652.
138. D. H. Evans, The fish gill: site of action and model for toxic effects of environmental pollutants, *Environ. Health Perspect.*, 1987, **71**, 47–58.
139. R. J. Erickson and J. M. McKim, A simple flow-limited model for gas exchange of organic chemicals at fish gills, *Environ. Toxicol. Chem.*, 1990, **9**, 159–165.
140. A. Quaranta, V. Bellantuono, G. Cassano and C. Lippe, Why amphibians are more sensitive than mammals to xenobiotics, *PLoS One*, 2009, **4**, e7699.
141. L. Weltje, P. Simpson, M. Gross, M. Crane and J. R. Wheeler, Comparative acute and chronic sensitivity of fish and amphibians: a critical review of data, *Environ. Toxicol. Chem.*, 2013, **32**, 984–994.
142. W. A. Battaglin, K. C. Rice, M. J. Focazio, S. Salmons and R. X. Barry, The occurrence of GLY, atrazine, and other pesticides in vernal pools and adjacent streams in Washington, DC, Maryland, Iowa, and Wyoming, 2005–2006, *Environ. Monit. Assess.*, 2009, **155**, 281–307.
143. N. Wagner and R. Hendler, Schutz von Amphibienlaichgewässern vor Pestizideinträgen durch Gewässerrandstreifen, *Nat. Landschaft.*, 2015, **90**, 224–229.
144. K. Greulich and S. Pflugmacher, Untersuchungen zur Wasserqualität eines Reproduktionsgewässers für Amphibien, *Herpetofauna*, 2002, **24**, 5–9.

CHAPTER 17

Developmental Stages of Rhinella arenarum *(Anura, Bufonidae) in Toxicity Studies: AMPHITOX, a Customized Laboratory Assay*

C. S. PÉREZ COLL,*[a,b] C. M. ARONZON[a,b] AND G. V. SVARTZ[a,b]

[a] Instituto de Investigación e Ingeniería Ambiental, Universidad Nacional de Gral. San Martín, Campus Miguelete, Av. 25 de Mayo y Martín de Irigoyen, 1650 San Martín, Prov. de Buenos Aires, Argentina; [b] Consejo Nacional de Investigaciones Científicas y Técnicas, Av. Rivadavia 1917, 1033 Ciudad Autónoma de Buenos Aires, Argentina
*Email: perezcoll@unsam.edu.ar; cristinaperezmilos@hotmail.com

17.1 Introduction

Increasing negative impacts on natural resources and life support systems result from continuing population growth and anthropic activities. This situation is worsened by increasing environmental pollution and inappropriate environmental management.[1–3] These facts are of major concern because the decline of environmental quality and biodiversity could severely affect ecosystem services that are essential to the quality of the biosphere and human health.[4,5] In present day conditions, it is well-documented that environmental degradation could result in lethal effects[6,7] while sublethal

Issues in Toxicology No. 33
Ecotoxicology and Genotoxicology: Non-traditional Aquatic Models
Edited by Marcelo L. Larramendy

Published by the Royal Society of Chemistry, www.rsc.org

impacts, from teratogenesis[8–11] to endocrine disruption,[12,13] could also be a threat to wildlife and human health. The aquatic environment is particularly important as a recipient, transporting medium and sink for the majority of xenobiotic chemicals.[1,14] In 1999, in order to cope with an increasing diversity of chemicals from different matrices and processes, AMPHITOX was developed[15] to cover the need for a sensitive, cost-effective and customized assay for different exposure times and developmental stage toxicity tests. The usefulness of the AMPHITOX test for toxicity assessment of environmental samples was shown in those early contributions.[16,17]

As stewardship for environmental and human health protection improves, the development of a more flexible and customized toxicity test for amphibians seems to be of high priority. FETAX, Frog Embryo Teratogenesis Assay-Xenopus,[18,19] represented the most important antecedent of toxicity test employing amphibian embryos for assessing lethal and teratogenic effects of xenobiotics within an acute exposure period. AMPHITOX has introduced additional possibilities to this test, such as: (1) The extension of the exposure time from acute to short-term chronic and chronic exposures, for the assessment of toxicity in adjusted bioassays, according to the characteristics of the sample.[16] (2) The use of two different developmental periods, embryos and larvae, for toxicity tests. (3) The plot of the LC_{10} (or LOEC), LC_{50} and LC_{90} (or LC_{100}), to describe the toxicity profile (TOP) curves from acute to chronic exposure, allowing the visualization of concentration- and time-exposure thresholds, as well as the range of concentrations that exert adverse effects in each case. These TOP curves provide, within a systemic toxicity approach, a more appropriate set of data for hazard and risk assessment purposes.[2,20–22] (4) Additionally, the use of *Rhinella arenarum* (ex *Bufo arenarum*) as a native amphibian specie distributed in a large area of South America also provides relevant information on the local impact of pollutants.[2,17] This species is highly sensitive to toxic agents and at the same time is placed among the "least concern" species according to the International Union for Conservation of Nature.[23] AMPHITOX is a standardized test employing amphibian embryos that can be used to evaluate toxicity for acute, short-term chronic, chronic, and early life stage exposure to hazardous chemicals and complex mixtures.[15] By means of AMPHITOX the toxicity of both environmental samples and physicochemical agents singly[8,24] or in combination[25,26] can be evaluated by adjusting the exposure period to the toxicity of the sample and/or the assessment purpose.[16,17,27]

The growing concern about chronic effects of xenobiotics at environmentally relevant concentrations led us to expand the exposure conditions of classic, acute (up to 96 h of exposure), short-term chronic (168 h) and chronic (14 days) AMPHITOX tests with young larvae to evaluate noxious effects on specific stages of the early life cycle (pulse exposures) and exposure periods covering whole development (embryo–larval and metamorphosis). This last test has even included post-exposure effects *e.g.* 60 days post-metamorphosis, to assess adverse effects on gonadogenesis and sexual differentiation.

It is useful to highlight that a worldwide decline of amphibian populations linked to the increasing environmental degradation has been widely documented.[28–32] Amphibians play a key role in food webs, living near or in water reservoirs, were they can be direct or indirectly affected by xenobiotic exposure. Moreover, as amphibians are considered keystone members of ecosystems and vital links in food chains, contaminants might be magnified through them.[33] Furthermore, contaminants might be transported by metamorphic amphibians to terrestrial communities.[34,35] This taxonomic group includes extremely sensitive species because they have permeable skin and eggs that readily absorb chemicals from the environment.[29,36] Moreover, many species complete their life cycle in shallow or ephemeral water bodies in which pollutants might be concentrated by seasonal temperature increments, coincident with breeding and larval development.[37–39]

In this contribution, the basic conditions of the AMPHITOX test and some case studies were included to illustrate both the basic and customized potential of this toxicity test. TOP curves, scanning electron microscopy (SEM) studies and histological analysis accompanying the toxicity assessment of different matrices were added.

17.2 Experimental

17.2.1 Aquisition of *Rhinella arenarum* Embryos and Larvae

R. arenarum adults were acquired in a non-impacted site in Lobos (Buenos Aires, Argentina: 35° 11′ S; 59° 05′ W). Ovulation of females was induced by means of an *i.p.* injection of a suspension of one homogenized homologous pituitary gland in 1 mL of AMPHITOX solution (AS) per female plus 2500 IU human chorionic gonadotropin (hCG). The composition of AS was: NaCl 36 mg L^{-1}, KCl 0.5 mg L^{-1}, $CaCl_2$ 1 mg L^{-1} and $NaHCO_3$ 2 mg L^{-1}, prepared in distilled water. Oocytes were fertilized *in vitro* using a 10% sperm suspension also prepared in AS. The sperm viability was confirmed by optical microscope. The egg quality and fertility were considered acceptable if the fertility rate was greater than 75% and embryo survival at the neurula stage was greater than 70%. For embryos used before hatching, the jelly coat was dissolved by immersing egg ribbons in a solution of 2% thioglycolic acid neutralized at pH 7.2–7.4 with 1.35 mL of saturated NaOH solution in 100 mL of AS, followed by thorough washing. Embryo and larval developmental stages were classified according to Del Conte and Sirlin[40] and Echeverria and Fiorito de Lopez,[41] respectively.

17.2.2 Toxicity Tests

The customized AMPHITOX tests include the following options according to the main purpose of the study: (1) Acute, short-term chronic and chronic toxicity tests employing young larvae at complete operculum stage (S.25). (2) Stage-dependent sensitivity assessments comparing different embryonic and/or larval stages. (3) Early Life Stage (ELS) studies treating embryos from two-cell blastomeres (S.4) up to complete operculum stages (S.25). (4) Complete development tests employing specimens from blastula to metamorphosis/60 days post-metamorphosis.

According to the purpose of the study, batches of 10 embryos/larvae, in triplicate, were placed in 10 cm Ø glass Petri dishes containing 40 mL of medium and maintained at 20 ± 1 °C. In cases when sample volume was critical, five embryos were placed in 5 cm Ø glass Petri dishes containing 10 mL of medium. In the case of growing larvae after 21 days, 10 individuals in triplicate were placed in 20 cm Ø glass Petri dishes containing 150 mL of test solution. Juveniles were maintained without exposure up to 60 days post-metamorphosis in glass flasks with 100 g of sand embedded on AS and kept wet by spraying with AS daily. Juveniles were fed with micro-crickets, *Acheta domestica* (~3 mm).

In all cases, control groups were simultaneously maintained in AS without additions. The conductivity and pH of the samples were measured and proper control bioassays for these parameters were conducted. In the case of continuous exposure, solutions were renewed every 48 hours. In 24 h pulse exposures the embryos were thoroughly washed after treatment and kept in AS. Specimens reaching the complete operculum stage (S.25), were fed with 3 granules (6 ± 0.5 mg) of balanced fish food Tetra Color® per Petri dish every 48 hours.

17.2.3 Effects Assessment

Lethal and sublethal (morphological and neurotoxic) effects were recorded daily by comparing the alterations observed in exposed individuals with the normal development and behaviour of control ones. Dead individuals were removed. Morphological and neurotoxic effects were evaluated by observing specimens under a binocular stereoscopic microscope (Zeiss Stemi DV4), which were photographed with a Sony DSC-S90 digital camera, and identified according to Bantle *et al.*[42] Embryos in Petri dishes showing significant adverse effects and control embryos were fixed in 4% formalin, dehydrated in a gradient of ethanol, prepared for scanning electron microscopy (SEM) by means of the critical point drying technique and observed using a Philips XL-30 operated at 10 kW for ultrastructure evaluation.

In metamorphosis studies, juveniles were euthanized at 60 days post-metamorphosis by submersion in a lethal concentration of benzocaine solution. Individuals were dissected under a Zeiss Stemi DV4 stereomicroscope and gonads were measured and examined for sex classification and gross gonadal anomalies. Gonad-kidney complexes were dissected and fixed in Bouin's solution for 24 h. Fixed organs were dehydrated in a graded ethanol series and embedded in paraffin. Serial cross-sections (7 μm) were obtained and then stained with hematoxylin and eosin. All slides of the sectioned ovaries and testes were observed and photographed under a Nikon Microphot FX microscope.

17.2.4 Data Analysis

Lethal (LC) and effective concentrations (EC) were statistically estimated by means of PROBIT analysis.[43] To examine statistical differences, a comparison was made, with the difference considered statistically significant when the higher LC/lower LC ratio exceeded the critical value (95%) established by APHA.[18,44] We conducted a one-way analysis of variance (ANOVA) to evaluate the NOEC value, and Tukey's tests were used to compare treatment means and establish significant differences ($P<0.05$). Sex ratios were statistically analyzed by Fisher's exact test, and Pearson's chi-square was used in the search for significant differences in the incidence of gonadal anomalies. Analyses were performed using GraphPad Prism software version 6.03 and differences were considered to be significant when $P<0.05$.

Additional information on the significance and use of the AMPHITOX test, including methodologies, water for culturing embryos, test material and organisms, procedures, test acceptability, records, safety precautions, references and information on amphibian species can be consulted in the standardized protocol for AMPHITOX.[15]

17.3 Results and Discussion

17.3.1 Acute, Short-term Chronic and Chronic Toxicity Studies

Individuals at complete operculum stage (S.25) provide the following advantages: (i) Increased maturity to assess neuromuscular, integument, respiratory, cardiovascular, digestive and excretory systems alterations. (ii) Relatively constant sensitivity to xenobiotics for about 21 days. (iii) Very low mortality baseline (less than 10%) registered throughout this period, which facilitates extension of toxicity assessment from acute to chronic period according to the purpose of the study. (iv) From a practical point of view, larval mortality is an endpoint to evaluate toxicity that just requires a quick and easy analysis and does not need as much observer expertise as teratological studies.

To illustrate this AMPHITOX test, we selected the toxicity study of endosulfan (ES). This compound is an organochlorine pesticide, identified as a persistent organic pollutant (POP) that was banned worldwide. However, it is still largely used, particularly in some developing countries, such as Argentina, where it has been phased out since July 2013.[45] Figure 17.1 exhibits the TOP curves of ES for *Rhinella arenarum* larvae, based on LCs, and shows a huge increase in the pesticide toxicity, from acute (96 h $LC_{10}=0.45$ mg ES L^{-1}), to chronic exposure (504 h $LC_{10}=0.004$ mg ES L^{-1}).[46]

It is important to note that ES environmental levels reported are in the range of 0.1–100 μg L^{-1},[47] reaching exceptional levels of 500 μg L^{-1}.[48] Taking into account the 504 h LC10 for *R. arenarum* as low as 4 μg ES L^{-1}, the measured and predicted ES concentrations exceed the levels that allow the

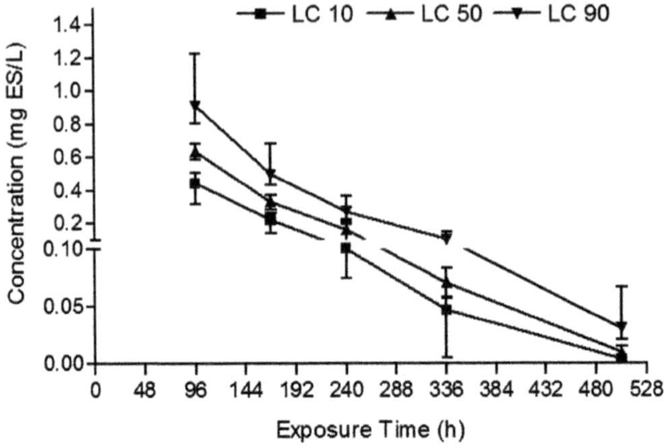

Figure 17.1 Toxicity profile (TOP curves) of *R. arenarum* continuously exposed to endosulfan (ES) starting at the complete operculum stage (S.25). The LC_{10}, LC_{50}, and LC_{90} and their corresponding confidence limits (95%) are plotted.
Reprinted from G. Svartz, I. R. Hutler Wolkowicz and C. S. Pérez-Coll, *Environ. Toxicol. Chem.*[46] © 2013 SETAC, with permission from John Wiley and Sons.

survival of this amphibian. Based on the analysis of a risk assessment scenario performed for this pesticide, Hazard Quotient approach (HQ) ES concentrations could be greater than ten times the level of concern, thus representing an important threat for this species.[49,50] By assessing the toxicity of the pesticide at environmentally relevant concentrations, the occurrence of lethal effects is of great concern, highlighting the significance and relevance of evaluating chronic effects for risk assessment studies.

This study shows the usefulness of this larval AMPHITOX test to reveal the toxicity of xenobiotics that affect organ systems in which maturation takes place late, such as the neuromuscular system, the main target organ of this pesticide. *R. arenarum* embryo survival, which was less affected than the larvae by the exposure not only to this pesticide but also to other xenobiotics, fell when organs and organ systems, including integument, respiratory, cardiovascular, digestive and excretory systems, were developed as maturity advanced (see Section 17.4.3.2).

17.3.2 Stage-dependent Sensitivity Studies

Stage-dependent sensitivity studies include basically two possibilities: (i) the comparative toxicity assessment of a xenobiotic on different periods of the life cycle, such as between complete embryo development and larval development, and (ii) comparative toxicity bioassays with embryos exposed at different stages (from S.4 to S.25).

To illustrate the first possibility (i), Figure 17.2 shows the comparative toxicity of a cypermethrin commercial formulation (CF) with 25% active

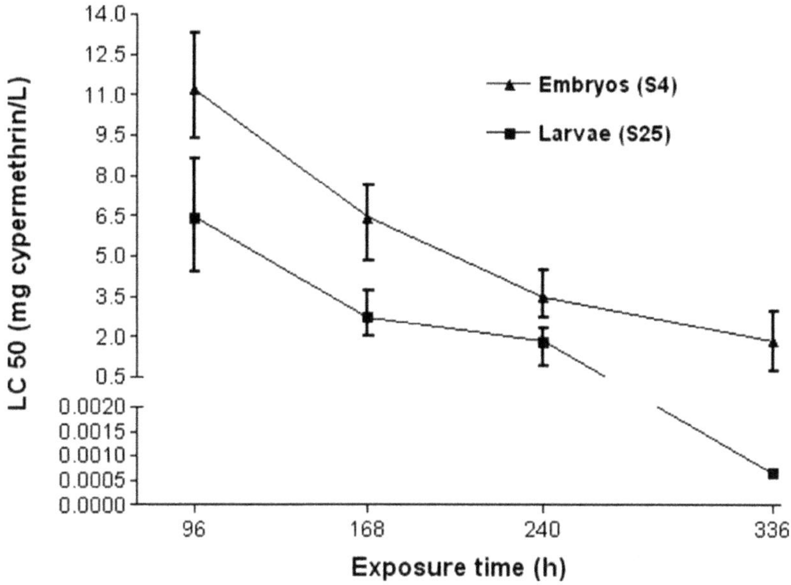

Figure 17.2 Toxicity profile (TOP curves) based on LC_{50} values of a commercial formulation with 25% active ingredient cypermethrin for *R. arenarum* embryos and larvae continuously exposed to the pesticide. The corresponding confidence limits (95%) are plotted.
Reprinted from *Environ. Sci. Pollut. Res.*, Comparative sensitivity among early life stages of the South American toad to cypermethrin-based pesticide, **23**, 2016, 2906–2913, G. Svartz, C. Aronzon and C.S. Pérez Coll (© Springer-Verlag Berlin Heidelberg 2015) With permission of Springer.[15]

ingredient (AI) on the overall *R. arenarum* embryo and larval development from complete operculum stage.[51]

The figure shows the TOP curves based on the LC_{50}, and their corresponding confidence limits (95%) at different exposure times. A huge differential sensitivity to the pesticide between embryos and larvae, with the early embryo development less susceptible to the CF cypermethrin (336 h NOEC = 0.5 mg L^{-1}) than larvae (336 h NOEC = 0.0005 mg L^{-1}) was obtained. Then, cypermethrin toxicity on larvae was almost 2800-fold higher than on embryos. Since pyrethroids act primarily on the Na^{+} and Ca^{++}channels in nerve tissues, the highest sensitivity of growing larvae could be related to the maturation of the nervous system. Another important advantage of TOP curves is that they provide relevant information at a quick glance. This simple AMPHITOX test allows also the possibility of highlighting the toxicity differences between the active ingredients (AI) and the commercial formulations (CF) of the pesticides. Thus, by assessing the comparative toxicity of cypermethrin AI and CF, the higher toxicity of the latter was demonstrated (336 h LC_{50} of 0.0048 mg CY L^{-1} and 0.00065 mg CY L^{-1}, for AI and CF, respectively).[52] On the other hand, the comparative toxicity of the herbicide 2,4-D (butyl ester of 2,4-dichlorophenoxyaceticacid) AI and CF on *R. arenarum* embryos and larvae also showed differential sensitivity to the herbicide, but in this case the early embryo development was most susceptible to the AI, 2,4-D (168 h NOEC = 6.4 mg L^{-1}) than larvae development (168 h NOEC = 12.4 mg L^{-1}). The CF NOEC for lethality was about 2 mg L^{-1} for both periods.[53] In both cases the CF was significantly more toxic, from 2 to 3 times higher than AI ($P<0.05$) for embryos and about four times more toxic for larvae ($P<0.05$). The elevated toxicity of the CF compared to that of the AI is frequently observed and may be related to the ability of adjuvants to facilitate the incorporation of AI into different tissues.[54] These results are ecologically relevant considering that commercial products are actually applied to farmlands. If risk assessment analysis is based only on the toxicity of AI, the real toxicity of commercial products on non-target organisms might be completely undervalued. Other pesticides that caused higher lethal effects on *R. arenarum* larval than on embryo developmental periods were endosulfan,[46] atrazine[55] and diazinon.[56] Other non-pesticides, such as bisphenol A,[57] nonylphenol,[58] copper[5] and the physical agent UV B irradiation,[60] also exerted higher lethal toxicity in larvae than in embryos. On the contrary, as with 2, 4-D and other xenobiotics, such as the pharmaceutical naringenin;[61] one of the precursors of the epoxide resin BADGE (bisphenol A diglycidyl ether), the epichlorohydrin[62] and the epoxy BADGE itself,[63] rendered higher lethal toxicity in embryos than in larvae.[63]

For the stage-dependent sensitivity studies (ii), we selected the assessment of copper toxicity. [59] Thus, Figure 17.3 shows the markedly stage-dependent sensitivity of *R. arenarum* to this metal.

Embryos at the beginning of development exhibited the highest resistance to Cu (LC_{50} = 154 µg Cu^{2+} L^{-1}).There was a significant increase in Cu toxicity from neural fold (S.14) to muscular response (S.18) until open mouth (S.21), with these stages being the most susceptible to Cu during the whole embryonic development (S.20, LC_{50} = 17 µg Cu^{2+} L^{-1}). From the opercular folds stage (S.23) onward, the resistance to Cu gradually increased until the hind limb bud (S.28) stage, showing greater sensitivity to Cu than the early stage embryos. It is noteworthy that 24 h pulse exposure toxicity bioassays provide very ecologically relevant toxicity data from toxic spills or any other emergent environmental accident. These last stage-dependent toxicity tests are also useful to identify the most critical developmental period for regulatory/conservation purposes.

17.3.3 Early Life Stage (ELS) Studies

For this test, embryos were exposed during their overall development. That is, from the two-cell blastomere stage (S.4) up to complete operculum stage (S.25). This is a more complex test, unlike tests with larvae, which requires a more careful and detailed analysis by an expert researcher. However, it provides very valuable information on teratogenesis and other early failures, such as delayed development, reduced body size and the first ethological

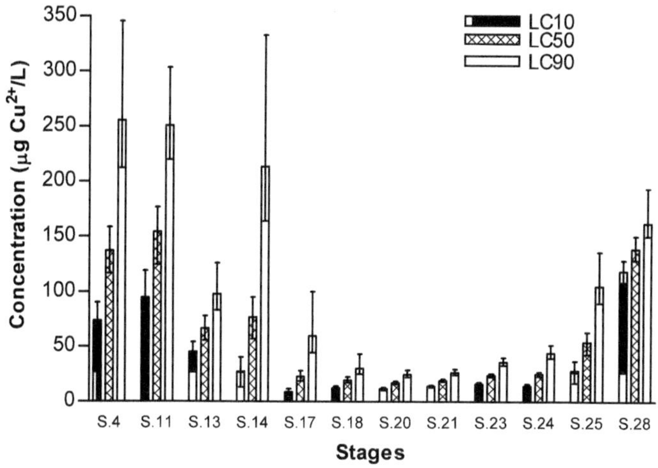

Figure 17.3 24 h LC_{10}, LC_{50}, and LC_{90} for Cu pulse-exposed *R. arenarum* at different developmental stages. The corresponding confidence limits (95%) are plotted.
Reprinted from C. M. Aronzon, M. T. Sandoval, J. Herkovits and C. S. Pérez-Coll, *Environ. Toxicol. Chem.*[59] Copyright © 2011 SETAC, with permission from John Wiley and Sons.

alterations as consequences of xenobiotic exposure. Rather few standardized tests are in use for the evaluation of early developmental failures in amphibians, even though fertilization and development take place externally in these aquatic organisms, thus allowing easy observation and monitoring of abnormalities. This is not a minor issue since early developmental disorders can make individuals more vulnerable to environmental challenges. Figure 17.4 shows different malformations in *R. arenarum* exposed to atrazine during the embryonic period (A–E) observed under scanning electron (SEM) and stereoscopic microscopy (SM).[64]

Furthermore, Figure 17.4F and G show the morphological alterations caused by the herbicide on larval development. A very relevant parameter for the comparison of substances' teratogenic potential is the Teratogenic Index (TI),[18] calculated as the LC_{10}/EC_{10} ratio. $TI > 1.5$ establishes a high risk for the occurrence of embryonic malformations in the absence of significant embryonic lethality. For example, atrazine's TI was 3.28. It was demonstrated that a large number of physicochemical agents assessed by AMPHITOX test were teratogenic, for example UVB irradiation,[60] cypermethrin,[51] endosulfan,[46] diazinon,[56] bisphenol A,[57] nonylphenol,[58] narigenin,[61] epichlorohydrine,[62] copper,[59] and the fungicide Maxim.[65] In contrast, other xenobiotics, such as the epoxide resin BADGE, showed low TI. The explanation for this finding resides in the fact that this compound was able to induce extremely severe abnormalities, which were incompatible with life. In addition, early ethological disturbances were observed as the consequence of embryo exposure to a diversity of adverse physicochemical agents.[46,51,53,56–58,60–63,65] Larvae that developed from atrazine-exposed[64] embryos exhibited abnormal and erratic

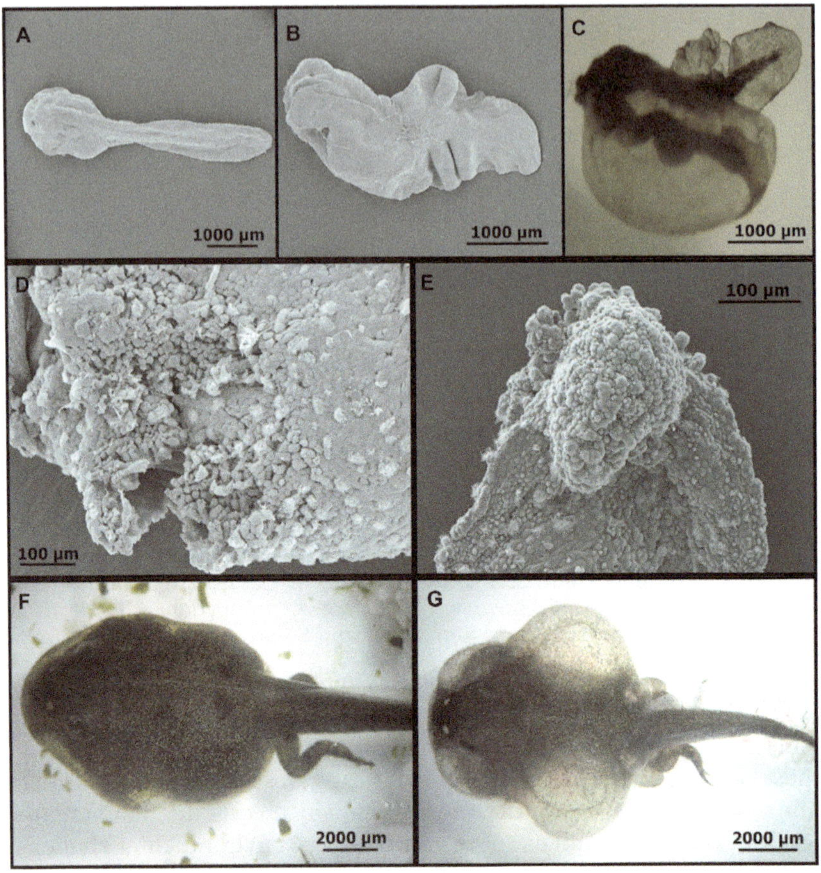

Figure 17.4 Scanning electron and stereoscopic microscopy pictures of *R. arenarum*. (A) Control larvae at S.25; (B and C) Larvae continuously treated with 20 $mg\,L^{-1}$ atrazine (ATR) from two-cell blastomere stage (S.4) onwards. The main anomalies were irregular border, flexures, pronounced edema, gut miscoiling, microcephaly, wavy and stunted tail. (D and E) Details of malformed mouth and sucker; the stunted tail exhibits cellular dissociation. (F) Control prometamorphosis larvae (S.28). (G) Larvae continuously exposed to 10 $mg\,L^{-1}$ ATR from S.28 onwards exhibit pronounced edema, tail flexures and reduced development of forelimbs. Reprinted from *Ecotoxicology*, Sublethal effects of atrazine on embryo-larval development of *Rhinella arenarum* (Anura: Bufonidae), **21**, 2012, 1251–1259, G. Svartz, J. Herkovits and C.S. Pérez Coll, (© Springer Science + Business Media, LLC 2012) With permission of Springer.[64]

swimming, spasmodic contractions, reduced food intake and general weakness. All of these disorders drastically alter the performance of organisms, making them more vulnerable (with decreased ability to escape from unfavorable conditions such as predators) and unable to compete with invasive species, eventually migrating toward better habitats. Both morphological and behavioural alterations can severely influence the health status of the animals and their offspring, with direct and indirect consequences on future population viability, as well as possible consequences at ecological levels.[66]

17.3.4 Studies on the Complete Development, From Blastula to Metamorphosis

This test is useful to perform a very complete toxicity analysis from almost the egg laying up to sexual differentiation, whose results enable us to inquire into the future viability of the exposed population. The treatment continues even after 60 days post-metamorphosis, which is the estimated time to reach sexual differentiation. Thus, AMPHITOX is also a very useful test for evaluating alterations in the metamorphic process, gonadogenesis and sexual differentiation as the consequence of toxicant exposure.

Figure 17.5 shows the effects of the commercial fungicide Maxim® XL on gonads of post-metamorphic *R. arenarum* juveniles after chronic exposure of early embryos for 75 days.[65]

Although in this case Maxim® XL affected neither the time required to complete metamorphosis nor sex proportions, gonadal development and differentiation were impaired. Gross gonadal analysis revealed a significant proportion of exposed individuals with underdevelopment of one or both gonads. Histological analysis confirmed that 18% of the individuals exposed to 0.25 mg L^{-1} Maxim® XL exhibited undifferentiated gonads characterized by a reduced number (or absence) of germ cells. It would be of great interest to continue the assessment of the effects of this and other fungicides on the endocrine system because a recent study has demonstrated that many fungicides with previously unknown endocrine activity were revealed as endocrine-disrupting chemicals.[67] Moreover, the assessment of histological alterations could be considered an "early warning" signal to assess environmental quality as they commonly occur at lower exposure concentrations than other endpoints, such as behavioural alterations and mortality.[68]

17.4 Conclusion

The results presented here show how the AMPHITOX test can be customized according to the toxicity of the xenobiotic and more relevant endpoints. The versatility of the AMPHITOX test for toxicity evaluation of a wide spectrum of adverse agents—chemicals either singly or in combinations, metals, organic compounds, UVB irradiation, *etc.*—was demonstrated. By extending the exposure from acute to short-term chronic and chronic it was possible to reveal the toxicity of xenobiotics at environmentally relevant concentrations, highlighting the significance and meaning of evaluating chronic exposure for risk assessment studies. Both the pulse exposure experiments and studies comparing different embryonic and/or larval stages are useful tools for regulation/conservation purposes thanks to their ability to show stage-dependent sensitivity to xenobiotics. In spite of the more complex ELS studies, teratogenesis and behavioural disorders provide very relevant

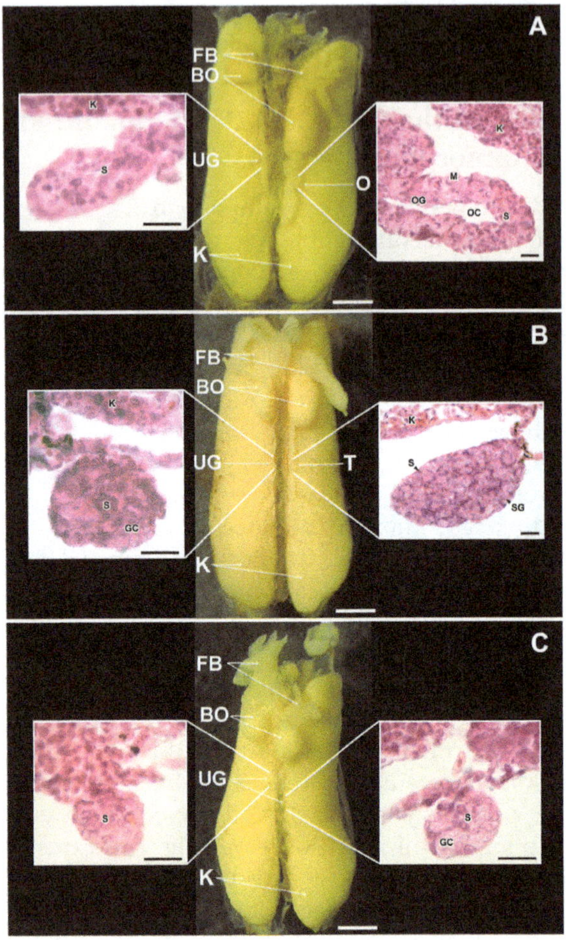

Figure 17.5 Gonadal development in 60 d post-metamorphic *R. arenarum* from Maxim® XL treatment. Stereomicroscope photographs of the kidney-gonad complex and inset showing light-microscope photographs of representative histological sections of the gonads. (A) Female from 0.25 $mg\,L^{-1}$ Maxim® XL treatment with underdeveloped right gonad; (B) Male from 2 $mg\,L^{-1}$ Maxim®XL treatment with underdeveloped right gonad; (C) Juvenile from 0.25 $mg\,L^{-1}$ Maxim® XL treatment showing underdevelopment of both gonads. Underdeveloped gonads exhibited scarce to complete absence of germ cells, so that frequently only somatic cells were observed. Scale bars: 500 μm (A–C), 20 μm (insets A–C). BO: Bidder's organ; FB: fat body; GC: germ cells; K: kidney; M: mitotic figures; O: ovary; OC: ovarian cavity; OG: oogonia; S: somatic cells; SG: spermatogonia; T: testis; UG: underdeveloped gonad.
Reprinted from *Environ. Toxicol. Pharmacol.*, **45**, G. Svartz, F. Meijide and C.S. Perez Coll, Effects of a fungicide formulation on embryo-larval development, metamorphosis, and gonadogenesis of the South American toad *Rhinella arenarum*, 1–7, Copyright (2016) with permission from Elsevier.[65]

information on the direct and indirect effects of xenobiotics at ecological levels. Finally, the AMPHITOX test covering complete development, from blastula to juvenile organisms, allows detection of xenobiotic toxicity on metamorphosis, gonadogenesis and sexual differentiation at ecologically relevant concentrations. AMPHITOX tests could be conducted with other amphibian species, although modifications might be necessary and the species' sensitivity range to xenobiotics should be taken into account when comparing data.

Acknowledgements

This study was supported by research Grants PICT-2013-0245 from Agencia de Promoción Científica y Tecnológica, Argentina and PIP 112 201301 00140 from Consejo Nacional de Investigaciones Científicas y Técnicas, Argentina. The authors are grateful to Dr Teresa M. Fonovich for her helpful criticism of the manuscript. We wish to thank Instituto Massone S. A. for providing the human chorionic gonadotropin.

References

1. J. Cairns Jr., Third wave ecotoxicology, *Ecotoxicology*, 1994, **3**, 1–10.
2. J. Herkovits, C. S. Pérez-Coll and F. D. Herkovits, Ecotoxicity in the Reconquista River, province of Buenos Aires, Argentina: a preliminary study, *Environ. Health Perspect.*, 1996, **104**, 186–189.
3. D. A. Vazquez Brust and C. Liston-Heyes, Environmental management intentions: An empirical investigation of Argentina's polluting firms, *J. Environ. Manage.*, 2010, **91**, 1111–1122.
4. R. Costanza, R. d'Arge, R. de Groot, S. Farber, M. Grasso, B. Hannon, K. Limburg, S. Naeem, R. O'Neill, J. Paruelo, R. G. Raskion, P. Sutton and M. van der Bel, The value of the world's ecosystem services and natural capital, *Nature*, 1997, **387**, 253–260.
5. C. Raudsepp-Hearne, G. D. Peterson, M. Tengö, E. M. Bennett, T. Holland, K. Benessaiah, G. K. MacDonald and L. Pfeifer, Untangling the environmentalist's paradox: why is human well-being increasing as ecosystem services degrade? *BioScience*, 2010, **60**, 576–589.
6. M. K. Hill, *Understanding Environmental Pollution*, Cambridge University Press, UK, 2010.
7. F. Pernet, J. Barret, C. Marty, J. Moal, P. Le Gall and P. Boudry, Environmental anomalies, energetic reserves and fatty acid modifications in oysters coincide with an exceptional mortality event, *Mar. Ecol.*, 2010, **401**, 129–146.
8. C. S. Pérez-Coll, J. Herkovits and A. Salibián, Teratogenic effects of cadmium on *Bufo arenarum* during gastrulation, *Experientia*, 1986, **42**, 1174–1176.
9. C. S. Pérez-Coll, J. Herkovits and A. Salibián, , Embryotoxicity of lead on *Bufo arenarum*, *Bull. Environ. Contam. Toxicol.*, 1988, **41**, 247–252.

10. P. G. Wells, Y. Bhuller, C. S. Chen, W. Jeng, S. Kasapinovic, J. C. Kennedy, P. M. Kim, R. R. Lapos, G. P. McCallum, C. J. Nicol, T. Parman, M. J. Wiley and A. W. Wong, Molecular and biochemical mechanisms in teratogenesis involving reactive oxygen species, *Toxicol. Appl. Pharmacol.*, 2005, **207**(Suppl.), 354–366.
11. V. Pašková, K. Hilscherová and L. Bláha, Teratogenicity and embryotoxicity in aquatic organisms after pesticide exposure and the role of oxidative stress, in *Reviews of Environmental Contamination and Toxicology*, **211**, ed. D. M. Whitacre, Springer, NY, 2011, pp. 25–61.
12. O. Fridman, L. Corro and J. Herkovits, Estradiol uptake, toxicity, metabolism, and adverse effects on cadmium-treated amphibian embryos, *Environ. Health Perspect.*, 2004, **112**, 862–866.
13. J.-H. Kang, D. Aasi and Y. Katayama, Bisphenol A in the aquatic environment and its endocrine-disruptive effects on aquatic organisms, *Crit. Rev. Toxicol.*, 2007, **37**, 607–625.
14. D. Hering, A. Borja, J. Carstensen, L. Carvalho, M. Elliott, C. K. Feld, A.-S. Heiskanen, R. K. Johnson, J. Moe, D. Pont, A. Solheim and W. van de Bund, The European Water Framework Directive at the age of 10: a critical review of the achievements with recommendations for the future, *Sci. Total Environ.*, 2010, **408**, 4007–4019.
15. J. Herkovits and C. S. Pérez-Coll, Bioensayos para test de toxicidad con embriones de anfibios ("ANFITOX"), *Ing. Sanit. Amb.*, 1999, **42**, 24–30 and **43**, 50–55.
16. J. Herkovits, C. S. Pérez-Coll and F. D. Herkovits, Ecotoxicological studies of environmental samples from Buenos Aires area using a standardized amphibian embryo toxicity test (AMPHITOX), *Environ. Pollut.*, 2002, **116**, 177–183.
17. J. Herkovits and C. S. Perez-Coll, Symposium on multiple stressor effects in relation to declining amphibian populations. AMPHITOX: A customized set of toxicity tests employing amphibian embryos, in *Multiple Stressor Effects in Relation to Declining Amphibian Populations*, ed. G. L. Linder, S. Crest, D. Sparling and E. E. Little, ASTM International, USA, STP 1443, 2003, pp. 46–60.
18. ASTM, Standard guide for conducting the Frog Embryo Teratogenesis Assay-Xenopus (FETAX), in *Standards on Aquatic Toxicology and Hazard Evaluation*, American Society for Testing and Materials, Philadelphia, PA, 1993, pp. 1199–1209.
19. J. A. Bantle, J. N. Dumont, R. A. Finch, G. Linder and D. J. Fort, *Atlas of Abnormalities, A Guide for the Performance of FETAX*, Oklahoma State University, Stillwater, 2nd edn, 1998.
20. J. Herkovits and A. L. Helguero, Copper toxicity and copper–zinc interactions in amphibian embryos, *Sci. Total Environ.*, 1998, **221**, 1–10.
21. J. Herkovits, F. D. Herkovits and C. S. Pérez-Coll, Identification of aluminium toxicity and Al-Zn interaction in amphibian *Bufo arenarum* embryos, *Environ. Sci.*, 1997, **5**, 57–64.

22. J. Herkovits, C. S. Pérez-Coll and F. D. Herkovits, Evaluation of Nickel-Zinc interactions by means of bioassays with amphibian embryos, *Ecotoxicol. Environ. Saf.*, 2000, **45**, 266–273.
23. International Union for Conservation of Nature (IUCN), Red List of Threatened Species. http://www.iucnredlist.org/ (accessed July 2016).
24. C. S. Pérez-Coll and J. Herkovits, Stage dependent susceptibility to lead in *Bufo arenarum* embryos, *Environ. Pollut.*, 1990, **63**, 239–245.
25. G. V. Svartz, C. M. Aronzon and C. S. Pérez Coll, Combined endosulfan and cypermethrin-induced toxicity to embryo–larval development of *Rhinella arenaru*, *J. Toxicol. Environ. Health*, 2016, **79**, 197–209.
26. C. M. Aronzon, G. V. Svartz and C. S. Pérez Coll, Synergy between diazinon and nonylphenol in toxicity during the early development of the *Rhinella arenarum* toad, *Water, Air, Soil Pollut.*, 2016, **227**, 1–10.
27. J. Herkovits, P. Cardellini, C. Pavanati and C. Pérez-Coll, Susceptibility of early life stages of *Xenopus laevis* to cadmium, *Environ. Toxicol. Chem.*, 1997, **16**, 312–316.
28. C. Simms, Indications of the decline of breeding amphibians at an isolated pond in marginal land, 1954–1967, *Br. J. Herpetol.*, 1969, **4**, 93–96.
29. R. Boyer and C. E. Grue, The need for water quality criteria for frogs, *Environ. Health Perspect.*, 1995, **103**, 352–357.
30. D. B. Wake and V. T. Vredenburg, Are we in the midst of the sixth mass extinction? A view from the world of amphibians, *Proc. Natl. Acad. Sci.*, 2008, **105**, 11466–11473.
31. M. A. Stapanian, M. Micacchion and J. V. Adams, Wetland habitat disturbance best predicts metrics of an amphibian index of biotic integrity, *Ecol. Indic.*, 2015, **56**, 237–242.
32. H. Cayuela, J. Lambrey, J. P. Vacher and C. Miaud, Highlighting the effects of land-use change on a threatened amphibian in a human-dominated landscape, *Popul. Ecol.*, 2015, **57**, 433–443.
33. M. R. Whiles, K. R. Lips, C. M. Pringle, S. S. Kilham, R. J. Bixby, R. Brenes, S. Connelly, J. Checo Colon-Gaud, M. H. Brown, A. D. Huryn, C. Montgomery and S. Peterson, The effects of amphibian population declines on the structure and function of Neotropical stream ecosystems, *Front. Ecol. Environ.*, 2006, **4**, 27–34.
34. G. W. Suter II, *Ecological Risk Assessment*, Boca Raton, FL, Lewis, 1993.
35. J. Unrine, W. Hopkins, C. Romanek and B. Jackson, Bioaccumulation of trace elements in omnivorous amphibian larvae: implications for amphibian health and contaminant transport, *Environ. Pollut.*, 2007, **149**, 182–192.
36. J. L. Kerby, K. L. Richards-Hrdlicka, A. Storfer and D. K. Skelly, An examination of amphibian sensitivity to environmental contaminants: are amphibians poor canaries? *Ecol. Lett.*, 2010, **13**, 60–67.
37. T. Hayes, K. Haston, M. Tsui, A. Hoang, C. Haefelle and A. Vonk, Atrazine-induced hermaphroditism at 0.1 ppb in American leopard frogs (*Rana pipiens*): laboratory and field evidence, *Environ. Health Perspect.*, 2003, **111**, 568–575.

38. R. M. Mann, R. V. Hyne, C. B. Choung and S. P. Wilson, Amphibians and agricultural chemicals: review of the risks in a complex environment, *Environ. Pollut.*, 2009, **157**, 2903–2907.
39. J. W. Allran and W. H. Karasov, Effects of atrazine and nitrate on northern leopard frog (*Rana pipiens*) larvae exposed in the laboratory from posthatch through metamorphosis, *Environ. Toxicol. Chem.*, 2000, **19**, 2850–2855.
40. E. Del Conte and L. Sirlin, The first stages of *Bufo arenarum* development, *Acta. Zool. Lilloana*, 1951, **12**, 495–499.
41. D. D. Echeverria and L. E. Fiorito de Lopez, Estadios de la metamorfosis en *Bufo arenarum* (Anura), *Physis*, 1981, **40**, 15–23.
42. J. A. Bantle, D. T. Burton, D. A. Dawson, J. N. Dumont, R. A. Finch, D. J. Fort, G. Linder, J. R. Rayburn, D. Buchwalter, A. M. Gaudet-Hull, M. A. Maurice and S. D. Turley, FETAX interlaboratory validation study: Phase II testing, *Environ. Toxicol. Chem.*, 1994, **13**, 1629–1637.
43. U. S. EPA, Users guide for a computer program for Probit analysis of data from acute and short-term chronic toxicity test with aquatic organisms. Biological Methods, Environmental 49. Monitoring and Support Laboratory, Cincinnati, OH, USA, 1988.
44. APHA, American Water Works Association, Water Pollution Control Federation, *Standard Methods for the Examination of Water and Wastewater*, American Public Health Association, Washington, DC, 21th edn, 2005.
45. SENASA, Res 511/11 https://viejaweb.senasa.gov.ar/contenido.php?to=n&in=1792&io=26869. (accessed July 2016).
46. G. Svartz, I. R. Hutler Wolkowicz and C. S. Pérez Coll, Toxicity of endosulfan on embryo-larval development of the South American toad *Rhinella arenarum*, *Environ. Toxicol. Chem.*, 2014, **33**, 875–881.
47. M. A. Dalvie, E. Cairncross, A. Solomon and L. London, Contamination of rural surface and ground water by endosulfan in farming areas of the Western Cape, South Africa, 2003. http://ehjournal.biomedcentral.com/articles/10.1186/1476-069X-2-1 (accessed July 2016).
48. U. S. EPA, *Overview of endosulfan: Risk assessment*, U.S.EPA, Office of Prevention, Pesticides, and Toxic Substances, 2001.
49. U.S. EPA, https://www.epa.gov/pesticide-science-and-assessing-pesticide-risks/technical-overview-ecological-risk-assessment-risk (accessed July 2016).
50. C. Boutin, K. E. Freemark and C. J. Keddy, *Proposed Guidelines for Registration of Chemical Pesticides: nontarget plant Testing and Evaluation*, Technical Report Series 145, Canadian Wildlife Service (Headquarters), Environment Canada, Ottawa, 1993.
51. G. Svartz, C. Aronzon and C. S. Pérez Coll, Comparative sensitivity among early life stages of the South American toad to cypermethrin-based pesticide, *Environ. Sci. Pollut. Res.*, 2016, **23**, 2906–2913.
52. G. V. Svartz and C. S. Pérez-Coll, Comparative toxicity of cypermethrin and a commercial formulation on *Rhinella arenarum* larval development (Anura: Bufonidae), *Int. J. Environ. Health*, 2013, **6**, 320–329.

53. C. M. Aronzon, M. T. Sandoval, J. Herkovits and C. S. Pérez-Coll, Stage-dependent toxicity of 2,4-Dichlorophenoxyacetic on the embryonic development of a South American toad, *Rhinella arenarum*, *Environ. Toxicol.*, 2011, **26**, 373–381.
54. T. Katagi, Surfactant effects on environmental behavior of pesticides, in *Reviews of Environmental Contamination and Toxicology*, ed. D. M. Whitacre, Springer, NY, 2008, vol. 194, pp. 71–177.
55. J. C. Brodeur, G. Svartz, C. S. Perez-Coll, D. J. G. Marino and J. Herkovits, Comparative susceptibility to atrazine of three developmental stages of *Rhinella arenarum* and influence on metamorphosis: non-monotonous acceleration of the time to climax and delayed tail resorption, *Aquat. Toxicol.*, 2009, **91**, 161–170.
56. C. M. Aronzon, D. J. G. Marino, A. E. Ronco and C. S. Pérez Coll, Differential toxicity and uptake of diazinon on embryo-larval development of *Rhinella arenarum*, *Chemosphere*, 2014, **100**, 50–56.
57. I. R. Hutler Wolkowicz, J. Herkovits and C. S. Pérez Coll, Stage-dependent toxicity of bisphenol A on *Rhinella arenarum* (Anura, Bufonidae) embryos and larvae, *Environ. Toxicol.*, 2014, **29**, 146–154.
58. C. M. Aronzon, P. A. Babay and C. S. Pérez Coll, Developmental toxicity and risk assessment of nonylphenol to the South American toad, *Rhinella arenarum*, *Environ. Toxicol. Pharmacol.*, 2014, **38**, 634–642.
59. C. M. Aronzon, M. T. Sandoval, J. Herkovits and C. S. Pérez-Coll, Stage-dependent susceptibility to copper in *Rhinella arenarum* embryos and larvae, *Environ. Toxicol. Chem.*, 2011, **30**, 2771–2777.
60. L. A. Castañaga, C. M. Asorey, M. T. Sandoval, C. S. Pérez-Coll, T. I. Argibay and J. Herkovits, Stage-dependent teratogenic and lethal effects exerted by ultraviolet B radiation on *Rhinella* (*Bufo*) *arenarum* embryos, *Environ. Toxicol. Chem.*, 2009, **28**, 427–433.
61. C. S. Pérez-Coll and J. Herkovits, Lethal and teratogenic effects of naringenin evaluated by means of an amphibian embryo toxicity test (AMPHITOX), *Food Chem. Toxicol.*, 2004, **42**, 299–306.
62. I. R. Hutler Wolkowicz, C. Aronzon and C. S. Perez Coll, Lethal and sublethal toxicity of the industrial chemical epichlorohydrin on *Rhinella arenarum* (Anura, Bufonidae) embryos and larvae, *J. Hazard. Mater.*, 2013, **263**, 784–791.
63. I. Hutler Wolkowicz, G. Svartz, C. Aronzon and C. S. Pérez Coll, Developmental toxicity of bisphenol A diglycidyl ether (epoxide resin badge) during the early life cycle of a native amphibian species. *Environ. Toxicol. Chem.*, 2016, **35**(12), 3031–3038.
64. G. Svartz, J. Herkovits and C. S. Pérez Coll, Sublethal effects of atrazine on embryo-larval development of *Rhinella arenarum* (Anura: Bufonidae), *Ecotoxicology*, 2012, **21**, 1251–1259.
65. G. Svartz, F. Meijide and C. S. Perez Coll, Effects of a fungicide formulation on embryo-larval development, metamorphosis, and gonadogenesis of the South American toad *Rhinella arenarum*, *Environ. Toxicol. Pharmacol.*, 2016, **45**, 1–7.

66. A. Egea-Serrano, R. A. Relyea, M. Tejedo and M. Torralva, Understanding of the impact of chemicals on amphibians: a meta-analytic review, *Ecol. Evol.*, 2012, **2**, 1382–1397.
67. F. Orton, E. Rosivatz, M. Scholze and A. Kortenkamp, Widely used pesticides with previously unknown endocrine activity revealed as in vitro antiandrogens, *Environ. Health Perspect.*, 2011, **119**, 794–800.
68. A. E. D. H. Sayed, U. M. Mahmoud and I. A. Mekkawy, Reproductive biomarkers to identify endocrine disruption in *Clarias gariepinus* exposed to 4-nonylphenol, *Ecotoxicol. Environ. Saf.*, 2012, **78**, 310–319.

CHAPTER 18

Hypsiboas pulchellus *(Anura, Hylidae) Tadpoles, a Novel Amphibian Experimental Model in Aquatic Pollution Research*

MARCELO L. LARRAMENDY*[a,b] AND SONIA SOLONESKI[a,b]

[a] Cátedra de Citología, Facultad de Ciencias Naturales y Museo, Universidad Nacional de La Plata, Calle 64 N° 3, B1904AMA La Plata, Argentina; [b] Consejo Nacional de Investigaciones Científicas y Técnicas (CONICET), Buenos Aires, Argentina

*Email: marcelo.larramendy@gmail.com

18.1 Introduction

It is accepted worldwide that amphibians are the evolutionary step that happened when vertebrates left the oceans, lakes, and other water bodies and came onto land. Today, amphibians comprise three lineages, including frogs and toads (Anura), salamanders (Urodela or Caudata), and caecilians (Gymnophiona, Apoda, or Caecilia). Amphibians are ectotherms and their skin is permeable to water, ions, and respiratory gases. The essential characteristics of their environment include appropriate levels of humidity, temperature, and lighting, as well as retreat sites. Terrestrial and arboreal species require moist substrates, water dishes, and high relative humidity.

Anurans are the largest group of amphibians, including approximately 7500 species,[1] and they are characterized by their ability to exploit both

Issues in Toxicology No. 33
Ecotoxicology and Genotoxicology: Non-traditional Aquatic Models
Edited by Marcelo L. Larramendy

Published by the Royal Society of Chemistry, www.rsc.org

aquatic and terrestrial habitats, residing on all of the major continents except Antarctica and on many oceanic islands.[2] Compared with other tetrapods, anurans have short trunks, broad, flat heads with extensive reduction of bone, and hind limbs that are substantially longer than the forelimbs in most species. Anurans have comparatively longer legs, particularly their strong hind limbs, than any other vertebrate group. Usually, they are associated with jumping, but they are also effective for swimming, walking, running, and climbing, and all of these forms of locomotion are represented among frogs and toads.

In recent years, increased interest in the biology of the organism and in environmental and evolutionary physiology has brought novel species into the laboratory, including amphibians, reptiles, birds, and mammals, and employed them in basic and applied scientific research. Among amphibians and reptiles, there are at least 12 280 distinct species with very varied ecological, environmental, and phylogenetic characteristics. Although the acquisition of basic scientific knowledge can justify research with amphibians and reptiles, the use of these animals in scientific research can produce effects that cannot always be predicted. The discovery of new species as well as novel attributes of known species is to be expected as a consequence of research studies. It is well known that amphibians have repeatedly been used through history as research animals, especially as models for studies of embryonic development, regeneration, and physiological function.[1] Laboratory studies are generally conducted under relatively controlled conditions with the purpose of testing particular hypotheses within the framework of a broader scientific investigation. Several salamanders and frogs have contributed much to these studies, and several methods for maintaining them in laboratory conditions have been well established.[3]

The use of amphibians as research models has been particularly successfully compared to the use of many other currently popular animal models to discover new information and/or obtain clues that could relate to other, if not all, species, particularly humans.[3] Amphibians offer many advantages, which include a relatively well-understood basic physiology that, in physiological studies, models not only for mammals, but also for several vertebrates that have key positions in the evolutionary path. Amphibians present a huge range of diversity in terms of morphology, physiology, and life histories that includes frogs and toads, salamanders, and caecilians, which makes them a rich source of experimental material for testing all types of physiological questions. They have phylogenies that are both well understood and complex, and therefore lend themselves to experimental approaches in phylogeneticy studies. Amphibians are adapted to a great diversity of habitats, including aquatic, sylvan, xeric, marine, arboreal, and burrowing, and high altitudes and latitudes, and they accordingly show a wide range of morphological and physiological adaptations. As a consequence of these adaptations, amphibians show a high tolerance to temperature and oxygen variation, and they present sufficient physiological traits in common with all vertebrates, including mammals.

Besides their place in the rich history of delineating basic physiological principles, amphibians as animal models have long been employed as indicators of environmental quality, *e.g.*, in water quality studies employing mortality analyses in both *in situ* and *in vivo* conditions to detect the toxicity risk of environmental pollutants.[4,5] Additionally, amphibian populations have been reported to suffer significant declines worldwide,[6,7] a phenomenon in most cases attributable not only to urban contamination but also to pollution of agricultural areas with pesticides.[8] However, other factors, *e.g.*, overexploitation, diseases, habitat loss and/or modification, introduced species, and climate change, could also contribute to their decline.[8] Negative effects against wild anuran populations exerted by industrial and urban discharges containing potentially xenoestrogenic endocrine disruptors have been well established and include major deleterious effects on amphibian reproduction and development.[9–11] In addition to interfering with normal amphibian growth, development, and susceptibility to disease, some of the chemicals of environmental pollution possess genotoxic, carcinogenic, or endocrine disrupting capacities that entail risks to the health of living organisms, especially amphibians.[12,13] Many amphibians have also been used as models to analyze the effects of several environmental pollutants, as pesticides, including insecticides, as well as herbicides are particularly detrimental to amphibian species. Several factors contribute to this, *e.g.*, their aquatic habitat, unprotected eggs, and sensitive, highly permeable skin, which is involved in gas, water, and electrolyte exchange with the environment.[14–17]

Hylidae is an extremely large, diverse family of American, Australopapuan, and temperate Eurasian tree frogs. As a family, it contains four subfamilies comprising approximately 38 genera with more than 870 known species described so far.[18] Hylids range from small to large in size specimens, and usually have distinct adhesive toe discs that contain a cartilage offsetting the terminal phalanx, which may aid in climbing.[18]

Hypsiboas is a genus of frogs in the family Hylidae. Their common name is either gladiator frogs or Wagler Neotropical tree frogs. Actually, the genus includes 93 species, and they are distributed in the tropical regions of Central and South America from Nicaragua to Argentina as well as in the Caribbean (Trinidad and Tobago). This genus was resurrected in 2005 after a major revision of Hylidae family.[18] Approximately 70 species previously placed in the *Hyla* genus were moved to this genus, and many new species have also been discovered.[1,18]

The common tree frog, *Hypsiboas pulchellus* (Duméril and Bibron, 1841), has an extensive distribution in Neotropical America, including Argentina, Brazil, Paraguay, and Uruguay.[1] *H. pulchellus* represents a very widespread and abundant species in the Argentinean region, including Misiones, Chaco, Buenos Aires, Córdoba, Corrientes, Entre Ríos, La Pampa, and Santa Fe provinces; all departments in Uruguay; Rio Grande do Sul and Santa Catarina in Brazil; and Southern Paraguay. This species lives in open habitats including forests, grasslands, and flooded savannahs.

Figure 18.1 Photograph showing an adult (A) and tadpoles (B) of *Hypsiboas pulchellus* kept in laboratory conditions. Bar represents 1 cm.

The species has an altitudinal range of 0–1250 m asl. It breeds in permanent ponds and flooded grasslands. It is frequently found in both natural and altered lentic water bodies of agricultural and urban areas, with marginal vegetation composed of small shrubs and riparian trees, interspersed with assemblages of Poaceae, Polygonaceae, and Cyperaceae.[19] It is possible that it adapts well to anthropogenic disturbance. However, this species is threatened by agricultural soil and water pollution (specifically pesticide runoff) in cropping areas from Córdoba in Argentina.[20] *H. prasinus* and *H. joaquini* are often considered to be subspecies of *H. pulchellus*. Similarly, *H. cordobae* and *H. riojanus* were previously considered subspecies, but were recently separated by Faivovich *et al.*[18]

H. pulchellus is a slender, medium-sized Hylidae (37–50 mm) that feeds on flies and mosquitoes and spends most of the day concealed among the vegetation, clinging on to leafs or branches (Figure 18.1, A: adult and B: tadpoles). It can be found in a variety of subtropical or tropical habitats, including wetlands and farmlands, and in urban and rural areas. Its natural habitats are dry lowland grassland, subtropical or tropical seasonally wet or flooded lowland grassland, intermittent freshwater lakes and marshes, and pastureland.[6] *H. pulchellus* has the most intense reproductive activity in three distinct periods of the year: August–September, November–December, and March–April.[21]

18.2 *Hypsiboas pulchellus* as a Model in Aquatic Acute Toxic and Genotoxic Studies

The significance of environmental pollution in the toxic and genotoxic effects in aquatic organisms is still important for detecting damage exerted by many xenobiotics. Knowledge of the acute and chronic toxicity of a xenobiotic helps in predicting and preventing several types of damage to the aquatic organisms in receiving waters, and this information is then useful in regulating toxic waste discharges.[22]

Genetic damage in somatic and germ cells is related to serious health effects, which in principle may occur even at low exposure levels. Accumulation of damage in DNA has been proposed to play a role in degenerative conditions, such as accelerated aging, immune dysfunction, and cardiovascular and neurodegenerative diseases.[23] The estimation of genotoxicity is an integral part of many global regulatory guidelines for assessing the risk of potentially toxic physical and chemical substances.

18.3 *Hypsiboas pulchellus* as a Biotic Matrix in Acute Lethality Studies

A common measure of the acute toxicity of a substance in organisms is the median lethal concentration, usually expressed as LC_{50}. This value represents the concentration required to kill half the members of a specific animal population when entering the animal's body by a particular route within a fixed time. The LC_{50} is a general indicator of the toxicity of a xenobiotic caused by small variations in tolerance between individuals of an assay and/or lack of knowledge of the toxicity of the substance. The data set is habitually characteristic for numerous routine toxicity experiments.

Several lethality studies have been conducted using *H. pulchellus* as a model organism and employing the LC_{50} estimation at different developmental larval stages. To date, the majority of studies examining the effects of xenobiotics on gladiator frog tadpoles as the model organism have been performed under laboratory conditions, whereas a smaller number have been conducted under field conditions. A summary of the results reported so far is presented in Table 18.1.

The lethal effect of the hexavalent chromium [Cr(VI)] contained in the potassium dichromate chemical reagent was studied using LC_{50} estimation in *H. pulchellus* tadpoles as a laboratory-exposed target amphibian.[24] The acute toxicity was measured for 96 h of exposure. Lethality results on *H. pulchellus* tadpoles showed a value of 29.60 mg L^{-1} Cr(VI) for the exposed

Table 18.1 Summary of median lethal concentrations (LC_{50}) exerted by several pollutants at different developmental larval stage of *Hypsiboas pulchellus* after different exposure times (h).

Developmental stage	Analyte	Pollutant/chemical analyzed	Exposure time	LC_{50} values mg L^{-1}	Ref.
25	Chromium(VI)	Potassium dichromate	96	29.60	24
25	Endosulfan	Thionex-L®	96	0.00013	25
25	Cypermethrin	Technical grade	96	0.4791	26
25	Cypermethrin	Sherpa®	96	0.1752	26
25	Imazethapyr	Pivot H®	96	1.48	30
26–30	Glufosinate ammonium	Liberty®	48	21.47	28
36	Imidacloprid	Technical grade	96	84.909	29
36	Imidacloprid	Glacoxan Imida®	96	52.622	29
37	Imazethapyr	Pivot H®	96	1.55	30

tadpole age group (Gosner Stage 25). Similarly, Agostini *et al.* reported 96 h LC_{50} values of 0.00013 $mg\,L^{-1}$ for premetamorphic tadpoles exposed to the insecticide endosulfan contained in the commercial formulation Thionex-L® (35% w/v) under laboratory conditions.[25] When the pyrethroid insecticide cypermethrin and the cypermethrin-based commercial formulation Sherpa® (25% w/v) were employed under laboratory conditions to test their acute toxicity in premetamorphic tadpoles, the results showed LC_{50} values of 0.4791 $mg\,L^{-1}$ and 0.1752 $mg\,L^{-1}$ for cypermethrin and the cypermethrin-based commercial formulation, respectively.[26] The lethal effect of the herbicide imazethapyr contained in the commercial formulation Pivot® was analyzed using LC_{50} estimation in premetamorphic *H. pulchellus* tadpoles laboratory exposed for 96 h.[27] Lethality results on *H. pulchellus* tadpoles showed a value of 1.48 $mg\,L^{-1}$ imazethapyr for the exposed tadpole age group (Gosner Stage 25). Lethality analyses performed on *H. pulchellus* tadpoles revealed a 48 h LC_{50} value of 21.47 $mg\,L^{-1}$ for the glufosinate ammonium-based herbicide Liberty®.[28] Recently, Ruiz de Arcaute *et al.* determined values of 84.909 and 52.622 $mg\,L^{-1}$ as the LC_{50} values (96 h) for the insecticide imidacloprid and the imidacloprid-based formulation Glacoxan Imida, respectively, on late-stage *H. pulchellus* larvae (Gosner Stage 36) exposed under laboratory conditions.[29] Finally, Pérez-Iglesias *et al.* reported an LC_{50} (96 h) value of 1.55 $mg\,L^{-1}$ for the imazethapyr-based formulation Pivot® on late-stage *H. pulchellus* larvae (Gosner Stage 36) exposed under laboratory conditions.[30] Accordingly, the reported studies indicate that the hylid frog tadpoles of *H. pulchellus* can be considered an adequate reference organism in toxicity risk assessment studies for different environmental pollutants, including pesticides.

18.4 *Hypsiboas pulchellus* as Biotic Matrix for the Single Cell Gel Electrophoresis or Comet Assay

In brief, the Comet assay is widely used in both *in vitro* and *in vivo* genotoxicity testing. It measures DNA strand breaks and alkali-labile sites in virtually any eukaryotic cell, including cells isolated from tissues.[31] The Comet assay is a simple, sensitive, and fast procedure that can detect primary DNA lesions and repair in any eukaryotic cell type after xenobiotic exposure, and can be used to identify and quantify short-lived DNA damage. This assay has well been applied in many animal and vegetal tissues to detect DNA damage at the level of the individual cells.[32–35]

Few genotoxicity studies employing *H. pulchellus* as a model organism have been conducted using the Comet assay methodology. To the best of our knowledge, this species has been employed to analyze the primary DNA damage exerted by the neonicotinoid insecticide imidacloprid[29] and an imidacloprid-based formulation[27] as well as by a commercial formulation of the imidazolinone herbicide imazethapyr[30] at developmental Gosner stage 36 (range 35–37) and continuously exposed for 96 h under laboratory

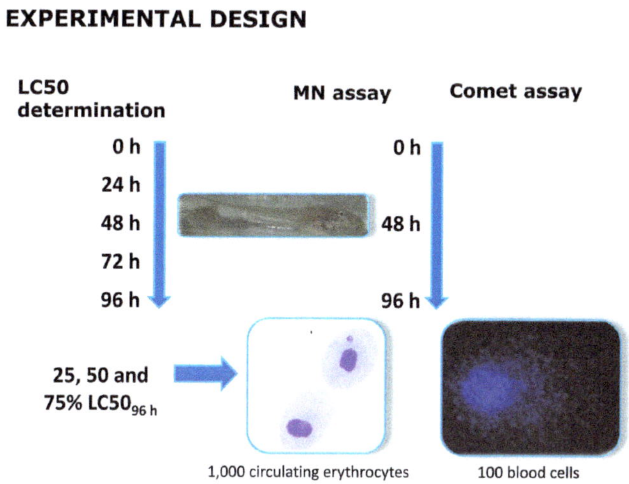

Figure 18.2 Image showing an experimental design for LC_{50} determination as a lethality endpoint and micronucleus (MN) and Comet assays as genotoxicity biomarkers.

conditions to sublethal concentrations equivalent to 25%, 50%, and 75% of the 96 h LC_{50}. In all these studies, a common general experimental design was employed, which is summarized in Figure 18.2.

Recently, Ruiz de Arcaute *et al.*[29] evaluated the genotoxic effect of the technical grade of the neonicotinoid insecticide imidacloprid on circulating blood cells of exposed tadpoles. The experiments were conducted over 48 h and 96 h of continuous exposure under laboratory conditions, and the results showed a significant increase in DNA single-strand breaks in tadpoles' circulating erythrocytes after 48 h exposure to a sublethal concentrations of 30–45 $mg L^{-1}$ imidacloprid and 15–45 $mg L^{-1}$ imidacloprid after 96 h exposure. After analyzing the genetic DNA damage values in those tadpoles treated with 30 $mg L^{-1}$ for 48 and 96 h, a significant increase in the genetic damage index was observed due to an enhancement, over negative control values, in the frequency of type III–IV nucleoids (damaged nucleoids) and a concomitant decrease of type 0–I nucleoids (non-damaged nucleoids).[29] A summary of the results is depicted in Figures 18.3 and 18.4.

Similar results were also reported by Pérez-Iglesias *et al.*[27] when *H. pulchellus* tadpoles were exposed to the commercial formulation Glacoxan Imida, containing 35% of the insecticide imidacloprid. They observed a significant increase in DNA single-strand breaks in erythrocytes of tadpoles exposed to 25.0 and 37.5 $mg L^{-1}$ Glacoxan Imida for both 48 and 96 h. An increase of the genetic damage index was observed at 48 h of treatment within the 12.5–37.5 $mg L^{-1}$ concentration range, whereas an increased frequency of DNA damage was observed only in tadpoles treated with 37.5 $mg L^{-1}$ of the imidacloprid-based commercial formulation for 96 h. A regression analysis demonstrated that whereas the genetic damage index

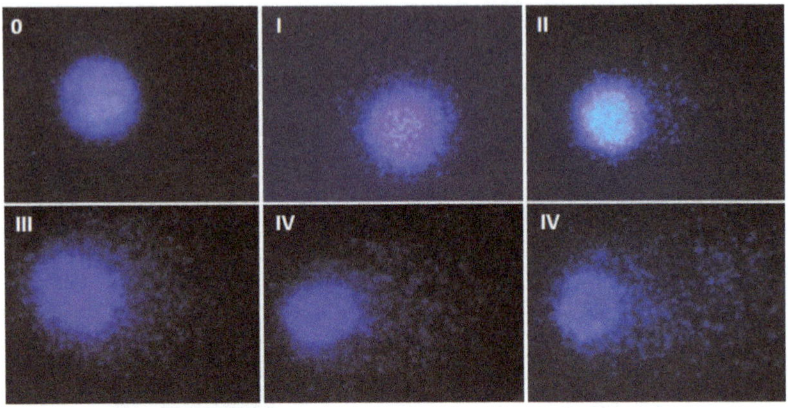

Figure 18.3 Digitized comet images showing undamaged (0–I) and damaged nucleoids (II–IV) of circulating blood cells from *Hypsiboas pulchellus* after *in vivo* exposure. They represent classes 0–IV as used for visual scoring (0–I: undamaged, II: minimum damage, III: medium damage, IV: maximum damage). Nucleoids were stained with 4′,6-diamidino-2-phenylindole (DAPI) and captured with a fluorescent microscope. Magnification, 1000×.

did not vary as a function of the imidacloprid concentration in tadpoles treated for 48 h, a significant dose-dependent increase in the genetic damage index was observed in tadpoles exposed for 96 h.[27] A summary of the results obtained by the authors is depicted in Figure 18.5.

Finally, the Comet assay methodology was employed to analyze the DNA damage exerted by the imidazolinone imazethapyr-based commercial herbicide formulation Pivot H®, containing 10.59% imazethapyr.[30] The authors revealed acute genotoxicity caused by the herbicide imazethapyr when 0.39, 0.78, and 1.17 mg L^{-1} concentrations of the based commercial formulation were employed. The results revealed a significant increase of DNA damage in blood peripheral erythrocytes following either 48 or 96 h of exposure time. When analyzing the genetic DNA damage values in those tadpoles exposed to Pivot H®, a significant increase of the genetic DNA damage was observed due to an enhancement, over negative control values, in the frequency of damaged nucleoids and a concomitant decrease of non-damaged nucleoids.[30] Figure 18.6 summarizes the results obtained in these experiments.

18.5 *Hypsiboas pulchellus* as a Biotic Matrix for the Micronucleus Assay and Assays of Other Nuclear Abnormalities

Regardless of the plausible clastogenic or aneugenic origin of micronuclei, the formation of micronuclei in circulating blood erythrocytes of *H. pulchellus* tadpoles has been employed as an indicator of chromosomal

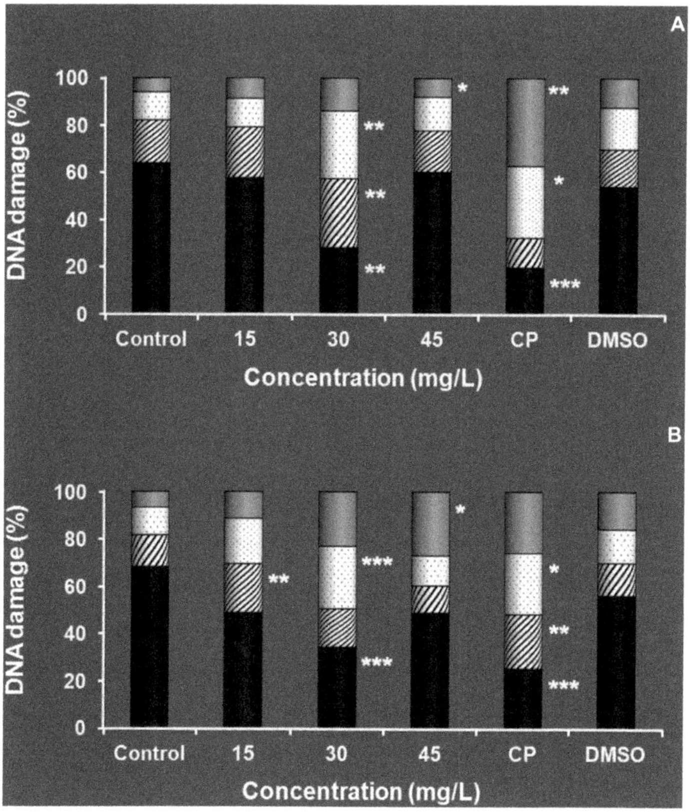

Figure 18.4 Imidacloprid-induced DNA damage measured by Comet assay in circulating blood cells from *Hypsiboas pulchellus* tadpoles exposed for 48 h (A) and 96 h (B). The frequencies of undamaged (type 0–I nucleoids; black column sections), type II (stripped column sections), type III (dotted column sections), and type IV (gray column sections) were determined by analyzing 100 nucleoids from each tadpole. Results are presented as percentages of pooled data from three independent experiments. Negative (untreated tadpoles), positive (CP, 40 mg L^{-1} cyclophosphamide tadpoles), and solvent (DMSO, 0.5% (v/v) dimethylsulfoxide L^{-1} treated tadpoles) controls were conducted and run simultaneously with imidacloprid-exposed larvae. *$P<0.05$; **$P<0.01$; ***$P<0.001$; significant differences with respect to control values. Reprinted from *Ecol. Indic.*, **45**, C. Ruiz de Arcaute, J. M. Pérez-Iglesias, N. Nikoloff, G. S. Natale, S. Soloneski and M. Larramendy, Genotoxicity evaluation of the insecticide imidacloprid on circulating blood cells of Montevideo tree frog *Hypsiboas pulchellus* tadpoles (Anura, Hylidae) by comet and micronucleus bioassays, 632–639. Copyright (2014), with permission from Elsevier.

injury after both the exposure and effect of a much reduced number of environmental pollutants compared to the Comet assay.

Micronucleus induction was reported in *H. pulchellus* tadpoles exposed to the 35% endosulfan-based insecticide Piastra after 96 h of exposure.[36] The

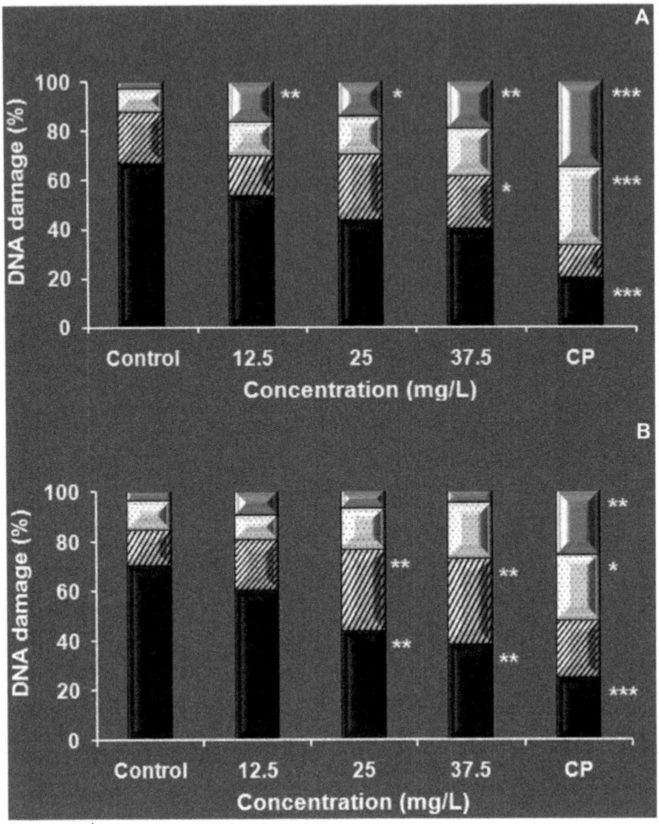

Figure 18.5 Glacoxan Imida-induced DNA damage measured by Comet assay in circulating blood cells from *Hypsiboas pulchellus* tadpoles exposed for 48 h (A) and 96 h (B). The frequencies of undamaged (type 0–I nucleoids; black bar sections), type II (stripped bar sections), type III (dotted bar sections), and type IV nucleoids (grey bar sections) were determined by analyzing 100 nucleoids from each tadpole. Results are presented as percentages of pooled data from three independent experiments. Negative (untreated tadpoles) and positive controls (CP, 40 mg L^{-1} cyclophosphamide-treated tadpoles) were conducted and run simultaneously with treatments for Glacoxan Imida-exposed tadpoles. *$P<0.05$; **$P<0.01$; ***$P<0.001$ (significant differences with respect to control values).
Reprinted from *Ecotoxicol. Environ. Saf.*, **104**, J. M. Pérez-Iglesias, C. Ruiz de Arcaute, N. Nikoloff, L. Dury, S. Soloneski, G. S. Natale and M. L. Larramendy, The genotoxic effects of the imidacloprid-based insecticide formulation Glacoxan Imida on Montevideo tree frog *Hypsiboas pulchellus* tadpoles (Anura, Hylidae), 120–126. Copyright (2014), with permission from Elsevier.

frequency of micronuclei was examined in blood smears obtained from tadpoles exposed to three different concentrations of 2.5, 5, and 10 μg L^{-1} commercial formulations after two sampling times, 48 and 96 h. The results

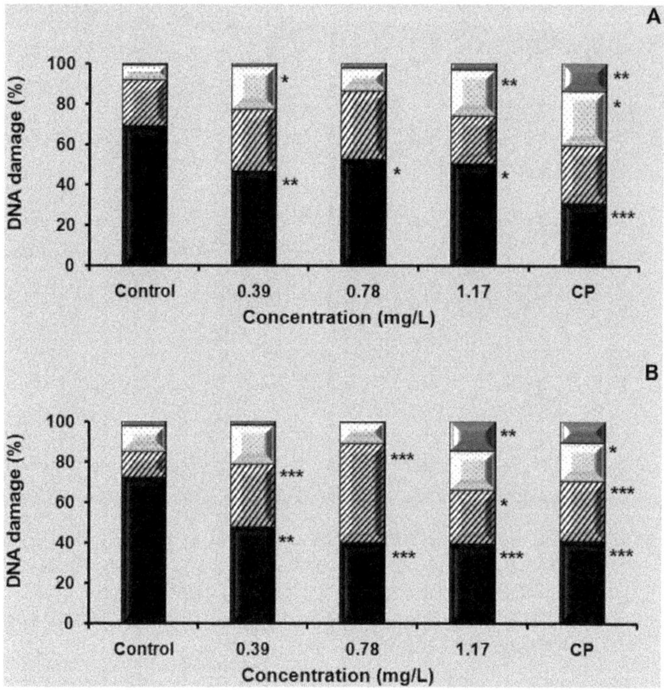

Figure 18.6 Imazethapyr-based commercial formulation herbicide Pivot H®-induced DNA damage measured by Comet assay in circulating blood cells from *Hypsiboas pulchellus* tadpoles exposed for 48 h (A) and 96 h (B). The frequencies of undamaged (type 0–I nucleoids; black bar sections), type II (stripped bar sections), type III (dotted bar sections), and type IV nucleoids (grey bar sections) were determined by analyzing 100 nucleoids from each tadpole. Results are presented as percentages of pooled data from three independent experiments. Negative (untreated tadpoles) and positive controls (CP, 40 mg L^{-1} cyclophosphamide-treated tadpoles) were conducted and run simultaneously with treatments for Pivot H®-exposed tadpoles. *$P < 0.05$; **$P < 0.01$; ***$P < 0.001$ (significant differences with respect to control values).
Reprinted from *Ecotoxicol. Environ. Saf.*, **119**, J. M. Pérez-Iglesias, S. Soloneski, N. Nikoloff, G. S. Natale and M. L. Larramendy, Toxic and genotoxic effects of the imazethapyr-based herbicide formulation Pivot H® on montevideo tree frog *Hypsiboas pulchellus* tadpoles (Anura, Hylidae), 15–24. Copyright 2015, with permission from Elsevier.

showed a significant increase in the frequency of micronuclei in circulating erythrocytes in those tadpoles exposed to 5 and 10 μg L^{-1} endosulfan.[36]

When the micronucleus methodology was employed to analyze the DNA damage exerted by the neonicotinoid insecticide imidacloprid in *H. pulchellus* tadpoles exposed to the 35% imidacloprid-based commercial formulation Glacoxan Imida, a significant increase in micronucleus frequency was reported.[27] Pérez-Iglesias *et al.* revealed acute genotoxicity caused by the insecticide imidacloprid when 25.0 mg L^{-1} was assayed for 96 h[27] (Figure 18.7).

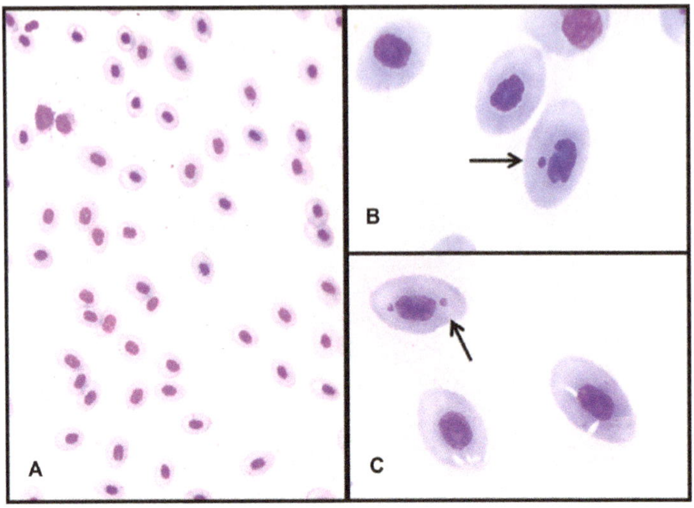

Figure 18.7 Photomicrographs from blood smear of *Hypsiboas pulchellus* showing erythrocytes with normal nucleus (A), a mature erythrocyte exhibiting one (B, arrow) or two micronucleus (C, arrow). Cells were stained with 5% Giemsa and viewed at 1000× magnification. Cells are approximately 20 μm along the long axis.

In amphibians and fish, there are several types of nuclear DNA lesions other than micronuclei whose origins are not completely understood.[37–39] In general, these nuclear abnormalities have been used as a signal of cytogenetic damage in many groups of organisms, and therefore they may complement micronucleus scoring in routine genotoxicity bioassays.[39,40] Nuclear abnormalities other than micronuclei were evaluated on erythrocytes of *H. pulchellus* tadpoles to monitor genetic damage after exposure to an insecticide such as imidacloprid.[29] Positive results were reported in *H. pulchellus* tadpoles exposed to the technical grade of the insecticide imidacloprid,[29] but not when the same organisms were exposed to the imidacloprid-based commercial formulation for both 48 and 96 h.[27] The experiments were conducted for 48 and 96 h under laboratory conditions, and the results showed a significant increase in the frequency of micronuclei when 15 and 30 mg L^{-1} concentrations of pure imidacloprid were assayed for 48 h; only 15 mg L^{-1} increased the frequency of micronuclei in tadpoles exposed for 96 h. Furthermore, other nuclear abnormalities, *i.e.*, binucleated cells and blebbed and notched nuclei, were induced in tadpoles exposed for 48 h to the 15 mg L^{-1} concentration and for 96 h to the 15 and 30 mg L^{-1} concentrations of imidacloprid.[29]

Finally, positive results were also observed when the frequency of micronuclei was estimated in *H. pulchellus* exposed to 1.17 mg L^{-1} of the imidazolinone imazethapyr-based commercial herbicide formulation Pivot H®, containing 10.59% imazethapyr, for 48 h, and in tadpoles exposed to 0.39, 0.78, and 1.17 mg L^{-1} Pivot H® for 96 h.[30] When the other nuclear abnormalities were analyzed, an increase in the frequency of lobbed and notched nuclei was observed after 48 h only in tadpoles exposed to 1.17 mg L^{-1} imazethapyr, and after 96 h in tadpoles exposed to 0.39, 0.78, and 1.17 mg L^{-1} Pivot H®.[30]

18.6 *Hypsiboas pulchellus* as Biotic Matrix for the Other Sublethal Endpoints

Antioxidant enzymes and nonenzymatic systems are essential for the conversion of reactive oxygen species into harmless metabolites and they are implicated in the protection and the restoration of normal metabolism and cellular function.[41] Peltzer *et al.*[28] demonstrated experimentally that the glufosinate ammonium-based herbicide Liberty® can induce oxidative stress in *H. pulchellus* and impair their health owing to its redox potential.[28] They investigated brain tissue of *H. pulchellus* exposed to different concentrations of glufosinate ammonium in a short-term static toxicity bioassay lasting 48 h. The results of this study clearly showed that the tadpoles experienced oxidative stress as characterized by significant modulation of the activity of antioxidant enzymes such as acetylcholinesterase and butyrylcholinesterase after exposure to a 3.55 to 15 mg L^{-1} concentration range of the glufosinate ammonium-based insecticide. The results revealed that sublethal concentrations of Liberty® significantly inhibited both acetylcholinesterase and butyrylcholinesterase activities in tadpoles with respect to the control, showing a concentration-dependent inhibitory effect.[28]

It is know that the study of behavioral responses of aquatic organisms provides multiple approaches to quantify the responses of contaminated organisms, and they can be precursors of other deleterious effects in other, parameters such as survival, growth, or reproduction.[42,43] Changes in behavioral responses due to exposure to a xenobiotic are among the most sensitive indicators of environmental stress, often between 10 and 100 times more sensitive when compared, for example, to survival response.[44] In amphibian tadpoles, the use of swimming activity as a behavioral endpoint has been well established as a sensitive endpoint of sublethal xenobiotic exposure. In this regard, Peltzer *et al.*[28] showed experimentally that the glufosinate ammonium-based herbicide Liberty® can induce alterations in the swimming speed of *H. pulchellus* tadpoles after acute exposure to a 15 mg L^{-1} concentration of the ammonium-based insecticide. At this concentration, the authors reported a significant increase in the swimming speed with respect to the control as well as a negative and significant correlation between swimming speed and butyrylcholinesterase activity, thus suggesting that inhibition of this enzyme is related to an increase in swimming speed.[28] Agostini *et al.*[26] analyzed sublethal effects of the insecticide cypermethrin as the technical grade and its commercial formulation Sherpa® on *H. pulchellus* tadpoles, assessing behavioral changes and growth during 96 h of exposure under standardized laboratory conditions. They analyzed changes in several parameters, such as regular swimming, delayed response, slow swimming, and immobility. Similarly, growth was assessed by measuring body length after 96 h of exposure. All effects on behavioral parameters showed an increasing degree of injury as the insecticide concentration increased in both the technical grade of cypermethrin and the cypermethrin-based commercial formulation Sherpa®. Both presentations of the analyzed insecticide caused abnormalities when 0.34–4.18 $\mu g\,L^{-1}$ cypermetrhin was employed, and malformation was detected in 100% of tadpoles after exposure to 34.4 $\mu g\,L^{-1}$ of the pure compound and the technical grade, and after exposure to 8.36 $\mu g\,L^{-1}$ of the cypermethrin-based commercial formulation Sherpa®. Organisms exposed to the technical grade insecticide showed lower body length with respect to controls when 3.44 $\mu g\,L^{-1}$ cypermetrhin and higher concentrations were employed, whereas those exposed to 0.83 $\mu g\,L^{-1}$ Sherpa® exhibited lower growth with respect to control values.[26]

Similarly, Pérez Iglesias *et al.*[30] analyzed sublethal effects of the herbicide imazethapyr contained in its commercial formulation Pivot H® after an acute exposure within the 3.55–15 $mg\,L^{-1}$ imazethapyr-based insecticide concentration range for 96 h. They reported changes in behavioral parameters estimated as irregular swimming and immobility in *H. pulchellus* tadpoles after an acute 48 h exposure to a 1.02 $mg\,L^{-1}$ concentration of the imazethapyr-based herbicide Pivot H®.[30]

18.7 Perspectives

The increasing attention to the use of non-traditional amphibian species in ecotoxicology and genotoxicology research lies in their ability not only to reveal toxic and genotoxic effects induced by many potential environmental pollutants, but also to help researchers understand the real performance of such ecosystems. *H. pulchellus* tadpoles are proving to be a useful animal model system for studying aquatic environmental toxicity. As a versatile tool *in vivo* and *in situ*, *H. pulchellus* tadpoles have a number of features that make them attractive as laboratory and *in situ* model organisms for assessing the toxicity impacts of environmental pollutants.

Acknowledgements

The support from the National University of La Plata (Grants 11/N699, 11/N746 and 11/N817) and the National Council for Scientific and Technological Research (CONICET, PIP N° 0344) from Argentina are acknowledged. The authors wish to thank MSci. Juan Manuel Pérez-Iglesias for the anuran images.

References

1. D. R. Frost, *Amphibian Species of the World: an Online Reference. Version 6.0 (Date of access)*. American Museum of Natural History, New York, USA, 2016.

2. F. Pough, R. M. Andrews, J. E. Cadle, M. L. Crump, A. H. Savitzky and K. D. Wells, *Herpetology*, Prentice Hall, Saddle River NJ, 3rd edn, 2004.
3. W. W. Burggren and S. Warburton, Amphibians as animal models for laboratory research in physiology, *ILAR J.*, 2007, **48**, 260–269.
4. R. Boyer and C. E. Grue, The need for water quality criteria for frogs, *Environ. Health Perspect.*, 1995, **103**, 352–357.
5. A. Venturino, E. Rosenbaum, A. Caballero, de Castro, O. L. Anguiano, L. Gauna, T. Fonovich, de Schroeder and A. M. Pechen de D'Angelo, Biomarkers of effect in toads and frogs, *Biomarkers*, 2003, **8**, 167–186.
6. J. Brodeur, J. Vera-Candioti, S. Soloneski, M. L. Larramendy and A. E. Ronco, Evidence of reduced feeding and oxidative stress in common tree frogs (*Hypsiboas pulchellus*) from an agroecosystem experiencing severe drought, *J. Herpetol.*, 2012, **46**, 72–78.
7. J. C. Brodeur, C. M. Asorey, A. Sztrum and J. Herkovits, Acute and subchronic toxicity of arsenite and zinc to tadpoles of *Rhinella arenarum* both alone and in combination, *J. Toxicol. Environ. Health*, 2009, **72**, 884–890.
8. R. M. Mann, R. V. Hyne, C. B. Choung and S. P. Wilson, Amphibians and agricultural chemicals: review of the risks in a complex environment, *Environ. Pollut.*, 2009, **157**, 2903–2927.
9. V. H. Hutchison and R. K. Dupré, in *Environmental Physiology of the Amphibians*, ed. M. E. Feder and W. K. Burggren, University of Chicago Press, Chicago, 1992, vol. 1, pp. 206–249.
10. T. Iguchi, H. Watanabe and Y. Katsu, Developmental effects of estrogenic agents on mice, fish, and frogs: a mini-review, *Horm. Behav.*, 2001, **40**, 248–251.
11. W. Kloas, Amphibians as a model for the study of endocrine disruptors, *Int. Rev. Cytol.*, 2002, **216**, 1–57.
12. T. J. Beebee, Conservation genetics of amphibians, *Heredity*, 2005, **95**, 423–427.
13. D. K. Jones, J. I. Hammond and R. A. Relyea, Very highly toxic effects of endosulfan across nine species of tadpoles: lag effects and family-level sensitivity, *Environ. Toxicol. Chem.*, 2009, **28**, 1939–1945.
14. A. R. Blaustein, D. B. Wake and W. P. Sousa, Amphibian declines: Judging stability, persistence, and susceptibility of populations to local and global extinctions, *Conserv. Biol.*, 1994, **8**, 60–71.
15. D. F. Bradford, R. A. Knapp, D. W. Sparling, M. S. Nash, K. A. Stanley, N. G. Tallent-Halsell, L. L. McConnell and S. M. Simonich, Pesticide distributions and population declines of California, USA, alpine frogs, *Rana muscosa* and *Rana sierrae*, *Environ. Toxicol. Chem.*, 2011, **30**, 682–691.
16. C. A. Brühl, S. Pieper and B. Weber, Amphibians at risk? Susceptibility of terrestrial amphibian life stages to pesticides, *Environ. Toxicol. Chem.*, 2011, **30**, 2465–2472.
17. D. W. Sparling and G. M. Fellers, Toxicity of two insecticides to California, USA, anurans and its relevance to declining amphibian populations, *Environ. Toxicol. Chem.*, 2009, **28**, 1696–1703.

18. J. Faivovich, C. F. B. Haddad, P. C. A. Garcia, D. R. Frost, J. A. Campbell and W. C. Wheeler, Systematic review of the frog family Hylidae, with special reference to Hylinae: phylogenetic analysis and taxonomic revision, *Bull. Am. Mus. Nat. Hist.*, 2005, **294**, 1–240.
19. P. M. Peltzer, R. C. Lajmanovich, A. M. Attademo and A. Beltzer, Diversity of anurans across agricultural ponds in Argentina, *Biodivers. Conserv.*, 2006, **15**, 3499–3513.
20. A. Kwet, S. Reichle, D. Silvano, C. Úbeda, D. Baldo and I. Di Tada, *Hypsiboas pulchellus, The International Union for Conservation of Nature. List of Threatened Species 2004: e.e.T55619A11340514*, 2004.
21. J. M. Gallardo, *Anfibios de los alrededores de Buenos Aires*, Editorial Universitaria de Buenos Aires, Buenos Aires, 1974.
22. USEPA, *Methods for Measuring the Acute Toxicity of Effluents and Receiving Waters to Freshwater and Marine Organisms*, USEPA, 5th edn, 2002, **821-R-02-012**.
23. M. Valko, D. Leibfritz, J. Moncol, M. T. Cronin, M. Mazur and J. Telser, Free radicals and antioxidants in normal physiological functions and human disease, *Int. J. Biochem. Cell Biol.*, 2007, **39**, 44–84.
24. G. S. Natale, L. L. Ammassari, N. G. Basso and A. E. Ronco, Acute and chronic effects of Cr(VI) on *Hypsiboas pulchellus* embryos and tadpoles, *Dis. Aquat. Org.*, 2006, **72**, 261–267.
25. M. G. Agostini, G. S. Natale and A. E. Ronco, Impact of endosulphan and cypermethrin mixture on amphibians under field use for biotech soya bean production, *Int. J. Environ. Res. Public Health*, 2009, **3**, 379–389.
26. M. G. Agostini, G. S. Natale and A. E. Ronco, Lethal and sublethal effects of cypermethrin to *Hypsiboas pulchellus* tadpoles, *Ecotoxicology*, 2010, **19**, 1545–1550.
27. J. M. Pérez-Iglesias, C. Ruiz de Arcaute, N. Nikoloff, L. Dury, S. Soloneski, G. S. Natale and M. L. Larramendy, The genotoxic effects of the imidacloprid-based insecticide formulation Glacoxan Imida on Montevideo tree frog *Hypsiboas pulchellus* tadpoles (Anura, Hylidae), *Ecotoxicol. Environ. Saf.*, 2014, **104**, 120–126.
28. P. M. Peltzer, C. M. Junges, A. M. Attademo, A. Bassó, P. Grenón and R. C. Lajmanovich, Cholinesterase activities and behavioral changes in *Hypsiboas pulchellus* (Anura: Hylidae) tadpoles exposed to glufosinate ammonium herbicide, *Ecotoxicology*, 2013, **22**, 1165–1173.
29. C. Ruiz, de Arcaute, J. M. Pérez-Iglesias, N. Nikoloff, G. S. Natale, S. Soloneski and M. Larramendy, Genotoxicity evaluation of the insecticide imidacloprid on circulating blood cells of Montevideo tree frog *Hypsiboas pulchellus* tadpoles (Anura, Hylidae) by comet and micronucleus bioassays, *Ecol. Indic.*, 2014, **45**, 632–639.
30. J. M. Pérez-Iglesias, S. Soloneski, N. Nikoloff, G. S. Natale and M. L. Larramendy, Toxic and genotoxic effects of the imazethapyr-based herbicide formulation Pivot H® on montevideo tree frog *Hypsiboas pulchellus* tadpoles (Anura, Hylidae), *Ecotoxicol. Environ. Saf.*, 2015, **119**, 15–24.

31. N. P. Singh, M. T. McCoy, R. R. Tice and E. L. Schneider, A simple technique for quantitation of low levels of DNA damage in individual cells, *Exp. Cell Res.*, 1988, **175**, 184–191.
32. S. Bendler-Schwaab, A. Hartmann, S. Pfuhler and G. Speit, The *in vivo* comet assay: use and status in genotoxicity testing, *Mutagenesis*, 2005, **20**, 245–254.
33. Y. F. Sasaki, T. Nakamura and S. Kawaqhuchi, What is better experimental design for in vitro comet assay to detect chemical genotoxicity? *AATEX*, 2008, **14**, 499–504.
34. C. L. Santos, B. Pourrut and J. M. Ferreira de Oliveira, The use of comet assay in plant toxicology: recent advances, *Front. Genet.*, 2015, **6**, 1–18.
35. S. Cotelle and J. F. Férard, Comet assay in genetic ecotoxicology: a review, *Environ. Mol. Mutagen.*, 1999, **34**, 246–255.
36. R. Lajmanovich, M. Cabagna, P. M. Peltzer, G. A. Stringhini and A. M. Attademo, Micronucleus induction in erythrocytes of the *Hyla pulchella* tadpoles (Amphibia: Hylidae) exposed to insecticide endosulfan, *Mutat. Res.*, 2005, **587**, 67–72.
37. F. Ayllón and E. Garcia-Vazquez, Induction of micronuclei and other nuclear abnormalidades in European minnow *Phoxinus phoxinus* and mollie *Poecilia latipinna*: An assessment of the fish micronucleus test, *Mutat. Res.*, 2000, **467**, 177–186.
38. T. Cavaş and S. Ergene-Gözükara, Micronuclei, nuclear lesions and interphase silver-stained nucleolar organizer regions (AgNORs) as cyto-genotoxicity indicators in *Oreochromis niloticus* exposed to textile mill effluent, *Mutat. Res.*, 2003, **538**, 81–91.
39. R. C. Lajmanovich, M. C. Cabagna-Zenklusen, A. M. Attademo, C. M. Junges, P. M. Peltzer, A. Bassó and E. Lorenzatti, Induction of micronuclei and nuclear abnormalities in tadpoles of the common toad (*Rhinella arenarum*) treated with the herbicides Liberty® and glufosinate-ammonium, *Mutat. Res.*, 2014, **769**, 7–12.
40. M. Pacheco and M. A. Santos, Induction of EROD activity and genotoxic effects by polycyclic aromatic hydrocarbons and resin acids on juvenile eel *Anguilla anguilla* L, *Ecotoxicol. Environ. Saf.*, 1997, **38**, 252–259.
41. M. Selvi, T. Cavaş, A. Cağlan Karasu Benli, B. Koçak Memmi, N. Çinkiliç, A. S. Dinçel, O. Vatan, D. Yilmaz, R. Sarikaya, T. Zorlu and F. Erkoç, Sub-lethal toxicity of esbiothrin relationship with total antioxidant status and *in vivo* genotoxicity assessment in fish (*Cyprinus carpio* L., 1758) using the micronucleus test and comet assay, *Environ. Toxicol.*, 2013, **28**, 644–651.
42. S. D. Broomhall, Egg temperature modifies predator avoidance and the effects of the insecticide endosulfan on tadpoles of an Australian frog, *J. Appl. Ecol.*, 2004, **41**, 105–113.
43. M. Denoël, M. Bichot, G. F. Ficetola, J. Delcourt, M. Y. Ylieff, P. Kestemont and P. Poncin, Cumulative effects of a road de-icing salt on amphibian behavior, *Aquat. Toxicol.*, 2010, **99**, 275–280.
44. A. Gerhardt, Aquatic behavioral ecotoxicology-prospects and limitations, *Hum. Ecol. Risk Assess.*, 2007, **13**, 481–491.

CHAPTER 19

Chemical Threats to Sea Turtles

M. CAMACHO,*[a] O. P. LUZARDO[a,b] AND J. ORÓS[c]

[a] Toxicology Unit, Research Institute of Biomedical and Health Sciences (IUIBS), University of Las Palmas de Gran Canaria, Paseo Blas Cabrera Felipe s/n, 35016 Las Palmas de Gran Canaria, Spain; [b] Spanish Biomedical Research Centre in Physiopathology of Obesity and Nutrition (CIBEROBn), Paseo Blas Cabrera Felipe s/n, 35016 Las Palmas de Gran Canaria, Spain; [c] Veterinary Faculty, University of Las Palmas de Gran Canaria, Trasmontana s/n, 35416 Arucas (Las Palmas), Spain
*Email: maria.camacho@ulpgc.es

19.1 Introduction

Seven species of sea turtles belonging to two different families inhabit the oceans of the planet. On the one hand, the Dermochelyidae family includes only one species, which is the leatherback turtle (*Dermochelys coriacea*). On the other hand, the other six species belong to the Cheloniidae family: green turtle (*Chelonia mydas*), loggerhead (*Caretta caretta*), hawksbill (*Eretmochelys imbricate*), Kemp's ridley (*Lepidochelys kemp*), olive ridley (*Lepidochelys olivacea*), and flatback turtle (*Natator depressus*). Due to the multiple threats globally, all species of sea turtles are listed in the Red List of Threatened Species of the International Union for Conservation of Nature.[1]

Toxicological research in sea turtles is relatively scarce. However, this is an indispensable topic as is a valuable tool that favors the conservation efforts of these flagship marine species. The interest in this topic has increased in recent years and several studies have been carried out in all species of sea

Issues in Toxicology No. 33
Ecotoxicology and Genotoxicology: Non-traditional Aquatic Models
Edited by Marcelo L. Larramendy

Published by the Royal Society of Chemistry, www.rsc.org

turtles due to the increased awareness of the toxic effects of pollutants on marine species. Because of their known adverse health effects, specifically organic chemicals and metals have been the focus of numerous monitoring studies on sea turtles. These animals are long-living vertebrates that may bioaccumulate organic and inorganic pollutants from food, sediment and water. Thus, sea turtles have been proposed as good bioindicators for pollution in marine ecosystems.

The first two publications on toxicology of marine turtles date from 1974,[2,3] and since then the number of publications has steadily grown all over the world. According to the bibliography, *Caretta caretta* and *Chelonia mydas* seem to be the sea turtle species most often studied regarding their body burden of pollutants.[4]

Historically, research of pollutants in sea turtles has been focused on deceased animals or eggs. However, the use of tissues from dead turtles complicates the interpretation of the results in some cases, especially the ones regarding the real impact of chemicals on the specimens' health. More recently, with the aim of selecting a non-invasive tool for pollutant analysis in live sea turtles, blood became an important matrix. In addition, some authors have reported that blood samples allow reasonably good estimations of the exposure of the target tissues to chemicals.[5,6] The possibility of being able of analyze the contaminant load using a nonlethal method is mandatory in these protected species.

Is a well-known fact that anthropogenic changes in the environment may be a determinant factor in the development of certain diseases for these marine reptiles. In this sense, aquatic species such as sea turtles are sensitive to the toxicological effects of certain chemicals, including the large class of endocrine disrupting compounds (including many synthetic organic chemicals and toxic elements). In fact, evaluating the effects of anthropogenic pollution is now a top global research priority for the conservation of marine turtles.[7]

In this chapter, we present the state of the art of the known presence and impact of organic and inorganic pollutants in sea turtles. Firstly, and according to the literature, some generalities of organic and inorganic pollutants in different sea turtle species are briefly described. Secondly, some biological factors that can influence the load of these pollutants in the tissues or eggs are considered. Thirdly, we give an overview of the occurrence of toxic pollutants effects, and finally we detail the current knowledge on the use of biomarkers that are available to assess the toxicological effects of chemicals in these marine reptiles.

19.2 Body Burden of Chemical Pollutants in Sea Turtles

19.2.1 Persistent Organic Pollutants

Several studies monitoring organic chemicals in sea turtles have been reported around the world.[4] The Stockholm Convention classifies certain anthropogenic chemicals as persistent organic pollutants (POPs) due to their resistance to degradation. Among these, organochlorine contaminants such as polychlorinated biphenyls (PCBs), organochlorine pesticides (OCPs) and polybrominated diphenyl ethers (PBDEs) are the most frequently studied compounds, followed by perfluorinated compounds (PFCs). Other POPs, such as dioxins or furans (among some others that we do not include in this chapter), have been still much less studied in these reptiles, mainly due to the high amount of sample that is necessary to perform the analyses and other technical difficulties. However, the total body burden of dioxins is usually assessed indirectly, through the measurement of dioxin-like PCBs, which have been reported to represent 60–65% of all dioxin compounds in living beings.[8] All POPs share the common characteristics of being resistant to degradation in the environment and biota. Owing to their fat solubility (except PFCs) and resistance to chemical and biological degradation, exposure of animals to certain classes of POPs leads to bioaccumulation of chemicals throughout their lives.

According to the literature, the pattern of contamination by organic pollutants in sea turtles differs greatly depending on their geographical location and other biological and non-biological factors.[4]

Organochlorine contaminants (OCs), such as PCBs and OCPs, have been deeply studied by several researchers in different tissues of marine turtles,[9] and they are still frequently detected in sea turtles even though most of them have been banned for decades and are no longer produced or used around the world. In fact, all sea turtle species have shown detectable levels of these organochlorine pollutants in their organs,[10–12] blood[5,6,13,14] or eggs.[15–17] Most of the studies carried out to date examined OCs in sea turtles using dead animals. In general, the pattern of tissue distribution in dead sea turtles shows higher concentrations in adipose tissue followed by liver, kidney, lung and muscle.[18–23] In the last decade, the number of studies in live sea turtles has increased, with the first study being done by Keller *et al.* in blood samples of loggerhead sea turtles.[6]

The highest concentrations of OCs have been registered in adipose tissues of dead turtles from the Mediterranean region[11] and Eastern Atlantic Ocean.[20] In a study done in live loggerhead turtles (plasma samples), the authors observed much higher levels of these pollutants in turtles from the Adriatic Sea than those detected in turtles from the Eastern Atlantic Ocean.[24] The lower levels of OCs detected in turtles from the Pacific Ocean compared to others areas were explained by the unindustrialized status of the region.[25,26]

Specifically, PCBs are one of the most commonly measured OC groups in biota, and are usually among the most prevalent chemicals.[27] Thus, the PCB levels and profiles in sea turtles have been deeply investigated in different tissues or eggs around the world.[20,28–31] In general, PCB profiles reported in sea turtles have been dominated by the higher chlorinated congeners in different tissues or blood. Hexachlorobiphenyls and heptachlorobiphenyls are predominant in sea turtles worldwide, including green turtles in the

Pacific,[28] loggerhead turtles in the Mediterranean sea,[11,12,18,31] and loggerhead and Kemp's ridley turtles in the Atlantic.[6,13,22] However, Gardner *et al.* reported a profile that was dominated by lower-chlorinated congeners in green turtles.[10] In addition, a different pattern has been described in green and hawksbill turtles from Cape Verde,[32] in which the most abundant congener was PCB 52 (tetrachlorobiphenyl). This frequency pattern was also detected in loggerhead turtles from the same region.[14] Differences in PCB patterns in turtles may be attributed to regional differences in congener compositions and also to dietary factors, which seem to be species-specific.[10,32] Furthermore, some authors have suggested that differences among congener compositions and levels of contamination could be related to different abilities to metabolize and excrete them among species.[26,30]

Within OC pesticides (OCPs), the most frequently investigated pesticides in sea turtles are dichlorodiphenyltrichloroethane (DDT) and its metabolites, dieldrin, chlordane isomers, and hexachlorobenzene (HCB). Several biomonitoring studies have investigated concentrations of pesticides in sea turtles in different organs,[11,33] blood[6,14,34] or eggs[17] worldwide. Among the studies on baseline levels of OCPs, the majority of these studies have reported values for DDTs in organs and tissues.[35] In fact, *p,p′*-DDE is the pesticide that is generally detected at the highest concentrations in sea turtles.[6,11,21,23–25,33,34] Swarthout *et al.* observed a different ratio of *p,p′*-DDT to *p,p′*-DDE concentrations in two species of sea turtle (green and Kemp's ridley).[30] These authors suggested that the higher mean ratio detected in green turtles *vs.* Kemp's ridley turtles could be caused by differences in the ability to metabolizing the parent compound *p,p′*-DDT to *p,p′*-DDE.

In recent years, information on the contamination status of PBDEs in sea turtles has been reported around the world. Levels of PBDEs in the organs,[25] blood[29,30,36,37] or eggs[17,38] of sea turtles have been investigated, and some differences have been revealed. Some studies have shown the existence of a common PBDEs pattern in sea turtles (which is also reported in other wildlife species),[39] with PBDE 47 being the predominant congener followed by PBDE 99, 100, 153 and 154.[30,40] However, as previously mentioned, different PBDE patterns have also been reported in other studies in sea turtles.[36,37,41]

Few studies to date have shown PFC contamination in marine turtles, and all of them have been done in live animals using blood samples.[42–45] An interesting study conducted on five species of sea turtles showed perfluorooctanesulfonate (PFOS) as the predominant PFC in all species, except in hawksbill turtles.[45] Other authors also reported PFOS as the dominant compound.[42–44]

19.2.2 Inorganic Pollutants

The pollution of the aquatic environment with metals (and other toxic elements) has become a worldwide problem in recent years. Inorganic elements, including both essential and non-essential elements, have a particular significance in ecotoxicology because they are indestructible and most of them have potential toxicity to all living organisms.[46] Although some of these elements enter the marine environment naturally as a consequence of erosion and weathering to soil and rocks, increasing quantities can be released into the sea as a result of unnatural sources or processes, such as industrial activities or urban growth, and alter the natural biological equilibrium.[47]

Therefore, some concern has arisen regarding the toxic effects of inorganic elements on the populations of sea turtles. In fact, Day *et al.* recently reported subtle negative impacts of mercury on the immune function of sea turtles, even at low concentrations.[48] The fact that most trace metals can be bioaccumulated makes species at higher trophic levels such as sea turtles at risk of being exposed to high levels of these pollutants.[49,50] In addition, essential elements can also be toxic to biota when a certain threshold is reached.[51]

In the last few decades, numerous studies focusing on metals in all species of marine turtles around the world have been published.[52–59] Differences in the levels of contaminants in sea turtles in relation to environmental contamination (specific to each area) have been reported, as well as inter-specific differences related to diet or age.[50,55,57,60,61] According to the literature, most studies have been focused on deceased specimens showing the levels of trace elements in different tissues.[52–54,62–66] Muscle, liver and kidney have been the most frequently monitored tissues.

In addition, since 1974 when a study on loggerhead turtles from South Caroline and Georgia was published,[3] several other studies have monitored levels of inorganic elements in eggs of sea turtles populations in different parts of the world.[61,62,64,67–72]

Furthermore, some researchers have reported the levels of essential elements and inorganic pollutants levels in live sea turtles[58,73,74] or carapace tissue samples.[75] Sakai *et al.* reported for the first time that concentrations of trace metals in the carapace were well-correlated with whole body burdens, thus suggesting carapace as a potentially useful non-lethal indicator for monitoring metal levels in sea turtles.[61] In the same way, van de Merwe *et al.* reported strong correlations between blood and tissue concentrations for metals.[5] In this sense, while blood samples seem to indicate recent exposure to metals, carapace tissue provides a more accurate estimation of long-term metal accumulation.[76–78]

Owing to their toxic effects, the most commonly targeted non-essential elements in sea turtles are mercury, cadmium, and lead. The concentrations of heavy metals vary consistently among tissues, showing a certain degree of organotropism. In general, like in other marine vertebrates, the pattern of metal accumulation in sea turtles tissues show the highest levels of mercury, cadmium and lead in the liver, kidney and bone/carapace, respectively.[56] In addition, the available investigations on essential metal concentrations in these marine reptiles have monitored several elements, such as zinc, selenium, copper, chromium and manganese, among others. Furthermore, the inter-organ distribution varies for essential elements. Copper levels have been reported to be highest in the liver,[50,52,59,61] while considerable levels of zinc have been detected in the fat tissue, bone and carapace,[46,50,61,66] or kidney.[59] In addition, zinc has been found to be the element that exhibits the highest blood concentrations.[58,60,68,73]

Besides correlations with inorganic elements and tissues (including blood) and carapace concentrations in sea turtles, several studies have also evaluated metal–metal correlations in samples of these species.[52,55,58,60,63,68,73,77] Selenium and mercury correlations have been reported.[60,63,77] This is relevant since selenium can have a protective role against mercury toxicity in sea turtles, as Perrault *et al.* reported.[71] In addition, correlations between cadmium and different elements have been observed, suggesting that elements such as zinc and copper could play a role in the detoxification of this toxic metal.[52,58]

19.3 Biological Factors Influencing Pollutant Concentrations

There are many available surveys showing different levels of contamination in marine turtles, but sometimes the interpretation of the results is quite difficult because sea turtles are wild animals and they are not under controlled conditions. Furthermore, the research carried out has also been widely heterogeneous due to: (a) types of samples utilized; (b) methodology; (c) location; (d) healthy or rehabilitated turtles; (e) live or dead animals; *etc.* In addition, different biological factors may affect the levels of pollutants in their bodies, such as the diet (indicating the trophic level), sex, age, lipid content, body condition/diseases, and other external conditions, such as geographic contamination or seasonal changes.

19.3.1 Body Condition

Several studies have been carried out using dead marine turtles. However, it is necessary to take into account that the body condition of these animals could have been affected by periods of starvation previous to death. Therefore, studies monitoring lipophilic xenobiotic compounds should take into account the differences among lipid content (corporal composition) according to the tissue analyzed. That is, an organ will contribute more or less significantly to the total load of pollutants depending on its fat content. Another issue to take into account is that body lipids are also mobilized to meet energy demands in certain biological circumstances, such as egg production or tissue maintenance/regeneration. Other situations in the life cycle of turtles, such as migration, which implies high energy requirements, may also produce variations in the redistribution of lipophilic compounds in these aquatic animals. Orós *et al.* suggested that the higher concentrations of PCBs observed in the liver of stranded turtles could be due to the remobilization of the fat (and the pollutants) due to the poor physical condition (severe caquexia) of the turtles.[20] Thus, the health status plays an important role in the interpretation of these contaminant burdens.

A recent study on live loggerhead turtles undergoing rehabilitation tried to determine whether the recovery of these animals to the normal physiological condition also impacted on the redistribution of their blood levels of toxic pollutants.[34] In fact, in this study the authors demonstrated that the circulating levels of OCs of sick turtles were correlated with the degree of emaciation, probably due to the mobilization of stored lipids into the bloodstream to meet energy demands, and correspondingly a mobilization of associated lipophilic compounds. A previous study reported a strong correlation between lipids in fat stores and blood OC levels in sea turtles.[6] However, it is not clear that the nutritional status may influence the tissue concentration of non-lipophilic contaminants (some metals). Day *et al.* did not observe differences between debilitated and healthy turtles, indicating no elevation in Hg exposure before and during the progression of this condition.[79]

19.3.2 Diet and Trophic Level

As we have said before, it is known that animals at higher trophic positions in the food web in aquatic ecosystems have major potential for bioaccumulation of POPs and some metals. Diet habits of marine turtles vary depending on the species, or there are even intraspecific differences. So, food preferences vary from more generalist species, as is the case with the loggerhead turtle, to a large degree of specialization, as is the case of the leatherback turtle, whose feeding is based mainly on jellyfish. In addition, in some species, the shift in diet through the different stages of development (juvenile *vs.* adult) also influences their exposure to pollutants. For example, green turtles are omnivorous after hatching, and then in adult life they change their feeding to vegetarian habits. Another example is found in hawksbill turtles, known to be omnivorous in their juvenile stage, and then later they chiefly feed on sponges during their adult life. Thus, knowing that the largest exposure of these pollutants comes from food, the higher the trophic level the more intense the bioaccumulation of chemicals in their bodies will be. Some authors have reported higher concentrations of organic[19,25,30] and inorganic pollutants[57,61,80] in omnivorous than in herbivorous marine turtle species. However, according to the study of García-Besné *et al.*, the trophic level of the hawksbill turtles did not reflect higher concentrations of all of the OC compared with those levels in green turtles.[81] Conversely, the levels of PCBs,[25] PBDEs,[41] or PFOS[45] found in samples of hawksbill turtles were much higher than those from other species (including species at higher trophic levels). In addition, no differences were observed in PCB bioaccumulation between loggerhead, olive ridley, and green sea turtles.[26] Furthermore, Andreani *et al.* observed higher concentrations of some metals in green *vs.* loggerhead turtles.[52] Swarthout *et al.* suggested that the observed differences between herbivorous and omnivorous sea turtle (green *vs.* olive ridley) may be due not only to differences in exposure/type of diet but also to differences in the ability to metabolize certain pollutants.[30] In a study by Richardson and Schlenk, some green and all hawksbill sea turtles had higher rates of PCB 52 metabolism than olive ridley and loggerhead turtles, indicating species-specific PCB biotransformation in sea turtles.[82] This fact could be due to the higher exposure to natural toxins through their diet in the spongivorous hawksbills and vegetarian green turtles.

19.3.3 Size (as a Proxy for Age)

As has been stated above, persistent pollutants (except PFCs) accumulate in the fatty tissues of exposed animals throughout their life. Most vertebrate animals do not have the ability to effectively metabolize and eliminate these pollutants, thus bioaccumulation occurs through their lives.

Size in sea turtles is usually the only available approximation for their age, and most authors hypothesize that increasing concentrations of these chemicals should be observed in older animals, and therefore the bigger ones. Several studies have investigated the phenomenon of the bioaccumulation of organic and inorganic pollutants in relation to sea turtles' growth, and the results are contradictory.[6,11,19,36,49,55,75,76] Thus, a few studies have reported increased organic pollutant or metal levels in relation to size.[54,76] Other studies have found negative correlations with size for some organochlorine compounds, such as mirex, chlordane, some PCBs, and *p,p′*-DDE,[6,14] and similar results have been also obtained for inorganic elements.[49,59,61] Finally, other investigations reported no associations between pollutant body burden and size.[42,77,83]

Mckenzie *et al.* explained the lower concentrations observed in larger green turtles as a consequence of the diet switch in this species: the main exposure of chemical pollutants would occur through diet during the pelagic stage (omnivorous period), and this load would be subsequently diluted as the animal grows during the herbivorous period.[19] Other authors have also supported this hypothesis in omnivorous loggerhead turtles. The hypothesis that the higher rate of accumulation during the earlier life stages, and the lower exposure in the neritic stage coupled with somatic growth, as a possible explanation for the higher levels of pollutants observed in smaller turtles has been also pointed out by other authors in different sea turtles species.[6,14] A similar theory has also been pointed out for inorganic pollutants.[59]

To explain the opposite findings, that larger turtles exhibit higher levels of pollutants than smaller ones, Ragland *et al.* suggested that the higher levels of OCs detected in larger male migratory turtles could be due to the greater food intake they need to meet their energetic demands (and so greater exposure to pollutants), and also that in adults the growth rate is substantially lower than in juveniles, and then bioaccumulation would be stronger than the above-explained effect of dilution.[36]

Finally, some investigations in these reptiles have pointed to the specific nature of the species in relation to the concentration of metals and growth.[49,68,80]

Given the contradictory results found in the literature, the interpretation of the bioaccumulation effect in relation to the size of sea turtles should be used with caution. If a study is based only on a single stage of the turtles' development (only juveniles or only adults), an important bias could exist.[6,11]

19.3.4 Sex and Maternal Transfer

In oviparous species, females often transfer a part of their body burden of chemicals to the eggs. Therefore, sex differences are expected between female and male marine turtles when they reach maturity. While several studies have shown the exposure of organic and inorganic pollutants in adult female sea turtles in different parts of the world,[13,37,58,67,81] very few studies have measured environmental pollutants in adult males.[36] In addition, some of the studies evaluated sex differences in juvenile turtles in relation to concentrations of toxic pollutants (both organic and inorganic), and as expected, no differences were found.[6,11,49,52,63]

Maternal transfer of organic pollutants to the eggs has been confirmed in sea turtles.[29,37,81,84] However, an absence of parallelism between the profile of chemical contamination of the mothers and eggs has been observed in different marine turtles,[29,37] and the results suggest that the less lipophilic PCBs seem to be more readily transferred from females to their eggs.

Maternal transfer of inorganic elements has also been studied, and the results seem to indicate that this is not the major route for the elimination of many of them, such as Pb and others.[67,85]

In addition, a decreasing trend of OCs in clutches throughout the same breeding season has been reported in marine turtles,[71] with the eggs of the first ovoposition receiving the highest concentrations of the pollutants transferred from the mother, and therefore those that are most exposed to the harmful effects of lipophilic contaminants. Conversely, no changes in the concentration of metals were detected in eggs from different clutches throughout the nesting season.[67]

19.4 Effects of Toxic Pollutants in Sea Turtles

Although the knowledge about the actual toxicity of anthropogenic pollutants in sea turtles is very scarce, there is increasing interest in understanding their effects. Although *in vivo* experiments would have the advantage of evaluating effects in a more definitive way, ethical and legal issues prevent the use of live specimens of these endangered species for experimentation, and emphasize the use of non-invasive methods. Thus, despite its limitations, the *in vitro* approach offers an ethical alternative to evaluating the potential adverse effects of pollutants on the health of these marine species.

In the following paragraphs, we describe the current state of knowledge of the potential health effects in turtles that may be attributable to chemicals, as well as the approaches taken to support this knowledge.

There are very few publications evaluating potential relationships between contaminants and diseases. The first publication dates from 1994, and examined pollutant levels (including organic and metals) and their possible correlation with fibropapillomatosis (FP) in sea turtles.[86] FP is a chronic debilitating disease that afflicts sea turtles, especially *C. mydas*. Although the suspected cause of this disease is a herpesvirus, it has been hypothesized that environmental pollutants may contribute to FP due to the potential immunosuppressive effect of chemical contaminants. However, in this particular study the authors could not find any association between analysed contaminants and disease.

Recently, Keller *et al.* investigated the potential role of POPs in green sea turtles with FP in Hawaii.[87] In this study, turtles had low concentrations of organic compounds, and concentrations did not increase with the prevalence of FP. The authors suggested that contaminants do not play a role in the aetiology of FP. The higher levels of POPs found in emaciated turtles were explained by the mobilization of lipids into the bloodstream during the late-stage weight loss of the disease.[87]

Also recently, the reduced concentration of serum cholesterol observed in green turtles from Brazil heavily afflicted with FP was related to Cu- and Pb-induced inhibition of 3-hydroxy-3-methylglutaryl-CoA reductase activity paralleled by a higher lipid peroxidation rate induced by increased Fe and Pb concentrations. Because oxidative stress is implicated in the pathogenesis of viral infections, the authors suggested that metal contamination, especially by Cu, Fe, and Pb, could be implicated in the aetiology of FP through oxidative stress generation.[88]

It has been described that turtles with FP experience immunosuppression, as it also generally occurs in unhealthy animals.[89] However, it is still unclear whether the immunosuppression occurs as a result of or as a precursor to FP development.[90] It is a well-known fact that the immune system of many species may be adversely affected by low concentrations of contaminants.[91] However, very few studies have evaluated the effects of environmental contamination on the immune system in marine turtles. Positive correlations between OCs and heterophil : lymphocyte ratios and white blood cell counts suggested probable modulation of the immune system in loggerhead turtles from North Carolina, USA.[92] The first study examining the potential effects of chronic exposure to OCs on immune function in sea turtles was done by Keller *et al.*[93] In this work, based on both field studies and *in vitro* experiments, the authors observed significant negative correlations between OC pesticides and lysozyme activity. Furthermore, an immunoenhancement of certain parameters with OC concentrations was detected, as is the case with lymphocyte proliferation. However, the authors remarked that any alteration of immune function, even an enhancement, can be considered an adverse effect.[93] Similarly to many other wildlife studies, these results suggest that the marine turtle's immune system is modulated by environmental concentrations of OCs.[93]

In addition, a study monitoring blood mercury concentrations in loggerhead turtles revealed that this metal may have a negative impact on the immune function.[48] The immune system protects the organism from infectious diseases and neoplastic cells. A depression of the immune system can lead to an increase of diseases caused by infectious agents, or in the severity and effects of diseases. Although in the study of Orós *et al.* a clear association between PCB concentrations and causes of death was not established, the authors observed that almost all turtles with severe septicaemia showed very high levels of these organochlorine congeners.[20] Furthermore, in one turtle from the Canary Islands, pansteatitis was linked with high levels of PCB congeners in the celomic fat.[94] The authors suggested that the high levels of PCBs detected could have induced lipid peroxidation in adipocytes, resulting in cell damage, deposition of ceroid pigment and an inflammatory response.

Blood parameters are used clinically to determine the general health status of sea turtles. Biological factors such as turtle reproductive status, age, sex, season, diet, or species may influence clinical parameters. With the aim of evaluating the potential effects of pollutants in sea turtles, several studies have assessed correlations between organic[13,30,92] and inorganic contaminants in sea turtles[48,58,95] and their clinical health parameters. The first study correlating contaminants and health parameters was done in free-ranging juvenile loggerhead sea turtles.[92] Since then, different studies have been conducted in free-ranging sub-adult and adult loggerhead turtles,[48] juvenile Kemp's ridley turtles,[95] juvenile and sub-adult Kemp's ridley and green turtles,[30] and female nesting loggerhead turtles.[13,58] In all these studies, some correlations were found between pollutant levels and clinical parameters, such as cell counts, enzymes, ions and other overall health indicators. However, positive and negative correlations between contaminant concentrations and clinical parameters were reported. In addition, some authors found that correlations were dependent on the sample type.[77,92]

Owing to the potential endocrine disruptive effects of estrogenic compounds, studies based on reproductive toxicology in sea turtles have been done using different approaches. *In vivo* experiments have been performed exposing eggs to xenoestrogens. In one study the estrogenic insecticide metabolite *p,p′*-DDE was injected in the eggshell of green turtles to evaluate its possible effect on sex reversal.[96] Sea turtles have temperature-dependent sex determination. Podreka *et al.* incubated the eggs at "male temperatures" and they observed that *p,p′*-DDE failed to reverse the sex of embryonic green sea turtles.[96] However, in another study the exposure to estrogens reversed the sex of olive ridley turtles that were also incubated at male-promoting temperatures.[97] More recently, in an *in vivo* experiment, GST-TS cells (an inmortal cell line derived from testes of a green sea turtle) were exposed to three known aromatase inducers and two pesticides (atrazine and *p,p′*-DDE) to evaluate the effects on cytochrome P450 aromatase.[98] This enzyme is critical in the sexual differentiation of reptiles as it is responsible for the conversion of testosterone to estradiol. Keller and McClellan-Green observed that atrazine significantly induced aromatase activity following a 24 h exposure, and *p,p′*-DDE inhibited the activity but only at cytotoxic concentrations, suggesting that induction of cytochrome P450 aromatase activity by environmental pollutants may result in changes to sex determination in hatchlings.[98] Recently, Webb *et al.* also observed significant induction of this enzyme in skin fibroblasts following 72 and 96 h exposure to perfluorooctanoic acid (PFOA).[99] Furthermore, in an *in vitro* system, Ikonomopoulou *et al.* observed that environmental pesticides and heavy metals might alter the binding of steroids (such as testosterone and estradiol) to plasma proteins, and suggested that this may be a mechanism that could potentially disrupt the endocrine balance.[100]

As we discussed earlier in this chapter, maternal transfer of organic and inorganic pollutants to eggs and hatchling sea turtles occurs. However, very few studies have evaluated the real impact of organic[37] and inorganic[71] pollutants on hatchlings. The study by Perrault *et al.* was the first one correlating both maternal and hatchling contaminant loads (Hg and Se) with reproductive success in sea turtles.[71] Both Se and the Se : Hg ratio were positively correlated with leatherback hatching and emergence success. The physiological protection from Se against Hg may allow increasing reproductive success.[71] However, later Perrault *et al.* studied the hazard quotients and risk assessment of mercury and selenium in leatherback turtles, suggesting potential harm to reproductive success and hatchling health and survival.[101] A recent study done by De Andrés *et al.* in leatherback turtles showed a strong negative correlation between the total PBDE concentration and hatching success rate, suggesting potential harmful effects of these contaminants on the reproduction of this species.[102]

The potential developmental effects caused by contaminants was investigated by van de Merwe *et al.*, who reported that organic pollutant exposure in marine turtles may cause abnormalities in embryonic development.[37] Curiously, Alava *et al.* found the highest levels of total POPs in a nest (4940 ng g lipid^{-1}), which contained a two-headed embryo of loggerhead turtle, suggesting a potential teratogenic effect of these pollutants.[16] Furthermore, higher POP levels in eggs of green turtles were significantly correlated with lower hatchling mass : length ratio, which may compromise offshore dispersal and predator avoidance.[37]

19.5 Biomarkers of Exposure

In recent years, due to concern regarding the toxic effects of environmental contamination on populations of sea turtles, some researchers studied traditional toxicological biomarkers in these marine species. The development and application of relatively non-invasive biomarkers represents a valuable means of evaluating the adverse effects of toxic pollutants in sea turtle populations.

Most toxicological biomarker studies in sea turtles have focused on characterization of the expression of some proteins or enzymes and their utility as potential biomarkers. In this sense, vitellogenin (Vtg) is the most commonly examined protein. Vtg is an adult female-specific protein that is induced by estrogens, and that is necessary for the development of egg yolk.[103] Thus, the expression of this protein in male turtles would indicate exposure to endocrine-disrupting compounds. However, the use of Vtg as a biomarker in females is conditioned by the physiological role of this protein in vitellogenesis, and therefore it is crucial to determine the starting point of vitellogenesis in the lifespan of the animals (mainly based on the turtle's size). A previous study in loggerhead turtles from the southeast of the US observed that most females start producing Vtg at a straight carapace length of around 77 cm.[104] However, Zaccaroni *et al.* found Vtg in plasma samples (42 out of 55) in which protein production was not expected, but they could not conclude whether the expression of Vtg was due to the exposure to estrogen-mimicking chemical compounds.[105] Heck *et al.* also reported Vtg expression in juvenile Kemp's ridley sea turtles, but this occurred after estradiol injection.[106] *In vivo* experiments have been also performed to test the performance of Vtg as a biomarker in sea turtles, in which live turtles were exposed to estradiol and the induction of Vtg.[107,108] A recent investigation describes the circulating concentration of Vtg in nesting and non-nesting females, in males, and in juvenile loggerhead sea turtles from the Northwest Atlantic Ocean.[109] The authors detected Vtg levels in male and juvenile turtles, and suggested that this may indicate exposure to environmental xenoestrogenic compounds.

Metallothionein (MT) has been proposed as another useful biomarker of exposure, in this case to metallic elements, as this protein has the capacity to bind inorganic contaminants through the thiol group of its cysteine residues, and plays an important role in the detoxification of metals. Andreani *et al.* examined MT concentrations in green and loggerhead sea turtles, finding them to be significantly higher in green turtles from the North Caribbean coast of Costa Rica than those from *Caretta caretta* from the Adriatic Sea.[52] Furthermore, they found positive correlations between Cu–MT and Cd–MT in liver and kidney in both species, suggesting that MT could be a useful biomarker of environmental exposure to metals.

The use of enzymes as biomarkers has also an important relevance. The first description of oxidative metabolism in sea turtles was reported by Valdivia *et al.*, who provided the first data of baseline values of indicators of oxidative stress in different tissues.[110] In recent years, with the aim of characterizing oxidative stress, the activities of antioxidant enzymes [superoxide dismutase (SOD), catalase (CAT), and glutathione *S*-transferase (GST)], and levels of lipid peroxidation have been analysed in different sea turtles species.[26,83,110,111] Richardson *et al.* characterized GST activity in four species of sea turtles (loggerhead, green, olive ridley and hawksbill turtles).[111] The results of this investigation showed differences in GST activity among species, with hawksbill turtles presenting the highest values (3- to 4.5-fold higher activity). This suggested differences in the biotransformation potential among sea turtle species. However, few studies have correlated oxidative stress indicators with organic and inorganic concentrations in marine turtles.[26,83] Labrada-Martagón *et al.* reported several correlations between antioxidant enzyme activities and concentrations of xenobiotics (both metals and organochlorine pesticide) in blood samples.[83] These associations may suggest physiological sensitivity of East Pacific green turtles to the effects of environmental chemicals. However, Richardson *et al.* did not find any correlations with GST and PCB concentrations among the three species studied (loggerhead, green and olive ridley turtles).[26]

Richardson *et al.* also measured the expression patterns of cytochrome P450 (CYP) enzymes, confirming that sea turtle microsomal proteins cross-reacted with anti-CYP2K1 and anti-CYP3A27 antibodies, but not with the anti-CYP1A antibody.[26] The CYP enzyme system is critical to the metabolism of both exogenous and endogenous chemicals, and the CYP 1 family is the most important for xenobiotic metabolism. The presence of PCB congeners and the lack of CYP1A expression suggest a potential mechanism of accumulation of these OCs.[26]

In the last few years, *in vitro* assays focusing on the comparison and identification of established cell lines as sensitive biomarkers for environmental contamination have been reported.[99,112,113] Webb *et al.* reported that 3-(4,5-dimethylthiazol-2-yl)-2,5-diphenyltetrazolium bromide (MTT) cytotoxicity assays along with CYP1A gene expression measures are viable methods for investigating the impact of organic contaminant exposure on marine turtle cells.[99] The cytotoxic effect of inorganic mercury on *in vitro* green turtle cells was also studied.[113] Wise *et al.* tested the cytotoxicity and genotoxicity of hexavalent chromium in hawksbill sea turtle cells due to its known carcinogenicity and frequent use in products.[114] The data from this study demonstrate that Cr(VI) is cytotoxic and genotoxic to hawksbill sea turtle cells. These results were also observed in a study that compared both cytotoxicity and genotoxicity chromate in human and hawksbill sea turtle skin fibroblasts.[115] Because of the similar responses of sea turtles and human cells to the cytotoxic and genotoxic effects of Cr(VI), the authors suggest that sea turtles may be a useful sentinel for human health responses to marine pollution.[115]

Furthermore, Caliani *et al.* demonstrated for the first time that the Comet assay is a useful method to evaluate genotoxic effects in loggerhead sea turtles. The Comet assay has been used in biomonitoring studies to detect DNA damage in several species.[116]

19.6 Conclusions

In this chapter we have seen how the increasing levels of chemical pollution produced by man may have serious impacts on the conservation of all species of sea turtles, both directly through the toxic effects on their health, as well as indirectly by affecting the ecosystem and food they consume. We have described the major findings in the body of scientific literature describing how turtles are exposed to these contaminants and are able to accumulate them throughout their life. Although it has been traditionally assumed that the vastness of the oceans of the Earth dilutes anthropogenic pollutants to safe levels, the direct measurements and the use of various biomarkers are currently indicating that exposure levels of marine turtles to these chemicals may pose a risk for them, and could have deleterious effects on their physiology. We have seen how research on these species is frequently difficult, and often yields contradictory and difficult to explain results, because it is subjected to different types of bias, either related to the type of samples used in the studies, or derived from the big gaps in the knowledge of the toxicodynamics of these compounds in these species of marine reptiles. It has also been demonstrated that the transfer of chemical contaminants from mothers to their eggs could have an effect on viability rates in many cases, and that younger animals are those that are most at risk. In other cases, causality could not be clearly determined, but there is very strong evidence that contaminants could have a decisive influence on the development of diseases, either directly or indirectly (*i.e.* causing damage to the immune system of these animals). In addition, when pollution episodes destroy aquatic plants, this can also destroy sea turtle feeding grounds. Clearly, much more research is needed to clarify all the remaining obscurities regarding the threats to sea turtles from pollution and the mechanisms underlying the possible toxic effects. However, there is overwhelming evidence that rising pollution levels pose a real threat to sea turtles. It has been estimated that only one in every thousand small turtles born survives to become an adult. To do so it has to overcome many natural predators and other threats. We know today that these few survivors could also be facing the "invisible threat" of chemical pollutants, affecting their lives on a long-term basis—a threat to these living fossils that have inhabited our planet for millions of years that the scientific community is now working to make more understandable and more visible.

References

1. IUCN, IUCN Red List of Threatened Species. Version 2015.4. /http://www.iucnredlist.org/).
2. N. P. Thompson, P. W. Rankin and D. W. Johnston, Polychlorinated biphenyls and p,p' DDE in green turtle eggs from Ascension Island, South Atlantic Ocean, *Bull. Environ. Contam. Toxicol.*, 1974, **11**, 399–406.
3. H. O. Hillestad, R. J. Reimold, R. R. Stickney, H. L. Windom and J. Jenkins, Pesticides, heavy metals and radionuclide uptake in loggerhead sea turtles from South Carolina and Georgia, *Herpetol. Rev.*, 1974, **5**, 75.

4. J. M. Keller, in *The Biology of Sea Turtles*, ed. J. Wyneken, K. J. Lohmann and J. A. Musick, CRC Marine Biology Series, CRC Press, Boca Raton, Florida, USA, 2013, ch. 11, vol. III, pp. 285–328.
5. J. P. van de Merwe, M. Hodge, H. A. Olszowy, J. M. Whittier and S. Y. Lee, Using blood samples to estimate persistent organic pollutants and metals in green sea turtles (*Chelonia mydas*), *Mar. Pollut. Bull.*, 2010, **60**, 579–588.
6. J. M. Keller, J. R. Kucklick, C. A. Harms and P. D. McClellan-Green, Organochlorine contaminants in sea turtles: correlations between whole blood and fat, *Environ. Toxicol. Chem.*, 2004, **23**, 726–738.
7. M. Hamann, M. H. Godfrey, J. A. Seminoff, K. Arthur, P. C. R. Barata, K. A. Bjorndal, A. B. Bolten, A. C. Broderick, L. M. Campbell, C. Carreras, P. Casale, M. Chaloupka, S. K. F. Chan, M. S. Coyne, L. B. Crowder, C. E. Diez, P. H. Dutton, S. P. Epperly, N. N. FitzSimmons, A. Formia, M. Girondot, G. C. Hays, I. J. Cheng, Y. Kaska, R. Lewison, J. A. Mortimer, W. J. Nichols, R. D. Reina, K. Shanker, J. R. Spotila, J. Tomás, B. Wallace, T. M. Work, J. Zbinden and B. J. Godley, Global research priorities for sea turtles:informing management and conservation in the 21st century, *Endang. Species Res.*, 2010, **11**, 245–269.
8. M. Van den Berg, L. S. Birnbaum, M. Denison, M. De Vito, W. Farland, M. Feeley, H. Fiedler, H. Hakansson, A. Hanberg, L. Haws, M. Rose, S. Safe, D. Schrenk, C. Tohyama, A. Tritscher, J. Tuomisto, M. Tysklind, N. Walker and R. E. Peterson, The 2005 World Health Organization reevaluation of human and Mammalian toxic equivalency factors for dioxins and dioxin-like compounds, *Toxicol. Sci.*, 2006, **93**, 223–241.
9. S. D'Ilio, D. Mattei, M. F. Blasi, A. Alimonti and S. Bogialli, The occurrence of chemical elements and POPs in loggerhead turtles (*Caretta caretta*): an overview, *Mar. Pollut. Bull.*, 2011, **62**, 1606–1615.
10. S. C. Gardner, M. D. Pier, R. Wesselman and J. A. Juarez, Organochlorine contaminants in sea turtles from the Eastern Pacific, *Mar. Pollut. Bull.*, 2003, **46**, 1082–1089.
11. B. Lazar, L. Maslov, S. H. Romanic, R. Gracan, B. Krauthacker, D. Holcer and N. Tvrtkovic, Accumulation of organochlorine contaminants in loggerhead sea turtles, *Caretta caretta*, from the eastern Adriatic Sea, *Chemosphere*, 2011, **82**, 121–129.
12. M. M. Storelli, G. Barone and G. O. Marcotrigiano, Polychlorinated biphenyls and other chlorinated organic contaminants in the tissues of Mediterranean loggerhead turtle *Caretta caretta*, *Sci. Total Environ.*, 2007, **373**, 456–463.
13. M. Camacho, O. P. Luzardo, L. D. Boada, L. F. López Jurado, M. Medina, M. Zumbado and J. Orós, Potential adverse health effects of persistent organic pollutants on sea turtles: Evidences from a cross-sectional study on Cape Verde loggerhead sea turtles, *Sci. Total Environ.*, 2013, **458–460C**, 283–289.
14. M. Camacho, L. D. Boada, J. Orós, P. López, M. Zumbado, M. Almeida-González and O. P. Luzardo, Comparative study of organohalogen

contamination between two populations of Eastern Atlantic loggerhead sea turtles (*Caretta caretta*), *Bull. Environ. Contam. Toxicol.*, 2013, **91**, 678–683.

15. S. K. Alam and M. S. Brim, Organochlorine, PCB, PAH, and metal concentrations in eggs of loggerhead sea turtles (*Caretta caretta*) from northwest Florida, USA, *J. Environ. Sci. Health B*, 2000, **35**, 705–724.
16. J. J. Alava, J. M. Keller, J. R. Kucklick, J. Wyneken, L. Crowder and G. I. Scott, Loggerhead sea turtle (*Caretta caretta*) egg yolk concentrations of persistent organic pollutants and lipid increase during the last stage of embryonic development, *Sci. Total Environ.*, 2006, **367**, 170–181.
17. J. J. Alava, J. M. Keller, J. Wyneken, L. Crowder, G. Scott and J. R. Kucklick, Geographical variation of persistent organic pollutants in eggs of threatened loggerhead sea turtles (*Caretta caretta*) from southeastern United States, *Environ. Toxicol. Chem.*, 2011, **30**, 1677–1688.
18. S. Corsolini, S. Aurigi and S. Focardi, Presence of polychlorobiphenyls (PCBs) and coplanar congeners in the tissues of the Mediterranean loggerhead turtle *Caretta caretta*, *Mar. Pollut. Bull.*, 2000, **40**(11), 952–960.
19. C. Mckenzie, B. J. Godley, R. W. Furness and D. E. Wells, Concentrations and patterns of organochlorine contaminants in marine turtles from Mediterranean and Atlantic waters, *Mar. Environ. Res.*, 1999, **47**, 117–135.
20. J. Orós, O. M. González-Díaz and P. Monagas, High levels of polychlorinated biphenyls in tissues of Atlantic turtles stranded in the Canary Islands, Spain, *Chemosphere*, 2009, **74**, 473–478.
21. M. Perugini, A. Giammarino, V. Olivieri, S. Guccione, O. R. Lai and M. Amorena, Polychlorinated biphenyls and organochlorine pesticide levels in tissues of *Caretta caretta* from the Adriatic Sea, *Dis. Aquat. Org.*, 2006, **71**, 155–161.
22. M. J. Rybitski, R. C. Hale and J. A. Musick, Distribution of organochlorine pollutants in Atlantic sea turtles, *Copeia*, 1995, **2**, 379–390.
23. M. M. Storelli, G. O. Marcotrigiano and Chlorobiphenyls, HCB, and organochlorine pesticides in some tissues of *Caretta caretta* (Linnaeus) specimens beached along the adriatic sea, Italy, *Bull. Environ. Contam. Toxicol.*, 2000, **64**, 481–488.
24. M. Bucchia, M. Camacho, M. R. Santos, L. D. Boada, P. Roncada, R. Mateo, M. E. Ortiz-Santaliestra, J. Rodríguez-Estival, M. Zumbado, J. Orós, L. A. Henríquez-Hernández, N. García-Álvarez and O. P. Luzardo, Plasma levels of pollutants are much higher in loggerhead turtle populations from the Adriatic Sea than in those from open waters (Eastern Atlantic Ocean), *Sci. Total Environ.*, 2015, **523**, 161–169.
25. G. Malarvannan, S. Takahashi, T. Isobe, T. Kunisue, A. Sudaryanto, T. Miyagi, M. Nakamura, S. Yasumura and S. Tanabe, Levels and distribution of polybrominated diphenyl ethers and organochlorine compounds in sea turtles from Japan, *Mar. Pollut. Bull.*, 2011, **63**, 172–178.

26. K. L. Richardson, M. Lopez Castro, S. C. Gardner and D. Schlenk, Polychlorinated biphenyls and biotransformation enzymes in three species of sea turtles from the Baja California peninsula of Mexico, *Arch. Environ. Con. Toxicol.*, 2010, **58**, 183–193.
27. S. R. de Solla, in *Ecotoxicology of Amphibians and Reptiles*, ed. D. W. Sparling, G. Linder, C. A. Bishop and S. K. Krest, New York, USA, 2010, ch. 10, p. 298.
28. X. S. Miao, G. H. Balazs, S. K. Murakawa and Q. X. Li, Congener-specific profile and toxicity assessment of PCBs in green turtles (*Chelonia mydas*) from the Hawaiian Islands, *Sci. Total Environ.*, 2001, **281**, 247–253.
29. K. R. Stewart, J. M. Keller, R. Templeton, J. R. Kucklick and C. Johnson, Monitoring persistent organic pollutants in leatherback turtles (*Dermochelys coriacea*) confirms maternal transfer, *Mar. Pollut. Bull.*, 2011, **62**, 1396–1409.
30. R. F. Swarthout, J. M. Keller, M. Peden-Adams, A. M. Landry, P. A. Fair and J. R. Kucklick, Organohalogen contaminants in blood of Kemp's ridley (*Lepidochelys kempii*) and green sea turtles (*Chelonia mydas*) from the Gulf of Mexico, *Chemosphere*, 2010, **78**, 731–741.
31. M. M. Storelli and N. Zizzo, Occurrence of organochlorine contaminants (PCBs, PCDDs and PCDFs) and pathologic findings in loggerhead sea turtles, *Caretta caretta*, from the Adriatic Sea (Mediterranean Sea), *Sci. Total Environ.*, 2014, **472**, 855–861.
32. M. Camacho, L. D. Boada, J. Orós, P. López, M. Zumbado, M. Almeida-González and O. P. Luzardo, Monitoring organic and inorganic pollutants in juvenile live sea turtles: results from a study of *Chelonia mydas* and *Eretmochelys imbricata* in Cape Verde, *Sci. Total Environ.*, 2014, **481**, 303–310.
33. P. Monagas, J. Orós, J. Araña and O. M. González-Díaz, Organochlorine pesticide levels in loggerhead turtles (*Caretta caretta*) stranded in the Canary Islands, Spain, *Mar. Pollut. Bull.*, 2008, **56**, 1949–1952.
34. M. Camacho, J. Orós, L. A. Henríquez-Hernández, P. F. Valerón, L. D. Boada, A. Zaccaroni, M. Zumbado and O. P. Luzardo, Influence of the rehabilitation of injured loggerhead turtles (*Caretta caretta*) on their blood levels of environmental organic pollutants and elements, *Sci. Total Environ.*, 2014, **487**, 436–442.
35. J. Orós, M. Camacho, L. D. Boada and O. P. Luzardo, in *Pesticides: Characteristics, Uses and Health Implications*, ed. A. J. Abrego and E. M. Lugo, Nova Sciense Publisher, Inc., New York, 2012, ch. 4, pp. 66–73.
36. J. M. Ragland, M. D. Arendt, J. R. Kucklick and J. M. Keller, Persistent organic pollutants in blood plasma of satellite-tracked adult male loggerhead sea turtles (*Caretta caretta*), *Environ. Toxicol. Chem.*, 2011, **30**, 1549–1556.
37. J. P. van de Merwe, M. Hodge, K. I. Whittier and S. Y. Lee, Persistent organic pollutants in the green sea turtle *Chelonia mydas*: nesting population variation, maternal transfer, and effects on development, *Mar. Ecol. Prog. Ser.*, 2010, **403**, 269–278.

38. J. M. Keller, Forty-seven days of decay does not change persistent organic pollutant levels in loggerhead sea turtle eggs, *Environ. Toxicol. Chem.*, 2013, **32**, 747–756.
39. R. A. Hites, Polybrominated diphenyl ethers in the environment and in people: a meta-analysis of concentrations, *Environ. Sci. Technol.*, 2004, **38**, 945–956.
40. J. M. Keller, J. J. Alava, K. Aleksa, B. Young and J. R. Kucklick, Spatial trends of polybrominated diphenyl ethers (PBDEs) in loggerhead sea turtle eggs and plasma, *Organohalog. Compd.*, 2005, **67**, 610–611.
41. S. Hermanussen, V. Matthews, O. Papke, C. J. Limpus and C. Gaus, Flame retardants (PBDEs) in marine turtles, dugongs and seafood from Queensland, Australia, *Mar. Pollut. Bull.*, 2008, **57**, 409–418.
42. C. Guerranti, S. Ancora, N. Bianchi, G. Perra, E. L. Fanello, S. Corsolini, M. C. Fossi and S. E. Focardi, Perfluorinated compounds in blood of *Caretta caretta* from the Mediterranean Sea, *Mar. Pollut. Bull.*, 2013, **73**, 98–101.
43. J. M. Keller, K. Kannan, S. Taniyasu, N. Yamashita, R. D. Day, M. D. Arendt, A. L. Segars and J. R. Kucklick, Perfluorinated compounds in the plasma of loggerhead and Kemp's ridley sea turtles from the southeastern coast of the United States, *Environ. Sci. Technol.*, 2005, **39**, 9101–9108.
44. S. G. O'Connell, M. Arendt, A. Segars, T. Kimmel, J. Braun-McNeill, L. Avens, B. Schroeder, L. Ngai, J. R. Kucklick and J. M. Keller, Temporal and spatial trends of perfluorinated compounds in juvenile loggerhead sea turtles (*Caretta caretta*) along the East Coast of the United States, *Environ. Sci. Technol.*, 2010, **44**, 5202–5209.
45. J. M. Keller, L. Ngai, J. Braun McNeill, L. D. Wood, K. R. Stewart, S. G. O'Connell and J. R. Kucklick, Perfluoroalkyl contaminants in plasma of five sea turtle species: comparisons in concentration and potential health risks, *Environ. Toxicol. Chem.*, 2012, **31**, 1223–1230.
46. M. M. Storelli, A. Storelli, R. D'Addabbo, C. Marano, R. Bruno and G. O. Marcotrigiano, Trace elements in loggerhead turtles (*Caretta caretta*) from the eastern Mediterranean Sea: overview and evaluation, *Environ. Pollut.*, 2005, **135**, 163–170.
47. D. Haynes and J. E. Johnson, Organochlorine heavy metal and polyaromatic hydrocarbon pollutant concentrations in the Great Barrier Reef (Australia): a review, *Mar. Pollut. Bull.*, 2000, **41**, 267–278.
48. R. D. Day, A. L. Segars, M. D. Arendt, A. M. Lee and M. M. Peden-Adams, Relationship of blood mercury levels to health parameters in the loggerhead sea turtle (*Caretta caretta*), *Environ. Health Perspect.*, 2007, **115**, 1421–1428.
49. Y. Anan, T. Kunito, I. Watanabe, H. Sakai and S. Tanabe, Trace element accumulation in hawksbill turtles (*Eretmochelys imbricata*) and green turtles (*Chelonia mydas*) from Yaeyama Islands, Japan, *Environ. Toxicol. Chem.*, 2001, **20**, 2802–2814.

50. F. Caurant, P. Bustamante, M. Bordes and P. Miramand, Bioaccumulation of cadmium, copper and zinc in some tissues of three species of marine turtles stranded along the French Atlantic Coasts, *Mar. Pollut. Bull.*, 1999, **38**, 1085–1091.
51. T. J. O'Shea and J. R. Geraci, in *Zoo & Wild Animal Medicine: Current Therapy*, ed. M. E. Fowler and D. L. Miller, W.B. Saunder Company, Philadelphia, 1999, pp. 412–478.
52. G. Andreani, M. Santoro, S. Cottignoli, M. Fabbri, E. Carpene and G. Isani, Metal distribution and metallothionein in loggerhead (*Caretta caretta*) and green (*Chelonia mydas*) sea turtles, *Sci. Total Environ.*, 2008, **390**, 287–294.
53. S. Franzellitti, C. Locatelli, G. Gerosa, C. Vallini and E. Fabbri, Heavy metals in tissues of loggerhead turtles (*Caretta caretta*) from the northwestern Adriatic Sea, *Comp. Biochem. Phys. C*, 2004, **138**, 187–194.
54. A. J. García-Fernández, P. Gómez-Ramírez, E. Martinez-López, A. Hernández-García, P. María-Mojica, D. Romero, P. Jiménez, J. J. Castillo and J. J. Bellido, Heavy metals in tissues from loggerhead turtles (*Caretta caretta*) from the southwestern Mediterranean (Spain), *Ecotoxicol. Environ. Saf.*, 2009, **72**, 557–563.
55. S. C. Gardner, S. L. Fitzgerald, B. A. Vargas and L. M. Rodriguez, Heavy metal accumulation in four species of sea turtles from the Baja California peninsula, Mexico, *Biometals*, 2006, **19**, 91–99.
56. M. M. Storelli and G. O. Marcotrigiano, Heavy metal residues in tissues of marine turtles, *Mar. Pollut. Bull.*, 2003, **46**, 397–400.
57. K. Suzuki, J. Noda, M. Yanagisawa, I. Kawazu, K. Sera, D. Fukui, M. Asakawa and H. Yokota, Particle-induced X-ray emission analysis of elements in plasma from wild and captive sea turtles (*Eretmochelys imbricata*, *Chelonia mydas*, and *Caretta caretta*) in Okinawa, Japan, *Biol. Trace Elem. Res.*, 2012, **148**, 302–308.
58. M. Camacho, J. Orós, L. D. Boada, A. Zaccaroni, M. Silvi, C. Formigaro, P. López, M. Zumbado and O. P. Luzardo, Potential adverse effects of inorganic pollutants on clinical parameters of loggerhead sea turtles (*Caretta caretta*): results from a nesting colony from Cape Verde, West Africa, *Mar. Environ. Res.*, 2013, **92**, 15–22.
59. C. C. da Silva, A. S. Varela, Jr., I. F. Barcarolli and A. Bianchini, Concentrations and distributions of metals in tissues of stranded green sea turtles (*Chelonia mydas*) from the southern Atlantic coast of Brazil, *Sci. Total Environ.*, 2014, **466–467**, 109–118.
60. S. Jerez, M. Motas, R. A. Cánovas, J. Talavera, R. M. Almela and A. B. Del Río, Accumulation and tissue distribution of heavy metals and essential elements in loggerhead turtles (*Caretta caretta*) from Spanish Mediterranean coastline of Murcia, *Chemosphere*, 2010, **78**, 256–264.
61. H. Sakai, K. Saeki, H. Ichihashi, H. Suganuma, S. Tanabe and R. Tatsukawa, Species-specific distribution of heavy metals in tissues and organs of loggerhead turtle (*Caretta caretta*) and green turtle

(*Chelonia mydas*) from Japanese coastal waters, *Mar. Pollut. Bull.*, 2000, **40**, 701–709.

62. B. J. Godley, D. R. Thompson and R. W. Furness, Do heavy metal concentrations pose a threat to marine turtles from the Mediterranean Sea? *Mar. Pollut. Bull.*, 1999, **38**, 497–502.
63. F. Maffucci, F. Caurant, P. Bustamante and F. Bentivegna, Trace element (Cd, Cu, Hg, Se, Zn) accumulation and tissue distribution in loggerhead turtles (*Caretta caretta*) from the Western Mediterranean Sea (southern Italy), *Chemosphere*, 2005, **58**, 535–542.
64. H. Sakai, H. Ichihashi, H. Suganuma and R. Tatsukawa, Heavy metal monitoring in sea turtles using eggs, *Mar. Pollut. Bull.*, 1995, **30**, 347–353.
65. M. M. Storelli, E. Ceci and G. O. Marcotrigiano, Distribution of heavy metal residues in some tissues of *Caretta caretta* (Linnaeus) specimen beached along the Adriatic Sea (Italy), *Bull. Environ. Contam. Toxicol.*, 1998, **60**, 546–552.
66. A. Torrent, O. M. González-Díaz, P. Monagas and J. Orós, Tissue distribution of metals in loggerhead turtles (*Caretta caretta*) stranded in the Canary Islands, Spain, *Mar. Pollut. Bull.*, 2004, **49**, 854–860.
67. E. Guirlet, K. Das and M. Girondot, Maternal transfer of trace elements in leatherback turtles (*Dermochelys coriacea*) of French Guiana, *Aquat. Toxicol.*, 2008, **88**, 267–276.
68. M. P. Ikonomopoulou, H. Olszowy, C. Limpus, R. Francis and J. Whittier, Trace element concentrations in nesting flatback turtles (*Natator depressus*) from Curtis Island, Queensland, Australia, *Mar. Environ. Res.*, 2011, **71**, 10–16.
69. J. C. Lam, S. Tanabe, S. K. Chan, M. H. Lam, M. Martin and P. K. Lam, Levels of trace elements in green turtle eggs collected from Hong Kong: Evidence of risks due to selenium and nickel, *Environ. Pollut.*, 2006, **144**, 790–801.
70. F. Páez-Osuna, M. F. Calderón-Campuzano, M. F. Soto-Jiménez and J. R. Ruelas-Inzunza, Trace metals (Cd, Cu, Ni, and Zn) in blood and eggs of the sea turtle *Lepidochelys olivacea* from a nesting colony of Oaxaca, Mexico, *Arch. Environ. Con. Toxicol.*, 2010, **59**, 632–641.
71. J. Perrault, J. Wyneken, L. J. Thompson, C. Johnson and D. L. Miller, Why are hatching and emergence success low? Mercury and selenium concentrations in nesting leatherback sea turtles (*Dermochelys coriacea*) and their young in Florida, *Mar. Pollut. Bull.*, 2011, **62**, 1671–1682.
72. M. Ehsanpour, M. Afkhami, R. Khoshnood and K. J. Reich, Determination and maternal transfer of heavy metals (Cd, Cu, Zn, Pb and Hg) in the Hawksbill sea turtle (*Eretmochelys imbricata*) from a nesting colony of Qeshm Island, Irán, *Bull. Environ. Contam. Toxicol.*, 2014, **92**, 667–673.
73. C. Ley-Quiñónez, A. A. Zavala-Norzagaray, T. L. Espinosa-Carreón, H. Peckham, C. Marquez-Herrera, L. Campos-Villegas and A. A. Aguirre, Baseline heavy metals and metalloid values in blood of loggerhead

turtles (*Caretta caretta*) from Baja California Sur, Mexico, *Mar. Pollut. Bull.*, 2011, **62**, 1979–1983.

74. A. A. Zavala-Norzagaray, C. P. Ley-Quinónez, T. L. Espinosa-Carreón, A. Canizalez-Román, C. E. Hart and A. A. Aguirre, Trace elements in blood of sea turtles *Lepidochelys olivacea* in the Gulf of California, Mexico, *Bull. Environ. Contam. Toxicol.*, 2014, **93**, 536–541.
75. D. R. Faust, M. J. Hooper, G. P. Cobb, M. Barnes, D. Shaver, S. Ertolacci and P. N. Smith, Inorganic elements in green sea turtles (*Chelonia mydas*): relationships among external and internal tissues, *Environ. Toxicol. Chem.*, 2014, **33**, 2020–2027.
76. R. D. Day, S. J. Christopher, P. R. Becker and D. W. Whitaker, Monitoring mercury in the loggerhead sea turtle, *Caretta caretta*, *Environ. Sci. Technol.*, 2005, **39**, 437–446.
77. L. M. Komoroske, R. L. Lewison, J. A. Seminoff, D. D. Deheyn and P. H. Dutton, Pollutants and the health of green sea turtles resident to an urbanized estuary in San Diego, CA, *Chemosphere*, 2011, **84**, 544–552.
78. H. Wang, Trace metal uptake and accumulation pathways in Kemp's Ridley sea turtles (*Lepidochelys kempii*), PhD thesis, Texas A&M University, 2005, pp. 99–130.
79. R. D. Day, J. M. Keller, C. A. Harms, A. L. Segars, W. M. Cluse, M. H. Godfrey, A. M. Lee, M. Peden-Adams, K. Thorvalson, M. Dodd and T. Norton, Comparison of mercury burdens in chronically debilitated and healthy loggerhead sea turtles (*Caretta caretta*), *J. Wildl. Dis.*, 2010, **46**, 111–117.
80. R. Kampalath, S. C. Gardner, L. Méndez-Rodríguez and J. A. Jay, Total and methylmercury in three species of sea turtles of Baja California Sur, *Mar. Pollut. Bull.*, 2006, **52**, 1784–1832.
81. G. García-Besné, C. Valdespino and J. Rendón-von Osten, Comparison of organochlorine pesticides and PCB residues among hawksbill (*Eretmochelys imbricata*) and green (*Chelonia mydas*) turtles in the Yucatan Peninsula and their maternal transfer, *Mar. Pollut. Bull.*, 2015, **91**, 139–148.
82. K. L. Richardson and D. Schlenk, Biotransformation of 2,2',5,5'-tetrachlorobiphenyl (PCB 52) and 3,3',4,4'-tetrachlorobiphenyl (PCB 77) by liver microsomes from four species of sea turtles, *Chem. Res. Toxicol.*, 2011, **24**, 718–725.
83. V. Labrada-Martagón, P. A. Rodríguez, L. C. Méndez-Rodríguez and T. Zenteno-Savín, Oxidative stress indicators and chemical contaminants in East Pacific green turtles (*Chelonia mydas*) inhabiting two foraging coastal lagoons in the Baja California peninsula, *Comp. Biochem. Phys. C*, 2011, **154**, 65–75.
84. E. Guirlet, K. Das, J. P. Thome and M. Girondot, Maternal transfer of chlorinated contaminants in the leatherback turtles, *Dermochelys coriacea*, nesting in French Guiana, *Chemosphere*, 2010, **79**, 720–726.
85. F. Páez-Osuna, M. F. Calderón-Campuzano, M. F. Soto-Jiménez and J. R. Ruelas-Inzunza, Lead in blood and eggs of the sea turtle,

Lepidochelys olivacea, from the Eastern Pacific: concentration, isotopic composition and maternal transfer, *Mar. Pollut. Bull.*, 2010, **60**, 433–439.
86. A. A. Aguirre, G. H. Balazs, B. Zimmerman and F. D. Galey, Organic contaminants and trace metals in the tissues of green turtles (*Chelonia mydas*) afflicted with fibropapillomas in the Hawaiian islands, *Mar. Pollut. Bull.*, 1994, **28**, 109–114.
87. J. M. Keller, G. H. Balazs, F. Nilsen, M. Rice, T. M. Work and B. A. Jensen, Investigating the potential role of persistent organic pollutants in Hawaiian green sea turtle fibropapillomatosis, *Environ. Sci. Technol.*, 2014, **48**, 7807–7816.
88. C. C. da Silva, R. D. Klein, I. F. Barcarolli and A. Bianchini, Metal contamination as a possible etiology of fibropapillomatosis in juvenile female green sea turtles *Chelonia mydas* from the southern Atlantic Ocean, *Aquat. Toxicol.*, 2016, **170**, 42–51.
89. K. Jones, E. Ariel, G. Burgess and M. Read, A review of fibropapillomatosis in Green turtles (*Chelonia mydas*), *Vet. J.*, 2016, **212**, 48–57.
90. T. M. Work, R. A. Rameyer, G. H. Balazs, C. Cray and S. P. Chang, Immune status of free-ranging green turtles with fibropapillomatosis from Hawaii, *J. Wildl. Dis.*, 2001, **37**, 574–581.
91. J. M. Keller, J. N. Meyer, M. Mattie, T. Augsperger, M. Rau, J. Dong and E. D. Levin, Assessment of immunotoxicology in wild populations: review and recommendations, *Rev. Toxicol.*, 2000, **3**, 167–212.
92. J. M. Keller, J. R. Kucklick, M. A. Stamper, C. A. Harms and P. D. McClellan-Green, Associations between organochlorine contaminant concentrations and clinical health parameters in loggerhead sea turtles from North Carolina, USA, *Environ. Health Perspect.*, 2004, **112**, 1074–1079.
93. J. M. Keller, P. D. McClellan-Green, J. R. Kucklick, D. E. Keil and M. M. Peden-Adams, Effects of organochlorine contaminants on loggerhead sea turtle immunity: comparison of a correlative field study and *in vitro* exposure experiments, *Environ. Health Perspect.*, 2006, **114**, 70–76.
94. J. Orós, P. Monagas, P. Calabuig, O. P. Luzardo and M. Camacho, Pansteatitis associated with high levels of polychlorinated biphenyls in a wild loggerhead sea turtle *Caretta caretta*, *Dis. Aquat. Org.*, 2013, **102**, 237–242.
95. C. Innis, M. Tlusty, C. Perkins, S. Holladay, C. Merigo and E. S. Weber, Trace metal and organochlorine pesticide concentrations in cold-stunned juvenile kemp's ridley turtles (*Lepidochelys kempii*) from Cape Cod, Massachusetts, *Chelonian Conserv. Biol.*, 2008, **7**, 230–239.
96. S. Podreka, A. Georges, B. Maher and C. J. Limpus, The environmental contaminant DDE fails to influence the outcome of sexual differentiation in the marine turtle *Chelonia mydas*, *Environ. Health Perspect.*, 1998, **106**, 185–188.

97. H. Merchant-Larios, S. Ruiz-Ramirez, N. Moreno-Mendoza and A. Marmolejo-Valencia, Correlation among thermosensitive period, estradiol response, and gonad differentiation in the sea turtle *Lepidochelys olivacea*, *Gen. Comp. Endocrinol.*, 1997, **107**, 373–385.
98. J. M. Keller and P. McClellan-Green, Effects of organochlorine compounds on cytochrome P450 aromatase activity in an immortal sea turtle cell line, *Mar. Environ. Res.*, 2004, **58**, 347–351.
99. S. J. Webb, G. V. Zychowski, S. W. Bauman, B. M. Higgins, T. Raudsepp, L. S. Gollahon, K. J. Wooten, J. M. Cole and C. Godard-Codding, Establishment, characterization, and toxicological application of loggerhead sea turtle (*Caretta caretta*) primary skin fibroblast cell cultures, *Environ. Sci. Technol.*, 2014, **48**, 14728–14737.
100. M. P. Ikonomopoulou, H. Olszowy, M. Hodge and A. J. Bradley, The effect of organochlorines and heavy metals on sex steroid-binding proteins *in vitro* in the plasma of nesting green turtles, *Chelonia mydas*, *J. Comp. Physiol. B*, 2009, **179**, 653–662.
101. J. R. Perrault, D. L. Miller, J. Garner and J. Wyneken, Mercury and selenium concentrations in leatherback sea turtles (*Dermochelys coriacea*): Population comparisons, implications for reproductive success, hazard quotients and directions for future research, *Sci. Total Environ.*, 2013, **463–464**, 61–71.
102. E. De Andrés, B. Gómara, D. González-Paredes, J. Ruiz-Martín and A. Marco, Persistent organic pollutant levels in eggs of leatherback turtles (*Dermochelys coriacea*) point to a decrease in hatching success, *Chemosphere*, 2016, **146**, 354–361.
103. S. M. Ho, S. Kleis, R. McPherson, G. J. Heisermann and I. P. Callard, Regulation of vitellogenesis in reptiles, *Herpetologica*, 1982, **38**, 40–50.
104. J. M. Keller, Occurrence and effects of organochlorine contaminants in sea turtles, PhD thesis, Duke University, 2003, pp. 197–221.
105. A. Zaccaroni, M. Zucchini, L. Segatta, M. Gamberoni, D. Freggi, P. A. Accorsi, D. Scaravelli and S. C. Gardner, Vitellogenin (VTG) conservation in sea turtles: anti-VTG antibody in *Chelonia mydas* versus *Caretta caretta*, *Physiol. Biochem. Zool.*, 2010, **83**, 191–195.
106. J. Heck, D. S. MacKenzie, D. Rostal, K. Medler and D. Owens, Estrogen induction of plasma vitellogenin in the Kemp's ridley sea turtle (*Lepidochelys kempii*), *Gen. Comp. Endocrinol.*, 1997, **107**, 280–288.
107. L. H. Herbst, L. Siconolfi-Baez, J. H. Torelli, P. A. Klein, M. J. Kerben and I. M. Schumacher, Induction of vitellogenesis by estradiol-17beta and development of enzyme-linked immunosorbant assays to quantify plasma vitellogenin levels in green turtles (*Chelonia mydas*), *Comp. Biochem. Physiol., Part B: Biochem. Mol. Biol.*, 2003, **135**, 551–563.
108. I. Sifuentes-Romero, C. Vázquez-Boucard, A. P. Sierra-Beltrán and S. C. Gardner, Vitellogenin in black turtle (*Chelonia mydas agassizii*): purification, partial characterization, and validation of an enzyme-linked immunosorbent assay for its detection, *Environ. Toxicol. Chem.*, 2006, **25**, 477–485.

109. K. Smelker, L. Smith, M. Arendt, J. Schwenter, D. Rostal, K. Selcer and R. Valverde, Plasma Vitellogenin in Free-Ranging Loggerhead Sea Turtles (*Caretta caretta*) of the Northwest Atlantic Ocean, *J. Mar. Biol.*, 2014, **2014**, 1–10.
110. P. A. Valdivia, T. Zenteno-Savin, S. C. Gardner and A. A. Aguirre, Basic oxidative stress metabolites in eastern Pacific green turtles (*Chelonia mydas agassizii*), *Comp. Biochem. Phys. C*, 2007, **146**, 111–117.
111. K. L. Richardson, G. Gold-Bouchot and D. Schlenk, The characterization of cytosolic glutathione transferase from four species of sea turtles: Loggerhead (*Caretta caretta*), green (*Chelonia mydas*), olive ridley (*Lepidochelys olivacea*), and hawksbill (*Eretmochelys imbricata*), *Comp. Biochem. Phys. C*, 2009, **150**, 279–284.
112. F. Tan, M. Wang, W. Wang, A. Alonso Aguirre and Y. Lu, Validation of an *in vitro* cytotoxicity test for four heavy metals using cell lines derived from a green sea turtle (*Chelonia mydas*), *Cell Biol. Toxicol.*, 2010, **26**, 255–263.
113. H. Wang, J. Tong, Y. Bi, C. Wang, L. Guo and Y. Lu, Evaluation of mercury mediated *in vitro* cytotoxicity among cell lines established from green sea turtles, *Toxicol. In Vitro*, 2013, **27**, 1025–1030.
114. S. S. Wise, H. Xie, T. Fukuda, W. Douglas Thompson and J. P. Wise, Hexavalent chromium is cytotoxic and genotoxic to hawksbill sea turtle cells, *Toxicol. Appl. Pharmacol.*, 2014, **279**, 113–118.
115. J. L. Young, S. S. Wise, H. Xie, C. Zhu, T. Fukuda and J. P. Wise Sr., Comparative cytotoxicity and genotoxicity of soluble and particulate hexavalent chromium in human and hawksbill sea turtle (*Eretmochelys imbricata*) skin cells, *Comp. Biochem. Phys. C*, 2015, **178**, 145–155.
116. I. Caliani, T. Campani, M. Giannetti, L. Marsili, S. Casini and M. C. Fossi, First application of comet assay in blood cells of Mediterranean loggerhead sea turtle (*Caretta caretta*), *Mar. Environ. Res.*, 2014, **96**, 68–72.

CHAPTER 20

Biomarkers of Environmental Contamination in Reptile Species: The Effect of Pesticide Formulations on Broad-snouted Caiman Caiman latirostris *(Crocodilia, Alligatoridae)*

G. L. POLETTA,*[a,b,c,d] E. LÓPEZ GONZÁLEZ,[b,d] P. M. BURELLA,[a,d] M. L. ROMITO,[b,d] P. A. SIROSKI[b,d,e] AND M. D. MUDRY[c,d]

[a] Cátedra de Toxicología, Farmacología y Bioquímica Legal, Facultad de Bioquímica y Ciencias Biológicas, Universidad Nacional del Litoral, Ciudad Universitaria, Paraje el Pozo, CP 3000, Santa Fe, Argentina; [b] "Proyecto Yacaré"-Laboratorio de Zoología Aplicada: Anexo Vertebrados (FHUC-UNL/MASPyMA), Aristóbulo del Valle 8700, CP 3000, Santa Fe, Argentina; [c] Grupo Investigación Biología Evolutiva (GIBE), IEGEBA-DEGE (CONICET-UBA), FCEyN, Pab. II, Ciudad Universitaria, Intendente Güiraldes 2160, CP 1428EGA, CABA, Argentina; [d] Consejo Nacional de Investigaciones Científicas y Técnicas (CONICET), Godoy Cruz 2290, CP C1425FQB, CABA, Argentina; [e] Instituto de Ciencias Veterinarias del Litoral (ICiVet CONICET-UNL), R. P. Kreder 2805. S3080HOF - Esperanza, Santa Fe, Argentina
*Email: gpoletta@fbcb.unl.edu.ar; gisepoletta@hotmail.com

Issues in Toxicology No. 33
Ecotoxicology and Genotoxicology: Non-traditional Aquatic Models
Edited by Marcelo L. Larramendy

Published by the Royal Society of Chemistry, www.rsc.org

20.1 Introduction

In all fields of toxicology, specific species are being adopted to act as surrogate models for field and laboratory-based research.[1] Environmental pollution has been recognized as one of the main contributing factors for endangered or vulnerable reptile species.[2,3] Despite consistent calls for greater emphasis on reptile ecotoxicology research, there is still a lack of knowledge regarding the responses of reptiles to contaminants.[4,5] Reptiles, as non-target organisms, can be directly exposed to contaminants through various routes, including: ingestion of contaminated food, accidental or deliberate ingestion of soil, inhalation, maternal transfer to eggs/young, dermal exposure and absorption by eggs of contaminants from surrounding environments,[6,7] as well as being indirectly affected as a consequence of diminished food availability or habitat loss.[8]

Pesticides, in view of their biocide activity, volume of application, and the fact that they are mixtures of chemicals, are the major contaminant of concern for terrestrial environments worldwide.[8] Agricultural activities are introducing extensive amounts of pesticides into the environment daily, contributing to physical and chemical changes in water properties, which can be reflected in the biological integrity of organisms living there. Argentina is the tenth largest agricultural nation in the world, with 31 million ha devoted to agriculture.[9] Pesticide use increased in the last two decades as agriculture became gradually transformed into a system of high technology in order to satisfy growing demands. The adoption of transgenic crops engineered to tolerate the broad-spectrum herbicide glyphosate has gradually increased over the last two decades, with over 22 million ha of cultivated land being occupied by glyphosate-tolerant corn and soybean in the 2012–2013 growing season.[10,11]

Agricultural frontier expansion has led to the loss of pristine habitat, with the consequent impairment of biodiversity and the ability of the ecosystems to provide essential resources.[12] Consequently, several wild species living in fragmented environments are affected by the overuse of pesticides and fertilizers related to this activity. In this framework, the expanding areas of extensive agriculture overlap with the natural geographic distribution of *Caiman latirostris* (broad-snouted caiman; Crocodylia, Alligatoridae), one of the two crocodilian species living in Argentina.[13] As a consequence, caimans are exposed constantly and simultaneously to complex mixtures of several pesticides, including mainly glyphosate based-formulations, and other organophosphates, pyrethroids and organochlorines. What is more, the period of maximum pesticide application coincides with the reproductive season of this species (November to March), so that developing embryos and hatchlings can be particularly susceptible.[14]

The direct effect of pesticides can interfere with molecular mechanisms regulation in early stages of development[15] and the protective mechanisms available in adults, such as DNA repair mechanisms, a competent immune system, detoxifying enzymes, liver metabolism, and the blood–brain barrier, are not fully functional in these early stages.[16] Exposure to pollutants has been shown to influence an organism's morphology, physiology, metabolism, and/or DNA integrity, and thus can affect its ability to acquire resources, grow and reproduce.[15,16] In the present chapter we describe the effects caused by widely used pesticide formulations on *C. latirostris* embryos and yearlings under different conditions of exposure, including biomarkers of genotoxicity, oxidative stress and growth.

20.2 Biomarkers of Early Warning for Pesticide Contamination

20.2.1 Genotoxicity

The alterations of DNA integrity produced by toxic compounds such as pesticides have been thoroughly investigated in many organisms. These compounds can disrupt normal cellular processes and interact directly or indirectly with DNA, causing genetic instability.[19] Biomarkers of genotoxicity are early indicators of damage to genetic material and/or associated structures. Among them, the analysis of micronuclei (MN) and other nuclear abnormalities (NAs), as well as DNA damage detected by the Comet assay (CA), are the most frequently used and recommended endpoints for detecting genotoxicity in environmental toxicology.[20–23]

20.2.1.1 Micronuclei and Other Nuclear Abnormalities

The analysis of blood cell morphology, particularly erythrocytes, has become an important biomarker of pollution, and thereby provides important tools for the prediction of the potential long-term effects of xenobiotics in wild species. In recent years, alterations in erythrocyte nuclei morphology have been increasingly used to evaluate the genotoxic effects of different compounds.[24–26]

Among cytogenetic test systems, the assessment of MN is commonly used for evaluating structural and numerical chromosomal aberrations induced by clastogenic and/or aneugenic agents. MN are cytoplasmic chromatin masses with the appearance of small nuclei that arise from chromosome fragments or intact whole chromosomes lagging behind in the anaphase stage of cell division. Their presence in cells is a reflection of structural and/or numerical chromosomal aberrations arising during mitosis.[27]

In recent years, several studies have described the presence of other NAs, also considered to be induced by genotoxic agents. The formation mechanisms of these NAs are not yet fully understood but, nowadays, many studies have included these abnormalities in the evaluation of genotoxicity as a complementary assay to the MN test.[22,28–30]

In our studies, the MN test and NAs were applied on peripheral blood erythrocytes. The criteria adopted for MN identification were the same described in Poletta *et al.*:[20] (1) MN should be smaller than one-third of the

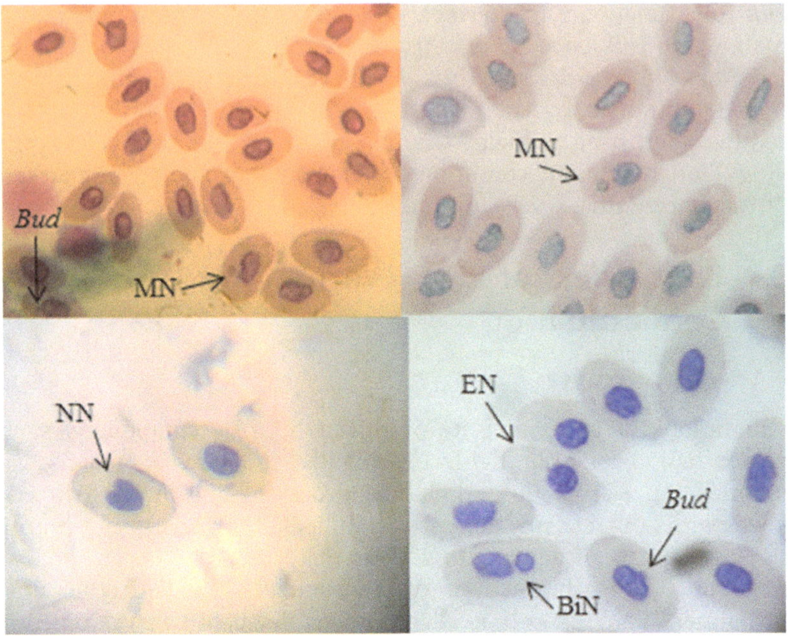

Figure 20.1 Images of *C. latirostris* erythrocytes with micronucleus (MN) and other nuclear abnormalities (NAs) analyzed: Buds, notched nuclei (NN), binuclei (BiN) and eccentric nuclei (EN) (arrows). 1000× magnification.

main nucleus, (2) MN should be separated from the main nucleus, and (3) MN should be the same color and intensity of the main nucleus. NAs were distinguished according to the classification proposed by Carrasco *et al.*[31] and Fenech:[32] notched nuclei (NN): appreciable depth into a nucleus that does not contain nuclear material; nuclear *buds*: nuclear evaginations related to DNA amplification in the S phase of the cell cycle; binucleated (BiN) cells with two nuclei, in division; and the presence of eccentric nuclei (EN): nucleus in a peripheral position. Besides these, we also included the category of total nuclear abnormalities (TNA), which is the sum of all the NAs observed (Figure 20.1).

20.2.1.2 Comet Assay

The alkaline Comet assay (single cell gel electrophoresis), also known as the Comet assay (CA), is the most widely used method for measuring DNA damage in eukaryotic cells. It is an effective biomarker for detecting DNA strand breaks, cross-links and alkali-labile sites in aquatic animals.[21] These types of damages, if not repaired, can initiate a cascade of biological consequences at the levels of cells, organs, whole animal and even populations through mutations, cancers, birth defects, reduced growth, abnormal

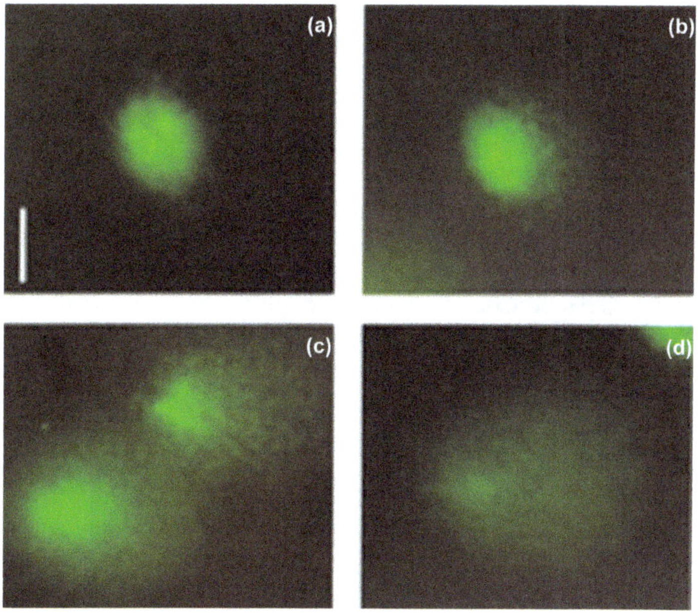

Figure 20.2 Comet images of different damage classes obtained from peripheral blood erythrocytes: (a) class 1, (b) class 2, (c) class 3, and (d) class 4. Bar = 10 μm.

development and the reduced survival of embryos, larvae and adults.[33] Cells with DNA damage display increased migration of DNA fragments from the nucleus, forming a "comet tail".

In all the studies presented here, the alkaline Comet assay was performed as described by Poletta *et al.*[20] for *C. latirostris* erythrocytes. All samples were coded for blind analysis, the slides were stained with Acridine orange, and Comet images were analyzed under a fluorescent microscope as previously described. Images of 100 randomly selected nucleoids (50 from each of two replicated slides) were scored from each animal and they were visually classified into five arbitrary classes according to tail size and intensity (from undamaged, class 0, to maximally damaged, class 4; Figure 20.2). A single DNA damage score (damage index, $DI = n1 + 2n2 + 3n3 + 4n4$) was calculated for each animal.[20]

20.2.2 Oxidative Stress Parameters

Several xenobiotics can induced oxidative stress by increasing reactive oxygen species (ROS) production or affecting antioxidant molecules. Chronic and accumulative oxidative stress induces deleterious modifications to DNA, lipid peroxidation products, and modifications in endogenous oxygen free radical scavengers, which are used as effective biomarkers to study pollutant-mediated oxidative stress.[34]

ROS can derive from subproducts in the mitochondrial respiratory chain, as well as from exogenous sources, such as smoke, radiation, UV light and contamination.[35] These reactive species can interact with DNA, proteins and lipids, causing oxidative damage, which leads to the alteration and destruction of membranes, enzymes and other proteins.[36]

In order to keep ROS at healthy levels in the cells, organisms have antioxidants defenses that work together: antioxidant molecules such as glutathione (GSH), vitamins C and E, and carotenoids, among others; and antioxidant enzymes: catalase (CAT), superoxide dismutase (SOD), GSH reductase (GR), GSH peroxidase (GPX) and glutathione *S*-transferase (GST). Antioxidants form stable complexes in order to avoid the action of reactive species in the cellular membrane and other cellular compounds.[36,37] The negative imbalance between ROS generation and the capacity of the biological systems to eliminate the reactive intermediaries or repair the damage is called oxidative stress (OS).[37]

Pesticides can act as prooxidants in a variety of tissues; they produce ROS accumulation, DNA damage, alteration of the antioxidants defenses and lipid peroxidation, causing large perturbations to intra- and inter-cellular homeostasis.[38,39] These xenobiotics interact with the plasmatic membrane and generate lipid peroxidation, resulting in phospholipid degradation, membrane damage and alteration of its functionality. Meanwhile, cellular protein damage is evidenced as an alteration of antioxidant enzymes, such as CAT and SOD, whose activity can increase or decrease depending on the intensity and duration of the exposure and susceptibility of the exposed species. Decreased CAT and SOD activities lead to OS and stimulate lipid peroxidation.[40] Different studies evaluated OS induced by pesticides in vertebrates,[28,41–43] but no information was available in reptiles. In a recent study we characterized a new set of OS biomarkers in *Caiman latirostris* blood so that OS induced by exogenous agents can be evaluated without any damage to the animals.[44] Biomarkers of OS applied in our studies were are described below.

20.2.2.1 Lipid Peroxidation in Erythrocytes (TBARS)

Malondialdehyde (MDA) as a marker of lipid peroxidation in red blood cells was determined by measuring the formation of the color produced during the reaction of TBA with MDA (tiobarbituric acid reactive species – TBARS–Assay), according to a modification of the method of Beauge and Aust[45] introduced by Poletta *et al.*, in order to adapt it for the broad snouted caiman:[44] Lysed erythrocyte dilution was mixed with potassium chloride buffer (0.154 M) plus protease inhibitors and distilled water. Then it was thoroughly mixed with four volumes of the solution: 15% w/v TCA, 0.375% w/v TBA, 0.25 mol l^{-1} HCl acid and 4% BHT. The mixture was heated in a dry bath at 95 °C for 45 min. After cooling, the flocculent precipitate was removed by centrifugation, the sample absorbance was determined at 535 nm and TBARS concentration was calculated using the extinction coefficient $1.56\times105\ M^{-1}\,cm^{-1}$. The MDA concentration in erythrocytes is expressed as nmol mg Hb^{-1} in the embryo experiment and as nmol mg $prot^{-1}$ in the hatchling experiment.

20.2.2.2 Catalase (CAT) Activity in Erythrocytes

CAT activity in lysed erythrocytes was measured spectrophotometrically by monitoring the decrease in H_2O_2 concentration over time with some modifications for this species.[44] The specific activity of each sample was calculated on the basis that one unit of enzyme activity is defined as the activity required to degrade 1 mole of hydrogen peroxide during 60 s g Hb^{-1}. H_2O_2 was added to a final concentration of 54 mM and the absorbance was measured at 240 nm, 25 °C during 60 s in the spectrophotometer. The results are expressed in arbitrary units as the activity of CAT (KU g Hb^{-1} or KU mg $prot^{-1}$).

20.2.2.3 Superoxide Dismutase (SOD) Activity in Erythrocytes

SOD was determined using the commercial kit 19160-1KT (Sigma).[44] SOD Assay Kit-WST utilizes a highly water-soluble tetrazolium salt, WST-1 (2-(4-iodophenyl)-3-(4-nitrophenyl)-5-(2,4-disulfophenyl)-2H-tetrazolium, monosodium salt), which produces a water-soluble formazan dye upon reduction with a superoxide anion. The rate of the reduction with O_2 is linearly related to the xanthine oxidase (XO) activity and is inhibited by SOD. Since the absorbance at 440 nm is proportional to the amount of superoxide anion, the SOD activity, as an inhibition activity, can be quantified by measuring the decrease in the color development at 440 nm. SOD activity is expressed as percentage of inhibition (%).

20.2.3 Growth

Growth is an integrated response to numerous physiological processes that influence the production budget of the individual. Thus, growth is a cumulative process integrating those factors that have positive and negative effects on the production balance. Growth rates are therefore often used as an index of overall individual health because an individual that has positive growth rates displays an ability to satisfy underlying survival (maintenance) costs while also allocating energy to the production of new tissues. Growth reduction can result from effects of contaminants on underlying bioenergetics processes, such as energy assimilation and metabolic expenditure. In some situations, chronic sublethal exposure to contaminants has been shown to result in an elevated standard metabolic rate (SMR) in reptiles and other taxa.[46] Given no compensatory increase in feeding or assimilation, individuals with SMRs elevated above normal would be expected to experience fitness costs associated with reduced growth or reproductive potential as a result of decreased energetic contributions to the production budget.[47]

Thus, a reduction in growth rate can serve as a biomarker of stressful environmental conditions, although the mechanisms responsible for the reductions may not be immediately evident.[17]

It is considered that body size is affected by processes that differ from those affecting others aspects of animal morphology. Factors known to influence body size are the size of the mother and conditions during egg incubation. Exchange of gasses and water with soil during egg incubation can expose them to chemical contaminants.[1]

In embryo exposure experiments, all hatchlings were individually identified, weighed (OHAUS® Compact scale CS200, precision 0.1 g) and measured in total length (TL) and snout–vent length (SVL) (tape measure, precision 0.5 cm). After that, they were maintained in plastic containers under common controlled conditions used in Proyecto Yacaré facilities. Food was supplied *ad libitum* three times a week, consisting of a mixture of 50% minced chicken head and 50% dry pellets for reptiles. At 3 and 6 months of age they were measured and weighed again in order to evaluate the effects of the treatments on the subsequent growth of the animals during the first months of life. In hatchling exposure experiments, animals were weighed and measured at the beginning and end of the experiment in order to compare their growth in each experimental group.

20.3 Effects of Pesticide Formulations on *Caiman latirostris* Exposed Under Controlled Conditions

20.3.1 Commercial Formulations

Pesticides formulations tested were: (1) Roundup® Full II (66.2% glyphosate [GLY] potassium salt [*N*-(phosphonomethyl)glycine monopotassium salt, $C_3H_7KNO_5P$] as its active ingredient, CAS No. 70901-12-1); (2) Panzer® Gold (60.2% isopropylamine salt of GLY [*N*-(phosphonomethyl) glycine; CAS No. 1071-83-6]; (3) Atanor® (25% mixture of different cypermethrin [CYP] isomers, $C_{22}H_{19}Cl_2NO_3$, CAS No. 52315-07-8); (4) Galgofan® (35% endosulfan [END], $C_8H_6Cl_6O_3S$, CAS No. 115-29-7); and (5) Lorsban® 48E (chlorpyrifos [CPF] 48% *O*,*O*-diethyl *O*-3,5,6-trichloro-2-pyridyl phosphorothioate).[48] Ethanol was used as a vehicle for END, CYP and CPF formulations while GLY formulations were diluted in distilled water for embryo experiments and tap water for hatchling exposure.

20.3.2 Exposure of Embryos to Pesticide Formulations by Topical Application on the Eggshell (*In Ovo* Exposure)

20.3.2.1 Egg Collection

C. latirostris eggs from different nests harvested in the Natural Managed Reserve El Fisco (30° 11′ 26″ S; 61° 0′27″ W; Dpto. San Cristobal, Santa Fe

Province, Argentina) as part of Proyecto Yacare ranching program activities[13] were used. This is a Protected Natural Area (Provincial Law 12,930; 2008), situated at least 20 km from any pesticide application area or other contaminant activity, which is part of the natural distribution of the species, and was chosen to ensure that eggs had not been environmentally exposed to any xenobiotic. All nests used in the experiments were collected within 5 days after oviposition, under the same conditions from harvest to treatment assignment, and egg viability was determined by analyzing the opaque eggshell banding.[49]

20.3.2.2 *Embryotoxicity of Glyphosate, Endosulfan and Cypermethrin Pesticide Formulations on* Caiman latirostris

In Argentina, many natural populations of *C. latirostris* overlap with areas of extensive agriculture, mostly soybean crops. The breeding season for broad-snouted caiman takes place during spring–summer, from late October–December (mating and copulation season) to March, when hatching occurs after an incubation period of approximately 65–80 days, depending on temperature.[13] During this period, females construct nests using surrounding vegetation so that embryos may be exposed not only to residual pesticides in the nest material, but also through direct spraying during applications when nests are close to crops, as this is the moment of maximum pesticide application.

Considerable scientific evidence demonstrates that early life stages of oviparous organisms often exhibit greater toxicological sensitivity to chemical contaminants than adult life stages. The texture of eggshells allows substantial air and water exchange during development and may allow other compounds to cross through the eggshell.[50]

Previous works demonstrated genotoxic and immunotoxic effects of Roundup formulation and its mixture with endosulfan and cypermethrin on *C. latirostris* embryos under controlled and seminatural exposure conditions.[14,51] Taking this into account, together with the lack of information on the effects of insecticides separately in this species, and about oxidative stress of any pesticide in any reptile species, we aim to assess the genotoxicity and oxidative stress of three pesticide formulations widely used in soybean crops (glyphosate, endosulfan, cypermethrin) in caiman embryos exposed by topical application through the eggshell.

20.3.2.2.1 Experimental Design and Treatments. This study was carried out at the Proyecto Yacaré, Laboratorio de Zoología Aplicada: Anexo Vertebrados (FHUC-UNL/MMA, Santa Fe, Argentina) facilities. Two different experiments were carried out:

Embryo Experiment 1 (EE_1):
Ninety-six eggs from three nests (32 eggs per nests) were randomly distributed into eight experimental groups (12 eggs per experimental group

with two replicates of six eggs each), as follows: (1) a negative control (NC) group, treated with distilled water; 2–4) three groups exposed to 500, 750, and 1000 μg egg^{-1} of GLY Panzer® Gold; 5–7) three groups exposed to 500, 750, 1000 μg egg^{-1} of GLY Roundup®; and 8), a positive control (PC) group treated with 700 μg egg^{-1} cyclophosphamide.[14]

Embryo Experiment 2 (EE_2):
One hundred and thirty two eggs from six nests (22 eggs per nest) were randomly distributed into eleven experimental groups ($N=132$; 12 eggs per experimental group with two replicates of six eggs each): (1) a negative control (NC) group treated with distilled water; (2) a vehicle control (VC) group treated with ethanol; 3–6) four groups exposed to 1, 10, 100, and 1000 μg egg^{-1} of END Galgofan®; 7–10) four groups exposed to 1, 10, 100, and 1000 μg egg^{-1} of CYP Atanor®; and 11) a PC treated with 700 μg egg^{-1} cyclophosphamide.[14]

Concentrations applied in our study were chosen by using previous studies carried out in *C. latirostris*[14,51,52] and other species, such as birds and mammals,[53–55] for reference and adapting them to the average weight of *C. latirostris* eggs (approximately 70 g) and to our experimental conditions. The cyclophosphamide concentration is based on previous studies carried out by our group in the same species.[14,51]

In both experiments, the pesticide solutions were applied to the eggshell (by topical application) at the embryo implantation zone within the first 5 days of incubation. Each experimental group was placed separately in a plastic container, using vermiculite as substrate and covering them with vegetal material similar to the nesting material, free of any exogenous substance. All eggs were incubated at a temperature of 31.5 ± 0.5 °C and 95% humidity in the Proyecto Yacaré incubator. They were controlled periodically during the experiment in order to identify and discard those that became non-viable.

When hatchlings started to call within the eggs, the corresponding eggs were removed from the incubator and if hatching did not occur spontaneously during the following 24 h, they were assisted.[13] If any eggs of the same clutch remain unhatched 72 h after the hatching of the first egg of that clutch, they were assisted. Immediately after hatching, blood samples were obtained (0.5 ml) from the spinal vein[56] with heparinized syringes and biomarkers of genotoxicity, oxidative stress and developmental parameters previously described were analyzed.

20.3.2.2.2 Results. Tables 20.1 and 20.2 show the results of MNs and NAs frequencies as mean ± SE per experimental group in EE_1 and EE_2, respectively.

In EE_1 we observed an increase in the FMN in Panzer® Gold 1000 and the PC, and also in the TNA for the latter, but no other NAs were observed (Table 20.1). The CA results showed significantly higher DNA damage in all treatments exposed to glyphosate-based formulations except Roundup® 750 when compared to NC ($p<0.01$ in all analyses performed; Figure 20.3).

Table 20.1 Micronucleus and nuclear abnormality frequencies observed in *C. latirostris* hatchlings exposed *in ovo* to different concentrations of the glyphosate-based formulations Roundup® and Panzer® Gold in EE_1.[a]

Experimental groups (EE_1)	FMN	*Buds*	Notched	Binuclei	Eccentric	TNA
NC	2.50 ± 0.62	89.35 ± 10.38	58.00 ± 7.45	1.21 ± 0.38	23.93 ± 8.11	172.50 ± 13.80
PC	5.80 ± 1.10^b	127.00 ± 11.64	56.20 ± 7.00	1.13 ± 0.31	32.00 ± 8.69	216.33 ± 16.12^b
Roundup® 500	4.20 ± 0.98	121.67 ± 10.77	53.06 ± 6.16	0.40 ± 0.13	28.47 ± 7.92	203.60 ± 11.90
Roundup® 750	3.44 ± 0.65	117.28 ± 9.22	58.33 ± 5.78	0.78 ± 0.27	26.11 ± 6.87	202.50 ± 12.31
Roundup® 1000	3.13 ± 0.68	97.60 ± 10.23	51.7 ± 5.46	0.73 ± 0.25	24.07 ± 7.92	173.67 ± 13.27
Panzer® 500	3.75 ± 1.03	138.50 ± 19.82	42.75 ± 9.31	0.75 ± 0.75	6.50 ± 2.63	188.50 ± 25.86
Panzer® 750	4.75 ± 1.11	162.25 ± 11.91	54.00 ± 9.86	0.50 ± 0.29	8.25 ± 2.29	225.00 ± 18.77
Panzer® 1000	5.80 ± 1.74^b	157.60 ± 20.86	43.80 ± 10.49	0.60 ± 0.60	11.00 ± 2.28	213.00 ± 29.76

[a]**EE_1: Embryo experiment 1.** All values are expressed as mean ± SEM. NC: negative control; VC: vehicle control; PC: positive control (cyclophosphamide 700 μg egg^{-1}); Roundup® 500, 750, 1000 (μg egg^{-1}): commercial formulation of glyphosate 66.2% (Roundup® Full II); Panzer® 500, 750, 1000 (μg egg^{-1}): commercial formulation of glyphosate 60.2% (Panzer® Gold).
[b]Statistically significant differences with respect to the NC: $p < 0.05$.

Table 20.2 Micronuclei and nuclear abnormality frequencies observed in *C. latirostris* hatchlings exposed *in ovo* to different concentrations of cypermethrin- and endosulfan-based formulations in EE_2.[a]

Experimental group (EE_2)	FMN	*Buds*	NN	BiN	EN	TNA
NC	2.90 ± 0.65	90.50 ± 11.45	61.20 ± 7.34	2.11 ± 0.45	25.03 ± 8.34	183.10 ± 14.50
VC	2.88 ± 0.84	96.23 ± 9.38	58.94 ± 7.11	0.70 ± 0.25	24.88 ± 7.27	180.76 ± 11.70
PC	6.33 ± 1.60^b	130.00 ± 12.24	51.20 ± 6.07	1.63 ± 0.55	36.08 ± 10.33	224.23 ± 23.02^b
Galgofan® 1	3.67 ± 0.80	133.67 ± 9.87	32.87 ± 3.56^c	1.07 ± 0.37	81.40 ± 13.41^c	249.00 ± 22.50^c
Galgofan® 10	3.81 ± 0.45^b	134.25 ± 11.08	33.69 ± 3.76^c	1.00 ± 0.30	64.87 ± 6.64^c	233.81 ± 14.52^c
Galgofan® 100	3.93 ± 0.76	142.27 ± 14.71^b	49.93 ± 6.54	0.93 ± 0.20	49.00 ± 10.25^b	242.13 ± 20.93^c
Galgofan® 1000	3.46 ± 0.52	140.53 ± 10.30^b	38.47 ± 3.28^b	0.87 ± 0.24	70.60 ± 8.10^c	250.47 ± 16.17^c
Atanor® 1	4.29 ± 0.62^b	107.71 ± 18.81	43.71 ± 6.75	0.93 ± 0.25	43.14 ± 11.90	195.50 ± 25.71
Atanor® 10	3.89 ± 0.62	142.16 ± 12.91^b	35.32 ± 4.69^b	0.68 ± 0.20	66.42 ± 7.38^c	244.58 ± 18.40^c
Atanor® 100	7.60 ± 0.62^b	138.13 ± 8.26	32.67 ± 4.36^b	0.87 ± 0.26	68.73 ± 6.87^c	240.40 ± 12.89^c
Atanor® 1000	3.06 ± 0.62	136.62 ± 12.18	36.50 ± 3.95^b	0.69 ± 0.18	67.75 ± 7.05^c	241.56 ± 15.08^c

[a]**EE_2: Embryo experiment 2.** All values are expressed as mean ± SEM. NC: negative control; VC: vehicle control; PC: positive control (cyclophosphamide 700 μg egg^{-1}); Galgofan® 1, 10, 100, 1000 (μg egg^{-1}): commercial formulation of endosulfan 35% (Galgofan®); Atanor® 1, 10, 100 and 1000 (μg egg^{-1}): commercial formulation of cypermethrin 25% (Atanor®).
[b]Statistically significant differences with respect to the NC: $p < 0.05$.
[c]Statistically significant differences with respect to the NC: $p < 0.01$.

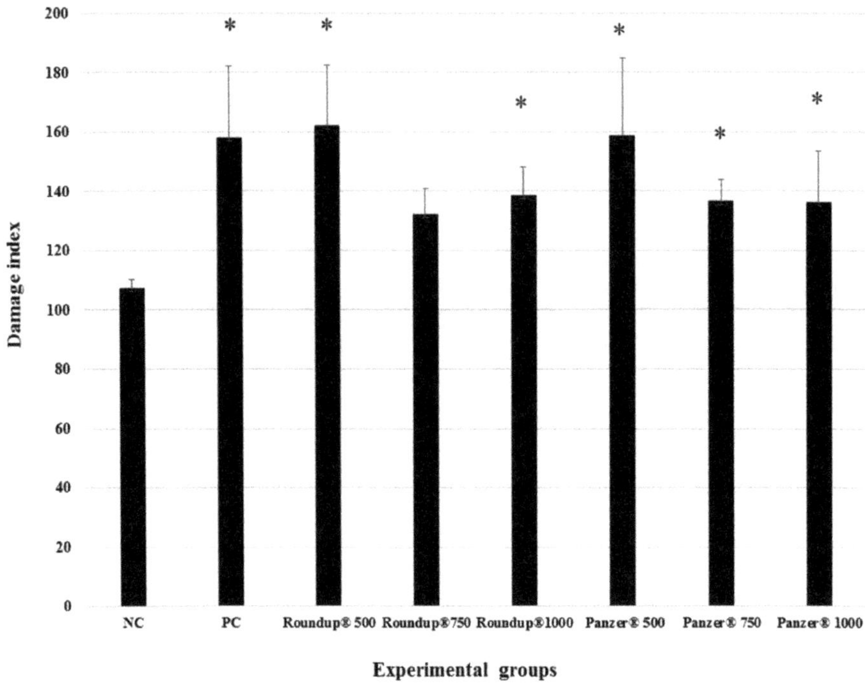

Figure 20.3 DNA damage observed in *C. latirostris* hatchlings exposed *in ovo* to different concentrations (500, 750, 1000 μg egg^{-1}) of glyphosate-based formulations Roundup® and Panzer® Gold in EE_1. *Statistically significant differences with respect to the NC.

In EE_2, NC and VC showed no differences in the FMN, FNA, or DI, indicating that ethanol caused no genotoxic damage ($p > 0.05$). There was an induction of MN caused by exposure to cyclophosphamide, Panzer® Gold, Galgofan® and Atanor® at different concentrations, compared to the NC (Table 20.2). Results obtained from the NAs showed significant differences with respect to the NC in *Buds*, EN, NN and TNA for Galgofan® and Atanor® at different concentrations, and for the PC in TNA (Table 20.2). CA exhibited an increase in the damage index with Galgofan® 1 and 1000, and Atanor® 100, compared to the NC ($p < 0.05$; Figure 20.4).

No significant differences were observed in growth between any of the treatments and the NC, in any of the experiments (EE_1 and EE_2; Table 20.3).

Regression analysis demonstrated a concentration-dependent effect only for Panzer® Gold on the FMN ($R^2 = 0.981$; $p < 0.01$) (Figure 20.5).

No relationship was found between MN, NAs or DI with caiman weight or length in EE_1 or EE_2 ($p > 0.05$ in all analyses performed).

Considering oxidative stress parameters, we observed a statistically significant increase in TBARS for GLY-based formulations Roundup® 500 and 1000, and Panzer® Gold 500 and 1000 (Figure 20.6); as well as for Galgofan® 10, 100 and 1000, and Atanor® 1, 10 and 1000 and in EE_2 (Figure 20.7), with

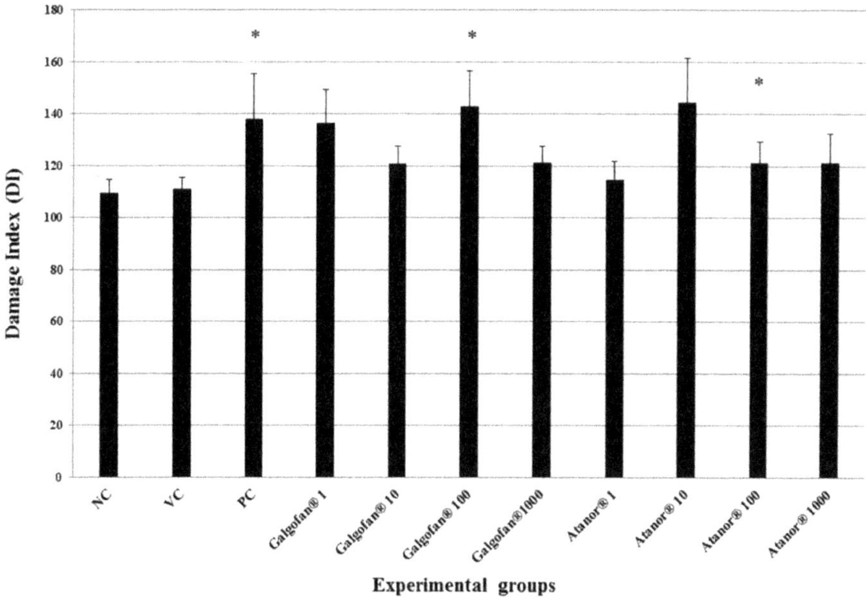

Figure 20.4 DNA damage observed in *C. latirostris* hatchlings exposed *in ovo* to different concentrations (1, 10, 100, 1000 μg egg^{-1}) of cypermetrhin formulation Atanor® and endosulfan formulation Galgofan® in EE_2. *Statistically significant differences with respect to the NC.

respect to the corresponding NC ($p < 0.05$ for all analyses performed). Antioxidant defenses showed an alteration in SOD activity for Galgofan® 1, and Atanor® 10 and 1000, compared with the NC ($p < 0.05$; Figure 20.8). CAT activity showed no statistically significant differences for any treatments ($p > 0.05$).

Figure 20.9 depicts the effect of Roundup® on SOD and CAT activities together, where an increase can be seen in both up to 750 μg egg^{-1}, followed by a decrease at higher concentrations. There was no relationship between OS parameters and the size of animals at birth ($p > 0.05$ in all cases).

20.3.2.3 Evaluation of Developmental Stage-dependent Genotoxic Effect of Roundup® Formulation on Embryos

Different studies have demonstrated the effects of a variety of pesticides during embryo development and the possible stage-dependent effects,[57–59] but there is no information concerning this for any pesticide in *C. latirotris.*

In previous works, we reported the genotoxic effects of Roundup® through the MN test and CA on *C. latirostris* hatchlings after *in ovo* exposure by topical application at the beginning of the incubation period.[14] Likewise, genetic, enzymatic and developmental alterations were observed in hatchlings after spraying of the nests with Roundup® alone, and in a mixture with endosulfan and cypermethrin formulations in a semi-natural experiment.[51]

Table 20.3 Growth parameters of *C. latirostris* hatchlings exposed *in ovo* to glyphosate-, endosulfan- and cypermethrin-based formulations in EE_1 and EE_2.[a]

Experimental group	TL	SVL	Weight
Embryo experiment 1 (EE_1)			
NC	23.42 ± 0.18	11.20 ± 0.07	44.19 ± 1.39
PC	22.57 ± 0.05	10.05 ± 0.05	30.62 ± 0.47
Panzer® 500	22.55 ± 0.55	10.87 ± 0.22	41.32 ± 1.21
Panzer® 750	23.55 ± 0.54	11.15 ± 0.26	44.42 ± 0.98
Panzer® 1000	23.52 ± 0.29	11.44 ± 0.07	46.58 ± 0.96
Roundup® 500	23.44 ± 0.26	11.23 ± 0.11	45.45 ± 0.62
Roundup® 750	23.14 ± 0.13	11.10 ± 0.07	44.90 ± 0.81
Roundup® 1000	23 ± 0.17	11.04 ± 0.08	43.85 ± 0.91
Embryo experiment 2 (EE_2)			
NC	22.17 ± 0.27	10.82 ± 0.12	38 ± 0.45
VC	23.02 ± 0.24	10.81 ± 0.12	44.45 ± 1.02
PC	23.04 ± 0.45	11.16 ± 0.11	43.01 ± 0.95
Galgofan® 1	23.46 ± 0.30	11.09 ± 0.19	43.92 ± 1.16
Galgofan® 10	23.49 ± 0.24	11.06 ± 0.11	44.76 ± 1.32
Galgofan® 100	22.42 ± 1.16	12.21 ± 1.03	44.77 ± 0.79
Galgofan® 1000	23.39 ± 0.14	11.32 ± 0.72	44.47 ± 0.50
Atanor® 1	23.45 ± 0.18	11.16 ± 0.18	44.86 ± 1.40
Atanor® 10	23.39 ± 0.16	11.09 ± 0.10	43.95 ± 1.01
Atanor® 100	23.42 ± 0.19	11.04 ± 0.09	44.30 ± 0.47
Atanor® 1000	23.17 ± 0.23	11.10 ± 0.10	44.52 ± 0.93

[a]All values are expressed as mean ± SEM. NC: negative control; VC: vehicle control; PC: positive control (cyclophosphamide 700 μg egg^{-1}); Roundup® 500, 750, 1000 (μg egg^{-1}): commercial formulation of glyphosate 66.2% (Roundup® Full II); Panzer® 500, 750, 1000 (μg egg^{-1}): commercial formulation of glyphosate 60.2% (Panzer® Gold); Galgofan® 1, 10, 100, 1000 (μg egg^{-1}): commercial formulation of endosulfan 35% (Galgofan®); Atanor® 1, 10, 100 and 1000 (μg egg^{-1}): commercial formulation of cypermethrin 25% (Atanor®).

Considering these results, we aimed to evaluate the stage-dependent genotoxic effect of the glyphosate-based formulation Roundup® through the CA, MN test and NA test in *C. latirostris* embryos exposed at three different stages of development (beginning, middle and end).

20.3.2.3.1 Experimental Design and Treatments. One hundred and thirty eggs (average weight 70 g) from five nests of the Proyecto Yacare Ranching Program, Santa Fe, Argentina, were used. They were equally and randomly distributed in 13 experimental groups of 10 eggs each: three for each effective concentration of Roundup® (750, 1250, 1750 μg egg^{-1}) and cyclophosphamide (700 μg egg^{-1}) as a PC, both determined in a previous study,[14] and a NC without exposure, treated with distilled water (Table 20.4).

To evaluate the sensitivity of embryos at different developmental stages, the total incubation period of approximately 70 ± 3 days (at 31 ± 1 °C) was divided into three stages of approximately 23 days each (first, second and third stages) and each concentration of Roundup® and cyclophosphamide was applied in triplicate at the beginning of these stages. Each group

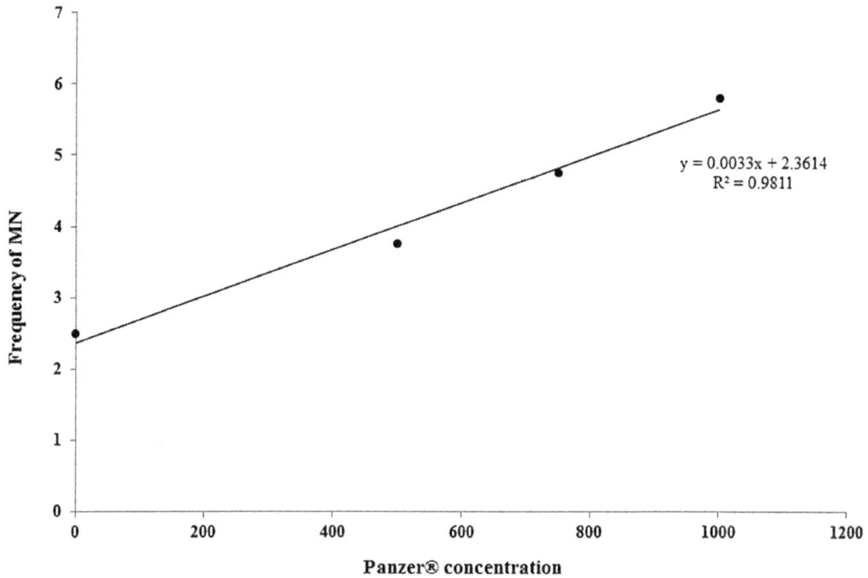

Figure 20.5 Concentration-dependent effect of Panzer® Gold (500, 750, 1000 μg egg^{-1}) on the FMN in EE_1 ($R^2 = 0.981$; $p < 0.01$).

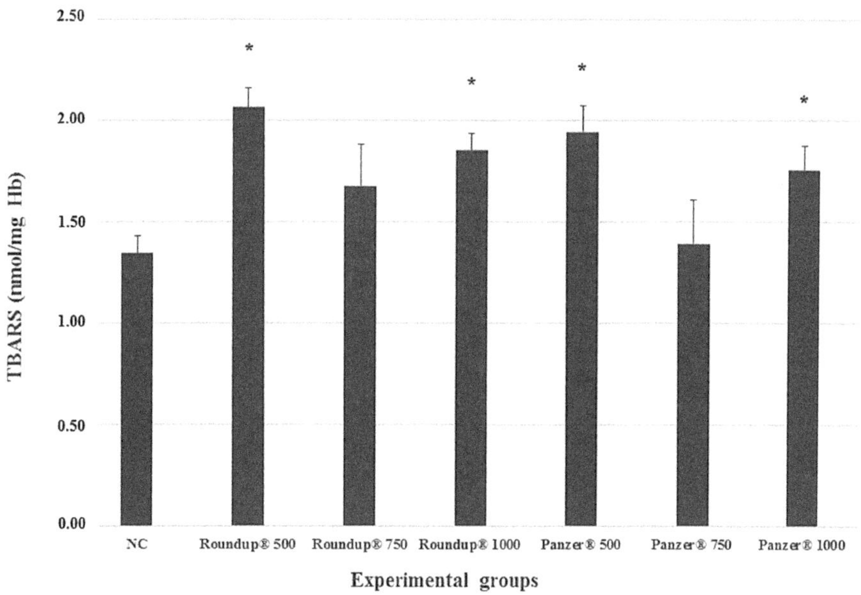

Figure 20.6 TBARS concentration (nmol mg Hb^{-1}) in erythrocytes of *C. latirostris* hatchlings exposed *in ovo* to different concentrations (500, 750, 1000 μg egg^{-1}) of glyphosate-based formulations Roundup® and Panzer® Gold in EE_1. *Statistically significant with respect to the NC (Dunnett Test, $p < 0.05$).

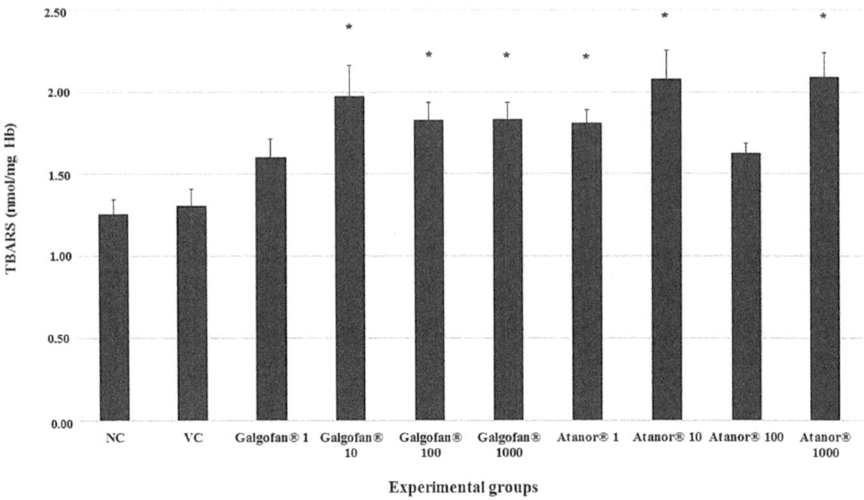

Figure 20.7 TBARS concentration (nmol mg Hb^{-1}) in erythrocytes of *C. latirostris* hatchlings exposed *in ovo* to different concentrations (1, 10, 100, 1000 μg egg^{-1}) of cypermethrin formulation Atanor® (CYP) and endosulfan formulation Galgofan® (END) in EE_2. *Statistically significant with respect to the NC and VC (Dunnett Test, $p<0.05$).

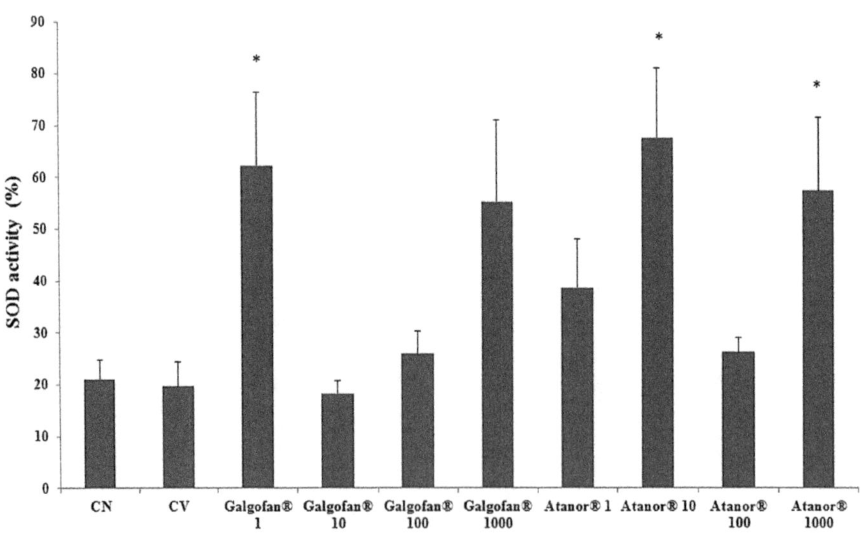

Figure 20.8 SOD activity (%) in erythrocytes of *C. latirostris* hatchlings exposed *in ovo* to different concentrations (1, 10, 100, 1000 μg egg^{-1}) of cypermethrin formulation Atanor® and endosulfan formulation Galgofan® in EE_2. *Statistically significant with respect to the NC and VC (Mann Whitey Test, $p<0.05$).

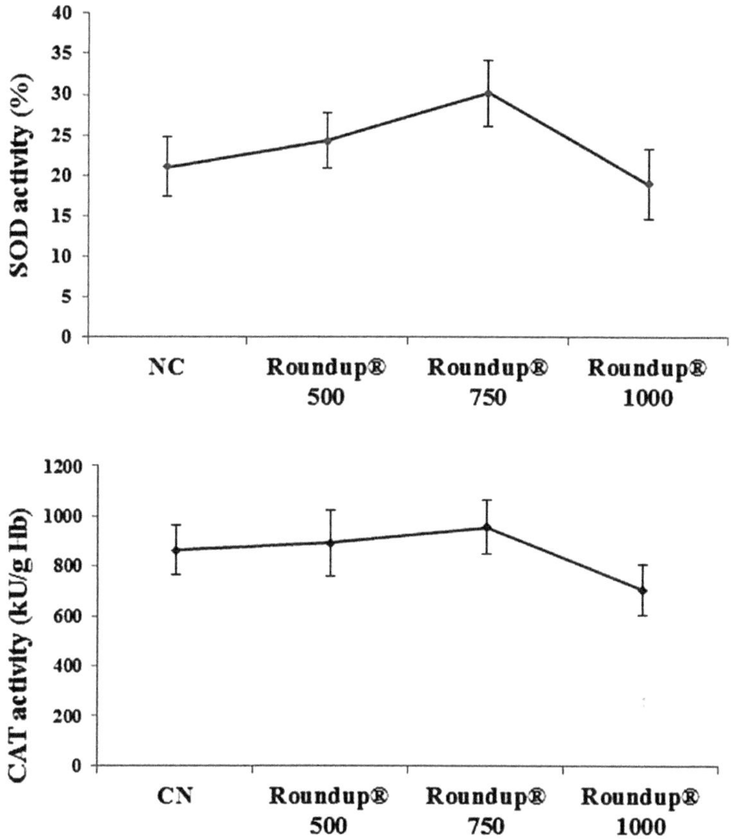

Figure 20.9 Comparison of SOD (above, %) and CAT (below, kU g Hb^{-1}) activities in erythrocytes of *C. latirostris* hatchlings exposed *in ovo* to different concentrations of Roundup® (500, 750, 1000 μg egg^{-1}). An increase can be seen in the activity of both enzymes and then a decrease at the higher concentration.

received the corresponding Roundup® or cyclophosphamide concentration by topical application on the eggshell, at a final volume of 50 μl, diluted in distilled water (Table 20.4).[14] Eggs from each single replicate of each treatment were placed separately in a plastic container (10 eggs), using vermiculite as the substrate and vegetal material to cover them, as described before. They were incubated in the Proyecto Yacaré artificial incubator under the conditions described before for the other experiments and viability was periodically controlled, as previously described.

Immediately after hatching, blood samples were obtained (0.5 ml) from the spinal vein, as described before, and genotoxicity (MN test, NA test and CA) and developmental parameters were analyzed.

20.3.2.3.2 Results. There were differences in DI between groups exposed to cyclophosphamide and to the different Roundup®

Table 20.4 Experimental groups and treatments applied at three different stages (S1, S2 and S3) of *C. latirostris* development.[a]

Experimental group	Treatment	Concentration ($\mu g\,egg^{-1}$)	Stage of exposure	No. of eggs exposed per clutch	No. of eggs exposed per stage	Total no. of eggs exposed per concentration
NC	Vehicle (distilled water)	0	Beginning	2	—	10
PC	Cyclophosphamide	700				
PC_{S1}			Beginning	2	10	
PC_{S2}			Middle	2	10	30
PC_{S3}			End	2	10	
T1	Roundup®	750				
$T1_{S1}$			Beginning	2	10	
$T1_{S2}$			Middle	2	10	30
$T1_{S3}$			End	2	10	
T2	Roundup®	1250				
$T2_{S1}$			Beginning	2	10	
$T2_{S2}$			Middle	2	10	30
$T2_{S3}$			End	2	10	
T3	Roundup®	1750				
$T3_{S1}$			Beginning	2	10	30
$T3_{S2}$			Middle	2	10	
$T3_{S3}$			End	2	10	

[a]NC: negative control; PC: positive control (700 $\mu g\,egg^{-1}$ cyclophosphamide); T1, T2 and T3: Treatments 1, 2 and 3 commercial formulation of glyphosate (Roundup®); S_1, S_2 and S_3: Moment of exposure in relation to first, second and third stage of development.

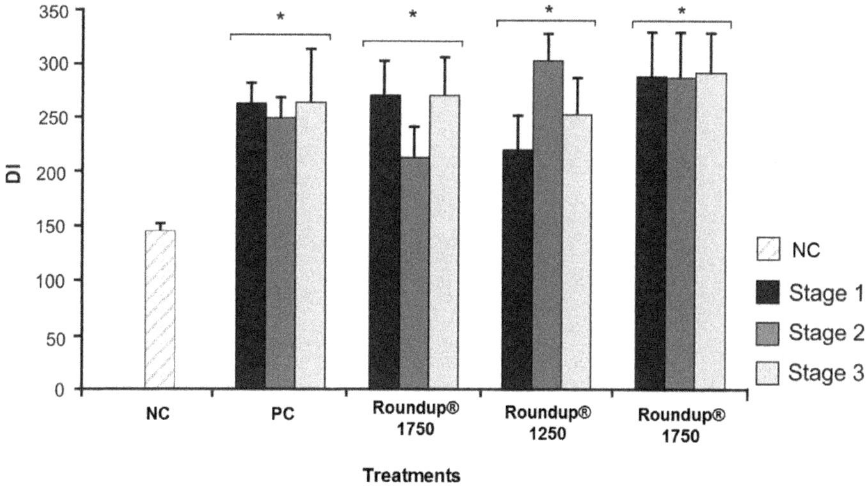

Figure 20.10 Damage index of the Comet assay for each experimental group (mean ± SEM) at three different stages of exposure to Roundup® (500, 1250, 1750 μg egg^{-1}) or the positive control (PC) cyclophosphamide (700 μg egg^{-1}). *Statistically significant compared to the NC (factorial ANOVA, $p < 0.05$).

concentrations with the NC ($p < 0.05$), but no differences were found among the other groups ($p > 0.05$). Similar damage was observed at the three stages of each treatment ($p > 0.05$), indicating that both Roundup® and cyclophosphamide produce DNA damage on *C. latirostris* embryos independently of the stage of development where the exposure occurs (Figure 20.10).

There was no differences in the frequency of MN or in TNA (Table 20.5) among the different treatment groups nor among the different stages within each treatment ($p > 0.05$).

No interaction was observed between treatment and stage for any of the biomarkers analyzed ($p > 0.05$). We found no differences in the size of animals among the different experimental groups at the moment of hatching ($p > 0.05$).

20.3.3 Effects of Pesticide Formulations and Mixtures on Caimans Exposed During Their First Months of Life

In natural environments in Argentina, *C. latirostris* can be exposed to pesticides as both embryos and hatchlings because the reproductive season of the species coincides with the moment of maximal pesticide application. During their first months of life, caimans spend most of their time in small water bodies. At this time of the year (February–May), short but heavy rainfalls occur frequently, which can cause intensive pesticide runoff to non-target compartments, such as aquatic ecosystems. Considering this, the aim

Table 20.5 Micronuclei and nuclear abnormality frequencies observed in *C. latirostris* hatchlings exposed *in ovo* to the glyphosate-based formulation Roundup® at different stages of development.[a]

Experimental groups	Biomarker	NC	PC	T1	T2	T3
Stage 1	FMN	0.86 ± 0.4	1 ± 0.58	0.75 ± 0.25	2 ± 1.24	0.67 ± 0.49
	FTNA	90 ± 20.79	95.5 ± 6.01	102.5 ± 8.92	106.17 ± 24.2	78 ± 5.27
Stage 2	FMN	—	0.83 ± 0.83	1.33 ± 0.44	1.43 ± 0.57	2.25 ± 1.65
	FTNA	—	115.17 ± 20.84	103.11 ± 16.26	91.29 ± 13.75	110.5 ± 22.87
Stage 3	FMN	—	1.33 ± 0.8	1.25 ± 0.62	1.22 ± 0.75	1.4 ± 0.98
	FTNA	—	115.17 ± 19.08	115.63 ± 24.28	122.8 ± 24.81	112.2 ± 28.56

[a] All values are expressed as mean ± SEM. NC: negative control; PC: positive control (700 μg egg^{-1} cyclophosphamide); T1: Treatment 1, Roundup® 750 μg egg^{-1}, T2: Treatment 1, Roundup® 1250 μg egg^{-1}; T3: Treatment 3, Roundup® 1750 μg egg^{-1}; FMN: number of cells with micronucleus found per 1000 erythrocytes observed. FTNA: sum of total nuclear abnormalities in 1000 erythrocytes observed, excluding MN.

of this study was to evaluate the genotoxic and oxidative effects on hatchlings under conditions similar to those occurring in small water bodies in the natural environments.

20.3.3.1 C. latirostris *Hatchlings*

We used 20 day old *C. latirostris* specimens, hatched from eggs harvested in different nests in the Natural Managed Reserve El Fisco, as previously described.

20.3.3.2 Experimental Design and Treatments

The study was carried out at the Proyecto Yacaré Laboratorio de Zoología Aplicada: Anexo Vertebrados (FHUC-UNL/MMA, Santa Fe) facilities and included two different experiments:

Hatchling Experiment 1 (*HE_1*)
One hundred and thirty two specimens, hatched from eggs coming from six different clutches, were randomly distributed into 11 experimental groups (12 specimens per experimental groups with two replicates): (1) a negative control (NC) treated with distilled water; (2) a vehicle control (VC) treated with ethanol; 2–3) two groups exposed to different concentrations of CYP Atanor® (Atanor® 1 and 2); 4–5) two groups exposed to different concentrations of END Galgofan® (Galgofan® 1 and 2); 6–7) two groups exposed to different concentrations of Roundup® (Roundup® 1 and 2); 8–9) two groups exposed to different concentrations of Panzer® Gold (Panzer® Gold 1 and 2); and 10) a group treated with the complex mixture (M_1): CYP Atanor® 1 + END Galgofan® 1 + GLY Roundup® 1 (Table 20.6).

Experiment 2 (*HE_2*)
Sixty-four specimens hatched from eggs coming from four different clutches were randomly distributed into four experimental groups of 16 specimens (two replicates of eight each): (1) A vehicle control (VC) treated with ethanol; 2–3) two groups exposed to different concentrations of CPF Lorsban® 48E; and 4) one group treated with the complex mixture (M_2): CYP Atanor® 1 + CPF Lorsban® 1 + GLY Roundup® 1 (Table 20.6). As in *HE_1* we tested both the NC and VC, but we avoided using a NC in *HE_2* in order to reduce the number of animals under experimentation.

Concentrations of GLY-based formulations were determined in relation to those recommended for their application in soybean crops (*i.e.*, 2% ha^{-1}) as described in previous works.[60,61] In the case of insecticide formulations (CYP, END and CPF), the amount recommended for their application in crops could not be used because when prepared to be applied in the bioassay containers, they gave concentrations higher than the toxic reference doses, so we decided to apply an equal and lower range for all three of them

Table 20.6 Experimental groups and treatments applied in hatchling experiments 1 (HE_1) and 2 (HE_2).[a]

Experimental groups	Compound	Initial concentration	Final concentration	No. of specimens exposed per nest	*N*
Hatchling experiment 1 (HE_1)					
NC	Distilled water	—	—	2	12
VC	Ethanol	200 µl L^{-1}	200 µl L^{-1}	2	12
Atanor® 1	CYP formulation	0.5 µg L^{-1}	0.05 µg L^{-1}	2	12
Atanor® 2	CYP formulation	1 µg L^{-1}	0.1 µg L^{-1}	2	12
Galgofan® 1	END formulation	0.5 µg L^{-1}	0.05 µg L^{-1}	2	12
Galgofan® 2	END formulation	1 µg L^{-1}	0.1 µg L^{-1}	2	12
Roundup® 1	GLY formulation Roundup®	2.5 mg L^{-1}	0.25 mg L^{-1}	2	12
Roundup® 2	GLY formulation Roundup®	5 mg L^{-1}	0.5 mg L^{-1}	2	12
Panzer® 1	GLY formulation Panzer® Gold	2.5 mg L^{-1}	0.25 mg L^{-1}	2	12
Panzer® 2	GLY formulation Panzer® Gold	5 mg L^{-1}	0.5 mg L^{-1}	2	12
M_1	CYP Atanor® + END Galgofan® + GLY Roundup®	0.5 µg L^{-1} + 0.5 µg L^{-1} + 2.5 mg L^{-1}	0.05 µg L^{-1} + 0.05 µg L^{-1} + 0.25 mg L^{-1}	2	12
Hatchling experiment 2 (HE_2)					
VC	Ethanol	200 µl L^{-1}	200 µl L^{-1}	2	16
Lorsban® 1	CPF formulation	0.5 µg L^{-1}	0.05 µg L^{-1}	2	16
Lorsban® 2	CPF formulation	1 µg L^{-1}	0.1 µg L^{-1}	2	16
M_2	CYP Atanor® +CPF Lorsban® +GLY Roundup®	0.5 µg L^{-1} + 0.5 µg L^{-1} + 2.5 mg L^{-1}	0.05 µg L^{-1} + 0.05 µg L^{-1} + 0.25 mg L^{-1}	2	16

[a]NC: negative control; VC: vehicle control; GLY: glyphosate; END: endosulfan; CYP: cypermethrin; CPF: chlorpyrifos; M_1 and M_2: complex pesticide mixtures.

(0.5–1 $\mu g L^{-1}$) based on the information available in other studies in fish.[62–65] These concentrations were progressively reduced in time during the experiment in order to simulate the degradation of the compounds in water (without additional entry of new material into the ecosystem (Table 20.6). The concentration at each point was based on compound degradation data previously determined by gas chromatography in the case of the insecticides, and by HPLC in the case of glyphosate, as thoroughly described in Poletta *et al.*[51] and Siroski *et al.*[66]

Sub-chronic exposure (60 days) was performed by immersion in plastic containers (75 cm long, 35 cm wide and 37 cm high, base surface = 0.2625 m^2), tilted to provide 60% dry and 40% water surface areas, with a maximum water depth of approximately 15 cm (fixed volume of 4 L). The temperature in the containers was maintained at 30 ± 2 °C and this was monitored with a Hobbo data logger (Onset Computer Corp., Pocasset, MA, USA). Food was supplied *ad libitum* three times a week, consisting of a mixture of 50% minced chicken head and 50% dry pellets for reptiles.

At the end of the experiment, blood samples (0.5 ml) were taken from the spinal vein of all animals with heparinized syringes and 25G×5/8″ needles. Samples were not taken at the beginning of the experiment to avoid any risk of death to caimans owing to their initial small size.

Animals were measured in snout–vent length (SVL) and total length (TL), and weighed at the beginning and at the end of the experiment to determine growth in each EG.

20.3.3.3 Results

The results of the genotoxicity tests are presented as mean±SE of FMN, FNAs, and DI of the CA per experimental group. In HE_1, data analysis showed that there were no significant differences in the DI, FMN or FNAs between the NC and VC, indicating that ethanol caused no genotoxic effect ($p > 0.05$).

There was a significant increase in the FMN in all exposed groups except for Galgofan® 2 compared to the NC ($p < 0.05$), while in HE_2 we observed similar results after exposure to both concentrations of Lorsban® ($p < 0.05$) and to M_2 ($p < 0.01$), with respect to the VC (Figure 20.11).

When we analyzed the FANs, the results in HE_1 showed significant differences in BiN for Galgofan® 2 ($p < 0.05$) and M_1 ($p < 0.01$); in EN for Galgofan® and Atanor® at both concentrations ($p < 0.01$); in TNA and *Buds* for the lower concentrations of Panzer® Gold ($p < 0.05$), while a tendency but no statistically significant increment was observed for Roundup® 1 ($p = 0.058$), compared to the NC. In HE_2, we observed a significant higher frequency of *Buds* in M_2 and lower EN for both concentrations of Lorsban® ($p < 0.05$) respect to the VC (Table 20.7).

In relation to the CA, the cell viability of all samples analyzed was in the range of 95–100%, indicating appropriate conditions for the application of the assay. In HE_1, the results demonstrated that pesticides induced a significant increase in DNA damage of caimans exposed to all pesticide

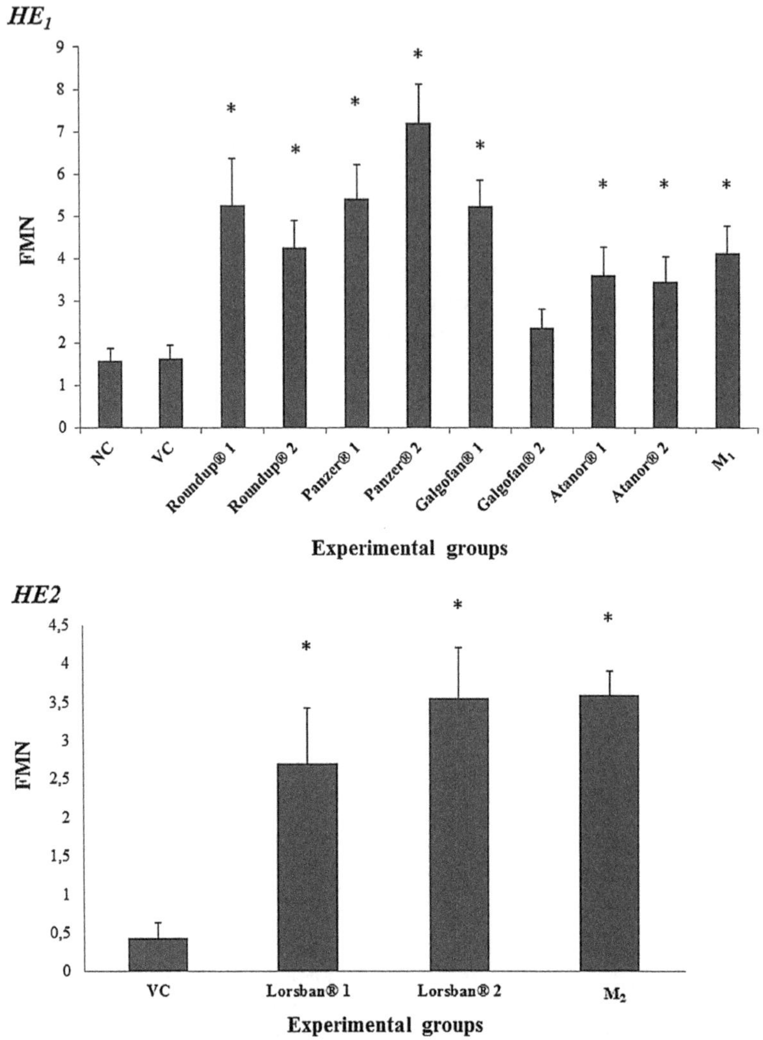

Figure 20.11 Micronucleus frequencies (FMN) in 1000 erythrocytes of *C. latirostris* observed in the different experimental groups. **Hatchling Experiment 1 (HE_1)**: NC: negative control; VC: vehicle control; Roundup® 1 and 2: commercial formulations of glyphosate 66.2% (Roundup® Full II, 2.5 and 5 mg L^{-1}, respectively); Panzer® 1 and 2: commercial formulations of glyphosate 60.8% (Panzer® Gold, 2.5 and 5 mg L^{-1}, respectively); Galgofan® 1 and 2: commercial formulations of endosulfan (Galgofan®, 0.5 and 1 μg L^{-1}, respectively); Atanor® 1 and 2: commercial formulations of cypermethrin (Atanor®, 0.5 and 1 μg L^{-1}, respectively); M_1: complex pesticide mixture of 2.5 mg L^{-1} GLY Roundup® + 0.5 μg L^{-1} CYP Atanor® + 0.5 μg L^{-1} END Galgofan®. **Hatchling Experiment 2 (HE_2)**: VC: vehicle control; Lorsban® 1 and 2: commercial formulations of chlorpyrifos (Lorsban® 48E, 0.5 and 1 μg L^{-1}, respectively); M_2: complex pesticide mixture of 2.5 mg L^{-1} GLY Roundup® + 0.5 μg L^{-1} CYP Atanor® + 0.5 μg L^{-1} CPF Lorsban® 48E. *Statistically significant with respect to the NC for HE_1 and VC for HE_2.

Table 20.7 Nuclear abnormalities frequency observed in *C. latirostris* hatchlings in different experimental groups.[a]

Experimental group	*Buds*	NN	BiN	EN	TNA
Hatchling experiment 1 (HE_1)					
NC	134.05 ± 7.51	31.26 ± 3.32	0.63 ± 0.16	54.16 ± 5.14	220.1 ± 11.67
VC	108.95 ± 7.52	24.26 ± 1.8	0.21 ± 0.1	52.8 ± 4.73	243.74 ± 10
Roundup® 1	145.69 ± 9.52[b]	38.95 ± 3.4	0.58 ± 0.28	50.47 ± 2.95	235.68 ± 11.56
Roundup® 2	150.25 ± 9.21	30.05 ± 2.75	1.05 ± 0.4	66.9 ± 4.55	248.25 ± 12.56
Panzer® 1	175.47 ± 11.49[b]	34.07 ± 3.4	0.6 ± 0.21	73.33 ± 9.23	283.5 ± 20.24[b]
Panzer® 2	147.71 ± 7.27	28 ± 2.43	0.5 ± 0.17	65.21 ± 5.18	241.43 ± 9.94
Atanor® 1	129.95 ± 7.84	28 ± 3.62	0.22 ± 0.15	94.83 ± 4.69[b]	253 ± 11.68
Atanor® 2	127.77 ± 7.81	23.24 ± 2.2	0.12 ± 0.12[b]	94 ± 6.2[b]	245.12 ± 13.4
Galgofan® 1	146.41 ± 9.36	27.53 ± 2.52	0.41 ± 0.21	90.82 ± 7.06[b]	265.18 ± 15.34
Galgofan® 2	133.9 ± 8.09	24.95 ± 2.25	0.47 ± 0.23	93.32 ± 6.44[b]	252.63 ± 12.04
M1	106.21 ± 4.29	33.72 ± 2.83	2.79 ± 0.41[b]	75.14 ± 5.12	217.86 ± 8.19
Hatchling experiment 2 (HE_2)					
VC	108 ± 4.33	36 ± 3.39	0.14 ± 0.14	112 ± 10.6	256 ± 11.05
Lorsban® 1	134.2 ± 10.2	58.9 ± 9.04	1.30 ± 0.45	84.7 ± 6.68[b]	279.1 ± 14.35
Lorsban® 2	128.1 ± 7.04	49.1 ± 5.7	0.55 ± 0.18	80.2 ± 4.95[b]	258 ± 12.12
M2	144.3 ± 8.36[b]	42.7 ± 2.85	1.20 ± 0.36	88.6 ± 6.5	276.8 ± 12.26

[a]All values are expressed as mean ± SEM; **Hatchling Experiment 1 (*HE_1*)**: NC: negative control; VC: vehicle control; Roundup® 1 and 2: commercial formulations of glyphosate 66.2% (Roundup® Full II, 2.5 and 5 mg L^{-1}, respectively); Panzer® 1 and 2: commercial formulations of glyphosate 60.8% (Panzer® Gold, 2.5 and 5 mg L^{-1}, respectively); Galgofan® 1 and 2: commercial formulations of endosulfan 35% (Galgofan®, 0.5 and 1 μg L^{-1}, respectively); Atanor® 1 and 2: commercial formulations of cypermethrin 25% (Atanor®, 0.5 and 1 μg L^{-1}, respectively); M_1: complex pesticide mixture of 2.5 mg L^{-1} GLY Roundup® + 0.5 μg L^{-1} CYP Atanor® + 0.5 μg L^{-1} END Galgofan®. **Hatchling Experiment 2 (*HE_2*)**: VC: vehicle control; Lorsban® 1 and 2: commercial formulations of chlorpyrifos 48% (Lorsban® 48E, 0.5 and 1 μg L^{-1}, respectively); M_2: complex pesticide mixture of 2.5 mg L^{-1} GLY Roundup® + 0.5 μg L^{-1} CYP Atanor® + 0.5 μg L^{-1} CPF Lorsban® 48E. NN: notched nuclei; BiN: binuclei; EN: eccentric nuclei; TNA: sum of total nuclear abnormalities per 1000 erythrocytes counted, excluding MN.
[b]<0.05 compared to NC (HE_1) or VC (HE_2).

formulations except Panzer® Gold, at both concentrations analyzed, compared to the NC (Figure 20.12). No differences were found with M_1 either (Atanor®–Roundup®–Galgofan®). Besides, in *HE_2*, Lorsban® at the higher concentration and the M_2 (Atanor®–Roundup®–Lorsban®) also induced a significantly higher DI with respect to the VC (Figure 20.12).

Considering oxidative stress parameters, in *HE_1* we found a significant increment in lipoperoxidation caused by all formulations except Panzer® Gold, a significantly higher activity of CAT caused by the higher concentration of Atanor® ($p = 0.014$) and higher activity of SOD induced by CYP and END formulations as well as by the M_1 ($p = 0.004$). In *HE_2* we observed only a significant lipoperoxidation with M_2 ($p = 0.032$) and an induction of SOD activity with the higher concentration of Lorsban® ($p = 0.022$) and M_2 ($p = 0.001$; Table 20.8).

Finally, no effects were observed on the length or weight of the caimans exposed to any pesticide formulation or complex mixtures tested in any of

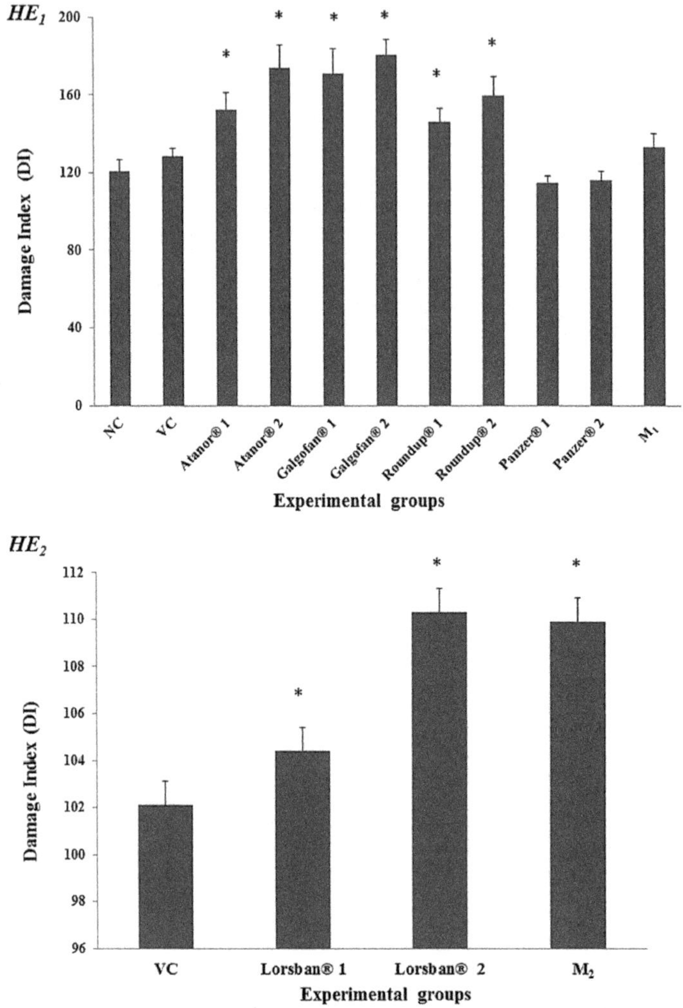

Figure 20.12 Damage index (DI) in *C. latirostris* exposed to different pesticide formulations and mixtures. **Hatchling Experiment 1 (*HE_1*)**: NC: negative control; VC: vehicle control; Roundup® 1 and 2: commercial formulations of glyphosate 66.2% (Roundup® Full II, 2.5 and 5 mg L^{-1}, respectively); Panzer® 1 and 2: commercial formulations of glyphosate 60.8% (Panzer® Gold, 2.5 and 5 mg L^{-1}, respectively); Galgofan® 1 and 2: commercial formulations of endosulfan (Galgofan®, 0.5 and 1 µg L^{-1}, respectively); Atanor® 1 and 2: commercial formulations of cypermethrin (Atanor®, 0.5 and 1 µg L^{-1}, respectively); M_1: complex pesticide mixture of 2.5 mg L^{-1} GLY Roundup® + 0.5 µg L^{-1} CYP Atanor® + 0.5 µg L^{-1} END Galgofan®. **Hatchling Experiment 2 (*HE_2*)**: VC: vehicle control; Lorsban® 1 and 2 commercial formulations of chlorpyrifos (Lorsban® 48E, 0.5 and 1 µg L^{-1}, respectively); M_2: complex pesticide mixture of 2.5 mg L^{-1} GLY Roundup® + 0.5 µg L^{-1} CYP Atanor® + 0.5 µg L^{-1} CPF Lorsban® 48E. *Statistically significant with respect to the NC for *HE_1* and VC for *HE_2*.

Table 20.8 Lipoperoxidation (TBARS) and antioxidant enzyme activities (Catalase and Superoxide dismutase) in *C. latirostris* hatchlings in the different experimental groups.[a]

Experimental group	TBARS (nmol mg prot^{-1})	CAT (KU mg prot^{-1})	SOD (% activity)
Hatchling experiment 1 (HE_1)			
NC	3.15 ± 0.2	216.77 ± 7.5	56.04 ± 2.99
VC	4.33 ± 0.46	219. ± 17.85	59.16 ± 2.10
Roundup® 1	5.78 ± 1.98[b]	229.04 ± 60.79	67.58 ± 4.5
Roundup® 2	3.27 ± 0.21	221.78 ± 31.43	62.99 ± 4.60
Panzer® 1	5.18 ± 1.47	197.31 ± 45.38	68.57 ± 3.72
Panzer® 2	4.74 ± 1.17	228.29 ± 39.33	63.89 ± 2.99
Atanor® 1	5.46 ± 1.29[b]	255.31 ± 16.72	70.35 ± 3.96[b]
Atanor® 2	5.15 ± 1.23[b]	262.83 ± 13.78[b]	73.73 ± 5.71[c]
Galgofan® 1	3.87 ± 0.16	192.04 ± 46.58	70.59 ± 3.75[b]
Galgofan® 2	5.00 ± 0.73[b]	211.33 ± 25.58	71.87 ± 1.58[b]
M1	5.59 ± 0.46	248.82 ± 23.02	90.64 ± 5.34[c]
Hatchling experiment 2 (HE_2)			
VC	7.97 ± 0.26	292.78 ± 17.64	54.79 ± 3.28
Lorbdan® 1	8.85 ± 0.52	348.53 ± 27.70	63.26 ± 2.23
Lorbdan® 2	8.21 ± 0.41	385.01 ± 38.71	68.38 ± 1.89[b]
M2	9.54 ± 0.45[b]	334.51 ± 28.19	84.66 ± 4.23[c]

[a]All values are expressed as mean ± SE. **Hatchling Experiment 1 (*HE_1*)**: NC: negative control; VC: vehicle control; Roundup® 1 and 2: commercial formulations of glyphosate 66.2% (Roundup® Full II, 2.5 and 5 mg L^{-1}, respectively); Panzer® 1 and 2: commercial formulations of glyphosate 60.8% (Panzer® Gold, 2.5 and 5 mg L^{-1}, respectively); Galgofan® 1 and 2: commercial formulations of endosulfan 35% (Galgofan®, 0.5 and 1 μg L^{-1}, respectively); Atanor® 1 and 2: commercial formulations of cypermethrin 25% (Atanor®, 0.5 and 1 μg L^{-1}, respectively); M_1: complex pesticide mixture of 2.5 mg L^{-1} GLY Roundup® + 0.5 μg L^{-1} CYP Atanor® + 0.5 μg L^{-1} END Galgofan®. **Hatchling Experiment 2 (*HE_2*)**: VC: vehicle control; Lorsban® 1 and 2: commercial formulations of chlorpyrifos 48% (Lorsban® 48E, 0.5 and 1 μg L^{-1}); M_2: complex pesticide mixture of 2.5 mg L^{-1} GLY Roundup® + 0.5 μg L^{-1} CYP Atanor® + 0.5 μg L^{-1} CPF Lorsban® 48E.

[b]Statistically significant differences with respect to the NC for HE_1 and VC for HE_2: $p < 0.05$.

[c]Statistically significant differences with respect to the NC for HE_1 and VC for HE_2: $p < 0.01$.

the experiments, compared with the NC or VC as appropriate ($p > 0.05$ in all analyses performed).

20.4 Final Considerations

Early life stages of oviparous organisms usually exhibit a greater toxicological sensitivity to chemical contaminants than adults, often because repair and detoxification systems are still immature. This can increase the susceptibility of some embryos and hatchlings to factors that affect their survival, such as the response to different types of infections or contaminant-induced stress.

Evaluation of pesticide effects through integral approaches, including complementary and non-destructive biomarkers of different endpoints, constitutes a more reliable tool for understanding real exposure scenarios. Genotoxicity and oxidative stress have been widely demonstrated to be sensitive groups of biomarkers, of rapid response and potentially related to each other.

Concerning mechanisms of nuclear abnormalities formation, MN can originate during the anaphase from lagging acentric chromosome or chromatid fragments caused by misrepair or unrepaired DNA breaks. Malsegregation of whole chromosomes in the anaphase may also lead to MN formation as a result of hypomethylation of repeated sequences in centromeric and pericentromeric DNA, defects in kinetochore proteins or assembly, dysfunctional spindles and defective anaphase checkpoint genes.[27,67] Enough information is known about the MN test applied in erythrocytes of different vertebrates exposed to genotoxic compounds. It has been widely used in biomonitoring studies to detect chromosome damage in different wild species, such as fish,[28,65] amphibians,[30,68,69] mammals,[70] and reptiles particularly by our group.[14,29,51,61,71] In all cases, this test demonstrated high sensitivity to detect the effects of genotoxic agents.

Several authors also recommended the analysis of other NAs, such as nuclear buds, eccentric nuclei, binucleated cells, notched nuclei and nucleoplasmic bridges, among others, as suitable biomarkers for genotoxicity screening, complementary to MN scoring. Although the mechanisms responsible for many of these nuclear abnormalities are not completely understood, they are considered as nuclear lesions analogous to MNs, which can be induced by genotoxic compounds even when MNs are not induced.[27,72] Some of them are interpreted as DNA amplifications, interruptions in the exocytosis mechanism, or stress factor exposure (*e.g.*, diet alterations, pathology, and metabolic damage), which could be primary manifestations preceding the formation of micronuclei.[22,27,67] Nuclear buds are proposed as the process of elimination of amplified DNA, DNA repair complexes and possibly excess chromosomes from aneuploid cells, while NPBs are said to originate from dicentric chromosomes, which may occur due to misrepair of DNA breaks, telomere end fusions, and also when defective separation of sister chromatids in the anaphase occurs due to failure of decatenation.[27,67]

Several studies have determined the induction of NAs by genotoxic agents in wild species.[22,73–75] Among Crocodilians, however, there were no previous data of NAs determination except for the MN test, so these were the first studies on the use of NAs as a biomarker of genotoxicity in *C. latirostris* and all other crocodilian species. NAs analyzed in broad-snouted caiman erythrocytes showed a significantly higher frequency in many of the treatments applied compared to the NC or VC, showing that they are good indicators of genotoxic damage.

On the other hand, the CA is the most widely used method for measuring DNA damage in eukaryotic cells. It detects strand breaks and alkali-labile sites at frequencies from a few hundred to several thousand breaks per cell—a biologically useful range—extending from low endogenous damage levels to the extent of damage that can be inflicted experimentally without killing cells. Cells from various tissues can be studied, in a wide variety of eukaryotic organisms.[76,77]

The results of our studies revealed that the commercial formulations of herbicides and insecticides widely applied in soy and other crops induce genotoxic effects in *C. latirostris* embryos at many of the concentrations tested. Other studies conducted by our research group in broad snouted caiman have demonstrated the genotoxic effect of Roundup® formulation and the active principle ingredient GLY after *in ovo* exposure by topical application at concentrations between 500 and 1750 $\mu g\,egg^{-1}$,[14,78] as well as immunotoxicity.[79] Besides, we reported the higher effects on genotoxicity and enzymatic alterations for the mixture of GLY–END–CYP formulations at concentrations recommended in agriculture for soybean crops, in a semi-natural exposure scenario for caiman nests, similar to that happening in natural environments near crops.[51] In another reptile species, the tegu lizard, we found higher DNA damage in erythrocytes of hatchlings exposed *in ovo* to 200–1600 μg Roundup® egg^{-1} but no significant differences were observed in the FMN and other NAs.[29]

In relation to the stage-dependent effect evaluated for this formulation, we found significantly higher DNA damage in embryos exposed to the three concentrations of Roundup® tested, compared to the NC, but no differences among the three stages of exposure. These results indicate that Roundup® produces genotoxicity at the beginning, middle and end of embryo development, which could be considered as evidence that the genotoxic effect is produced regardless of the moment in which exposure occurs. In their natural environment, embryos may receive pesticides during the whole developmental period, as the reproductive season for this species take place during the moment of the year with the maximum pesticide application rate, implying a higher probability of exposure. In studies conducted in agricultural environments in Argentina, values of GLY found in sediments and soils were from 1500 to 5000 $\mu g\,kg^{-1}$ after one application.[80,81] Recently, similar values were found in sediments of the Saladillo River, a tributary of the Parana River, in the central-east of Argentina,[11] which is part of the geographic distribution of the species. If this amount is applied on an egg of 70 g, on average, it would be approximately equivalent to 100–350 μg applied to the egg. Even when these concentrations are lower than those tested in the present study, it must be considered that caiman nests may receive repeated exposure to this and other compounds as pesticide spraying is done repeatedly on crops, even as a preventive action. Accordingly, the two most common crop rotations of the Argentine pampas (*i.e.*, corn *vs.* full-season soybeans and corn *vs.* wheat *vs.* short-season soybeans) usually require two or three applications of glyphosate.[82]

Different authors have studied the genotoxic capacity and lethality of Roundup® in species of mammals, fish, amphibians and reptiles.[5,74,83] Recently, based on a review of the literature available on GLY effects in different organisms, the World Health Organization (March, 2015) in the report of the International Agency for Research on Cancer (IARC) (WHO International Agency for Research on Cancer, 2015) classified GLY into Group 2A (*probably carcinogenic to humans*).

Others studies that evaluated the effects of toxic compounds on developing organisms found differences in the effects depending on the stage of development, suggesting that this is a consequence of different capacities of the repair processes.[84] These authors reported that grass shrimp (*Palaemonetes pugio*) embryos exposed to various concentrations of benzo[α]pyrene (BαP), Cr(VI) and hydrogen peroxide showed a lower capacity to repair DNA strand breaks, determined by the CA, in early-stage compared with later-stage embryos, which may contribute to the high sensitivity of early embryos to genotoxic compounds. Osman *et al.* studied DNA damage using the CA in African catfish (*Clarias gariepinus*) embryos and larvae exposed to three concentrations of lead nitrate at different times post-fertilization. A significant dose-related DNA damage response was observed in embryos exposed after 96 h post-fertilization, but after 144 h no remarkable genotoxic increase was observed at higher concentrations. The authors suggest that DNA repair processes took place after 144 h post-fertilization, preventing further DNA damage.[85] Considering those results, it can be supposed that the enzymatic repair mechanisms in *C. latirostris* embryos are still immature and, therefore, deficient during all its development. Accordingly, in a previous study we observed that embryos environmentally exposed to pesticides showed a delay in the time needed to repair the damage.[86] Oxidative stress has also been implicated in inducing mutations that debilitate the normal functioning of a variety of repair mechanisms (base excision, transcription-coupled repair, mismatch repair, among others) and thus allow for the loss-of-function of cell cycle control genes. Further efforts should be devoted in order to assess the DNA repair capacity of *C. latirostris* during embryo development.

Exposure during critical stages of development can have severe and permanent consequences that can be evident much later in life.[87] Caimans exposed to pesticides during the embryological stage showed developmental abnormalities, affected gonadal differentiation and, unlike the results obtained in our studies, decreased hatchling weight.[52,88,89] The pesticides evaluated in those studies are organochlorines, which are known to be highly toxic to organisms, so it is possible that they affect some metabolic function, which caused the lower hatchling weigh, which is not altered by GLY-based pesticides.

In the stage-dependent study, while the CA showed a positive response following Roundup® exposure, the MN and NA tests did not indicate any genotoxic effect of Roundup® in *C. latirostris* embryos. Differences in the results found by the three techniques applied (CA, MN and NA) can be explained by the endpoints they evaluated. While MN registers alterations at the chromosomal level and NA are considered complementary to it, CA identifies single strand breaks and maximizes the expression of alkali-labile sites in the DNA molecule. Therefore, CA is more sensitive than MN test to detect damage at DNA level.[19] In the previous study made by our group with the same formulation, both the MN frequency and DNA damage increased at 500–1750 μg egg^{-1}, compared to controls, and we observed a concentration-dependent effect after *in ovo* exposure of caimans to Roundup® at the beginning of development.[14] These results coincide with those reported here for the CA in embryo exposure to Roundup®, but without a dose-dependent effect. The observation of similar results in three independent experiments is an evidence of DNA damage caused by Roundup® on *C. latirostris* embryos and their supposed inefficient DNA repair mechanisms, as suggested by Osman *et al.*[85] and Hook and Lee[84] in other species. However, the lack of a dose-dependent effect in DNA damage in the study reported here is a difference that must be taken into account. On the other hand, differences observed in the MN results of Roundup®-exposed groups between the three works could be explained by a higher susceptibility of the clutches used in the first work, or by the lower sensitivity of the MN and NA tests to detect damage at these concentrations. Besides, it must be note that almost no difference was observed in the genotoxicity cause by the two GLY-based formulations Roundup® and Panzer® Gold, indicating that despite the different components that are included in the formulations together with the active principle, both of them, at least in this case, induce almost the same level of genotoxicity.

Similar to the present study, the positive control alkylating agent cyclophosphamide induced a significant difference compared to the NC in the CA in all assays conducted by our group in two different reptile species, the broad snouted caiman and the tegu lizard.[14,23,29] In the case of the MN and NA tests, results were less consistent, showing positive results in some studies[14,23] but not in others,[29] as it was observed in the *in ovo* studies presented in this chapter. A possible explanation for this could be the mechanism of action of this compound in relation to the kind of damage that these biomarkers detect. The mechanisms of action of cyclophosphamide, as an alkylating agent, generally induced DNA fragmentation, a kind of damage detected by the CA, but rather frequently; this may lead to clastogenic or aneugenic effects that are recognized by the MN or NA tests.

Concerning the effects of CYP (Atanor®) during development, we observed an increase in the FMN and other NAs from the lowest concentration tested, compared to the controls (NC and VC), while the CA showed only an effect at 100 $\mu g\, egg^{-1}$. Anwar[54] observed severe teratological abnormalities in chick embryos exposed to 100–400 $mg\, kg^{-1}$ of CYP, as a single sublethal dose (final volume 50 μl) at day "0" of incubation, with the lowest concentration being seven times higher than the highest concentration applied in our work. In amphibians, Cabagna *et al.*[73] reported that 5, 10, 20 and 40 $\mu g\, L^{-1}$ of the CYP formulation produced a significant increase in the frequency of micronucleated erythrocytes at 48 and 96 h of treatment in *Odontophrynus americanus* tadpoles.

In relation to the END formulation (Galgofan®), our results showed an increase in DNA damage in the lowest and higher concentrations tested, while the MN test only showed an effect at 100 $\mu g\, egg^{-1}$. However, we observed other NAs at all concentrations evaluated, especially eccentric and notched nuclei, which could indicate particular mechanisms of damage. Mobarak and Al-Asmsari[90] suggested embryotoxic and teratogenic effects on developing chick embryos incubated for 24 h with a single dose of 7, 14 and 24 mg egg^{-1} administered by topical application, which are, as in the case of the CYP formulation, doses at least seven times higher than those applied here. Considering their toxic properties, bioaccumulation and persistence, the United Nations Environment Programme (UNEP) added END to the list of prohibited organic pollutants (Stockholm Convention on POPs, 2011). In Argentina, a period of 5 years was established for phasing it out since the date of the effective Resolution 511/2011,[91] including total prohibition on using, importing, processing (synthesis), formulation, and marketing the active ingredient and formulated products, thus extending legal use until 2016. In spite of this, at the time when this study took place, END was one of the main insecticides used for pest control in Argentina, and residues and metabolites are still now found in the environment, and it is even still used in some places.[92]

In recent years, ecologists have taken to studying antioxidants and oxidative stress in free-ranging organisms and have integrated principles of oxidative stress into several core evolutionary concepts, such as life history trade-offs (*e.g.* survival *vs.* reproduction), senescence and sexual selection.[93] High levels of ROS do not necessarily result in oxidative stress if this can be balanced by an up regulation of defenses; nor does it follow that individuals with relatively high levels of antioxidants are necessarily in a better redox state than those with lower levels as this will depend on the levels of ROS that these defenses have to deal with.[94]

It is widely known that agrochemicals may produce ROS through different mechanisms: by interference in electron transport in the mitochondrial membrane with the consequent accumulation of reactive intermediates, by inactivation of antioxidant enzymes, or by the deterioration of non-enzymatic antioxidants and lipid peroxidation.[95] In early developmental stages oxidative stress levels are high owing to the presumed link between high metabolic activities required for growth and ROS generation.[96]

In the first instance, we evaluated oxidative stress parameters (CAT and SOD activities and lipid peroxidation) in caimans exposed *in ovo* to GLY, CIP and END pesticide formulations, as previously described. Many studies have reported oxidative stress of CIP, END and the GLY formulation Roundup® on different vertebrate species,[97–100] but information about the effect of the formulation Panzer® is limited, with some studies performed on the fish *Cnesterodon decemmaculatus,* but with a formulation containing a lower proportion of the active ingredient (48%)[101,116] than the one used in our studies (Panzer® Gold, 66.2%).

In Figure 20.9, we analyzed the enzymatic activity of CAT and SOD for Roundup®, where it showed an increase in both enzymes as concentration increases, up to 750 μg egg^{-1}, after which the activities decrease drastically. When comparing the results with those of TBARS (Figure 20.6) and DNA damage (Figure 20.4), we found significant differences at 500 and

1000 μg egg^{-1}, but not at 750 μg egg^{-1}. These values could demonstrate that the damage in TBARS exists because the increase in enzymatic activity is not enough to counteract the ROS increase at those concentrations, which indicates impaired ability of antioxidant defenses.[100] Moreover, ROS can react with cellular components, such as nucleic acids, thus a dysfunction in the antioxidant system can be the indirect result of an alteration in gene expression,[102] altering the enzyme synthesis in neonates exposed to Roundup®. Unlike these results, a study conducted in the Nile crocodile (*Crocodylus niloticus*) exposed to aquatic contaminants reported an efficient antioxidant system evaluated by GST, GPx, GR GSH, in which the enzymatic activities of CAT, GSH and GPx particularly suggest an elevated capability to reduce lipid hydroperoxides and hydrogen peroxide.[103]

The results observed in TBARS for the GLY-based formulation Panzer® Gold were similar to those obtained for Roundup®, with a significantly higher level with Panzer® Gold 500 and 1000 μg egg^{-1} compared to the controls (Figure 20.6). These results could suggest that, though there were no significant changes in the enzymatic activity, there is a consistent response in TBARS.

Damage to lipids measured in caimans exposed to Galgofan® was significant with respect to the controls with concentrations of 10 μg egg^{-1} and higher, while the activity of SOD was 62% higher than the NC at 1 μg egg^{-1}, showing no differences at higher concentrations. These results could indicate that at lower concentrations there is a significant increase of SOD activity that counteracts the ROS increase, avoiding lipid peroxidation. At higher concentrations, large amounts of ROS could alter the activity of erythrocyte antioxidant defenses, stimulating lipid peroxidation (Figures 20.7 and 20.8). Similarly, Crupkin *et al.*[28] reported inhibition of SOD, CAT and GSH enzyme activity in the fish *Australoherus facetus* exposed to different concentrations of END formulation (0.02, 0.5, 5 and 10 μg L^{-1}) for 24 h, with a consequent increase in H_2O_2 and lipid peroxidation in the liver and brain.

As mentioned, in the present work, we found a significant increase in lipid peroxidation with different concentrations of Roundup® (500 and 1000) and Panzer® Gold (500 and 1000), Galgofan® (10, 100 and 1000) and Atanor® (1, 10 and 1000). Lipid peroxidation is considered one of the best biomarkers of oxidative damage because it indicates indirectly the incidence of free radicals on the biological membranes. The exposure to GLY, END and CYP formulations was done *in ovo*; thus, we can expect high susceptibility to oxidative damage in embryos owing to their high unsaturated fatty acid content and deficiencies in antioxidant enzymes as a consequence of immaturity. This was explained by Hilscherova *et al.*[102] when they studied oxidative stress induced by 2,3,7,8-tetrachlorodibenzo-*p*-dioxin (TCDD) on embryos of *Gallus domesticus*.

During their evolutionary history, reptiles have had to endure variations in the availability of atmospheric oxygen.[104] Additionally, some of these organisms, such as turtles and crocodiles, experience long immersion periods, where the blood circulation is intended primarily for vital organs, resulting in hypoxia and ischemia in others. After immersion, blood flows back to all the organs normally. In fact, these animals have a particular tolerance to changes in the oxygen availability that involves the generation of ROS, and OS may occur after an abrupt reoxygenation, for which they must have an effective antioxidant system.[105] In this context, Furtado Fhilo *et al.*[105] applied oxidative stress biomarkers in various organs of *C. yacare* comparing the embryonic, juvenile and adult stage, and the differences in the summer and winter seasons to assess the influence of immersion on OS. They found that the liver is one of the organs with the highest levels of TBARS, which were lower for hatchlings than embryos, but they increase in adults. Such higher levels in embryos could be due to the liver's metabolic demands and its specific role during embryonic development.

Furthermore, Hermes-Lima *et al.*,[106] continuing with the previous research, evaluated the effect of season (summer and winter) and development on GSH content and the enzymatic activity of CAT, GPX, GST and Glucose 6-phosphate dehydrogenase (G6PDH) in *C. yacare.* They observed increased GSH values in the liver, lung, kidney and brain from embryo–hatchling stages to juveniles, while in the youth–adult transition there were no significant changes.

Recently, Poletta *et al.*[44] characterized a new set of OS biomarkers in the blood of juvenile *C. latirostris* and determined basal values of lipid peroxidation (827.45 ± 196.54 nmol g Hb^{-1}) and antioxidant defenses: CAT (591.87 ± 132.74 KU g Hb^{-1}), SOD ($31.12 \pm 7.34\%$ activity) and GSH (1.66 ± 0.02 μmol mg Hb^{-1}). Values of TBARS, CAT and SOD are slightly lower than those of the NC obtained in the embryo experiment: 1255.87 nmol g Hb^{-1}, 863.6 KU g Hb^{-1} and 21.05% activity, respectively. A possible explanation for these differences may be the fact that hatchlings have to deal with the stress generated during birth related to changes in the environment from inside to outside the egg, causing higher levels of lipid peroxidation and antioxidant enzyme activity.

According to the literature available on reptiles, it is important to highlight that the knowledge about OS on these animals is still very limited. Although there have been advances in the study of OS baseline values in certain organs, and the influence of temperature, seasonality, and development on them,[105–108] there are no previous reports on the application of these parameters as biomarkers of OS produced by pesticides in any reptile.

For *C. latirostris*, the adaptation of the techniques to be applied in peripheral blood, avoiding any damage to the animals,[44] represents a major advance over previous studies reported in other reptile species, where organs or tissues were used, with the consequent slaughter of animals. In this work, it was possible to use these techniques as markers of OS in caimans exposed *in ovo* to different pesticides, and this is the first report of oxidative damage and impaired antioxidant defenses produced by pesticides in all crocodilians and even in all reptile species.

Considering the particularity of reptiles in terms of their antioxidant capacity systems, it is expected that there is variability in responses of individuals to different stressors because of sensitivity or resistance. Besides, the different responses found in this and the cited works could also be related to the intensity, duration and route of exposure and the type of xenobiotic studied. In our study, embryos were exposed only once within 5 days after oviposition and the evaluation was conducted at the time of birth, which could explain the absence of an effect on CAT. However, it was enough to affect the enzymatic activity of SOD in some of the exposed groups, as well as TBARS levels in most of them.

Concerning growth parameters, under controlled incubation conditions eggs generally undergo weight loss. Beldomenico *et al.*[52] measured egg weights at the beginning and at the end of the incubation period and determined the decrease in weight during embryonic development and its relation to pesticide exposure, as well as hatchling weight. They found that in exposed eggs, part of the egg weight loss was produced by a decrease in the mass of non-embryonic contents. Greater egg weight loss during incubation was observed in those treated with an environmentally relevant dose of atrazine (0.2 mg kg^{-1}) and relatively low doses of END formulation (2 and 20 mg kg^{-1}, similar to concentrations applied in our study) with respect to the NC. They suggest that egg weight loss is related to some metabolic aspect of the embryo involved in the process of water evaporation through of eggshell, and that the *in ovo* exposure to pesticides affects the functioning of several organs and/or disrupts signals that control development. Most reptiles display indeterminate growth, accruing new biomass throughout their lifetimes. However, in most species, growth rates decline after the rapidly growing juvenile period and, whereas the growth rate remains positive, it can become quite low in old individuals. Because growth rate depends on the initial size of the individual, especially during the rapidly growing phase, environmental factors that lead to a period of reduced growth can be particularly important during the juvenile period, when animals grow at their maximum rate, attaining a size at which certain predators can be avoided. Therefore, smaller hatchlings would have less chance of survival during their early years, thus affecting the population dynamic of the species.[17] In our study only the initial eggs weight were measured and then the weight of hatchlings at birth, so no analysis of egg weight variation among experimental groups could be done. Besides, we found no significant differences among experimental groups with respect to hatchling weight at birth or during the first months of life (3 and 6 month measurements), so possibly the concentrations applied in our study were too low to induce effects on growth.

In conclusion, these results add further evidence to those obtained with the MN and other NA tests and the CA in erythrocytes of broad-snouted caiman, which are demonstrated to be useful parameters in the evaluation of the genotoxic effects of END-, CYP- and GLY-based formulations.

Our data also indicated that different concentrations of different pesticide formulations induce different NAs, including MN, and that neither of the biomarkers showed a concentration-dependent effect, except for Panzer® Gold in MN, unlike what we observed in previous studies at least for Roundup®.[14] Further studies are necessary to better understand the mechanisms of damage that induce these abnormalities in broad-snouted caiman embryos.

In a second instance of evaluation, juvenile caimans exposed to sub-lethal concentrations of CYP-, END-, CPF- and GLY-based formulations, as well as to complex mixtures of them, revealed significant increases in MN, other NAs, DNA damage of the CA, and oxidative stress through damage to lipids and enzyme alterations.

Considering the commercial Roundup® formulation, this study confirmed genotoxicity evidenced by the FMN in *C. latirostris* previously reported by our group under the same experimental conditions.[61]

Our findings agree with other reports of genotoxic effects after *in vivo* exposure to similar concentrations and formulations in fish species, but in short-term exposure studies (up to 96 h). Vera-Candioti *et al.*[101] tested the toxicity of 48% GLY-based formulations, Panzer® and Credit® on *C. decemmaculatus* (Pisces, Poeciliidae). Credit® exposure showed an increase in the MN frequency after 96 h of treatment, whereas a similar increase was observed with Panzer® between 48 and 96 h exposure time at 3.9 and 7.8 $mg L^{-1}$. Cavas and Könen[109] observed high FMN, NAs and strand breaks in *Carassius auratus* after *in vivo* exposure to Roundup® at 5–15 $mg L^{-1}$.

In a study conducted by Guilherme *et al.*[75] in the European eel (*Anguilla anguilla*), the authors evaluated the genotoxic potential of Roundup® in gills and liver following short-term (1 and 3 day) exposure to environmentally relevant concentrations (58 and 116 $\mu g L^{-1}$), also addressing the possible association with oxidative stress. Both concentrations induced DNA strand breaks, but the modified Comet assay showed the occurrence of FPG-sensitive sites in the liver only after a 3 day exposure to the higher concentration. The antioxidant defenses were in general unresponsive, despite a single increment of catalase activity in the gills (116 $\mu g L^{-1}$, 3 day) and a decrease of superoxide dismutase activity in the liver (58 $\mu g L^{-1}$, 3 day). The authors stated that the mechanisms involved in Roundup®-induced DNA strand breaks should be similar in both organs but the type of DNA damage varies with the concentration and exposure duration. Hence, after 1 day of exposure, an increase in the pro-oxidant state is not a necessary condition for the induction of the DNA-damaging effects of Roundup®, while when the duration of exposure increased to three days, ROS-dependent processes gained preponderance as a mechanism of DNA damage induction at the higher concentration. Unlike this, we found no concentration-dependent effect in genotoxicity either in hatchlings or in embryos exposed to Roundup® or Panzer® Gold, but we observed a similar lack of response in antioxidant enzymes SOD and CAT.

As these authors proposed, we observed that different formulations at different concentrations seem to induce different mechanisms of damage, showing different types of NAs, as well as DNA damage and oxidative stress in some cases but not in others, without consistency in the results.

De Castilhos *et al.*[110] observed negative results with the MN test but high levels of DNA damage with the CA in blood and hepatic cells of *Corydoras paleatus* exposed *in vivo* to 6.67 μg L^{-1} of Roundup® for 3, 6 and 9 days. Moreno *et al.*[111] reported significantly higher DNA damage scores in erythrocytes and gill cells of the fish *P. lineatus* exposed to two nominal concentrations of GLY (1 mg L^{-1} and 5 mg L^{-1}) for 6, 24 and 96 h. The same species exposed to 10 mg L^{-1} of Roundup® showed DNA damage in erythrocytes, but no significant difference in the FMN and FNAs after 6 and 96 h of exposure.[74] In contrast, we observed genotoxicity induced by this compound with both biomarkers, and even the presence of *Buds* at the lowest concentration, together with lipoperoxidation. The greater effect observed in our study could possibly be to do with the higher concentrations applied in relation to the first work,[110] and the longer period of exposure considering the other two.[74,111] However, the sensitivity of the species is a factor that must always be taken into account.

In addition, our data indicated negative results in DI for the GLY formulation Panzer® Gold, in contrast with results obtained by the MN test with the same treatments. These data disagree with those obtained in *C. decemmaculatus* after *in vivo* exposure to a concentration of 3.90 mg L^{-1},[109] which is similar to the concentration tested by us. However, is important to say that even when the formulation used in our study has a higher percentage of the active component in the formula (60.2%), in the experimental design we made a progressive decrease in the concentration with time, finishing with 10-fold less than the initial one. Thus, it is possible that DNA damage generated at the higher concentration had been repaired, and damage is no longer generated when the concentration decreases, while in the case of the MN test those MN induced at higher concentration at the beginning of the experiment cannot be repaired and cells remain in circulation, thus exerting an increase in the FMN in these groups.

In a meta-analytical review of experimental studies on the relationship between exposure to GLY formulations and the formation of MN, different responses are seen according to the test system, the group of animals tested, and the type of cells analyzed (polychromatic erythrocytes or all erythrocytes). A negative relationship between exposure time and effects was found so that over the course of exposure time, MN formation is decreased. This can be explained by the adaptation of detoxification mechanisms, the metabolization of xenobiotics and the repair of DNA damage over the time of exposure. This was not observed in our study so we concluded that the effect generated is one that cannot be repaired.[112] The difference observed between both GLY-based formulations in genotoxicity and oxidative stress can be explained by the various adjuvants and surfactants these formulas have, most of which are unknown.

Besides herbicides, numerous insecticides are widely used in agricultural practices, especially associated with soybean crops. Results obtained in this study show increased DNA damage in caimans exposed to different insecticide formulations. Similar results were observed in the freshwater fish *Channa punctatus* exposed for 96 h to three different concentrations of CPF formulation (203 $\mu g L^{-1}$, 406 $\mu g L^{-1}$ and 609 $\mu g L^{-1}$) under laboratory-controlled conditions.[113] On the other hand, Yin *et al.*[114] demonstrated the genotoxicity of a CPF formulation in the Chinese toad *Bufo bufo gargarizands* at five concentrations (0.32, 0.64, 0.72, 1.08 and 2.56 $\mu g L^{-1}$). More recently, Ismail *et al.*[115] observed an increase in DI induced by sublethal concentrations of CPF formulation (221.4, 110.7 and 73.8 $\mu g L^{-1}$) in erythrocytes and gill cells of the fish *Labeo rohita* after 96 h of exposure. The commercial formulations of CPF Lorsban® were also tested in *C. decemmaculatus*[65] within the concentration range of 8–25 $\mu g L^{-1}$, demonstrating an increase in the FMN with 48 and 96 h of treatment.[116]

In relation to END Galgofan®, both concentrations tested induced significant DNA damage compared to the control. Sharma *et al.*[62] reported significant higher DNA damage in erythrocytes and other tissues of fish *Mystus vittatus* exposed *in vivo* to two sub-lethal concentrations of END formulation (0.50 and 0.25 $\mu g L^{-1}$) and a non-lethal concentration (0.20 $\mu g L^{-1}$), one of them being equal to the lowest concentration tested in our study. Similar results were observed in zebrafish (*Danio rerio*) exposed under laboratory controlled conditions to four different concentrations of END formulation, also covering the same range tested by us (0.01, 0.1, 1 and 10 $\mu g L^{-1}$), and sampled every 7 days for 28 days. The genotoxic effects of another formulation were detected by Crupkin *et al.*[28] in *Australoheros facetus* exposed for 24 h, with an increased frequency of NAs at 0.02 $\mu g L^{-1}$ and MN at 5 $\mu g L^{-1}$. END has been reported as an oxidant creating oxidative stress.[121] A decrease of catalase activity has been observed after exposure in fish.[122]

Our data clearly show that when caimans are exposed to sublethal concentrations of CYP (Atanor®) an increase in DNA DI is induced in both treatments. The results are consistent with other reports where the insecticide caused DNA damage in different cells of *P. lineatus.* Simoniello *et al.*[64] found significantly higher levels of DI in erythrocytes of fish exposed *in vivo* to all concentrations tested (0.300, 0.150 and 0.075 $\mu g L^{-1}$) for 96 h. The same was corroborated in gill cells exposed *in vivo* to a CYP formulation that indicated an induction of genotoxic damage at 0.300 and 0.150 $\mu g L^{-1}$ with respect to the control evidenced by the CA.[117] Ansari *et al.*[100] reported an increase in FMN correlated with increased oxidative stress and disturbance of antioxidant enzymes in *Channa punctata* exposed to CYP formulation at 0.4, 0.8 and 1.2 $mg L^{-1}$ for 48 and 72 h. In our study, we also observed a correlation between genotoxicity and oxidative stress biomarkers for groups exposed to both concentrations of CYP Atanor®.

The application of pesticide mixtures is a common agricultural practice. In this study we examined each pesticide formulation alone and a mixture of different combinations of them, M_1 (GLY–CYP–END) and M_2 (GLY–CYP–CPF) formulations. Our data show no significant difference in the DI in individuals exposed to M_1. On the contrary, the same mixture demonstrated a high level of damage in *C. latirostris* exposed *in ovo* under semi-natural conditions.[51] However, we found genotoxicity damage in yearlings exposed to M_2, as well as in those exposed to the compounds separately. A mixture of pesticides can produce a synergy or antagonism and the response depends on the chemical properties and modes of toxic action of the pesticides in addition to the biological characteristics of the species.[118,119]

Teratological abnormalities were reported in chick embryos exposed to 0.005–0.5 $\mu g\,egg^{-1}$ of the mixture of CYP + CPF formulations, as a single sublethal dose (final volume 50 μl) at day 0 of incubation.[16]

Contaminants with similar or different modes of action can influence each other's toxicity, resulting in an almost unlimited number of possible additive, synergistic or antagonistic combinations. In addition, non-chemical factors may also act as stressors and add to the complexity of multiple stressor situations.[120] In this context, pesticides have received much attention as possible combined toxicity stressors. Contrarily to what was expected in the present study, no additive or synergistic effect was observed for the mixtures in any of the parameters analyzed.

Species may greatly differ in their tolerance to stressors generating oxidative stress, so it is suggested that a marker of antioxidant capacity should always be associated with at least a marker of oxidative damage when the aim is to make inferences about oxidative stress.[123] The consequent loss-of-function and structural integrity of modified bio-molecules through oxidative stress can have a wide range of downstream functional consequences and may be the cause of subsequent cellular dysfunctions and tissue damage.[124] Lipids are one of the major targets of oxidative stress. As mentioned earlier, lipid peroxidation gives rise to a number of secondary, highly damaging products, known to further perpetuate ROS production. In our studies, lipid peroxidation was evident with most of the formulations and concentrations tested, showing to be a very consistent biomarker of oxidative stress.

Methods utilizing blood, like in our studies, or other body fluids are favored in studies on wild species because they are less invasive and do not require terminal sampling. Besides, all the biomarkers we applied routinely to the evaluation of these compounds are made in blood, so we can ensure a battery of biomarkers of different endpoints for the evaluation of environmental contamination, especially pesticides, in this and other related species.

20.5 Concluding Remarks

The four formulations tested in the different experiments, as well as the complex mixtures of them, were shown to induce genotoxicity and oxidative stress, but no synergistic effects were observed for the mixtures. Exposure of *C. latirostris* embryos to different concentrations of Roundup® produces DNA damage, independently of the time of development in which exposure occurs. This could be a consequence of immaturity and deficiency of the repair mechanisms in *C. latirostris* embryos during development.

The variations found in the results of the different biomarkers analyzed in all the studies made by our group demonstrated that all of them are sensitive biomarkers but it is important to make integral approaches in biomonitoring as the use of more parameters with different endpoints ensures more reliable and better interpretations of the results. Some biomarkers indicate potential toxicity among caimans exposed to pesticides alone and in mixtures, even though no individual biomarker has provided categorical evidence for toxic effect. However, analyzed together, genotoxicity and oxidative stress provide a coherent profile of animals under metabolic stress.

It is surprising the lack of information that still exists on the physiological consequences on animal development after or during the exposure to different genotoxic compounds. This approach encourages further investigations to understand the final consequence of these alterations in *C. latirotris* as a sentinel of environmental pesticide contamination.

20.6 The Major Aim: Assessing Environmental Exposure of Natural Populations

The extraordinary growth in the chemical industry during the second half of the twentieth century has led to the appearance in nature of thousands of new products every year, a large percentage of which have significant biological effects. The presence in the environment of xenobiotics that are biologically active and difficult to break down represents a degree of stress that is frequently unacceptable for living organisms and that is also expressed at the ecosystem level. Both direct and indirect toxic activity can, in certain circumstances, be an important risk factor for the human population as well. The usual way to approach ecotoxicity testing, according to relevant EPA and OECD guidelines for the testing of chemicals, is the use of well-defined tests in which selected species are exposed to a single pollutant under controlled laboratory conditions. Such a standardized approach is necessary to acquire information in a relatively short time in order to gather data that is easy to compare and to interpret. However, extrapolation to the real world is challenging, if at all feasible.

Environmental biomonitoring makes use of 'sentinel' organisms living in their natural habitat and reflecting long-term, continuous exposure. The primary hypothesis for these kinds of studies is that the physiological effects of exposure to a cocktail of pesticides will be manifested as changes in individual biomarkers of exposure and effect within the exposed populations of *Caiman latirostris.* In natural environments, if the added stress of pesticide toxicity takes animals closer to their survival thresholds, then they are likely to be more vulnerable to events such as unusually cold weather, drought, disease, *etc.* so that they might be vulnerable to chemical contamination in combination with other natural stressors.[18]

Although field situations are ecologically complex and factors other than pesticides may be acting, the absence of observable effects on field subpopulations is probably indicative that animals are coping or compensating for this level of exposure.[1]

Models to study environmental toxicity are a necessary compromise between the control of experimental parameters (through the use of laboratory-reared substitute species and the setting of a thoroughly controlled exposure scenario) and realism (field or semi-field studies). An entirely different approach is based on the use of native species, which essentially considers pollution as a complex situation and therefore implies a more holistic interpretation of the real conditions of exposure in the field. This kind of study includes the capture and sampling of animals in the field or collection of exposed and controls eggs.

This approach allows consideration of interactions between pollutants as well as homeostasis. Interpretation of the results, on the other hand, may be particularly difficult in the face of the many constraints and confounding factors of the natural environment.[77]

Studies that convincingly demonstrate whether environmental contaminants have detrimental effects, or not, on natural reptile subpopulations are still lacking. The main reasons for this knowledge gap are probably related with the nonexistence of protocols and model species for this group as there are for other vertebrates.[1]

Acknowledgements

These studies were supported by the Agencia Nacional de Promoción Científica y Tecnológica (ANPCyT: PICT 2011-1349 to GLP), Universidad Nacional del Litoral (CAI + D 2011 – 50120110100189 to GLP) and Proyecto Yacaré and Yacarés Santafesinos (Gob. Sta. Fe/MUPCN).

References

1. M. J. Amaral, M. A. Carretero, R. C. Bicho, A. M. V. M. Soares and R. M. Mann, The use of a lacertid lizard as a model for reptile ecotoxicology studies - Part 1 Field demographics and morphology, *Chemosphere*, 2012, **87**, 757–764.
2. H. J. D. Lange, J. Lahr, J. J. Van der Pol, Y. Wessels and J. H. Faber, Ecological vulnerability in wildlife: an expert judgment and multicriteria analysis tool using ecological traits to assess relative impact of pollutants, *Environ. Toxicol. Chem.*, 2009, **28**, 2233–2240.
3. B. D. Todd, J. D. Wilson and J. W. Gibbons, The global status of reptiles and causes of their decline, in *Ecotoxicology of Amphibians and Reptiles*,

ed. D. W. Sparling, G. Linder, C. A. Bishop and S. K. Krest, CRC Press, New York, 2010, pp. 69–104.

4. S. C. Gardner, Introduction to Reptilian Toxicology in *Toxicology of Reptiles. New perspectives: Toxicology and the Environment*, ed. S. C. Gardner and E. Oberdörster, Taylor & Francis Group, Florida, 2005, pp. 1–8.
5. D. W. Sparling, G. Linder, C. A. Bishop and S. K. Krest, Recent advancements in amphibian and reptile ecotoxicology in *Ecotoxicology of Amphibians and Reptiles*, ed. D. W. Sparling, G. Linder, C. A. Bishop and S. K. Krest, CRC Press, New York, 2010, pp. 1–14.
6. R. H. Rauschenberger, M. S. Sepúlveda, J. J. Wiebe, N. J. Szabo and T. S. Gross, Predicting maternal body burdens of organochlorine pesticides from eggs and evidence of maternal transfer *in Alligator mississippiensis*, *Environ. Toxicol. Chem.*, 2004, **23**, 2906–2915.
7. C. N. Rich and L. G. Talent, Soil ingestion may be an important route for the uptake of contaminants by some reptiles, *Environ. Toxicol. Chem.*, 2009, **28**, 311–315.
8. F. Sánchez-Bayo, Impacts of agricultural pesticides on terrestrial ecosystems, in *Ecological Impacts of Toxic Chemicals*, ed. F. Sánchez-Bayo, P. J. van den Brink and R. M. Mann, Bentham Publishers, 2011, pp. 63–87.
9. A. Leguizamón, Modifying Argentina: GM soy and socio-environmental change, *Geoforum*, 2014, **53**, 149–160.
10. S. López, D. Aiassa, S. Benitez-Lite, R. Lajmanovich, F. Mañas, G. L. Poletta, N. Sánchez, M. F. Simoniello and A. Carrasco, Pesticides Used in South American GMO-Based Agriculture: A Review of Their Effects on Humans and Animal Models in *Advances in Molecular Toxicology*, ed. J. C. Fishbein and J. M. Heilman, Elsevier, Amsterdam, 2012, pp. 41–75.
11. A. E. Ronco, D. J. G. Marino, M. Abelando, P. Almada and C. D. Apartin, Water quality of the main tributaries of the Paraná Basin: glyphosate and AMPA in surface water and bottom sediments, *Environ. Monit. Assess.*, 2016, **188**(8), 458–470.
12. M. A. Aizen, L. A. Garibaldi and M. Dondo, Expansión de la soja y diversidad de la agricultura argentina, *Ecol. Aust.*, 2009, **19**, 45–54.
13. A. Larriera, A. Imhof and P. A. Siroski, Estado actual de los programas de conservación y manejo de género *Caiman* en Argentina in *Contribución al conocimiento del Genero Caiman de Suramerica*, ed. J. Castroviejo, J. Ayarzaguena and A. Velasco, Public. Asoc. Amigos de Doña Ana, vol. 18, Sevilla, 2008, pp. 139–179.
14. G. L. Poletta, A. Larriera, E. Kleinsorge and M. D. Mudry, Genotoxicity of the herbicide formulation Roundup® (glyphosate) in broad-snouted caiman (*Caiman latirostris*) evidenced by the Comet assay and the Micronucleus test, *Mutat. Res.*, 2009, **672**, 95–102.
15. Paganelli, V. Gnazzo, H. Acosta, S. L. López and A. E. Carrasco, Glyphosate-based herbicides produce teratogenic effects on vertebrates by impairing retinoic acid signaling, *Chem. Res. Toxicol.*, 2010, **23**, 1586–1595.

16. K. Uggini Gowri, P. V. Patel and S. Balakrishnan, Embryotoxic and teratogenic effects of pesticides in chick embryos: A comparative study using two commercial formulations, *Environ. Toxicol.*, 2012, **27**(3), 166–174.
17. C. L. Mitchelmore, C. L. Rowe, A. R. Place, Tools for assessing contaminant exposure and effects in reptiles in *Toxicology of Reptiles*, ed. S. C. Gardner and E. Oberdorster, Taylor & Francis Group, Boca Raton, 2006, pp. 63–122.
18. M. J. Amaral, R. C. Bicho, M. A. Carretero, J. C. Sánchez-Hernández, A. M. R. Faustino, A. M. V. M. Soares and R. M. Mann, The use of a lacertid lizard as a model for reptile ecotoxicologystudies: Part 2 – Biomarkers of exposure and toxicity amongpesticide exposed lizards, *Chemosphere*, 2012, **87**, 765–774.
19. M. A. Carballo and M. D. Mudry, Indicadores y marcadores biológicos, in *Genética Toxicológica*, ed. M. D. Mudry and M. A. Carballo, De los Cuatro Vientos Editorial, Buenos Aires, 2006, pp. 83–108.
20. G. L. Poletta, A. Larriera, E. Kleinsorge and M. D. Mudry, *Caiman latirostris* (broad-snouted caiman) as a sentinel organism for genotoxic monitoring: Basal values determination of micronucleus and comet assay, *Mutat. Res.*, 2008, **650**, 202–209.
21. G. Frenzilli, M. Nigro and B. P. Lyons, The Comet assay for the evaluation of genotoxic impact in aquatic environments, *Mutat. Res.*, 2009, **681**(1), 80–92.
22. R. C. Lajmanovich, M. C. Cabagna-Zenklusen, A. M. Attademo, C. M. Junge, P. M. Peltzer, A. Bassó and E. Lorenzatti, Induction of micronuclei and nuclear abnormalities in tadpoles of the common toad (*Rhinella arenarum*) treated with the herbicides Liberty® and glufosinate-ammonium, *Mutat. Res.*, 2014, **769**, 7–12.
23. E. C. López González, A. Larriera, P. A. Siroski and G. L. Poletta, Micronuclei and other nuclear abnormalities on *Caiman latirostris* (Broad-snouted caiman) hatchlings after embryonic exposure to different pesticide formulations, *Ecotoxicol. Environ. Saf.*, 2017, **136**, 84–91.
24. I. Strunjak-Perovic, D. Lisicic, R. Coz-Rakovac, N. Topic Popovic, M. Jadan, V. Benkovic and Z. Tadic, Evaluation of micronucleus and erythrocytic nuclear abnormalities in Balkan whip snake *Hierophis gemonensis*, *Ecotoxicology*, 2010, **19**, 1460–1465.
25. G. N. A. Furnus, J. D. Caffetti, E. M. García, M. F. Benítez, M. C. Pastori and A. S. Fenocchio, Baseline micronuclei and nuclear abnormalities frequencies in native fishes from the Paraná River (Argentina), *Braz. J. Biol.*, 2014, **74**(1), 217–221.
26. J. M. Pérez-Iglesias, C. Ruiz de Arcaute, N. Nikoloff, L. Dury, S. Soloneski, G. S. Natale and M. L. Larramendy, The genotoxic effects of the imidacloprid-based insecticide formulation Glacoxan Imida on Montevideo tree frog *Hypsiboas pulchellus* tadpoles (Anura, Hylidae), *Ecotoxicol. Environ. Saf.*, 2014, **104**, 120–126.

27. M. Fenech, M. Kirsch-Volders, A. T. Natarajan, J. Surralles, J. W. Crott, J. Parry, H. Norppa, D. A. Eastmond, J. D. Tucker and P. Thomas, Molecular mechanisms of micronucleus, nucleoplasmic bridge and nuclear bud formation in mammalian and human cells, *Mutagenesis*, 2011, **26**(1), 125–132.
28. A. C. Crupkin, P. Carriquiriborde, J. Mendieta, A. M. Panzeri, M. L. Ballesteros, K. S. B. Miglioranza and M. L. Menone, Oxidative stress and genotoxicity in the South American cichlid, *Australoheros facetus*, after short-term sublethal exposure to endosulfan, *Pest. Biochem. Physiol.*, 2013, **105**, 102–110.
29. G. L. Schaumburg, G. L. Poletta, P. A. Siroski and M. D. Mudry, Genotoxicity induced by Roundup® (Glyphosate) in Tegu lizard (*Salvator merianae*) embryos, *Pest. Biochem. Physiol.*, 2016, **130**, 71–78.
30. F. E. Pollo, C. L. Bionda, Z. A. Salinas, N. E. Salas, A. L. Martino, Common toad *Rhinella arenarum* (Hensel, 1867) and its importance in assessing environmental health: test of micronuclei and nuclear abnormalities in erythrocytes, *Environ. Monit. Assess.*, 2015, **187**, 587–595.
31. K. R. Carrasco, K. L. Tilbury and M. S. Myers, Assessment of the piscine micronucleus test as an in situ biological indicator of chemical contaminant effects, *Can. J. Fish. Aquat. Sci.*, 1990, **47**, 2123–2136.
32. M. Fenech, The *in vitro* micronucleus technique, *Mutat. Res.*, 2000, **455**, 81–95.
33. R. F. Lee and S. Steinert, Use of the single cell gel electrophoresis/comet assay fordetecting DNA damage in aquatic (marine and freshwater) animals, *Mutat. Res.*, 2003, **544**, 43–64.
34. D. Ziech, R. Franco, A. G. Georgakilas, S. Georgakila, V. Malamou-Mitsi, O. Schoneveld, A. Pappa and M. I. Panayiotidis, The role of reactive oxygen species and oxidative stress in environmental carcinogenesis and biomarker development, *Chem. -Biol. Interact.*, 2010, **188**, 334–339.
35. B. Halliwell and S. Chirico, Lipid peroxidation: its mechanism, measurement, and significance, *Am. J. Clin. Nutr.*, 1993, **57**, 715S–724S.
36. B. Halliwell, Free radicals and antioxidants: updating a personal view, *Nutr. Rev.*, 2012, **70**, 257–265.
37. J. Limón-Pacheco and M. E. Gonsebatt, The role of antioxidants and antioxidant-related enzymes in protective responses to environmentally induced oxidative stress, *Mutat. Res.*, 2009, **674**, 137–147.
38. R. Franco, R. Sánchez-Olea, E. M. Reyes-Reyes and M. I. Panayiotidis, Environmental toxicity oxidative stress and apoptosis: Ménage à Trois, *Mutat. Res.*, 2009, **674**, 3–22.
39. S. Mena, A. Ortega and J. M. Estrela, Oxidative stress in environmental-induced carcinogenesis, *Mutat. Res.*, 2009, **674**, 36–44.
40. E. Ö. Oruç and D. Usta, Evaluation of oxidative stress responses and neurotoxicity potential of diazinon in different tissues of *Cyprinus carpio*, *Environ. Toxicol. Pharmacol.*, 2007, **23**, 48–55.
41. M. J. Costa, D. A. Monteiro, A. L. Oliveira-Neto, F. T. Rantin and A. L. Kalinin, Oxidative stress biomarkers and heart function in

bullfrog tadpoles exposed to Roundup Original, *Ecotoxicology*, 2008, **17**, 153–163.

42. R. Cattaneo, B. Clasen, V. L. Loro, C. C. de Menezes, A. Pretto, B. Baldisserotto, A. Santi and L. A. de Avila, Toxicological responses of *Cyprinus carpio* exposed to a commercial formulation containing glyphosate, *Bull. Environ. Contam. Toxicol.*, 2011, **87**, 597–602.
43. S. C. Rossi, M. D. da Silva, L. D. S. Piancini, C. A. Oliveira Ribeiro, M. M. Cestari and H. C. Silva de Assis, Sublethal effects of waterborne herbicides in tropical freshwater fish, *Bull. Environ. Contam. Toxicol.*, 2011, **87**, 603–607.
44. G. L. Poletta, M. F. Simoniello and M. D. Mudry, Biomarkers of oxidative damage and antioxidant defense capacity in *Caiman latirostris* blood, *Comp. Biochem. Physiol.*, 2016, **79**, 29–36.
45. J. A. Beauge and S. D. Aust, Microsomal lipid peroxidation, *Methods Enzymol.*, 1978, **52**, 302–310.
46. W. A. Hopkins, C. L. Rowe and J. D. Congdon, Elevated trace element concentrations and standard metabolic rate in banded water snakes (*Nerodia fasciata*) exposed to coal combustion wastes, *Environ. Toxicol. Chem.*, 1999, **18**, 1258–1263.
47. J. D. Congdon, A. E. Dunham, W. A. Hopkins, C. L. Rowe and T. G. Hinton, Resource allocation-based life histories: A conceptual basis for studies of ecological toxicology, *Environ. Toxicol. Chem.*, 2001, **20**, 1698–1703.
48. EXTOXNET: The Extension Toxicology Network. Pesticide Information Profiles (PIPs) [Internet; access July 2016]. Available from: /http://extoxnet.orst.edu/pips/ghindex.htmlS.
49. P. Donayo, C. Piña and A. Larriera, Período de incubación, banda de calcificación, peso de los huevos y desarrollo embrionario de *Caiman latirostris* a tres temperaturas diferentes in *Conservação e manejo de jacarés e crocodilos de América Latina*, ed. L. M. Verdade and A. Larriera, C. N. Editoria, Piracicaba, vol. 2, 2002, pp. 79–90.
50. S. R. de Solla, E. Palonen and P. A. Martin, Toxicity of pesticides associated with potato production, including soil fumigants, to snapping turtle eggs (*Chelydra serpentina*), *Environ. Toxicol. Chem.*, 2014, **33**(1), 102–106.
51. G. L. Poletta, E. Kleinsorge, A. Paonessa, M. D. Mudry, A. Larriera and P. A. Siroski, Genetic enzymatic and developmental alterations observed in *Caiman latirostris* exposed *in ovo* to pesticide formulations and mixtures in an experiment simulating environmental exposure, *Ecotoxicol. Environ. Saf.*, 2011, **74**, 852–859.
52. P. M. Beldomenico, F. Rey, W. S. Prado, J. C. Villarreal, M. Muñoz, del Toro and E. H. Luque, *In ovum* exposure to pesticides increases the egg weight loss and decreases hatchlings weight of *Caiman latirostris* (Crocodylia: Alligatoridae), *Ecotoxicol. Environ. Saf.*, 2007, **68**, 246–251.
53. N. Sinha, R. Narayan and D. K. Saxena, Effect of endosulfan on the testis of growing rats, *Bull. Environ. Contam. Toxicol.*, 1997, **58**, 79–86.

54. K. Anwar, Toxic effects of cypermethrin on the biochemistry and methodology of 11th day chik embryo (*Gallus domesticus*), *Pak. J. Appl. Sci.*, 2003, **3**(6), 432–445.
55. S. Patel, A. K. Pandey, M. Bajpayee, D. Parmar and A. Dhawan, Cypermethrin-induced DNA damage in organs and tissues of the mouse: Evidence from the comet assay, *Mutat. Res.*, 2006, **607**, 176–183.
56. J. G. Myburgh, R. M. Kirberger, J. C. A. Steyl, J. T. Soley, D. G. Booyse, F. W. Huchzermeyer, R. H. Lowers and L. J. Guillette, The post-occipital spinal venous sinus of the Nile crocodile (*Crocodylus niloticus*): Its anatomy and use for blood sample collection and intravenous infusions, *J. S. Afr. Vet. Assoc.*, 2014, **85**, 1–10.
57. C. S. Bentivegna and T. Piatkowski, Effects of tributyltin on medaka (*Oryzias latipes*) embryos at different stages of development, *Aquat. Toxicol.*, 1998, **44**, 117–128.
58. C. M. Howe, M. Berrill, B. D. Pauli, C. C. Helbing, K. Werry and N. Veldhoen, Toxicity of glyphosate-based pesticides to four North American frog species, *Environ. Toxicol. Chem.*, 2004, **23**, 1928–1938.
59. P. K. Mensah, W. J. Muller and C. G. Palmer, Acute toxicity of Roundup® herbicide to three life stages of the freshwater shrimp *Caridina nilotica* (Decapoda: Atyidae), *Phys. Chem. Earth*, 2011, **36**, 905–909.
60. M. A. Latorre, E. C. López González, A. Larriera, G. L. Poletta and P. A. Siroski, Effects of in vivo exposure to Roundup® on immune system of *Caiman latirostris*, *J. Immunotoxicol.*, 2013, **10**(4), 349–354.
61. E. C. López González, M. A. Latorre, A. Larriera, P. A. Siroski and G. L. Poletta, Induction of micronucleiin broad snouted caiman (*Caiman latirostris*) hatchlings exposed *in vivo* to Roundup® (glyphosate) concentrations used in agriculture, *Pest. Biochem. Physiol.*, 2013, **105**, 131–134.
62. S. Sharma, N. S. Nagpure, R. Kumar, S. Pandey, S. K. Srivastava, P. J. Singh and P. K. Mathur, Studies on the genotoxicity of endosulfan in different tissues of fresh water fish *Mystus vittatus* using the comet assay, *Arch. Environ. Contam. Toxicol.*, 2007, **53**(4), 617–623.
63. B. Shao, L. Zhu, M. Dong, J. Wang, J. Wang, H. Xie and S. Zhu, DNA damage and oxidative stress induced by endosulfan exposure in zebrafish (*Danio rerio*), *Ecotoxicology*, 2012, **21**(5), 1533–1540.
64. M. F. Simoniello, F. Gigena, G. Poletta, A. Loteste, E. Kleinsorge, M. Campana, J. Scagnetti and M. J. Parma, Alkaline comet assay for genotoxic effect detection in neotropical fish *Prochilodus lineatus* (Pisces, Curimatidae), *Bull. Environ. Contam. Toxicol.*, 2009, **83**, 155–158.
65. J. Vera-Candioti, S. Soloneski and M. L. Larramendy, Chlorpyrifos-based insecticides induced genotoxic and cytotoxic effects in the ten spotted live-bearer fish, *Cnesterodon decemmaculatus* (Jenyns, 1842), *Environ. Toxicol.*, 2013, **29**, 1390–1398.
66. P. A. Siroski, G. L. Poletta, M. A. Latorre, M. E. Merchant, H. H. Ortega and M. D. Mudry, Immunotoxicity of commercial-mixed glyphosate in

broad snoutedcaiman (*Caiman latirostris*), *Chem. -Biol. Interact.*, 2016, **244**, 64–70.

67. R. Seriani, M. J. Tavares Ranzani-Paiva, A. T. Silva-Souza and S. Roseli Napoleão, Hematology, micronuclei and nuclear abnormalities in fishes from São Francisco river, Minas Gerais state, Brazil, *Acta Sci., Biol. Sci.*, 2011, **33**(1), 107–112.
68. B. Bosch, F. Mañas, N. Gorla and D. Aiassa, Micronucleus test in post metamorphic *Odontophrynus cordobae* and *Rhinella arenarum* (Amphibia: Anura) for environmental monitoring, *J. Toxicol. Environ. Health Sci.*, 2011, **3**(6), 155–163.
69. M. S. Babini, C. L. Bionda, N. E. Salas and A. L. Martino, Health status of tadpoles and metamorphs of *Rhinella arenarum* (Anura,Bufonidae) that inhabit agroecosystems and its implications forland use, *Ecotoxicol. Environ. Saf.*, 2015, **118**, 118–125.
70. V. D. Heuser, J. da Silva, H. Moriske, J. F. Dias, M. L. Yoneama and T. R. O. de Freitas, Genotoxicity biomonitoring in regions exposed to vehicle emissions using the comet assay and the micronucleus test in native rodent *Ctenomys minutus*, *Environ. Mol. Mutagen.*, 2002, **40**, 227–235.
71. M. A. Latorre, E. C. López González, P. A. Siroski and G. L. Poletta, Basal frequency of micronuclei and hematological parameters in the side-necked turtle, *Phrynops hilarii* (Duméril & Bibron, 1835), *Acta Herpetol.*, 2015, **10**(1), 31–37.
72. M. Oliveira, I. Ahmad, V. L. Maria, C. S. S. Ferreira, A. Serafim, M. J. Bebianno, M. Pacheco and M. A. Santos, Evaluation of oxidative DNA lesions in plasma and nuclear abnormalities in erythrocytes of wild fish (*Liza aurata*) as an integrated approach to genotoxicity assessment, *Mutat. Res.*, 2010, **703**, 83–89.
73. M. C. Cabagna, R. C. Lajmanovich, P. M. Peltzer, A. M. Attademo and E. Ale, Induction of micronuclei in tadpoles of *Odontophrynus americanus* (Amphibia: Leptodactylidae) by the pyrethroid insecticide Cypermethrin, *Toxicol. Environ. Chem.*, 2006, **88**(4), 729–737.
74. D. G. S. M. Cavalcante, C. B. R. Martinez and S. H. Sofia, Genotoxic effects of Roundup® on the fish *Prochilodus lineatus*, *Mutat. Res.*, 2008, **655**, 41–46.
75. S. Guilherme, I. Gaivao, M. A. Santos and M. Pacheco, European eel (*Anguilla anguilla*) genotoxic and pro-oxidant responses following short-term exposure to Roundup®—a glyphosate-based herbicide, *Mutagenesis*, 2010, **25**(5), 523–530.
76. S. A. S. Langie, A. Azqueta and A. R. Collins, The comet assay: past, present, and future, *Front. Genet.*, 2015, **6**, 266–271.
77. J. de Lapuente, J. Lourenço, S. A. Mendo, M. Borràs, M. G. Martins, P. M. Costa and M. Pacheco, The Comet assay and its applications in the field of ecotoxicology: a mature tool that continues to expand its perspectives, *Front. Genet.*, 2015, **6**(180), 21–40.
78. G. L. Poletta, A. Larriera, P. Siroski, E. Kleinsorge and M. D. Mudry, Integral aproach of glyphosate-induced alterations in a South American

caiman species, in *Herbicides: Properties, Crop Protection and Environmental Hazards*, ed. K. D. Piotrowski, Nova Sci. Pub., New York, 2011, pp. 189–210.

79. M. A. Latorre, M. L. Romito, A. Larriera, G. L. Poletta and P. A. Siroski, Total and differential white blood cells count in *Caiman latirostris* after *in ovo* and *in vivo* exposure to insecticides, *J. Immunotoxicol.*, 2016, **13**(6), 903–908.
80. P. J. Peruzzo, A. A. Porta and A. E. Ronco, Levels of glyphosate in surface waters, sediments and soils associated with direct sowing soybean cultivation in north pampasic region of Argentina, *Environ. Pollut.*, 2008, **156**, 61–66.
81. V. C. Aparicio, E. De Gerónimo, D. Marino, J. Primost, P. Carriquiriborde and J. L. Costa, Environmental fate of glyphosate and aminomethylphosphonic acid in surface waters and soil of agricultural basins, *Chemosphere*, 2013, **93**(9), 1866–1873.
82. P. S. Bindraban, A. C. Franke, D. O. Ferrar, C. M. Ghersa, L. A. P. Lotz, A. Nepomuceno, M. J. M. Smulders and C. C. M. van de Wiel, *GM-related Sustainability: Agroecological Impacts, Risks and Opportunities of Soy Production in Argentina and Brazil*, Plant Research International B.V., Wageningen, 2009, vol. 259, pp. 1–45.
83. R. A. Relyea, The lethal impact of Roundup on aquatic and terrestrial amphibeans, *Ecol. Appl.*, 2005, **15**, 1118–1125.
84. S. E. Hook and R. F. Lee, Genotoxicant induced DNA damage and repair in early and late developmental stages of the grass shrimp *Paleomonetes pugio* embryo as measured by the comet assay, *Aquat. Toxicol.*, 2004, **66**, 1–14.
85. A. G. M. Osman, A. M. Imam, J. Verreth, S. Wuertz, W. Kloas and F. Kirschbaum, Monitoring of DNA breakage in embryonic stages of the African catfish *Clarias gariepinus* (Burchell, 1822) after exposure to lead nitrate using alkaline comet assay, *Environ. Toxicol.*, 2008, **23**, 679–687.
86. G. L. Poletta, P. A. Siroski, P. S. Amavet, H. H. Ortega and M. D. Mudry, Reptiles as animal models: Examples of their utility in genetics, immunology and toxicology in *Reptiles in Research: Investigations of Ecology, Physiology and Behavior from Desert to Sea*, ed. W. Lutterschmidt, Nova Science Publishers, New York, 2013, pp. 406–446.
87. H. J. Hamlin and L. J. Guillette, Embryos as targets of endocrine disrupting contaminants in wildlife, *Birth Defects Res., Part C*, 2008, **93**, 19–33.
88. M. R. Milnes, D. Allen, T. A. Bryan, C. D. Sedacca and L. J. Guillette Jr., Developmental effects of embryonic exposure to toxaphene in the American alligator (*Alligator mississippiensis*), *Comp. Biochem. Physiol.*, 2004, **138**, 81–87.
89. M. R. Milnes, T. A. Bryan, J. G. Medina, M. P. Gunderson and L. J. Guillette Jr., Developmental alterations as a result of *in ovo* exposure to the pesticide metabolite *p,p*'-DDE in *Alligator mississippiensis*, *Gen. Comp. Endocrinol.*, 2005, **144**, 257–263.

90. Y. M. Mobarak and M. A. Al-Asmsari, Endosulfan impacts on the developing chick embryos: Morphological, morphometric and skeletal changes, *Int. J. Zool. Res.*, 2011, 7(2), 107–127.
91. SENASA, 2011. Resol 511/11 del Servicio Nacional de Sanidad y Calidad Agroalimentaria-Ministerio de Agricultura, Ganadería y Pesca de la Nación. [Internet; access June 2016]. Available at: http://www.senasa.gov.ar/senasa-comunica/noticias/el-senasa-controla-que-se-respete-la-prohibicion-de-elaborar-y-comercializar-productos.
92. M. L. Ballesteros, K. S. B. Miglioranza, M. González, G. Fillmann, D. A. Wunderlin and M. A. Bistoni, Multimatrix measurement of persistent organic pollutants in Mar Chiquita, a continental saline shallow lake, *Sci. Total Environ.*, 2014, **490**, 73–80.
93. D. Costantini, M. Rowe, M. W. Butler and K. J. McGraw, From molecules to living systems: historical and contemporary issues in oxidative stress and antioxidant ecology, *Funct. Ecol.*, 2010, **24**, 950–959.
94. K. B. Storey, Oxidative stress: animal adaptations in nature, *Braz. J. Med. Biol. Res.*, 1996, **29**, 1715–1733.
95. G. W. Winston and R. T. Di Giulio, Prooxidant and antioxidant mechanisms in aquatic organisms, *Aquat. Toxicol.*, 1991, **19**(2), 137–161.
96. P. Monaghan, N. B. Metcalfe and T. Torres, Oxidative stress as a mediator of life history trade-offs: mechanisms measurements and interpretation, *Ecol. Lett.*, 2009, **12**(1), 75–92.
97. M. Kale, N. Rathore, S. John and D. Bhatnagar, Lipid peroxidative damage on pyrethroid exposure and alterations in antioxidant status in rat erythrocytes: a possible involvement of reactive oxygen species, *Toxicol. Lett.*, 1999, **105**, 197–205.
98. C. Beuret, F. Zirulnik and M. Gimenez, Effect of the herbicide glyphosate on liver lipoperoxidation in pregnant rats and their fetuses, *Reprod. Toxicol.*, 2005, **19**, 501–504.
99. L. Glusczak, D. dos Santos, Miron, B. S. Moraes, R. R. Simões, M. R. Chitolina Schetinger, V. M. Morsch and V. L. Loro, Acute effects of glyphosate herbicide on metabolic and enzymatic parameters of silver catfish (*Rhamdia quelen*), *Comp. Biochem. Physiol.*, 2007, **146**, 519–524.
100. R. A. Ansari, S. Rahman, M. Kaur, S. Anjum and S. Raisuddin, In vivo cytogenetic and oxidative stress-inducing effects of cypermethrin in freshwater fish *Channa punctata* Bloch, *Ecotoxicol. Environ. Saf.*, 2011, **74**, 150–156.
101. J. Vera-Candioti, S. Soloneski and M. L. Larramendy, Evaluation of the genotoxic and cytotoxic effects of glyphosate-based herbicides in the ten spotted live-bearer fish *Cnesterodon decemmaculatus* (Jenyns, 1842), *Ecotoxicol. Environ. Saf.*, 2013, **89**, 166–173.
102. K. Hilscherova, A. L. Blankenship, M. Nie, K. K. Coady, B. L. Upham, J. E. Trosko and J. P. Giesy, Oxidative stress in liver and brain of the hatchling chicken (*Gallus domesticus*) following in ovo injection with TCDD, *Comp. Biochem. Physiol.*, 2003, **136**, 29–45.

103. A. Arukwe, R. Røsbak, A. O. Adeogun, H. A. Langberg, A. Venter, J. Myburgh, C. Botha, M. Benedetti and F. Regoli, Biotransformation and oxidative stress responses in captive Nile crocodile (*Crocodylus niloticus*) exposed to organic contaminants from the natural environment in South Africa, *Plos One*, 2015, **10**(6), 1–19.
104. P. E. Bickler and L. T. Buck, Hypoxia tolerance in reptiles, amphibians, and fishes: Life with variable oxygen availability, *Annu. Rev. Physiol.*, 2007, **69**, 145–170.
105. O. V. Furtado-Filho, C. Polcheira, D. P. Machado, G. Mourão and M. Hermes-Lima, Selected oxidative stress markers in a South American crocodilian species, *Comp. Biochem. Physiol., Part C: Toxicol. Pharmacol.*, 2007, **146**(1–2), 241–254.
106. M. Hermes-Lima, C. Carreiro, D. C. Moreira, C. Polcheira, D. P. Machado and E. G. Campos, Glutathione status and antioxidant enzymes in a crocodilian species from the swamps of the Brazilian Pantanal, *Comp. Biochem. Physiol., Part A: Mol. Integr. Physiol.*, 2012, **163**, 189–198.
107. Y. Voituron, S. Servais, C. Romestaing, T. Douki and H. Barre, Oxidative DNA damageand antioxidant defenses in the European common lizard *(Lacerta vivipara)* insupercooled and frozen state, *Cryobiology*, 2006, **52**, 74–82.
108. P. A. Valdivia, T. Zenteno-Savín, S. C. Gardner and A. A. Aquirre, Basic, oxidative stress metabolites in eastern Pacific green turtles (*Chelonia mydas agassizii*), *Comp. Biochem. Physiol., Part C: Toxicol. Pharmacol.*, 2007, **146**, 111–117.
109. T. Cavas and S. Könen, Detection of cytogenetic and DNA damage in peripheral erythrocytes of goldfish (*Carassius auratus*) exposed to a glyphosate formulation using the micronucleus test and the comet assay, *Mutagenesis*, 2007, **22**(4), 263–268.
110. N. De Castilhos Ghisi and M. M. Cestari, Genotoxic effects of the herbicide Roundup® in the fish *Corydoras paleatus* (Jenyns 1842) after short-term, environmentally low concentration exposure, *Environ. Monit. Assess.*, 2013, **185**, 3201–3207.
111. N. C. Moreno, S. H. Sofia and C. B. Martinez, Genotoxic effects of the herbicide Roundup Transorb® and its active ingredient glyphosate on the fish *Prochilodus lineatus*, *Environ. Toxicol. Pharmacol.*, 2014, **37**(1), 448–454.
112. N. C. Ghisi, E. C. de Oliveira and A. J. Prioli, Does exposure to glyphosate lead to an increase in the micronuclei frequency? A systematic and meta-analytic review, *Chemosphere*, 2016, **145**, 42–54.
113. D. Ali, N. S. Nagpure, S. Kumar, R. Kumar and B. Kushwaha, Genotoxicity assessment of acute exposure of chlorpyrifos to freshwater fish *Channa punctatus* (Bloch) using micronucleus assay and alkaline single-cell gel electrophoresis, *Chemosphere*, 2008, **71**(10), 1823–1831.

114. X. Yin, G. Zhu, X. B. Li and S. Liu, Genotoxicity evaluation of chlorpyrifos to amphibian Chinese toad (Amphibian: Anura) by comet assay and micronucleus test, *Mutat. Res.*, 2009, **680**(1), 2–6.
115. M. Ismail, Q. M. Khan, R. Ali, T. Ali and A. Mobeen, Genotoxicity of chlorpyrifos in freshwater fish *Labeo rohita* using Alkaline Single-Cell Gel Electrophoresis (Comet) assay, *Drug Chem. Toxicol.*, 2014, **37**(4), 466–471.
116. J. Vera-Candioti, S. Soloneski and M. L. Larramendy, Single-cell gel electrophoresis assay in the ten spotted live-bearer fish, *Cnesterodon decemmaculatus* (Jenyns, 1842), as bioassay for agrochemical-induced genotoxicity, *Ecotoxicol. Environ. Saf.*, 2013, **98**, 368–373.
117. G. L. Poletta, F. Gigena, A. Loteste, M. J. Parma, E. C. Kleinsorge and M. F. Simoniello, Single-cell gel electrophoresis assay in the ten spotted live-bearer fish, *Cnesterodon decemmaculatus* (Jenyns, 1842), as bioassay for agrochemical-induced genotoxicity, *Pestic. Biochem. Physiol.*, 2013, **107**(3), 385–390.
118. J. C. Brodeur, M. B. Poliserpi and M. Sánchez, Synergy between glyphosate- and cypermethrin-based pesticides during acute exposures in tadpoles of the common South American Toad *Rhinella arenarum*, *Chemosphere*, 2014, **112**, 70–76.
119. M. Lydy, J. Belden, C. Wheelock, B. Hammock and D. Denton, Challenges in regulating pesticide mixtures, *Ecol. Soc.*, 2004, **9**(6), 1–15.
120. J. Beyer, K. Petersen, Y. Song, A. Ruus, M. Grung, T. Bakke and K. E. Tollefsen, Environmental risk assessment of combined effects in aquatic ecotoxicology: A discussion paper, *Mar. Environ. Res.*, 2014, **96**, 81–91.
121. R. Saxena, P. Garg and D. K. Jain, In Vitro Anti-oxidant Effect of Vitamin E on Oxidative Stress Induced due to Pesticides in Rat Erythrocytes, *Toxicol. Int.*, 2011, **18**(1), 73–76.
122. A. Slaninova, M. Smutna, H. Modra and Z. Svobodova, A review: oxidative stress in fish induced by pesticides, *Neuro Endocrinol. Lett.*, 2009, **30**(1), 2–12.
123. D. Costantini and S. Verhulst, Does high antioxidant capacity indicate low oxidativestress? *Funct. Ecol.*, 2009, **23**, 506–509.
124. R. Pamplona, Membrane phospholipids, lipoxidative damage and molecular integrity: a causal role in aging and longevity, *Biochim. Biophys. Acta*, 2008, **1777**, 1249–1262.

CHAPTER 21

Epilogue and Final Remarks

MARCELO L. LARRAMENDY*[a,b] AND GUILLERMO ELI LIWSZYC[c]

[a] Cátedra de Citología, Facultad de Ciencias Naturales y Museo, Universidad Nacional de La Plata, Calle 64 N° 3, B1904AMA La Plata, Argentina; [b] Consejo Nacional de Investigaciones Científicas y Técnicas (CONICET), Buenos Aires, Argentina; [c] Physician, Specialist in Internal Medicine, former Guest Scientist at the Department of Pathology, University of Helsinki, Helsinki, Finland

*Email: marcelo.larramendy@gmail.com

Nowadays, the growing impact of anthropogenic activities generates huge quantities of toxic residues that can have direct or indirect detrimental effects upon the quality of our environment. The increase in their jeopardizing effects can exert short-, medium- and long-term consequences affecting human and environmental health in general, but are also capable of reducing the biodiversity of native flora and fauna, which will, in turn, encourage the resistance and emergence of new pests and diseases.

This book, *Ecotoxicology and Genotoxicology: Non-traditional Aquatic Models*, intends to provide an overview and relevant examples to stimulate practical discussions on the use of non-conventional biotic matrices within the scientific challenges faced by the Ecotoxicology and Genotoxicology academic world. Furthermore, the book endows relevant tools that may be of use in the implementation of decisions leading to actions that will hopefully reduce environmental health risk against environmental factors that may adversely impact human health or ecological balances.

We aimed to compile information from a diversity of sources into a single volume. The rationale is to give some real-life examples in order to widen the

Issues in Toxicology No. 33
Ecotoxicology and Genotoxicology: Non-traditional Aquatic Models
Edited by Marcelo L. Larramendy

Published by the Royal Society of Chemistry, www.rsc.org

concept that the use of a non-conventional animal model, far from being a scientific conundrum, may instead provide real answers to some of the actual problems the whole planet is dealing with. Concomitantly, these real examples extend concepts of hazardous factors to living species that may stimulate new research ideas and trends in the relevant fields.

In *Aquatic Invertebrates as Experimental Models*, readers will find ten chapters (Chapters 1–10) with background information about the nature of some environmental pollutants and some of the most versatile and validated methods of analyses employing different organisms. Specific examples of the use of several taxa from microscopic and near-microscopic pseudocoelomate organisms, such as Rotifers, to macroscopic marine and freshwater Molluscs and Crustaceans, including prawns and crabs, are listed as well. The first chapter provides an overview of the single cell gel electrophoresis methodology (also called the Comet assay) as currently one of the most important tools for analyzing genotoxicity at the molecular DNA level, underpinning its sensitivity, reliability and cost-effectiveness in different taxa. Chapter 2 provides a review of the adverse effects of pharmaceutical products upon marine environmental ecosystems. The use of several different taxa, including Echinoderms, Molluscs and Crustaceans, among others, stresses the necessity of adopting a tiered approach to determine the potential risks associated with these compounds. The third chapter documents the use of Rotifers, a pseudocoelomata aquatic taxon, as models for both ecotoxicology and genotoxicology. The authors have carried out a complete search in the mainstream scientific literature describing the types of studies, perspectives and use of rotifers in both areas. Chapter 4 reviews the health status monitoring of aquatic ecosystems employing molluscan immunomarkers. The authors stress that several cellular and molecular immunological parameters of freshwater molluscs have been established as immunomarkers of waterborne toxic substances. Among them, haemocyte density dynamics, lysosomal dye retention, apoptosis and generation of intrahaemocytic nitric oxide and superoxide anion are included. The filter feeding mode of digestion, limited mobility, prolonged life span, optimal sensitivity and resilience of immunological parameters are postulated as advantages of molluscs to function as both indicators and sources of effective immunomarkers of water pollution. The fifth chapter highlights the suitability of the zebra mussel (*Dreissena polymorpha*) as a non-conventional mollusc for the surveillance of emerging aquatic pollutants of widely used pharmaceutical and personal care products. These include triclosan, trimethoprim, ibuprofen, diclofenac and paracetamol, with a battery of different biomarkers suggested. The authors demonstrate that the sensitivity of zebra mussel in response to low concentrations of pharmaceutical and personal care products confers to this bivalve the status of a reliable model organism for aquatic ecotoxicology. Chapter 6 reviews the use of benthic aquatic macroinvertebrates, namely crayfish. These are considered to be a keystone species owing to their pivotal role in the aquatic habitat, transferring energy within the aquatic food web and between the aquatic and terrestrial food chains. The chapter is focused on presenting these organisms as a bioindicator species to measure contamination as they have low genetic and ecological variability, a large distribution, a long life-span, large body size, relatively low mobility and are easily reared in a laboratory setting. Furthermore, it is pointed out that crayfish exhibit measureable changes in population size, behaviour, tissue accumulation, histological and other physiological responses when exposed to contaminants. These types of changes qualify these organisms as optimal sentinel species. The following chapter also highlights crayfish as appropriate biomonitors in ecotoxicological and genotoxicological studies. *Cambarellus montezumae*, a particular Mexican crayfish, has most of the appropriate characteristics required for a biomonitor and it lives in a variety of freshwater bodies in México. This type of crayfish has shown dose- and time-dependent responses with respect to the effect of pesticides, suggesting that it can be proposed as a non-conventional freshwater biomonitor. The eighth chapter emphasizes the potential use of the freshwater crabs *Potamonautes* spp., prevalent Crustaceans in rivers of the South-Western region of the Western Cape of South Africa, as important model organisms for nanotoxicology studies. Promising results were obtained when these species were used for nanotoxicity assessment following exposure to silver nanoparticles and climatic stressors. Advantages of the *Potamonautes* spp. in identifying endpoints of toxicity and in elucidating mechanisms of toxicity are presented as well as addressing gaps in nanotoxicological research and recommendations for future research lines. The ninth chapter presents two other Crustaceans, the freshwater prawns, *Macrobrachium borellii* and *Palaemonetes argentinus*, major representatives of the South American freshwater environment. The scarce data available related to the basal activities of biomarker enzymes show that they can still be used as tools for ecotoxicological studies and/or monitoring. The authors give a clear example of the biomarking role played by different enzymes, *e.g.*, cholinesterase, superoxide dismutase, catalase and glutathione *S*-transferases, when these two crustaceans are exposed to an organophosphorate substance, such as fenitrothion. Chapter 10, the last one of this section of the book, also highlights the fact that some species of macroinvertebrates are used worldwide as bioindicators. As an example, the authors present results obtained when the mangrove crab *Ucides cordatus* is used as a model to monitor the conservation status of mangrove ecosystems. Owing to the wide interaction of this crab with different environmental matrices, the use of biomarkers in haemolymph samples, including neutral Red retention time as well as micro or macro DNA damage by Comet and micronucleus assays, allows an evaluation of the effect of stressors promoted by a set of pollutants.

In the second section of this book, *Aquatic Vertebrates as Experimental Models*, readers will find ten chapters (Chapters 11–20) dealing with background information about the nature of some environmental pollutants, some of the most useful and validated worldwide methods of analyses employing different organisms, as well as specific examples of the use of a several taxa including fish, amphibians and reptiles. The eleventh chapter presents an overview of the use of fish as model organisms in aquatic genotoxicity studies and provides a guide for measuring genotoxic damage in different fish cells. The most commonly used genotoxicity endpoints, namely the Comet assay and micronucleus test in different fish tissues as well as erythrocyte nuclear abnormalities, are described in detail, including the respective methodology. Study designs for both laboratory and field conditions when fish are used as experimental models are quoted as well. Chapter 12 offers a practical example of the use of two different endemic fish species from the Orinoco and Amazon River basins: the characid "tambaqui" (*Colossoma macropomum*), and the syngnathid seahorse (*Hippocampus reidi*). These serve as models to assess water pollution induced by the presence of polycyclic aromatic hydrocarbons in freshwater as well as marine and estuary environments, respectively. In this regard, the authors clearly address the fact that both "tambaqui" and seahorses can be considered excellent bioindicators, at least for polycyclic aromatic hydrocarbons, since they are extremely responsive to these strong organic pollutants. In the thirteenth chapter, information on the distribution, habitat, life history and reproduction of two families of teleost fish, *Blenniidae* and *Syngnathidae*, is presented. This chapter focuses on marine and estuarine species that have already been shown to adequately respond to xenobiotic insults. The use as non-conventional vertebrate species of the blenniid "shanny" (*Lipophrys pholis*) as well as several signatids is well documented. Special emphasis is given to the revision of literature that supports the use of both families as sentinels for pollution monitoring. Finalizing the series of chapters committed to the employment of fish species as experimental models, a fourteenth chapter is included. Data using the "ten spotted live-bearer fish" (*Cnesterodon decemmaculatus*) as a target organism in ecotoxicological and genotoxicological studies are presented. This species is an endemic fish member of the family Poeciliidae with an extensive distribution in Neotropical America, which attains high densities in a large variety of water-bodies within the whole River Plate (Río de La Plata) and other South American basins. Furthermore, several reports found this species suitable as a test organism in acute and chronic toxicity riparian bioassays. This chapter presents an overview of selected studies that have led to the use of *C. decemmaculatus* as a reliable and valid model of ecotoxicology and genotoxicology both *in vivo* and *in situ*. A battery of several different endpoints for lethality, cytotoxicity and genotoxicity induced by different widely used agrochemicals is presented, including glyphosate, chlorpyrifos, pirimicarb, dicamba, 2,4-D, and paraquat. Next comes a series of five chapters employing aquatic life-stage of different amphibians, including urodele and anuran species. Chapter 15 describes the potential use of two experimental biotic matrices employing urodele amphibians such as *Pleurodeles waltl* and *Ambystoma mexicanum* for the design and development of a sensitive method for the detection of genotoxic effects. This chapter highlights that testing methods provide a standardized protocol and are used worldwide in different research areas, such as public health, wastewater treatment, ecotoxicology, environmental genotoxicology and functional ecology. The sixteenth chapter shows that within European amphibians, anuran larvae are especially more suitable for toxicity testing than urodele larvae in respect of their biology. Embryos as well as larvae are sensitive to toxic agents. Ontogeny provides measurable meaningful endpoints that are representative of toxic effects. *Rana temporaria* larvae are proposed as a European anuran test species, both in order to achieve basic knowledge on the impairment of developmental steps in amphibians and for developmental toxicity studies. Finally, the chapter highlights that anurans of the genera *Bombina* and *Discoglossus* could be used as laboratory model species. Few pesticide studies use European anuran larvae and toxicological experience with larval stages of urodeles is limited. Chapter 17 describes the potential of AMPHITOX, a standardized test, employing anuran amphibian embryos to evaluate the toxicity for acute, short-term, chronic and early life stage exposure to hazardous chemicals and complex mixtures of xenobiotics. By means of this bioassay the toxicity of both environmental samples and physicochemical agents in a single or combined form can be evaluated. The exposure period can be adjusted to the toxicity of the sample and/or the assessment purpose. The growing concern about chronic effects of xenobiotics at environmentally relevant concentrations has led researchers to expand the exposure conditions of classic acute, short-term chronic and chronic AMPHITOX tests with young larvae to evaluate noxious effects on specific developmental stages of the early life cycle (pulse-exposures) and exposure periods covering the whole development (embryo-larval and metamorphosis). This last condition can even include post-exposure effects to assess adverse effects on gonadogenesis and sexual differentiation. The eighteenth chapter presents an overview of selected investigations that have led to the use of the common tree frog *Hypsiboas pulchellus* tadpoles as a reliable and valid novel model for *in vivo* and *in situ* studies of aquatic pollution. In order to validate the use of this hylid species as a non-conventional biotic matrix for ecotoxicological and genotoxicological studies, results obtained on both early and late-stage larvae of acute lethal and acute sublethal effects induced after exposure to different xenobiotics, mostly pesticides, are presented. Finally, two chapters employing other Chordata, such as Reptiles, as non-conventional models are presented. Chapter 19 emphasizes that sea turtles are considered of special interest as potential bioindicators for organic and inorganic pollutants in marine ecosystems. It includes generalities of these pollutants in sea turtle species as well as describing some biological factors that can influence the load of these pollutants in the tissues or eggs of the reptiles. The occurrence of toxic pollutant effects and detailed current knowledge on the use of different suitable biomarkers to assess the toxicological effects of chemicals in these marine reptiles are also overviewed. Finally, the last chapter of this compilation, Chapter 20, stresses the fact that several wild species can be affected by the overuse of pesticides and fertilizers in agricultural activities. The authors describe the employment of another reptile sentinel species, the broad snouted caiman (*Caiman latirostris*), for pesticide contamination by different pesticide formulations (glyphosate-, endosulfan-, cypermethrin- and chlorpyrifos-based formulations) and mixtures widely used in Argentina and other neighbouring countries. This chapter includes several instances of evaluations conducted in embryos and hatchlings, under controlled and semi-natural conditions employing biomarkers of genotoxicity (micronuclei, other nuclear abnormalities and Comet assay) and oxidative stress (oxidation of lipids and DNA as well as the activity of antioxidant enzymes) on blood cells. Not surprisingly, all pesticides tested as well as the complex mixtures generally applied induced genotoxicity and oxidative stress, as revealed by several biomarkers of different endpoints.

Despite dealing with many diverse topics, we have tried to compile this "wealth of information" into two major parts for the sake of clarity and order. First, *Aquatic Invertebrates as Experimental Models*, and second, *Aquatic Vertebrates as Experimental Models*, taking into consideration whether invertebrate or vertebrate taxa of one or more populations of a living organism(s) have been selected as experimental matrices, respectively.

Without running the risk of being repetitive, we would like to recap on some important concepts as previously mentioned in *Ecotoxicology and Genotoxicology: Non-traditional Terrestrial Models*, by the same editor. We strongly recommend the perusal of both volumes, which are not overlapping subjects, in order to gain the full benefit of this series and have a holistic and stimulating approach to the matter.

Many scientists feel that the time has come for a shift in emphasis away from basic research towards applied science. While basic research is needed in order to shed more light on the fundamentals, applied research is required in order to find solutions to the many problems the world faces, *e.g.*, overpopulation, excessive use of the Earth's natural resources and pollution.

New studies in several reputable publications [*e.g.*, Worldwide Trends 1975–2015, *The Lancet*, 2016, **384**(10064), 37–55] of common worldwide human diseases, such as arterial hypertension, among others, seem to suggest that not only classical and lifestyle factors should be addressed in the search for adequate treatment. From a medical perspective, it is not only a disease of affluence. One in eight deaths worldwide are due to this condition and its corresponding factors (heart disease, kidney disease and stroke). These studies seem to suggest that a closer look at other, up till now unrelated, factors is required. Early-life nutrition and exposure to air pollution, heavy metals and even noise are factors that may push blood pressure up later in life. Thinking out of silos and cross-linking medical with applied science will lead to a faster pace of understanding all the factors related to the emergence of a disease and thus finding a cure.

The chapters included in this book are a mere enumeration of some practical examples. There are many more species that can be used as experimental models and the list should be expanded by different research groups and academics all over the world. We hope that many more scientists will realise that it is important to tackle subjects related to the status of the environment using autochthonous, non-target species, which are truly exposed to locally used xenobiotic agents. In addition, this research plan would also increase the chances of independent scientists and research institutions getting access to grants and attracting the attention of local authorities, who would be very interested in financing projects that really are within their sphere of political interest and pride.

Many researchers have contributed to the publication of this book. We hope that it serves as a herald in order to bolster enthusiasm for the use of native, easily available local species in order to widen our knowledge on the subject. Last but not least, we would like to especially thank the authors for their positive responses, their time, contributions and feedback, which made the compilation of this book possible.

Subject Index